BARRON'S
THE TRUSTED NAME IN TEST PREP

2024

AP®

Calculus
PREMIUM

Dennis Donovan, M.S.,
David Bock, M.S.,
and Shirley O. Hockett, M.A.

AP® is a registered trademark of the College Board, which is not affiliated with
Barron's and was not involved in the production of, and does not endorse, this product.

AP® is a registered trademark of the College Board, which is not affiliated with Barron's and was not involved in the production of, and does not endorse, this product.

© Copyright 2023, 2022, 2019, 2017, 2015, 2013, 2012, 2010, 2008 by Kaplan North America, LLC, d/b/a Barron's Educational Series
Prior editions © Copyright 2005, 2002, 1998 by Kaplan North America, LLC, d/b/a Barron's Educational Series under the title *How to Prepare for the AP Advanced Placement Examination in Calculus* and 1995, 1992, 1987, 1983, 1971 under the title *How to Prepare for the Advanced Placement Examination: Mathematics*.

Published by Kaplan North America, LLC, d/b/a Barron's Educational Series
1515 W. Cypress Creek Road
Fort Lauderdale, FL 33309
www.barronseduc.com

ISBN: 978-1-5062-8783-6

10 9 8 7 6 5 4 3 2 1

Kaplan North America, LLC, d/b/a Barron's Educational Series print books are available at special quantity discounts to use for sales promotions, employee premiums, or educational purposes. For more information or to purchase books, please call the Simon & Schuster special sales department at 866-506-1949.

About the Authors

Dennis Donovan has been a math teacher at Xaverian Brothers High School for more than 25 years, teaching AP Calculus, both AB and BC, for more than 20 years. He has served as an AP Calculus Reader and as one of nine national Question Leaders for the AP Calculus exam. Dennis leads professional development workshops for math teachers as a consultant for the College Board and a T^3 Regional Instructor for Texas Instruments.

David Bock taught AP Calculus during his 35 years at Ithaca High School, and served for several years as an Exam Reader for the College Board. He also taught mathematics at Tompkins-Cortland Community College, Ithaca College, and Cornell University. A recipient of several local, state, and national teaching awards, Dave has coauthored five textbooks, and now leads workshops for AP teachers.

IN MEMORIAM
Shirley O. Hockett
(1920–2013)

Shirley Hockett taught mathematics for 45 years, first at Cornell University and later at Ithaca College, where she was named Professor Emerita in 1991. An outstanding teacher, she won numerous awards and authored six mathematics textbooks. Shirley's experiences as an Exam Reader and Table Leader for AP Calculus led her to write the first ever AP Calculus review book, published in 1971. Her knowledge of calculus, attention to detail, pedagogical creativity, and dedication to students continue to shine throughout this book. On behalf of the thousands of AP Calculus students who have benefited from her tireless efforts, we gratefully dedicate this review book to Shirley.

Table of Contents

How to Use This Book...ix

Barron's Essential 5...x

Introduction...1

    The Courses...1

    Topic Outline for the AB and BC Calculus Exams...1

    The Examinations...7

    The Graphing Calculator: Using Your Graphing Calculator on the AP Exam..........8

    Grading the Examinations..12

DIAGNOSTIC TESTS

    Diagnostic Test Calculus AB...17

    Diagnostic Test Calculus BC...41

TOPICAL REVIEW AND PRACTICE

1   Functions...63

    A. Definitions..63

    B. Special Functions..66

    C. Polynomial and Other Rational Functions...69

    D. Trigonometric Functions...69

    E. Exponential and Logarithmic Functions..72

    BC ONLY F. Parametrically Defined Functions...73

    G. Polar Functions..76

    Practice Exercises..79

2   Limits and Continuity..89

    A. Definitions and Examples..89

    B. Asymptotes...94

    C. Theorems on Limits...96

    D. Limit of a Quotient of Polynomials..98

    E. Other Basic Limits...99

    F. Continuity...99

    Practice Exercises..105

3   Differentiation..113

    A. Definition of Derivative...113

    B. Formulas...115

    C. The Chain Rule: The Derivative of a Composite Function........................116

    D. Differentiability and Continuity...121

    E. Estimating a Derivative...122

        E1. Numerically..122

        E2. Graphically..125

    BC ONLY F. Derivatives of Parametrically Defined Functions.........................126

G. Implicit Differentiation 128

H. Derivative of the Inverse of a Function 130

I. The Mean Value Theorem 131

J. Indeterminate Forms and L'Hospital's Rule 133

K. Recognizing a Given Limit as a Derivative 136

Practice Exercises 139

4   Applications of Differential Calculus **159**

A. Slope; Critical Points 159

B. Tangents to a Curve 161

C. Increasing and Decreasing Functions 163

Case I. Functions with Continuous Derivatives 163

Case II. Functions Whose Derivatives Have Discontinuities 163

D. Maximum, Minimum, Concavity, and Inflection Points: Definitions ... 164

E. Maximum, Minimum, and Inflection Points: Curve Sketching 165

Case I. Functions That Are Everywhere Differentiable 165

Case II. Functions Whose Derivatives May Not Exist Everywhere 169

F. Global Maximum or Minimum 171

Case I. Differentiable Functions 171

Case II. Functions That Are Not Everywhere Differentiable 171

G. Further Aids in Sketching 171

H. Optimization: Problems Involving Maxima and Minima 173

I. Relating a Function and Its Derivatives Graphically 177

J. Motion Along a Line 180

`BC ONLY` K. Motion Along a Curve: Velocity and Acceleration Vectors 182

L. Tangent-Line Approximations 186

M. Related Rates 188

`BC ONLY` N. Slope of a Polar Curve 190

Practice Exercises 193

5   Antidifferentiation **213**

A. Antiderivatives 213

B. Basic Formulas 213

`BC ONLY` C. Integration by Partial Fractions 220

D. Integration by Parts 221

E. Applications of Antiderivatives; Differential Equations 224

Practice Exercises 227

6   Definite Integrals **241**

A. Fundamental Theorem of Calculus (FTC); Evaluation of Definite Integrals ... 241

B. Properties of Definite Integrals 241

C. Definition of Definite Integral as the Limit of a Riemann Sum 246

D. The Fundamental Theorem Again 247

E. Approximations of the Definite Integral; Riemann Sums 248

E1. Using Rectangles 248

E2. Using Trapezoids 250

E3. Comparing Approximating Sums 252

F. Graphing a Function from Its Derivative; Another Look 253

G. Interpreting ln *x* as an Area 260

H. Average Value 261

Practice Exercises 270

7   Applications of Integration to Geometry **283**

A. Area 283

A1. Area Between Curves 285

A2. Using Symmetry 286

`BC ONLY` | A3. Region Bounded by Polar Curve 288

B. Volume 290

B1. Solids with Known Cross Sections 290

B2. Solids of Revolution 292

`BC ONLY` | C. Length of Curve (Arc Length) 297

D. Improper Integrals 299

Practice Exercises 309

8   Further Applications of Integration **337**

A. Motion Along a Straight Line 337

`BC ONLY` | B. Motion Along a Plane Curve 339

C. Other Applications of Riemann Sums 342

D. FTC: Definite Integral of a Rate Is Net Change 344

Practice Exercises 346

9   Differential Equations **357**

A. Basic Definitions 357

B. Slope Fields 358

`BC ONLY` | C. Euler's Method 363

D. Solving First-Order Differential Equations Analytically 367

E. Exponential Growth and Decay 369

Case I: Exponential Growth 369

Case II: Restricted Growth 373

`BC ONLY` | Case III: Logistic Growth 376

Practice Exercises 381

10   Sequences and Series **399**

`BC ONLY` | A. Sequences of Real Numbers 399

B. Infinite Series 400

B1. Definitions 400

B2. Theorems About Convergence or Divergence of Infinite Series 402

B3. Tests for Convergence of Infinite Series 403

B4. Tests for Convergence of Nonnegative Series 404

B5. Alternating Series and Absolute Convergence 408

C. Power Series 411

C1. Definitions; Convergence 411

C2. Functions Defined by Power Series 413

C3. Finding a Power Series for a Function: Taylor and Maclaurin Series 415

BC ONLY | C4. Approximating Functions with Taylor and Maclaurin Polynomials 418

C5. Taylor's Formula with Remainder; Lagrange Error Bound 423

C6. Computations with Power Series 425

C7. Power Series over Complex Numbers 429

Practice Exercises 431

11 Miscellaneous Multiple-Choice Practice Questions **443**

12 Miscellaneous Free-Response Practice Exercises **475**

AB PRACTICE TESTS

AB Practice Test 1 505

AB Practice Test 2 529

AB Practice Test 3 555

BC PRACTICE TESTS

BC Practice Test 1 583

BC Practice Test 2 605

BC Practice Test 3 625

Appendix: Formulas and Theorems for Reference **645**

Index **653**

How to Use This Book

Introduction and Diagnostic Tests

TStart with the Introduction, which provides an overview of the AP Calculus AB and BC courses and exams. Then, proceed to the diagnostic tests, which allow you to assess your strengths and areas for improvement. There is one AB diagnostic and one BC diagnostic, and both are followed by solutions keyed to the corresponding review chapter.

Review and Practice

"Topical Review and Practice" includes 10 chapters with notes on the main topics of the Calculus AB and BC syllabi and numerous carefully worked-out examples. Each chapter concludes with a set of multiple-choice questions, usually divided into calculator and no-calculator sections, followed immediately by answers and solutions.

This review is followed by further practice: (1) Chapter 11, which includes a set of multiple-choice questions on miscellaneous topics and an answer key and answer explanations; (2) Chapter 12, a set of miscellaneous free-response problems that are intended to be similar to those in Section II of the AP examinations. They are followed by detailed solutions.

In this book, review material on topics covered only in Calculus BC is indicated by a BC icon and a vertical line in the margin, as are both multiple-choice questions and free-response-type problems that are likely to occur only on a BC examination.

Practice Tests

This book concludes with three full-length AB practice tests and three full-length BC practice tests, all of which mirror the actual exams in format, content, and level of difficulty. Each test is followed by detailed answers and explanations.

Online Practice

There are also four additional full-length practice tests online. You may take these tests in practice (untimed) mode or in timed mode. Two of these online tests mirror the AB test, and the other two mirror the BC test. All questions are answered and explained.

For Students

Students who use this book independently will improve their performance by studying the illustrative examples carefully and trying to complete practice problems before referring to the solutions

For Teachers

The teacher who uses this book with a class may profitably do so in several ways. If the book is used throughout a year's course, the teacher can assign all or part of each set of multiple-choice questions and some miscellaneous exercises after the topic has been covered. These sets can also be used for review purposes shortly before exam time. The practice tests will also be very helpful when reviewing toward the end of the year. Teachers may also assemble exams by choosing appropriate problems from the miscellaneous practice questions in Chapters 11 and 12.

BARRON'S ESSENTIAL 5

As you review the content in this book to work toward earning that **5** on your AP CALCULUS **AB** exam, here are five things that you **<u>MUST</u>** know above everything else:

Learn the basic facts:

- derivatives (p. 115) and antiderivatives (p. 213) of common functions;
- the Product (p. 115), Quotient (p. 115), and Chain Rules (p. 116) for finding derivatives;
- the midpoint, left and right rectangle, and trapezoid approximations for estimating definite integrals (p. 248);
- finding antiderivatives by substitution (p. 224);
- the important theorems: Rolle's Theorem (p. 132), the Mean Value Theorem (p. 131), and especially the Fundamental Theorem of Calculus (p. 241).

Understand that a derivative is an instantaneous rate of change, and be able to apply that concept to:

- use L'Hospital's Rule to find limits of indeterminate forms (only $\frac{0}{0}$ and $\frac{\infty}{\infty}$) (p. 133);
- find equations of tangent lines (p. 161);
- determine where a function is increasing/decreasing (p. 163), is concave up/down (p. 164), or has maxima, minima, or points of inflection (pp. 165, 170);
- analyze the speed, velocity, and acceleration of an object in motion (p. 180);
- solve related rates problems (p. 188), using implicit differentiation (p. 128) when necessary.

Understand that integrals represent accumulation functions based on antiderivatives, and be able to apply those concepts to:

- the average value of a function (p. 261);
- area (p. 283) and volume (p. 290);
- the position of objects in motion and distance traveled (p. 337);
- total amount when given the rate of accumulation (p. 344);
- differential equations, including solutions and slope fields (p. 357).

4 Be able to apply any of the aforementioned calculus concepts to functions defined algebraically, graphically, or in tables.

5 Be able to maximize your score on the exam by:

- answering *all* the multiple-choice questions;
- knowing how and when to use your calculator, and when not to;
- understanding what work you need to show;
- knowing how to explain, interpret, and justify answers when a question requires that. (The free-response solutions in this book model such answers.)

BARRON'S ESSENTIAL 5

As you review the content in this book to work toward earning that **5** on your AP CALCULUS **BC** exam, here are five things that you **<u>MUST</u>** know above everything else:

Master the Essential 5 listed for the AB Calculus Exam. These form the core for questions that determine your AB subscore and provide the essential knowledge base you'll need for questions related to the additional BC topics.

Understand how to extend AB Calculus concepts to more advanced situations, including:

- using limits to analyze improper integrals (p. 299);
- solving logistic differential equations (p. 377) and estimating solutions using Euler's method (p. 363);
- finding antiderivatives using integration by parts (p. 221) or partial fractions (p. 220);
- finding length of curve (arc length) (p. 297).

Be able to apply calculus concepts to parametrically defined functions (pp. 73, 126, 339) and polar functions (pp. 76, 190, 288).

Know how to analyze the position, velocity, speed, acceleration, and distance traveled for an object in motion in two dimensions by applying calculus concepts to vectors (p. 182, 339).

Understand infinite series. You must be able to:

- determine whether a series converges or diverges (p. 399);
- use Taylor's theorem to represent functions as power series (p. 415);
- determine the interval of convergence for a power series (p. 411);
- find bounds on the error for estimates based on series (pp. 410, 423).

Introduction

This book is intended for students who are preparing to take either of the two Advanced Placement examinations in Mathematics offered by the College Board, and for their teachers. It is based on the latest Course and Exam Description published by the College Board and covers the topics listed there for both Calculus AB and Calculus BC.

The Courses

Calculus AB and BC are both full-year courses in the calculus of functions of a single variable. Both courses emphasize:

(1) student understanding of concepts and applications of calculus over manipulation and memorization;

(2) developing the student's ability to express functions, concepts, problems, and conclusions analytically, graphically, numerically, and verbally and to understand how these are related; and

(3) using a graphing calculator as a tool for mathematical investigations and for problem solving.

Both courses are intended for those students who have already studied college-preparatory mathematics: algebra, geometry, trigonometry, analytic geometry, and elementary functions (linear, polynomial, rational, exponential, logarithmic, trigonometric, inverse trigonometric, and piecewise).

Topic Outline for the AB and BC Calculus Exams

The AP Calculus course topics can be arranged into four content areas: 1. Limits, 2. Derivatives, 3. Integrals and the Fundamental Theorem, and 4. Series. The AB exam tests content areas 1, 2, and 3. The BC exam tests all four content areas. There are BC-only topics in content areas 2 and 3, as well. Roughly 40 percent of the points available for the BC exam are BC-only topics.

Content Area 1: Limits

Limits are used in many calculus concepts to go from the discrete to the continuous case. Students must understand the idea of limits so that a deeper understanding of definitions and theorems can be achieved. Students must be presented with different representations of functions when calculating limits. Working with tables, graphs, and algebraically defined functions with and without the calculator are essential skills that students need to master.

I. Understanding the behavior of a function
   A. Limit—writing and interpreting
      1. Limit definition—existence versus nonexistence
      2. Writing limits using correct symbolic notation
         a. One-sided limits
         b. Limits at infinity
         c. Infinite limits
         d. Nonexisting limits
   B. Estimating limits
      1. Graphical and numerical representations of functions may be used to estimate limits

    C. Calculating limits

        1. Use theorems of limits to calculate limits of sums, differences, products, quotients, and compositions of functions

        2. Use algebraic manipulation, trigonometric substitution, and Squeeze Theorem

        3. L'Hospital's Rule may be used to evaluate limits of indeterminate forms $\frac{0}{0}$ and $\frac{\infty}{\infty}$

    D. Function behavior

        1. Limits can be used to explain asymptotic (vertical and horizontal) behavior of functions

        2. Relative rates of growth of functions can be compared using limits

II. Continuity of functions

    A. Intervals of continuity and points of discontinuity

        1. Definition of continuity

        2. Some functions are continuous at all points in their domain

            a. Polynomials

            b. Rational functions

            c. Power functions

            d. Exponential functions

            e. Logarithmic functions

            f. Trigonometric functions

        3. Types of discontinuities

            a. Removable

            b. Jump

            c. Vertical asymptotes

    B. Continuity allows the application of important calculus theorems

        1. Continuity is a condition for the application of many calculus theorems

            a. Intermediate Value Theorem

            b. Extreme Value Theorem

            c. Mean Value Theorem

Content Area 2: Derivatives

The derivative describes the rate of change. The concept of the limit helps us develop the derivative as an instantaneous rate of change. Many applications rely on the use of the derivative to help determine where a function attains a maximum or a minimum value. Again, students must be presented with different representations of functions and use different definitions of the derivative when calculating and estimating derivatives. Working with tables, graphs, and algebraically defined functions with and without the calculator are essential skills that students need to master.

I. The derivative as the limit of a difference quotient

    A. Identifying the derivative

        1. Difference quotients give the average rate of change on an interval; common forms include $\dfrac{f(a+h)-f(a)}{h}$ and $\dfrac{f(x)-f(a)}{x-a}$

        2. Instantaneous rate of change at a point is the limit of the difference quotient, provided it exists; $f'(a)=\lim\limits_{h\to 0}\dfrac{f(a+h)-f(a)}{h}$ or $f'(a)=\lim\limits_{x\to a}\dfrac{f(x)-f(a)}{x-a}$

        3. The derivative of the function f is given by $f'(x)=\lim\limits_{h\to 0}\dfrac{f(x+h)-f(x)}{h}$

        4. Various notations for the derivative of a function $y=f(x)$ include $\dfrac{dy}{dx}$, $f'(x)$, and y'

        5. The derivative can be given using any of the representations in the rule of four: graphically, numerically, analytically, and verbally

    B. Estimating the derivative

        1. Tables and graphs allow the estimation of the derivative at a point

    C. Calculating the derivative

        1. Apply the rules for differentiating families of functions

            a. Polynomial, rational, power, exponential, logarithmic, trigonometric, and inverse trigonometric

        2. Differentiation rules can be used to find the derivatives of sums, differences, products, and quotients of functions

        3. Chain Rule

            a. Composite functions can be differentiated with the Chain Rule

            b. Implicit differentiation

            c. The derivative of an inverse function

        4. Parametric, vector, and polar functions can be differentiated using the methods described above   `BC ONLY`

    D. Higher-order derivatives

        1. Differentiating the first derivative produces the second derivative, differentiating the second derivative produces the third derivative, and so on, provided these derivatives exist

        2. Notations for higher-order derivatives: second derivatives $\frac{d^2y}{dx^2}$, $f''(x)$, and y''; third derivatives $\frac{d^3y}{dx^3}$, $f'''(x)$, and y'''; higher than third derivatives $\frac{d^ny}{dx^n}$ or $f^{(n)}(x)$, where n is the number of the derivative

II. Using the derivative of a function to determine the behavior of the function

    A. Analyzing the properties of a function

    B. Connecting differentiability and continuity

III. Interpreting and applying the derivative

    A. The meaning of a derivative

        1. Instantaneous rate of change with respect to the independent variable

        2. The units for the derivative of a function are the units of the function over the units for the independent variable

    B. Using the slope of the tangent line

        1. The slope of the line tangent to a graph at a point is the derivative at that point

        2. The tangent line provides a local linear approximation of function values near the point of tangency

    C. Solving problems

        1. Related rates

            a. Find the rate of change of one quantity by knowing the rate(s) of change of related quantities

        2. Optimization

            a. Finding the maximum or minimum value of a function on an interval

        3. Rectilinear motion

            a. Using the derivative to determine velocity, speed, and acceleration for particles moving along a line

        4. Planar motion   `BC ONLY`

            a. Using the derivative to determine velocity, speed, and acceleration for particles moving along a curve defined by parametric or vector functions

    D. Differential equations

        1. Verify a function is a solution to a given differential equation using derivatives

        2. Estimate solutions to differential equations

            a. Slope fields allow the visualization of a solution curve to a differential equation; students may be asked to draw a solution curve through a given point on a slope field

BC ONLY

            b. Euler's method provides a numerical method to approximate points on the solution curve for a differential equation

    IV. The Mean Value Theorem (MVT)

    A. If a function is continuous on the closed interval $[a,b]$ and is differentiable on the open interval (a,b), then MVT guarantees the existence of a point in the open interval (a,b) where the instantaneous rate of change is equal to the average rate of change on the interval $[a,b]$

Content Area 3: Integrals and the Fundamental Theorem

The definition of a definite integral comes from a Riemann Sum. Students should be able to estimate a definite integral using various methods and be able to calculate definite integrals using geometry and analytic techniques using the Fundamental Theorem of Calculus (FTC). Once again, students must be presented with different representations of functions and use different methods when calculating and estimating definite integrals. Working with tables, graphs, and algebraically defined functions with and without the calculator are essential skills that students need to master. Students must be able to work with functions defined by an integral and understand the connection between integral and differential calculus as demonstrated in the FTC. Students will see many applications of integral calculus; for example, area, volume, accumulation functions, and motion problems. Students should be able to interpret the meaning of an integral expression in the context of a problem.

    I. Antidifferentiation

    A. The antiderivative of basic functions follows from the differentiation rules

    II. The definite integral

    A. Riemann Sums and definite integrals

        1. The Riemann Sum is a sum of products, height times width, where the height is a value of the function in a subinterval and the width is the length of the subinterval

        2. The limit of a Riemann Sum is the definite integral

$$\int_a^b f(x)\,dx = \lim_{n \to \infty} \sum_{k=1}^{n} f(a + k \cdot \Delta x) \cdot \Delta x, \text{ where } \Delta x = \frac{b-a}{n}$$

    B. Approximation of definite integrals can be computed with partitions of uniform length or varying lengths by using two methods

        1. Riemann Sum—left, right, or midpoint

        2. Trapezoidal sum

    C. Geometric area formulas can be used to evaluate a definite integral of a function

    D. Properties of integrals can be employed when the integral meets certain criteria

        1. The sum or difference of two functions

        2. A constant times a function

        3. Involving the reversal of limits

        4. Over adjacent intervals

BC ONLY

    E. Improper integrals

III. The Fundamental Theorem of Calculus

    A. Functions defined as an integral, for example, $g(x) = \int_a^x f(t)\,dt$

        1. Differentiate functions defined by integrals, such as

$$g'(x) = \frac{d}{dx}\left(\int_a^x f(t)\,dt\right) = f(x)$$

        2. Determine information about the function g by using graphical, numerical, analytical, or verbal information about the function f

    B. Calculate antiderivatives and evaluate definite integrals

        1. A definite integral can be evaluated by $\int_a^b f(x)\,dx = F(b) - F(a)$, where $F(x)$ is an antiderivative of $f(x)$

        2. Techniques for determining antiderivatives

            a. Algebraic manipulation, for example, long division and completing the square

            b. Substitution of variables

            c. Integration by parts   `BC ONLY`

            d. Partial fractions, only nonrepeating linear

IV. Applications involving the definite integral

    A. Interpreting a definite integral

        1. $\int_a^b f'(t)\,dt$ represents the accumulation of the rate of change over the interval $[a,b]$

        2. An accumulation function can be defined as $f(t) = f(a) + \int_a^t f'(q)\,dq$

    B. Average value of a function

    C. Motion problems

        1. Rectilinear motion—displacement is the integral of velocity over an interval of time, and total distance traveled is the integral of speed over an interval of time

        2. Planar motion—displacement, distance, and position of a particle moving on a curve, defined by a parametric or vector function, can be determined using a definite integral   `BC ONLY`

    D. Area and volume

        1. Definite integrals are used to calculate certain areas, volumes, and lengths

            a. Areas of regions between curves in the plane

            b. Volumes of solids with known cross sections; this includes disks, washers, and given cross-sectional shapes; *NOTE:* volume by cylindrical shells is not tested on the AP exam

            c. Areas bounded by polar curves   `BC ONLY`

            d. Length of curve (arc length)—both planar curves and those defined by parametric functions

V. Separable differential equations

    A. General and particular (specific) solutions to differential equations

        1. Using antidifferentiation, a general solution involves the constant of integration

        2. Using antidifferentiation, a particular solution can be found for a differential equation with a given initial condition

            a. Linear motion

            b. Exponential growth and decay

            c. Logistic growth   `BC ONLY`

        3. Separation of variables must be used prior to antidifferentiation on some differential equations

        4. Domain restrictions may apply to the solutions to differential equations

   B. Modeling, interpreting, and solving differential equations from context

      1. Exponential growth and decay models come from the statement, "The rate of change of a quantity is proportional to the amount of the quantity"; the model is $\frac{dy}{dt} = ky$

BC ONLY

      2. Logistic growth model comes from the statement, "The rate of change of a quantity is jointly proportional to the amount of the quantity and the difference between the quantity and the carrying capacity"; the model is $\frac{dy}{dt} = ky(M - y)$

BC ONLY

Content Area 4: Series

Students will study series of numbers and power series. Several tests will be introduced to determine the convergence or divergence of a series of numbers and to determine the radius of and interval of convergence of a power series. Students will become familiar with Maclaurin series and general Taylor series representations of functions. The concept of using a Taylor polynomial to approximate a function near a specific value is an extension of using the tangent line for such approximations from earlier in the course and allows for better approximations.

  I. Convergence

    A. Does a series converge or diverge?

      1. A series converges if and only if the limit of the sequence of its partial sums exists

      2. Series are absolutely convergent, conditionally convergent, or divergent

      3. Common series

         a. Geometric series

         b. Harmonic series

         c. p-series

      4. Tests for convergence (these are the only tests that are assessed on the AP exam)

         a. nth Term Test

         b. Comparison Test

         c. Limit Comparison Test

         d. Integral Test

         e. Ratio Test

         f. Alternating Series Test

    B. Find or estimate the sum of a series

      1. The exact sum of a geometric series can be found as $\frac{a}{1 - r}$, where a is the first term in the series and r is the common ratio, such that $|r| < 1$

      2. The Alternating Series Error Bound may be used to determine the maximum difference between the actual value of the infinite series and the partial sum used to estimate, provided the alternating series converges by the Alternating Series Test

      3. If a series converges absolutely, the commutative property of addition applies on the sum

  II. Power series of a function

    A. Taylor polynomials

      1. The nth degree term in the Taylor polynomial centered at $x = a$ has a coefficient equal to $\frac{f^{(n)}(a)}{n!}$

      2. Taylor polynomials can be used to approximate the function value near the center

3. Error bound
   a. The error of an approximation can be bound using a Taylor polynomial by using the Lagrange error bound
   b. If the Taylor polynomial has alternating signs, the error of an approximation might be bound using a Taylor polynomial by the Alternating Series Error Bound

B. Power series

1. Power series have the form $\displaystyle\sum_{n=0}^{\infty} c_n(x-a)^n$, where the coefficients $\{c_n\}$ are a sequence of real numbers and the series is centered at $x = a$
   a. Students must know the Maclaurin series for $\sin(x)$, $\cos(x)$, e^x, and the geometric series, $\dfrac{1}{1-x}$

2. A power series for a function can be found using various methods
   a. Algebra
   b. Substitution
   c. Properties of geometric series
   d. Term-by-term differentiation
   e. Term-by-term integration

C. Interval of convergence

1. The radius of convergence of a power series can be found with the Ratio Test
   a. If the radius is zero, the series converges only at a single point, the center
   b. If the radius is positive, the series converges on an interval and the series is the Taylor series for the function

2. If a power series is obtained via term-by-term differentiation or integration, the radius of convergence is the same as that of the original power series

The Examinations

The Calculus AB and BC examinations and the course descriptions are prepared by committees of teachers from colleges or universities and from secondary schools. The examinations are intended to determine the extent to which a student has mastered the subject matter of the course.

Each examination is 3 hours and 15 minutes long, as follows:

Section I has two parts. Part A has 30 multiple-choice questions for which 60 minutes are allowed. The use of calculators is *not* permitted in Part A.

Part B has 15 multiple-choice questions for which 45 minutes are allowed. Some of the questions in Part B require the use of a graphing calculator.

Section II, the free-response section, has a total of six questions in two parts:

Part A has 2 questions, of which some parts *require* the use of a graphing calculator. After 30 minutes, however, *you will no longer be permitted to use a calculator.* If you finish Part A early, you will not be permitted to start work on Part B.

Part B has 4 questions and you are allotted an additional 60 minutes, but *you are not allowed to use a calculator.* You may work further on the Part A questions (without your calculator).

The section that follows gives important information on the use (and misuse!) of the graphing calculator.

The Graphing Calculator: Using Your Graphing Calculator on the AP Exam

The Four Calculator Procedures

Each student is expected to bring a graphing calculator to the AP exam. Different models of calculators vary in their features and capabilities; however, there are four procedures you must be able to perform on your calculator:

C1. Produce the graph of a function within an arbitrary viewing window.

C2. Solve an equation numerically.

C3. Compute the derivative of a function numerically.

C4. Compute definite integrals numerically.

Guidelines for Calculator Use

1. On multiple-choice questions in Section I, Part B, *you may use any feature or program on your calculator.* Warning: Don't rely on it too much! Only a few of these questions require the calculator, and in some cases using it may be too time-consuming or otherwise disadvantageous.

2. On the free-response questions in Section II, Part A:

(a) You may use the calculator to perform any of the four listed procedures. When you do, you need only write the equation, derivative, or definite integral (called the "setup") that will produce the solution and then write the calculator result to the required degree of accuracy (three places after the decimal point unless otherwise specified). Note that a setup must be presented in standard algebraic or calculus notation, not in calculator syntax. For example, you *must* include in your work the setup $\int_0^\pi \cos(t)\,dt$ even if you use your calculator to evaluate the integral.

(b) For a solution for which you use a calculator capability other than the four listed previously, you must write down the mathematical steps that yield the answer. A correct answer alone will not earn full credit and will likely earn no credit.

(c) You must provide *mathematical reasoning* to support your answer. Calculator results alone will not be sufficient.

The Procedures Explained

Here is more detailed guidance for the four allowed procedures.

C1. "Produce the graph of a function within an arbitrary viewing window." More than likely, you will not have to produce a graph on the exam that will be graded. However, you must be able to graph a wide variety of functions, both simple and complex, and be able to analyze those graphs. Skills you need include, but are not limited to, typing complex functions correctly into your calculator including correct notation, which will ensure that the graph on the screen is what the question writer intended you to see, and finding a window that accurately represents the graph and its features. Note that on rare occasions you may wish to draw a graph in your exam booklet to justify an answer in the free-response section; such a graph must be clearly labeled as to what is being graphed, and there should be an accompanying sentence or two explaining why the graph you produced justifies the answer.

C2. "Solve an equation numerically" is equivalent to "Find the zeros of a function" or "Find the point of intersection of two curves." Remember: you must first show your setup—write the equation out algebraically; then it is sufficient just to write down the calculator solution.

C3. "Compute the derivative of a function numerically." When you seek the value of the derivative of a function at a specific point, you may use your calculator. First, indicate what you are finding—for example, $f'(6)$—then write the numerical answer obtained from your calculator. Note that if you need to find the derivative of the function, rather than its value at a particular point, you must write the derivative symbolically. Note that some calculators are able to perform symbolic operations.

C4. "Compute definite integrals numerically." If, for example, you need to find the area under a curve, you must first show your setup. Write the complete integral, including the integrand in terms of a single variable, with the limits of integration. You may then simply write the calculator answer; you need not compute an antiderivative.

Accuracy

Calculator answers must be correct to three decimal places. To achieve this required accuracy, never type in decimal numbers unless they came from the original question. Do *not* round off numbers at intermediate steps, as this is likely to produce error accumulations resulting in a loss of credit. If necessary, store intermediate answers in the calculator's memory. Do *not* copy them down on paper; storing is faster and avoids transcription errors. Round off, or truncate, only after your calculator produces the final answer.

Sample Solution for a Free-Response Question

The following example question illustrates proper use of your calculator on the AP exam. This example has been simplified (compared to an actual free-response question); it is designed to illustrate just the procedures **(C1–C4)** that you can use by supplying the setup and the value from your calculator.

> **Example**

For $0 \leq t \leq 3$, a particle moves along the x-axis. The velocity of the particle at time t is given by $v(t) = 2 + 3 \cos\left(\frac{t^3}{8}\right)$. The particle is at position $x = 5$ at time $t = 2$.

(a) Find the acceleration of the particle at $t = 2$.

✓ **Solution**

The acceleration is the derivative of the velocity—that connection must be made in your work. **(C3)** Since the velocity function is defined, you can use the derivative at a point function on your calculator to find $v'(2)$. The calculator gives the value as $v'(2) = -3.78661943164$, which you may write as either $v'(2) = -3.787$ or $v'(2) = -3.786$ under the decimal presentation rules.

Your work should look like this:

$$a(2) = v'(2) = -3.787$$

(b) At what time(s) is the velocity of the particle equal to zero?

✓ Solution

(C2) You will need to solve the equation $v(t) = 0$. Some calculators have a solve function on the calculator/home screen, but sometimes they are a little difficult to work with. **(C1)** Our suggestion is to graph the function, $v(t)$, and use the calculator's root/zero function on the graph page. There is only one zero for $v(t)$ on the interval $0 \leq t \leq 3$, and it occurs at $t = 2.64021$.

Your work should look like this:

$$v(t) = 0$$

$$t = 2.640$$

(c) Find the position of the particle at $t = 1$.

✓ Solution

(C4) You will use a definite integral by using the Fundamental Theorem of Calculus (FTC) to find the position. The form of the FTC you use is $f(b) = f(a) + \int_a^b f(x)dx$. Using this form of the FTC, you need to know a value of the function, $f(a)$, and the rate of change of the function, $f'(x)$. Here we know the position at $t = 2$ (i.e., $x(2) = 5$), and the rate of change of the position is the velocity, $x'(t) = v(t)$. We want to find $x(1)$, so our setup using the FTC is $x(1) = x(2) + \int_2^1 x'(t)dt$, and our calculator gives a value of 0.4064274888. Notice in the work below that we left the integrand as $v(t)$; you may also do this on the AP Calculus exam since it is a defined function.

Your work should look like this:

$$x(1) = x(2) + \int_2^1 x'(t)dt$$

$$x(1) = 5 + \int_2^1 v(t)dt$$

$$x(1) = 0.406$$

A Note About Solutions in This Book

Students should be aware that in this book we sometimes do *not* observe the restrictions cited previously on the use of the calculator. When providing explanations for solutions to illustrative examples or to exercises, we often exploit the capabilities of the calculator to the fullest. Indeed, students are encouraged to do just that on any question in Section I, Part B of the AP examination for which they use a calculator. However, to avoid losing credit, you must carefully observe the restrictions imposed on when and how the calculator may be used when answering questions in Section II of the examination.

Additional Notes and Reminders

- SYNTAX. Learn the proper syntax for your calculator: the correct way to enter operations, functions, and other commands. Parentheses, commas, variables, or parameters that are missing or entered in the wrong order can produce error messages, waste time, or (worst of all) yield wrong answers.

- **RADIANS.** Keep your calculator set in radian mode. Almost all questions about angles and trigonometric functions use radians. If you ever need to change to degrees for a specific calculation, return the calculator to radian mode as soon as that calculation is complete.

- **TRIGONOMETRIC FUNCTIONS.** Many calculators do not have keys for the secant, cosecant, or cotangent function. To obtain these functions, use their reciprocals.

    For example, $\sec\left(\dfrac{\pi}{8}\right) = 1/\cos\left(\dfrac{\pi}{8}\right)$.

    Evaluate inverse functions such as arcsin, arccos, and arctan on your calculator. Those function keys are usually denoted as $\sin^{-1}$, $\cos^{-1}$, and $\tan^{-1}$.

    Don't confuse reciprocal functions with inverse functions. For example:

    $$\cos^{-1}\left(\frac{1}{2}\right) = \arccos\left(\frac{1}{2}\right) = \frac{\pi}{3}$$
    $$(\cos 2)^{-1} = \frac{1}{\cos 2} = \sec 2 \approx -2.403$$
    $$\cos 2^{-1} = \cos\left(\frac{1}{2}\right) \approx 0.878$$
    $$\cos^{-1}(2) = \arccos 2, \text{ which does not exist}$$

- **NUMERICAL DERIVATIVES.** You may be misled by your calculator if you ask for the derivative of a function at a point where the function is not differentiable because the calculator evaluates numerical derivatives using the difference quotient (or the symmetric difference quotient). For example, if $f(x) = |x|$, then $f'(0)$ does not exist. Yet the calculator may find the value of the derivative as

    $$\frac{f(x+0.001) - f(0)}{0.001} = 1 \quad \text{or} \quad \frac{f(x+0.001) - f(x-0.001)}{0.002} = 0$$

    Remember: always be sure f is differentiable at a before asking the calculator to evaluate $f'(a)$.

- **IMPROPER INTEGRALS.** Most calculators can compute only definite integrals. Avoid using yours to obtain an improper integral, such as

    $$\int_0^\pi \frac{1}{x^2}\,dx \quad \text{or} \quad \int_0^2 \frac{dx}{(x-1)^{2/3}}$$

- **FINAL ANSWERS TO SECTION II QUESTIONS.** Although we usually express a final answer in this book in simplest form (often evaluating it on the calculator), this is hardly ever necessary for Section II questions on the AP examination. According to the directions printed on the exam, "unless otherwise specified" (1) you need not simplify algebraic or numerical answers and (2) answers involving decimals should be correct to three places after the decimal point. However, be aware that if you try to simplify, you must do so correctly or you will lose credit.

- **USE YOUR CALCULATOR WISELY.** Bear in mind that you will not be allowed to use your calculator at all in Part A of Section I. In Part B of Section I and Part A of Section II, *only a few questions* will require one. The questions that require a calculator *will not be identified.* You will have to be sensitive not only to when it is necessary to use the calculator but also to when it is efficient to do so.

    The calculator is a marvelous tool, capable of illustrating complicated concepts with detailed pictures and of performing tasks that would otherwise be excessively time consuming—or even impossible. But the completion of calculations and the displaying of graphs on the calculator can be slow. Sometimes it is faster to find an answer using arithmetic, algebra, and analysis without recourse to the calculator. Before you start pushing buttons, take a few seconds to decide on the best way to solve a problem.

Grading the Examinations

Each completed AP examination paper receives a grade according to the following five-point scale:

5. Extremely well qualified
4. Well qualified
3. Qualified
2. Possibly qualified
1. No recommendation

Many colleges and universities accept a grade of 3 or better for credit or advanced placement or both; some also consider a grade of 2, while others require a grade of 4. (Students should check AP credit policies at individual colleges' websites.) More than 51 percent of the candidates who took the 2021 Calculus AB examination earned grades of 3, 4, or 5. Nearly 75 percent of the 2021 BC candidates earned a 3 or better. Altogether, about 375,000 students took the 2021 AP Calculus examinations.

The multiple-choice questions in Section I are scored by a machine. Students should note that *the score will be the number of questions answered correctly*. Since no points can be earned if answers are left blank and there is no deduction for wrong answers, *students should answer every question*. For questions they cannot do, students should try to eliminate as many of the choices as possible and then pick the best remaining answer.

The problems in Section II are graded by college and high school teachers called "readers." The answers in any one examination booklet are evaluated by different readers, and for each reader, all scores given by preceding readers are concealed, as are the student's name and school. Readers are provided sample solutions for each problem, with detailed scoring scales and point distributions that allow partial credit for correct portions of a student's answer. Problems in Section II are all counted equally.

In the determination of the overall grade for each examination, the two sections are given equal weight. The total raw score is then converted into one of the five grades listed previously. Students should not think of these raw scores as percents in the usual sense of testing and grading. A student who averages 6 out of 9 points on the Section II questions and performs similarly well on Section I's multiple-choice questions will typically earn a 5. Many colleges offer credit for a score of 3, historically awarded for earning over 40 of 108 possible points.

Students who take the BC examination are given not only a Calculus-BC grade but also a Calculus-AB subscore grade. The latter is based on the part of the BC examination that deals with topics in the AB syllabus.

In general, students will not be expected to answer all the questions correctly in either Section I or Section II.

Great care is taken by all involved in the scoring and reading of papers to make certain that they are graded consistently and fairly so that a student's overall AP grade reflects as accurately as possible his or her achievement in calculus.

Diagnostic Tests

ANSWER SHEET
Diagnostic Test Calculus AB

SECTION I

Part A

1. Ⓐ Ⓑ Ⓒ Ⓓ   9. Ⓐ Ⓑ Ⓒ Ⓓ   17. Ⓐ Ⓑ Ⓒ Ⓓ   25. Ⓐ Ⓑ Ⓒ Ⓓ
2. Ⓐ Ⓑ Ⓒ Ⓓ   10. Ⓐ Ⓑ Ⓒ Ⓓ   18. Ⓐ Ⓑ Ⓒ Ⓓ   26. Ⓐ Ⓑ Ⓒ Ⓓ
3. Ⓐ Ⓑ Ⓒ Ⓓ   11. Ⓐ Ⓑ Ⓒ Ⓓ   19. Ⓐ Ⓑ Ⓒ Ⓓ   27. Ⓐ Ⓑ Ⓒ Ⓓ
4. Ⓐ Ⓑ Ⓒ Ⓓ   12. Ⓐ Ⓑ Ⓒ Ⓓ   20. Ⓐ Ⓑ Ⓒ Ⓓ   28. Ⓐ Ⓑ Ⓒ Ⓓ
5. Ⓐ Ⓑ Ⓒ Ⓓ   13. Ⓐ Ⓑ Ⓒ Ⓓ   21. Ⓐ Ⓑ Ⓒ Ⓓ   29. Ⓐ Ⓑ Ⓒ Ⓓ
6. Ⓐ Ⓑ Ⓒ Ⓓ   14. Ⓐ Ⓑ Ⓒ Ⓓ   22. Ⓐ Ⓑ Ⓒ Ⓓ   30. Ⓐ Ⓑ Ⓒ Ⓓ
7. Ⓐ Ⓑ Ⓒ Ⓓ   15. Ⓐ Ⓑ Ⓒ Ⓓ   23. Ⓐ Ⓑ Ⓒ Ⓓ
8. Ⓐ Ⓑ Ⓒ Ⓓ   16. Ⓐ Ⓑ Ⓒ Ⓓ   24. Ⓐ Ⓑ Ⓒ Ⓓ

Part B

31. Ⓐ Ⓑ Ⓒ Ⓓ   35. Ⓐ Ⓑ Ⓒ Ⓓ   39. Ⓐ Ⓑ Ⓒ Ⓓ   43. Ⓐ Ⓑ Ⓒ Ⓓ
32. Ⓐ Ⓑ Ⓒ Ⓓ   36. Ⓐ Ⓑ Ⓒ Ⓓ   40. Ⓐ Ⓑ Ⓒ Ⓓ   44. Ⓐ Ⓑ Ⓒ Ⓓ
33. Ⓐ Ⓑ Ⓒ Ⓓ   37. Ⓐ Ⓑ Ⓒ Ⓓ   41. Ⓐ Ⓑ Ⓒ Ⓓ   45. Ⓐ Ⓑ Ⓒ Ⓓ
34. Ⓐ Ⓑ Ⓒ Ⓓ   38. Ⓐ Ⓑ Ⓒ Ⓓ   42. Ⓐ Ⓑ Ⓒ Ⓓ

Diagnostic Test Calculus AB

Section I

Part A

TIME: 60 MINUTES

The use of calculators is *not* permitted for this part of the examination.
There are 30 questions in Part A, for which 60 minutes are allowed.

DIRECTIONS: Choose the best answer for each question.

1. $\lim\limits_{x \to \infty} \dfrac{3x^2 - 4}{2 - 7x - x^2}$ is

   (A) -3

   (B) 0

   (C) 3

   (D) ∞

2. $\lim\limits_{h \to 0} \dfrac{\cos\left(\dfrac{\pi}{2} + h\right) - \cos\left(\dfrac{\pi}{2}\right)}{h}$ is

   (A) 1

   (B) nonexistent

   (C) 0

   (D) -1

3. If, for all x, $f'(x) = (x - 2)^4(x - 1)^3$, it follows that the function f has

   (A) a relative minimum at $x = 1$

   (B) a relative maximum at $x = 1$

   (C) both a relative minimum at $x = 1$ and a relative maximum at $x = 2$

   (D) relative minima at $x = 1$ and at $x = 2$

4. Let $F(x) = \displaystyle\int_0^x \dfrac{10}{1 + e^t}\,dt$. Which of the following statements is (are) true?

   I. $F'(0) = 5$

   II. $F(2) < F(6)$

   III. F is concave upward

   (A) I only

   (B) II only

   (C) I and II only

   (D) I and III only

5. If $f(x) = 10^x$ and $10^{1.04} \approx 10.96$, which is closest to $f'(1)$?

   (A) 0.92

   (B) 0.96

   (C) 10.5

   (D) 24

6. If f is differentiable, we can use the line tangent to f at $x = a$ to approximate values of f near $x = a$. Suppose that for a certain function f this method always <u>underestimates</u> the correct values. If so, then in an interval surrounding $x = a$, the graph of f must be

(A) increasing

(B) decreasing

(C) concave upward

(D) concave downward

7. If $f(x) = \cos x \sin 3x$, then $f'\left(\dfrac{\pi}{6}\right)$ is equal to

(A) $\dfrac{1}{2}$

(B) $-\dfrac{\sqrt{3}}{2}$

(C) $\dfrac{\sqrt{3}}{2}$

(D) $-\dfrac{1}{2}$

8. $\displaystyle\int_0^1 \dfrac{x}{x^2+1}\,dx$ is equal to

(A) $\dfrac{\pi}{4}$

(B) $\ln \sqrt{2}$

(C) $\dfrac{1}{2}(\ln 2 - 1)$

(D) $\ln 2$

9. The graph of f'' is shown below. If $f'(1) = 0$, then $f'(x) = 0$ at what other value of x on the interval $[0,8]$?

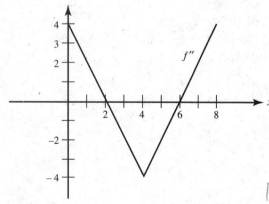

(A) 2

(B) 3

(C) 4

(D) 7

Questions 10 and 11. Use the following table, which shows the values of differentiable functions f and g.

x	f	f'	g	g'
1	2	$\dfrac{1}{2}$	-3	5
2	3	1	0	4
3	4	2	2	3
4	6	4	3	$\dfrac{1}{2}$

10. If $P(x) = (g(x))^2$, then $P'(3)$ equals

(A) 4

(B) 6

(C) 9

(D) 12

11. If $H(x) = f^{-1}(x)$, then $H'(3)$ equals

(A) $-\dfrac{1}{16}$

(B) $-\dfrac{1}{8}$

(C) $\dfrac{1}{2}$

(D) 1

12. The total area of the region bounded by the graph of $y = x\sqrt{1-x^2}$ and the x-axis is

(A) $\dfrac{1}{3}$

(B) $\dfrac{1}{2}$

(C) $\dfrac{2}{3}$

(D) 1

13. The graph of $y = \dfrac{1-x}{x-3}$ is concave upward when

(A) $x > 3$

(B) $1 < x < 3$

(C) $x < 1$

(D) $x < 3$

GO ON TO THE NEXT PAGE

14. As an ice block melts, the rate at which its mass, M, decreases is directly proportional to the square root of the mass. Which equation describes this relationship?

   (A) $\sqrt{M(t)} = kt$

   (B) $\dfrac{dM}{dt} = k\sqrt{t}$

   (C) $\dfrac{dM}{dt} = k\sqrt{M}$

   (D) $\dfrac{dM}{dt} = \dfrac{k}{\sqrt{M}}$

15. The average (mean) value of $\tan x$ on the interval from $x = 0$ to $x = \dfrac{\pi}{3}$ is

   (A) $\ln \dfrac{1}{2}$

   (B) $\dfrac{3}{\pi} \ln 2$

   (C) $\dfrac{\sqrt{3}}{2}$

   (D) $\dfrac{9}{\pi}$

16. If $y = x^2 \ln x$ for $x > 0$, then y'' is equal to

   (A) $3 + \ln x$

   (B) $3 + 2\ln x$

   (C) $3 + 3\ln x$

   (D) $2 + x + \ln x$

17. Water is poured at a constant rate into the conical reservoir shown in the figure. If the depth of the water, h, is graphed as a function of time, the graph is

   (A) constant

   (B) linear

   (C) concave upward

   (D) concave downward

18. If $f(x) = \begin{cases} x^2 & \text{for } x \le 1 \\ 2x - 1 & \text{for } x > 1 \end{cases}$, then

   (A) $f(x)$ is not continuous at $x = 1$

   (B) $f(x)$ is continuous at $x = 1$ but $f'(1)$ does not exist

   (C) $f'(1) = 2$

   (D) $\lim\limits_{x \to 1} f(x)$ does not exist

19. $\lim\limits_{x \to 2^-} \dfrac{|x - 2|}{x - 2}$ is

   (A) $-\infty$

   (B) -1

   (C) ∞

   (D) nonexistent

Questions 20 and 21. The graph below consists of a quarter-circle and two line segments and represents the velocity of an object during a 6-second interval.

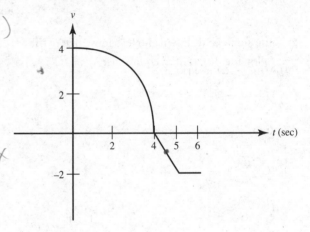

20. The object's average speed (in units/sec) during the 6-second interval is

   (A) $\dfrac{4\pi + 3}{6}$

   (B) $\dfrac{4\pi - 3}{6}$

   (C) 1

   (D) 1

21. The object's acceleration (in units/sec^2) at $t = 4.5$ is

(A) 0

(B) −1

(C) −2

(D) $-\frac{1}{4}$

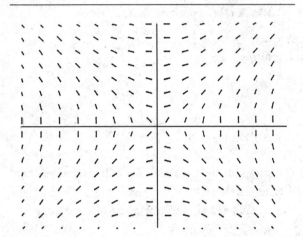

22. The slope field shown above is for which of the following differential equations?

(A) $\dfrac{dy}{dx} = -\dfrac{y}{x}$

(B) $\dfrac{dy}{dx} = \dfrac{1}{xy}$

(C) $\dfrac{dy}{dx} = -\dfrac{x}{y}$

(D) $\dfrac{dy}{dx} = \dfrac{x}{y}$

23. If y is a differentiable function of x, then the slope of the curve of $xy^2 - 2y + 4y^3 = 6$ at the point where $y = 1$ is

$x - 2 + 4 = 6$
$x + 2 = 6$
$x = 4$

(A) $\dfrac{1}{18}$

(B) $-\dfrac{1}{26}$

(C) $\dfrac{5}{18}$

(D) $-\dfrac{11}{18}$

$y^2 + x2yy' - 2y' + 12y^2y' = 0$

$x2yy' - 2y' + 12y^2y' = -y^2$

$y'(x2y - 2 + 12y^2) = -y^2$

$\underline{\qquad 2xy - 2 + 12y^2 \qquad}$

$y' = \dfrac{-y^2}{12y^2 + 2xy - 2}$

$y' = \dfrac{-1}{12 + 8 - 2} = \boxed{\dfrac{-1}{18}}$

24. In the following, $L(n)$, $R(n)$, $M(n)$, and $T(n)$ denote, respectively, left, right, midpoint, and trapezoidal sums with n equal subdivisions. Which of the following is *not* equal exactly to $\int_{-3}^{3} |x|\, dx$? $\dfrac{b-a}{n}$ $\dfrac{3+3}{n} = \dfrac{6}{n}$

(A) $L(2)$

(B) $T(3)$

(C) $M(4)$

(D) $R(6)$

25. The table shows some values of a differentiable function f and its derivative f':

x	0	1	2	3
$f(x)$	3	4	2	8
$f'(x)$	4	−1	1	10

Find $\int_{0}^{3} f'(x)\,dx.$

(A) 5 $8 - 3 = 5$

(B) 6

(C) 11.5

(D) 14

26. The solution of the differential equation $\dfrac{dy}{dx} = 2xy^2$ for which $y = -1$ when $x = 1$ is

(A) $y = -\dfrac{1}{x^2}$ for $x \neq 0$

(B) $y = -\dfrac{1}{x^2}$ for $x > 0$

(C) $\ln y^2 = x^2 - 1$ for all x

(D) $y = -\dfrac{1}{x}$ for $x > 0$

27. The base of a solid is the region bounded by the parabola $y^2 = 4x$ and the line $x = 2$. Each plane section perpendicular to the x-axis is a square. The volume of the solid is

(A) 8

(B) 16

(C) 32

(D) 64

GO ON TO THE NEXT PAGE

28. Which of the following could be the graph of $y = \dfrac{x^2}{e^x}$?

(A)

(B)

(C)

(D)

29. If $F(3) = 8$ and $F'(3) = -4$, then $F(3.02)$ is approximately

(A) 7.92
(B) 7.98
(C) 8.02
(D) 8.08

[handwritten: $y =$]

[handwritten: $y - y_1 = x - x_1$]

[handwritten: $y = (x - x_1 + y_1$]

[handwritten: $(-4 - 3.02) + 8$]

[handwritten: $-7.08 + 8$]

[handwritten: 1.02]

30. If $F(x) = \displaystyle\int_0^{2x} \dfrac{1}{1 - t^3}\, dt$, then $F'(x) =$

(A) $\dfrac{1}{1 - x^3}$

(B) $\dfrac{2}{1 - 2x^3}$

(C) $\dfrac{1}{1 - 8x^3}$

(D) $\dfrac{2}{1 - 8x^3}$

STOP   End of Part A, Section I

DIAGNOSTIC TEST CALCULUS AB

Part B

TIME: 45 MINUTES

Some questions in this part of the examination require the use of a graphing calculator. There are 15 questions in Part B, for which 45 minutes are allowed.

DIRECTIONS: Choose the best answer for each question. If the exact numerical value of the correct answer is not listed as a choice, select the choice that is closest to the exact numerical answer.

Questions 31 and 32. Refer to the graph of f' below.

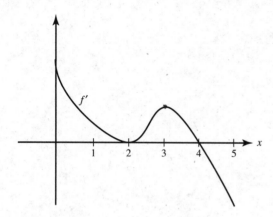

31. f has a local maximum at $x =$

    (A) 3 only
    (B) 4 only
    (C) 2 and 4
    (D) 3 and 4

32. The graph of f has a point of inflection at $x =$

    (A) 2 only
    (B) 3 only
    (C) 2 and 3 only
    (D) 2 and 4 only

33. For what value of c on $0 < x < 1$ is the tangent to the graph of $f(x) = e^x - x^2$ parallel to the secant line on the interval (0,1)?

    (A) 0.351
    (B) 0.500
    (C) 0.693
    (D) 0.718

34. Find the volume of the solid generated when the region bounded by the y-axis, $y = e^x$, and $y = 2$ is rotated around the y-axis.

    (A) 0.386
    (B) 0.592
    (C) 1.216
    (D) 3.998

35. The table below shows the "hit rate" for an Internet site, measured at various intervals during a day. Use a trapezoid approximation with 6 subintervals to estimate the total number of people who visited that site.

Time	Midnight	6 A.M.	8 A.M.
People per minute	5	2	3

Noon	5 P.M.	8 P.M.	Midnight
8	10	16	5

    (A) 5,280
    (B) 10,080
    (C) 10,440
    (D) 10,560

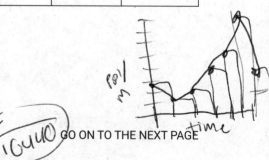

GO ON TO THE NEXT PAGE

36. The acceleration of a particle moving along a straight line is given by $a = 6t$. If, when $t = 0$, its velocity, v, is 1 and its position, s, is 3, then at any time t

    (A) $s = t^3 + 3$

    (B) $s = t^3 + t + 3$

    (C) $s = \dfrac{t^3}{3} + t + 3$

    (D) $s = \dfrac{t^3}{3} + \dfrac{t^2}{3} + 3$

37. If $y = f(x^2)$ and $f'(x) = \sqrt{5x - 1}$ then $\dfrac{dy}{dx}$ is equal to

    (A) $2x\sqrt{5x^2 - 1}$

    (B) $\sqrt{5x - 1}$

    (C) $2x\sqrt{5x - 1}$

    (D) $\dfrac{\sqrt{5x - 1}}{2x}$

38. Find the area of the first quadrant region bounded by $y = x^2$, $y = \cos(x)$, and the y-axis.

    (A) 0.292

    (B) 0.508

    (C) 0.547

    (D) 0.921

39. If the substitution $x = 2t + 1$ is used, which of the following is equivalent to $\displaystyle\int_0^3 \sqrt[4]{2t + 1}\, dt$?

    (A) $\displaystyle\int_0^3 \sqrt[4]{x}\, dx$

    (B) $\dfrac{1}{2}\displaystyle\int_0^3 \sqrt[4]{x}\, dx$

    (C) $\dfrac{1}{2}\displaystyle\int_{-1/2}^1 \sqrt[4]{x}\, dx$

    (D) $\displaystyle\int_1^7 \dfrac{1}{2} \sqrt[4]{x}\, dx$

40. An object moving along a line has velocity $v(t) = t\cos t - \ln(t + 2)$, where $0 \le t \le 10$. How many times does the object reverse direction?

    (A) one

    (B) two

    (C) three

    (D) four

41. A 26-foot ladder leans against a building so that its foot moves away from the building at the rate of 3 feet per second. When the foot of the ladder is 10 feet from the building, the top is moving down at the rate of r feet per second, where r is

    (A) 0.80

    (B) 1.25

    (C) 7.20

    (D) 12.50

$f(3) = 3$	$g(3) = -3$	$h(3) = 3$
$f'(3) = 2$	$g'(3) = -4$	$h'(3) = 4$
$f''(3) = 1/2$	$g''(3) = 8$	$h''(3) = 2$

42. The functions $f(x)$, $g(x)$, and $h(x)$ have derivatives of all orders. Listed above are values for the functions and their first and second derivatives at $x = 3$. Find $\displaystyle\lim_{x\to 3} \dfrac{(f(x))^2 - 3h(x)}{g(x) + h(x)}$.

    (A) $-\dfrac{1}{5}$

    (B) $\dfrac{1}{2}$

    (C) 1

    (D) nonexistent

43. The graph above shows an object's <u>acceleration</u> (in ft/sec²). It consists of a <u>quarter-circle</u> and two line segments. If the object was at rest at $t = 5$ seconds, what was its initial velocity?

    (A) -2 ft/sec

    (B) $3 - \pi$ ft/sec

    (C) $\pi - 3$ ft/sec

    (D) $\pi + 3$ ft/sec

44. Water is leaking from a tank at the rate of $R(t) = 5 \arctan\left(\dfrac{t}{5}\right)$ gallons per hour, where t is the number of hours since the leak began. To the nearest gallon, how much water will leak out during the first day?

24 hrs.

(A) 7
(B) 12
(C) 24
(D) 124

$R(24) =$

45. Find the y-intercept of the line tangent to $y = \left(x^3 - 4x^2 + 8\right)e^{\cos x^2}$ at $x = 2$.

calc? 4 to?

(A) 0
(B) 2.081
(C) 4.161
(D) 21.746

$y =$ $x_1 = 2$

$y - y_1 = mx - x_1$ $y_1 = 0$

$y = (mx - x_1) + y_1$

$3x^2 - 12x$

$(3x^2 - 8x)e^{\cos x^2} + (x^3 - 4x^2 + 8)$

STOP   End of Section I

Section II

Part A

TIME: 30 MINUTES
2 PROBLEMS

> A graphing calculator is required for some of these problems. See instructions on page 8.

1. When a faulty seam opened at the bottom of an elevated hopper, grain began leaking out onto the ground. After a while, a worker spotted the growing pile below and began making repairs. The following table shows how fast the grain was leaking (in cubic feet per minute) at various times during the 20 minutes it took to repair the hopper.

t (min)	0	4	5	7	10	12	18	20
$L(t)$ (ft^3/min)	4	7	9	8	6	5	2	0

   (a) Estimate $L'(15)$ using the data in the table. Show the computations that lead to your answer. Using correct units, explain the meaning of $L'(15)$ in the context of the problem.

(b) The falling grain forms a conical pile that the worker estimates to be 5 times as far across as it is deep. The pile was 3 feet deep when the repairs had been half-completed. How fast was the depth increasing then?

   NOTE: The volume of a cone with height h and radius r is given by: $V = \frac{1}{3}\pi r^2 h$.

(c) Use a trapezoidal sum with seven subintervals as indicated in the table to approximate $\int_0^{20} L(t)\,dt$. Using correct units, explain the meaning of $\int_0^{20} L(t)\,dt$ in the context of the problem.

2. An object in motion along the x-axis has velocity $v(t) = \left(t + e^t\right)\sin t^2$ for $1 \le t \le 3$.

   (a) At what time, t, is the object moving to the left?

   (b) Is the speed of the object increasing or decreasing when $t = 2$? Justify your answer.

   (c) At $t = 1$ this object's position was $x = 10$. What is the position of the object at $t = 3$?

STOP   End of Part A, Section II

Part B

TIME: 60 MINUTES

4 PROBLEMS

> No calculator is allowed for any of these problems.
>
> If you finish Part B before time has expired, you may return to work on Part A, but you may not use a calculator.

3. The graph of function f consists of the semicircle and line segment shown in the figure below.

   Define the area function $A(x) = \int_0^x f(t)\,dt$ for $0 \le x \le 18$.

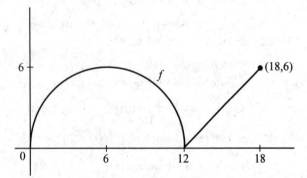

   (a) Find $A(6)$ and $A(18)$.

   (b) What is the average value of f on the interval $0 \le x \le 18$?

   (c) Write an equation of the line tangent to the graph of A at $x = 6$. Use the tangent line to estimate $A(7)$.

   (d) Give the coordinates of any points of inflection on the graph of A. Justify your answer.

4. Consider the curve: $2x^2 - 4xy + 3y^2 = 16$.

   (a) Show $\dfrac{dy}{dx} = \dfrac{2y - 2x}{3y - 2x}$.

   (b) Verify that there exists a point Q where the curve has both an x-coordinate of 4 and a slope of zero. Find the y-coordinate of point Q.

   (c) Find $\dfrac{d^2y}{dx^2}$ at point Q. Classify point Q as a local maximum, local minimum, or neither. Justify your answer.

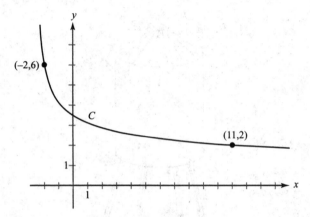

5. The graph above represents the curve C, given by $f(x) = \dfrac{6}{\sqrt[3]{2x + 5}}$ for $-2 \le x \le 11$.

   (a) Let R represent the region between C and the x-axis. Find the area of R.

   (b) Set up, but do not solve, an equation to find the value of k such that the line $x = k$ divides R into two regions of equal area.

   (c) Set up but do not evaluate an integral for the volume of the solid generated when R is rotated around the x-axis.

6. Let $y = f(x)$ be the function that has an x-intercept at $(2,0)$ and satisfies the differential equation $\dfrac{dy}{dx} = \dfrac{4}{x^2 e^y}$.

   (a) Write an equation for the line tangent to the graph of f at the point $(2,0)$.

   (b) Solve the differential equation, expressing y as a function of x and specifying the domain of the function.

   (c) Find an equation of each horizontal asymptote to the graph of $y = f(x)$.

STOP   **End of Test**

ANSWER KEY
Diagnostic Test Calculus AB

SECTION I

Part A

1. **A** (2)	9. **B** (6)	17. **D** (4)	25. **A** (6)
2. **D** (3)	10. **D** (3)	18. **C** (3)	26. **B** (9)
3. **A** (4)	11. **D** (3)	19. **B** (2)	27. **C** (7)
4. **C** (6)	12. **C** (7)	20. **A** (6)	28. **C** (2)
5. **D** (3)	13. **D** (4)	21. **C** (4)	29. **A** (4)
6. **C** (4)	14. **C** (9)	22. **D** (9)	30. **D** (4)
7. **D** (3)	15. **B** (6)	23. **A** (3)	
8. **B** (5)	16. **B** (3)	24. **B** (6)	

Part B

31. **B** (4)	35. **C** (6)	39. **D** (6)	43. **C** (6)
32. **C** (4)	36. **B** (8)	40. **B** (8)	44. **D** (8)
33. **A** (3)	37. **A** (3)	41. **B** (4)	45. **C** (4)
34. **B** (7)	38. **C** (7)	42. **B** (3)	

NOTE: Chapters that review and offer additional practice for each topic are specified in parentheses.

Answer Explanations

Section I Multiple-Choice

Part A

1. **(A)** Use the Rational Function Theorem (page 98); the ratio of the coefficients of the highest power of x is $\dfrac{3}{-1} = -3$.

2. **(D)** $\displaystyle\lim_{h\to 0}\dfrac{f\!\left(\dfrac{\pi}{2}+h\right)-f\!\left(\dfrac{\pi}{2}\right)}{h} = f'\!\left(\dfrac{\pi}{2}\right) \cdot f'(x) = -\sin(x)$, so $f'\!\left(\dfrac{\pi}{2}\right) = -\sin\!\left(\dfrac{\pi}{2}\right) = -1$.

3. **(A)** Since $f'(1) = 0$ and f' changes from negative to positive there, f reaches a minimum at $x = 1$. Although $f'(2) = 0$ as well, f' does not change sign there, and thus f has neither a maximum nor a minimum at $x = 2$.

4. **(C)** $F'(x) = \dfrac{10}{1 + e^x} > 0$, and $F''(x) = \dfrac{-10e^x}{\left(1 + e^x\right)^2} < 0$.

5. **(D)** $f'(1) \approx \dfrac{f(1.04) - f(1)}{1.04 - 1} = \dfrac{0.96}{0.04}$.

6. **(C)** The graph must look like one of these two:

7. **(D)** $f'(x) = 3\cos x \cos 3x - \sin x \sin 3x$.

$$f'\!\left(\dfrac{\pi}{6}\right) = 3\cos\!\left(\dfrac{\pi}{6}\right)\cos\!\left(\dfrac{\pi}{2}\right) - \sin\!\left(\dfrac{\pi}{6}\right)\sin\!\left(\dfrac{\pi}{2}\right) = 3 \cdot \dfrac{\sqrt{3}}{2} \cdot 0 - \dfrac{1}{2} \cdot 1$$

8. **(B)** $\displaystyle\int_0^1 \dfrac{x\,dx}{x^2 + 1} = \dfrac{1}{2}\ln\left(x^2 + 1\right)\Big|_0^1$

$$= \dfrac{1}{2}\left(\ln 2 - \ln 1\right)$$

$$= \ln \sqrt{2}$$

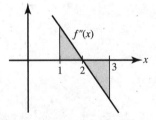

9. **(B)** Let $f'(x) = \displaystyle\int_1^x f''(t)\,dt$. Then f' increases for $1 < x < 2$, and then begins to decrease. In the figure above, the area below the x-axis, from 2 to 3, is equal in magnitude to that above the x-axis; hence, $\displaystyle\int_1^3 f''(t)\,dt = 0$.

10. **(D)** $P'(x) = 2g(x) \cdot g'(x)$; $P'(3) = 2g(3) \cdot g'(3) = 2 \cdot 2 \cdot 3 = 12$.

11. **(D)** Note that $H(3) = f^{-1}(3) = 2$. Therefore

$$H'(3) = \dfrac{1}{f'(H(3))} = \dfrac{1}{f'(2)} = 1$$

12. **(C)** Note that the domain of y is all x such that $|x| \leq 1$ and that the graph is symmetric to the origin. The area is given by

$$2\int_0^1 x\sqrt{1-x^2}\ dx \Rightarrow u = 1 - x^2; \ du = -2x\,dx$$

So, $2\displaystyle\int_0^1 x\sqrt{1-x^2}\ dx = 2\int_1^0 x\ \sqrt{u}\ \frac{du}{-2x} = -1\int_1^0 \sqrt{u}\ du = \int_0^1 u^{1/2}\ du = \frac{2}{3}u^{3/2}\Big|_0^1 = \frac{2}{3}(1-0)$

13. **(D)** Since

$$y' = 2(x-3)^{-2} \text{ and } y'' = -4(x-3)^{-3} = \frac{-4}{(x-3)^3}$$

y'' is positive when $x < 3$.

14. **(C)** $\dfrac{dM}{dt}$ represents the rate of change of mass with respect to time; y is directly proportional to the square root of x if $y = k\sqrt{x}$.

15. **(B)** $\dfrac{1}{\pi/3}\displaystyle\int_0^{\pi/3} \tan x\ dx = \frac{3}{\pi}\big[-\ln\cos x\big]\Big|_0^{\pi/3} = \frac{3}{\pi}\Big(-\ln\frac{1}{2}\Big)$.

16. **(B)** $y' = \Big(x^2 \cdot \dfrac{1}{x}\Big) + 2x\ln x = x + 2x\ln x$ and $y'' = 1 + \Big(2x \cdot \dfrac{1}{x} + 2\ln x\Big) = 3 + 2\ln x$.

17. **(D)** As the water gets deeper, the depth increases more slowly. Hence, the rate of change of depth decreases: $\dfrac{d^2h}{dt^2} < 0$.

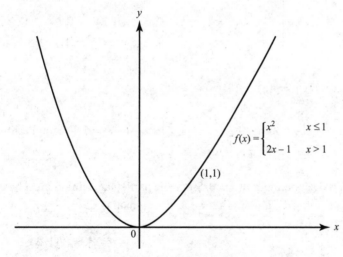

Figure for Solution 18

18. **(C)** The graph of f is shown in the figure above; f is defined and continuous at all x, including $x = 1$. Since

$$\lim_{x\to 1^-} f'(x) = 2 = \lim_{x\to 1^+} f'(x)$$

$f'(1)$ exists and is equal to 2.

19. **(B)** Since $|x-2| = 2 - x$ if $x < 2$, the limit as $x \to 2^-$ is $\dfrac{2-x}{x-2} = -1$.

20. **(A)** Average speed $= \dfrac{\text{distance covered in 6 sec}}{\text{time elapsed}}$

$$= \dfrac{\frac{1}{4}\pi\left(4^2\right) + \frac{1}{2}(1\cdot 2) + 1\cdot 2}{6}$$

Note that the distance covered in 6 seconds is $\displaystyle\int_0^6 |v(t)|\, dt$, the area between the velocity curve and the t-axis.

21. **(C)** Acceleration is the slope of the velocity curve, $\dfrac{-2-0}{5-4}$.

22. **(D)** Slopes are: 1 along $y = x$, -1 along $y = -x$, 0 along $x = 0$, and undefined along $y = 0$.

23. **(A)** Differentiating implicitly yields $2xyy' + y^2 - 2y' + 12y^2y' = 0$. When $y = 1$, $x = 4$. Substitute to find y'.

24. **(B)**

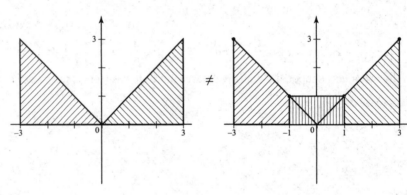

25. **(A)** $\displaystyle\int_0^3 f'(x)dx = f(x)\Big|_0^3 = f(3) - f(0) = 8 - 3 = 5.$

26. **(B)** Separate to get $\dfrac{dy}{y^2} = 2x\, dx$, $-\dfrac{1}{y} = x^2 + C$. Since $-(-1) = 1 + C$ implies that $C = 0$, the solution is $-\dfrac{1}{y} = x^2$ or $y = -\dfrac{1}{x^2}$.

This function is discontinuous at $x = 0$. Since the particular solution must be differentiable in an interval containing the initial value $x = 1$, the domain is $x > 0$.

27. **(C)**

$$\Delta V = (2y)^2\Delta x$$
$$V = 4\int_0^2 y^2 dx$$
$$= 4\int_0^2 4x\, dx$$
$$= 32$$

28. **(C)** Note that $\displaystyle\lim_{x\to\infty}\frac{x^2}{e^x} = 0$, $\displaystyle\lim_{x\to-\infty}\frac{x^2}{e^x} = \infty$, and $\dfrac{x^2}{e^x} \geq 0$ for all x.

29. **(A)** At $x = 3$, an equation of the tangent line is $y - 8 = -4(x - 3)$, so $f(x) \approx -4(x - 3) + 8$. $f(3.02) \approx -4(0.02) + 8$.

30. **(D)** Let $u = 2x$ and note that $F'(x) = \dfrac{1}{1 - u^3}$.

    Then $F'(x) = F'(u)u'(x) = 2F'(u) = 2 \cdot \dfrac{1}{1 - (2x)^3}$.

Part B

31. **(B)** The sign diagram shows that f changes from increasing to decreasing at $x = 4$

$$f \quad \underset{0 \qquad\quad 2 \qquad\quad 4 \qquad\quad 5}{\underbrace{\quad \text{inc} \quad | \quad \text{inc} \quad | \quad \text{dec} \quad}}$$

$$f' \qquad\qquad + \qquad\qquad + \qquad\qquad -$$

and thus f has a maximum there. Because f increases to the right of $x = 0$ and decreases to the left of $x = 5$, there are minima at the endpoints.

32. **(C)** Since f' decreases, increases, then decreases, f'' changes from negative to positive, then back to negative. Hence, the graph of f changes concavity at $x = 2$ and $x = 3$.

33. **(A)**

On the curve of $f(x) = e^x - x^2$, the two points labeled are $(0,1)$ and $(1, e - 1)$.

The slope of the secant line is $m = \dfrac{\Delta y}{\Delta x} = \dfrac{e - 2}{1} = e - 2$. Find c in $[0,1]$ such that $f'(c) = e - 2$,

or $f'(c) - (e - 2) = 0$. Since $f'(x) = e^x - 2x$, c can be calculated by solving $0 = e^x - 2x - (e - 2)$. The answer is 0.351.

34. **(B)**

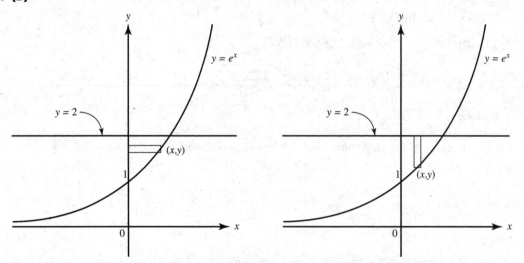

Use disks; then $\Delta V = \pi R^2 H = \pi(\ln y)^2 \,\Delta y$. Note that the limits of the definite integral are 1 and 2. Evaluate the integral

$$\pi \int_1^2 (\ln y)^2 \, dy = 0.592$$

Alternatively, use shells*; then $\Delta V = 2\pi RHT = 2\pi x(2 - e^x)\,\Delta x$. Here, the upper limit of integration is the value of x for which $e^x = 2$, namely, $\ln 2$. Now evaluate

$$2\pi \int_0^{\ln 2} x(2 - e^x) \, dx = 0.592$$

35. **(C)** Note that the rate is people *per minute*, so the first interval width from midnight to 6 A.M. is 360 minutes. The total number of people is estimated as the sum of the areas of six trapezoids:

$$T = 360\left(\frac{5+2}{2}\right) + 120\left(\frac{2+3}{2}\right) + 240\left(\frac{3+8}{2}\right) + 300\left(\frac{8+10}{2}\right) + 180\left(\frac{10+16}{2}\right) + 240\left(\frac{16+5}{2}\right) = 10{,}440$$

36. **(B)** $a = \dfrac{dv}{dt} = 6t$, so $v = 3t^3 + c$.

    Since $v = 1$ when $t = 0$, $c = 1$.

    Now $v = \dfrac{ds}{dt} = 3t^2 + 1$, so $s = t^3 + t + c$.

    Since $s = 3$ when $t = 0$, $c = 3$; then $s = t^3 + t + 3$.

37. **(A)** Let $u = x^2$. Then

$$\frac{dy}{dx} = \frac{dy}{du} \cdot \frac{du}{dx} = \frac{df}{du} \cdot f'(u)\frac{du}{dx} = \sqrt{5u - 1} \cdot 2x = 2x\sqrt{5x^2 - 1}$$

---

* Questions that require the use of the shells method will not appear on the AP exam.

38. **(C)**

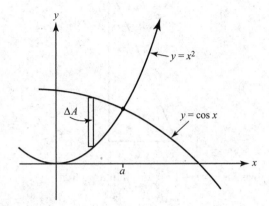

To find a, the point of intersection of $y = x^2$ and $y = \cos(x)$, use your calculator to solve the equation $x^2 - \cos(x) = 0$. (Store the value for later use; $a \approx 0.8241$.)

As shown in the diagram above, $\Delta A = (\cos(x) - x^2)\Delta x$.

Evaluate the area: $A = \displaystyle\int_0^a \left(\cos(x) - x^2 \right) dx \approx 0.547$.

39. **(D)** If $x = 2t + 1$, then $t = \dfrac{x - 1}{2}$, so $dt = \dfrac{1}{2} dx$. When $t = 0$, $x = 1$; when $t = 3$, $x = 7$.

40. **(B)**

The velocity is graphed in $[-1,11] \times [-15,5]$. The object reverses direction when the velocity changes sign, that is, when the graph crosses the x-axis. There are two such reversals—at $x = a$ and at $x = b$.

41. **(B)** See the figure above. Since $x^2 + y^2 = 26^2$, it follows that

$$2x\frac{dx}{dt} + 2y\frac{dy}{dt} = 0$$

at any time t. When $x = 10$, then $y = 24$ and it is given that $\frac{dx}{dt} = 3$.

Hence, $2(10)(3) + 2(24)\frac{dy}{dt} = 0$, so $\frac{dy}{dt} = -\frac{5}{4}$.

42. **(B)** Since this is the limit of a quotient, first try the quotient property of limits. Take the limits of the numerator and denominator separately, and since the functions have derivatives of all orders, the function and all its derivatives are continuous, so use substitution when finding limits.

$$\lim_{x \to 3}\left((f(x))^2 - 3h(x)\right) = 3^2 - 3 \cdot 3 = 0 \qquad \lim_{x \to 3}(g(x) + h(x)) = -3 + 3 = 0$$

Since the limit of the denominator is zero, we cannot use the quotient property of limits. Notice that the limit of both the numerator and the denominator are zero, which means that L'Hospital's Rule may work. Using L'Hospital's Rule, we get:

$$\lim_{x \to 3}\frac{(f(x))^2 - 3h(x)}{g(x) + h(x)} = \lim_{x \to 3}\frac{2f(x) \cdot f'(x) - 3h'(x)}{g'(x) + h'(x)}$$

By substitution, we find:

$$\lim_{x \to 3}(2f(x) \cdot f'(x) - 3h'(x)) = 2(3)(2) - 3(4) = 0 \text{ and } \lim_{x \to 3}(g'(x) + h'(x)) = -4 + 4 = 0$$

Again, the limits of both the numerator and the denominator are zero, so try L'Hospital's Rule a second time.

$$\lim_{x \to 3}\frac{2f(x) \cdot f'(x) - 3h'(x)}{g'(x) + h'(x)} = \lim_{x \to 3}\frac{2f(x) \cdot f''(x) + 2f'(x) \cdot f'(x) - 3h''(x)}{g''(x) + h''(x)}$$

By substitution, we find:

$$\lim_{x \to 3}\frac{2f(x) \cdot f''(x) + 2f'(x) \cdot f'(x) - 3h''(x)}{g''(x) + h''(x)} = \frac{2(3)(1/2) + 2(2)(2) - 3(2)}{8 + 2} = \frac{5}{10} = \frac{1}{2}$$

43. **(C)** $v(5) - v(0) = \int_0^5 a(t)\,dt = -\frac{1}{4}\pi \cdot 2^2 + \frac{1}{2}(3)(2) = -\pi + 3$. Since $v(5) = 0$, $-v(0) = -\pi + 3$; so $v(0) = \pi - 3$.

44. **(D)** $\int_0^{24} 5\arctan\left(\frac{t}{5}\right)dt = 124.102$.

45. **(C)** Let $y = (x^3 - 4x^2 + 8)e^{\cos(x^2)}$. An equation of the tangent at point $(2, y(2))$ is $y - y(2) = y'(2)(x - 2)$. Note that $y(2) = 0$. To find the y-intercept, let $x = 0$ and solve for y: $y = -2y'(2)$. A calculator yields $y = 4.161$.

Section II Free-Response

Part A

AB/BC 1. (a) $L'(15) \approx \dfrac{2-5}{18-12} = -0.5 \text{ ft}^3/\text{min}/\text{min}$

At 15 minutes, the rate at which grain is leaking is decreasing by one-half a cubic foot per minute per minute.   **(Review Chapter 3)**

(b) Let $h =$ the height of the cone and $r =$ its radius. The cone's diameter is given to be $5h$, so $r = \dfrac{5}{2}h$, and the cone's volume,

$$V = \frac{1}{3}\pi r^2 h = \frac{1}{3}\pi\left(\frac{5}{2}h\right)^2 h = \frac{25}{12}\pi h^3$$

Then

$$\frac{dV}{dt} = \frac{25}{4}\pi h^2 \frac{dh}{dt}$$

At $t = 10$ the table shows $L = \dfrac{dV}{dt} = 6$, and it is given that $h = 3$; thus:

$$6 = \frac{25}{4}\pi 3^2 \frac{dh}{dt}, \text{ so } \frac{dh}{dt} = \frac{8}{75\pi} \text{ ft/min}$$

(Review Chapter 4)

(c) $\displaystyle\int_0^{20} L(t)dt \approx (4-0)\left(\frac{7+4}{2}\right) + (5-4)\left(\frac{9+7}{2}\right) + (7-5)\left(\frac{8+9}{2}\right) + (10-7)\left(\frac{6+8}{2}\right) +$

$(12-10)\left(\dfrac{5+6}{2}\right) + (18-12)\left(\dfrac{2+5}{2}\right) + (20-18)\left(\dfrac{0+2}{2}\right) = 102 \text{ ft}^3$

The total amount of grain that leaked out of the elevated hopper while the repairs were underway was approximately 102 ft³.

(Review Chapter 6)

AB 2. Graph $y = (x + e^x)\sin(x^2)$ in $[1,3] \times [-15,20]$. Note that y represents velocity v and x represents time t.

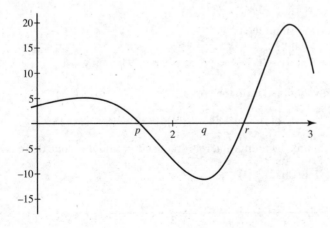

(a) The object moves to the left when the velocity is negative, namely, on the interval $p < t < r$. Use the calculator to solve $(x + e^x)\left(\sin\left(x^2\right)\right) = 0$; then $p = 1.772$ and $r = 2.507$. The answer is $1.772 < t < 2.507$.   **(Review Chapter 4)**

(b) At $t = 2$, the object is speeding up; because $v(2) = -7.106 < 0$ and $a(2) = -30.897 < 0$, the object is accelerating in the direction of motion. (*NOTE:* You can use your calculator to find both $v(2)$ and $a(2)$.)   **(Review Chapter 4)**

(c) $x(3) = x(1) + \int_1^3 x'(t)dt = x(1) + \int_1^3 v(t)\,dt = 10 + 4.491 = 14.491$

(Review Chapter 8)

Part B

AB/BC 3. (a) $A(6) = \frac{1}{4}\left(\pi 6^2\right) = 9\pi$, $A(18) = \frac{1}{2}\left(\pi 6^2\right) + \frac{1}{2}(6)(6) = 18\pi + 18$.   **(Review Chapter 6)**

(b) The average value of $f = \dfrac{\displaystyle\int_0^{18} f(x)dx}{18 - 0} = \dfrac{18\pi + 18}{18} = \pi + 1$.   **(Review Chapter 6)**

(c) The line tangent to the graph of A at $x = 6$ passes through point $(6, A(6))$ or $(6, 9\pi)$. Since $A'(x) = f(x)$, the graph of f shows that $A'(6) = f(6) = 6$. Hence, an equation of the line is $y - 9\pi = 6(x - 6)$.

Use the tangent line; then $A(x) \approx y = 6(x - 6) + 9\pi$, so $A(7) \approx 6(7 - 6) + 9\pi = 6 + 9\pi$.
(Review Chapter 4)

(d) Since f is increasing on $[0,6]$, f' is positive there. Because $f(x) = A'(x)$, $f'(x) = A''(x)$; thus A is concave upward for $[0,6]$. Similarly, the graph of A is concave downward for $[6,12]$ and upward for $[12,18]$. There are points of inflection on the graph of A at $(6, 9\pi)$ and $(12, 18\pi)$.   **(Review Chapter 4)**

AB 4.   (a) $4x - 4xy' - 4y + 6yy' = 0$

$(6y - 4x)\, y' = 4x - 4y$

$\dfrac{dy}{dx} = \dfrac{2y - 2x}{3y - 2x}$

(Review Chapter 3)

(b) $\dfrac{dy}{dx} = 0 \Rightarrow 2y - 2x = 0 \Rightarrow y = x$; when $x = 4$, then $y = 4$.

Verify that $(4,4)$ is on the curve:

$$2(4)^2 - 4(4)(4) + 3(4)^2 = 32 - 64 + 48 = 16$$

Therefore, the point $Q(4,4)$ is on the curve and the slope there is zero.

(Review Chapter 3)

(c) $y'' = \dfrac{(3y - 2x)(2y' - 2) - (2y - 2x)(3y' - 2)}{(3y - 2x)^2}$

$y''\Big|_{(4,4)} = \dfrac{(4)(-2) - (0)(-2)}{(4)^2} = -\dfrac{1}{2}$

At $Q = (4,4)$ the curve has a local maximum because $y' = 0$ and $y'' < 0$.

(Review Chapters 3 and 4)

AB/BC 5. (a) Draw elements as shown. Then

$$\Delta A = y\,\Delta x = \frac{6}{\sqrt[3]{2x+5}}\,\Delta x$$

$$A = \int_{-2}^{11} \frac{6}{\sqrt[3]{2x+5}}\,dx = \frac{6}{2}\int_{-2}^{11}(2x+5)^{-1/3}\,(2\,dx)$$

$$= 3\cdot\frac{3}{2}(2x+5)^{2/3}\Big|_{-2}^{11} = \frac{9}{2}(27^{2/3}-1^{2/3}) = 36$$

(b) $$\int_{-2}^{k}\frac{6}{\sqrt[3]{2x+5}}\,dx = \int_{k}^{11}\frac{6}{\sqrt[3]{2x+5}}\,dx$$

(c) Revolving the element around the x-axis generates disks. Then

$$\Delta V = \pi r^2\,\Delta x = \pi y^2\,\Delta x = \pi\left(\frac{6}{\sqrt[3]{2x+5}}\right)^2\Delta x,\ \text{so}$$

$$V = \pi\int_{-2}^{11}\frac{36}{(2x+5)^{2/3}}\,dx$$

(Review Chapter 7)

AB 6. (a) $\dfrac{dy}{dx}\Big|_{(x,y)=(2,0)} = \dfrac{4}{(2)^2\,e^0} = 1.$ An equation for the tangent line is

$y = 0 + 1(x-2) = x - 2.$

(Review Chapter 4)

(b) The differential equation $\dfrac{dy}{dx} = \dfrac{4}{x^2\,e^y}$ is separable:

$$\int e^y dy = \int \frac{4}{x^2}\,dx$$

$$e^y = -\frac{4}{x} + c$$

If $y = 0$ when $x = 2$, then $e^0 = -\dfrac{4}{2} + c$; thus $c = 3$, and $e^y = -\dfrac{4}{x} + 3.$

Solving for y gives the solution: $y = \ln\left(3 - \dfrac{4}{x}\right).$

Domain: The domain of a solution to a differential equation is the largest contiguous open interval containing the initial condition. For this solution, the initial condition is at $x = 2$. Since $x \neq 0$ from the differential equation (d.e.) (because x^2 is in the denominator), that means the largest the domain can be is $x > 0$.

Next, determine whether or not the solution adds any further restrictions.

Note that $y = \ln\left(3 - \dfrac{4}{x}\right)$ is defined only if $3 - \dfrac{4}{x} > 0$.

$\dfrac{3x - 4}{x} > 0$ only if the numerator and denominator have the same sign.

We already know that $x > 0$, so $3x - 4 > 0 \Rightarrow x > \dfrac{4}{3}$. Since the domain must contain $x = 2$, the domain is $x > \dfrac{4}{3}$.

(Review Chapter 9)

(c) Since $\lim\limits_{x \to \pm\infty} \ln\left(3 - \dfrac{4}{x}\right) = \ln 3$, the function $y = \ln\left(3 - \dfrac{4}{x}\right)$ has a horizontal asymptote at $y = \ln 3$.

(Review Chapter 2)

ANSWER SHEET
Diagnostic Test Calculus BC

SECTION I
Part A

1. Ⓐ Ⓑ Ⓒ Ⓓ
2. Ⓐ Ⓑ Ⓒ Ⓓ
3. Ⓐ Ⓑ Ⓒ Ⓓ
4. Ⓐ Ⓑ Ⓒ Ⓓ
5. Ⓐ Ⓑ Ⓒ Ⓓ
6. Ⓐ Ⓑ Ⓒ Ⓓ
7. Ⓐ Ⓑ Ⓒ Ⓓ
8. Ⓐ Ⓑ Ⓒ Ⓓ

9. Ⓐ Ⓑ Ⓒ Ⓓ
10. Ⓐ Ⓑ Ⓒ Ⓓ
11. Ⓐ Ⓑ Ⓒ Ⓓ
12. Ⓐ Ⓑ Ⓒ Ⓓ
13. Ⓐ Ⓑ Ⓒ Ⓓ
14. Ⓐ Ⓑ Ⓒ Ⓓ
15. Ⓐ Ⓑ Ⓒ Ⓓ
16. Ⓐ Ⓑ Ⓒ Ⓓ

17. Ⓐ Ⓑ Ⓒ Ⓓ
18. Ⓐ Ⓑ Ⓒ Ⓓ
19. Ⓐ Ⓑ Ⓒ Ⓓ
20. Ⓐ Ⓑ Ⓒ Ⓓ
21. Ⓐ Ⓑ Ⓒ Ⓓ
22. Ⓐ Ⓑ Ⓒ Ⓓ
23. Ⓐ Ⓑ Ⓒ Ⓓ
24. Ⓐ Ⓑ Ⓒ Ⓓ

25. Ⓐ Ⓑ Ⓒ Ⓓ
26. Ⓐ Ⓑ Ⓒ Ⓓ
27. Ⓐ Ⓑ Ⓒ Ⓓ
28. Ⓐ Ⓑ Ⓒ Ⓓ
29. Ⓐ Ⓑ Ⓒ Ⓓ
30. Ⓐ Ⓑ Ⓒ Ⓓ

Part B

31. Ⓐ Ⓑ Ⓒ Ⓓ
32. Ⓐ Ⓑ Ⓒ Ⓓ
33. Ⓐ Ⓑ Ⓒ Ⓓ
34. Ⓐ Ⓑ Ⓒ Ⓓ

35. Ⓐ Ⓑ Ⓒ Ⓓ
36. Ⓐ Ⓑ Ⓒ Ⓓ
37. Ⓐ Ⓑ Ⓒ Ⓓ
38. Ⓐ Ⓑ Ⓒ Ⓓ

39. Ⓐ Ⓑ Ⓒ Ⓓ
40. Ⓐ Ⓑ Ⓒ Ⓓ
41. Ⓐ Ⓑ Ⓒ Ⓓ
42. Ⓐ Ⓑ Ⓒ Ⓓ

43. Ⓐ Ⓑ Ⓒ Ⓓ
44. Ⓐ Ⓑ Ⓒ Ⓓ
45. Ⓐ Ⓑ Ⓒ Ⓓ

Diagnostic Test Calculus BC

Section I

Part A

TIME: 60 MINUTES

The use of calculators is *not permitted for this part of the examination.*
There are 30 questions in Part A, for which 60 minutes are allowed.

DIRECTIONS: Choose the best answer for each question.

1. A particle moves along the parametric curve given by $x(t) = e^{t^2+2}$ and $y(t) = (t^3+2)^2$. Which of the following is the velocity vector at time $t=1$?

   (A) $\langle 2e^3, 18 \rangle$
   (B) $\langle 2e^3, 6 \rangle$
   (C) $\langle e^3, 6 \rangle$
   (D) $\langle e^3, 18 \rangle$

2. $\displaystyle\lim_{h \to 0} \frac{\cos\left(\dfrac{\pi}{2} + h\right) - \cos\left(\dfrac{\pi}{2}\right)}{h}$ is

   (A) 1
   (B) nonexistent
   (C) 0
   (D) −1

3. If, for all x, $f'(x) = (x-2)^4(x-1)^3$, it follows that the function f has

   (A) a relative minimum at $x=1$
   (B) a relative maximum at $x=1$
   (C) both a relative minimum at $x=1$ and a relative maximum at $x=1$
   (D) relative minima at $x=1$ and at $x=2$

4. Let $F(x) = \displaystyle\int_0^x \frac{10}{1+e^t}\, dt$. Which of the following statements is (are) true?

   I. $F'(0) = 5$
   II. $F(2) < F(6)$
   III. F is concave upward

   (A) I only
   (B) II only
   (C) I and II only
   (D) I and III only

5. If $f(x) = 10^x$ and $10^{1.04} \approx 10.96$, which is closest to $f'(1)$?

   (A) 0.92
   (B) 0.96
   (C) 10.5
   (D) 24

GO ON TO THE NEXT PAGE

6. If f is differentiable, we can use the line tangent to f at $x = a$ to approximate values of f near $x = a$. Suppose that for a certain function f this method always underestimates the correct values. If so, then in an interval surrounding $x = a$, the graph of f must be

   (A) increasing
   (B) decreasing
   (C) concave upward
   (D) concave downward

7. The region in the first quadrant bounded by the x-axis, the y-axis, and the curve of $y = e^{-x}$ is rotated about the x-axis. The volume of the solid obtained is equal to

   (A) $\dfrac{1}{2}$

   (B) $\dfrac{\pi}{2}$

   (C) π

   (D) 2π

8. Of the following, which is the Maclaurin series for $x^2 \sin(x^2)$?

   (A) $x^3 - \dfrac{x^7}{3!} + \dfrac{x^{11}}{5!} - \dfrac{x^{15}}{7!} + \cdots$

   (B) $x^4 - \dfrac{x^8}{3!} + \dfrac{x^{12}}{5!} - \dfrac{x^{16}}{7!} + \cdots$

   (C) $x^2 - \dfrac{x^6}{2!} + \dfrac{x^{10}}{4!} - \dfrac{x^{14}}{6!} + \cdots$

   (D) $x - \dfrac{x^5}{2!} + \dfrac{x^9}{4!} - \dfrac{x^{13}}{6!} + \cdots$

9. Which series diverges?

   (A) $\displaystyle\sum_{n=1}^{\infty} \dfrac{(-1)^n}{n^5}$

   (B) $\displaystyle\sum_{n=1}^{\infty} \dfrac{(-1)^n}{\sqrt[5]{n}}$

   (C) $\displaystyle\sum_{n=1}^{\infty} \dfrac{(-1)^n}{5n+1}$

   (D) $\displaystyle\sum_{n=1}^{\infty} \dfrac{(-1)^n \cdot n}{5n+1}$

10. $\displaystyle\int_1^{\infty} \dfrac{12}{(x+8)^{3/2}}\, dx$ is

    (A) $\dfrac{4}{9}$

    (B) 4

    (C) 8

    (D) divergent

x	f	f'	g	g'
1	2	$\dfrac{1}{2}$	-3	5
2	3	1	0	4
3	4	2	2	3
4	6	4	3	$\dfrac{1}{2}$

11. The table above gives values of differentiable functions f and g. If $H(x) = f^{-1}(x)$, then $H'(3)$ equals

    (A) $-\dfrac{1}{16}$

    (B) $-\dfrac{1}{8}$

    (C) $\dfrac{1}{2}$

    (D) 1

12. $\displaystyle\int_0^1 x e^x\, dx$ equals

    (A) 1
    (B) -1
    (C) $\dfrac{e}{2}$
    (D) $e - 1$

13. The graph of $y = \dfrac{1-x}{x-3}$ is concave upward when

    (A) $x > 3$
    (B) $1 < x < 3$
    (C) $x < 1$
    (D) $x < 3$

14. As an ice block melts, the rate at which its mass, M, decreases is directly proportional to the square root of the mass. Which equation describes this relationship?

    (A) $\sqrt{M(t)} = kt$

    (B) $\dfrac{dM}{dt} = k\sqrt{t}$

    (C) $\dfrac{dM}{dt} = k\sqrt{M}$

    (D) $\dfrac{dM}{dt} = \dfrac{k}{\sqrt{M}}$

15. The length of the curve $y = 2x^{3/2}$ between $x = 0$ and $x = 1$ is equal to

    (A) $\dfrac{2}{27}\left(10^{3/2}\right)$

    (B) $\dfrac{2}{27}\left(10^{3/2} - 1\right)$

    (C) $\dfrac{2}{3}\left(10^{3/2}\right)$

    (D) $\sqrt{5}$

16. If $y = x^2 \ln x$, for $x > 0$, then y'' is equal to

    (A) $3 + \ln x$

    (B) $3 + 2 \ln x$

    (C) $3 + 3 \ln x$

    (D) $2 + x + \ln x$

17. Water is poured at a constant rate into the conical reservoir shown above. If the depth of the water, h, is graphed as a function of time, the graph is

    (A) constant

    (B) linear

    (C) concave upward

    (D) concave downward

18. A particle moves along the curve given parametrically by $x = \tan t$ and $y = 2 \sin t$. At the instant when $t = \dfrac{\pi}{3}$, the particle's speed equals

    (A) $\sqrt{3}$

    (B) $\sqrt{5}$

    (C) $\sqrt{6}$

    (D) $\sqrt{17}$

19. Suppose $\dfrac{dy}{dx} = \dfrac{10x}{x+y}$ and $y = 2$ when $x = 0$. Use Euler's method with two steps to estimate y at $x = 1$.

    (A) 1

    (B) 2

    (C) 3

    (D) 5

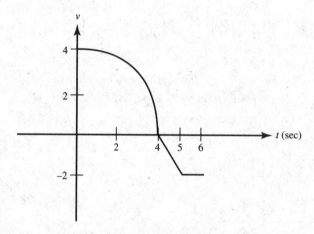

20. The graph above consists of a quarter-circle and two line segments and represents the velocity of an object during a 6-second interval. The object's average speed (in units/sec) during the 6-second interval is

    (A) $\dfrac{4\pi + 3}{6}$

    (B) $\dfrac{4\pi - 3}{6}$

    (C) -1

    (D) 1

21. Which of the following is the interval of convergence for the series $\sum_{n=1}^{\infty} \frac{(x-2)^n}{n \cdot 3^n}$?

    (A) $(-3,3)$
    (B) $[-5,-1)$
    (C) $[-1,5)$
    (D) $(-\infty,\infty)$

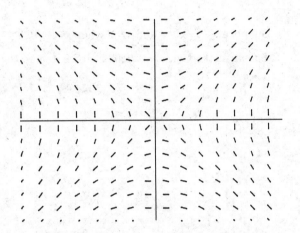

22. The slope field shown above is for which of the following differential equations?

    (A) $\frac{dy}{dx} = -\frac{y}{x}$

    (B) $\frac{dy}{dx} = \frac{1}{xy}$

    (C) $\frac{dy}{dx} = -\frac{x}{y}$

    (D) $\frac{dy}{dx} = \frac{x}{y}$

23. If y is a differentiable function of x, then the slope of the curve of $xy^2 - 2y + 4y^3 = 6$ at the point where $y = 1$ is

    (A) $-\frac{1}{18}$
    (B) 0
    (C) $\frac{5}{18}$
    (D) $\frac{1}{3}$

24. For the function f shown in the graph, which has the smallest value on the interval $2 \le x \le 6$?

    (A) $\int_2^6 f(x)dx$
    (B) the left Riemann Sum with 8 equal subintervals
    (C) the midpoint Riemann Sum with 8 equal subintervals
    (D) the trapezoidal approximation with 8 equal subintervals

25. The table shows some values of a differentiable function f and its derivative f':

x	0	1	2	3
$f(x)$	3	4	2	8
$f'(x)$	4	−1	1	10

Find $\int_0^3 f'(x)dx$.

    (A) 5
    (B) 6
    (C) 11.5
    (D) 14

GO ON TO THE NEXT PAGE

26. The solution of the differential equation $\dfrac{dy}{dx} = 2xy^2$ for which $y = -1$ when $x = 1$ is

   (A) $y = -\dfrac{1}{x^2}$ for $x \neq 0$

   (B) $y = -\dfrac{1}{x^2}$ for $x > 0$

   (C) $\ln y^2 = x^2 - 1$ for all x

   (D) $y = -\dfrac{1}{x}$ for $x > 0$

27. The base of a solid is the region bounded by the parabola $y^2 = 4x$ and the line $x = 2$. Each plane section perpendicular to the x-axis is a square. The volume of the solid is

   (A) 8
   (B) 16
   (C) 32
   (D) 64

28. What is the radius of the Maclaurin series for $\dfrac{3x}{1 - 4x^2}$?

   (A) $\dfrac{1}{3}$

   (B) $\dfrac{1}{2}$

   (C) 1

   (D) infinite

29. $\displaystyle\int \dfrac{1 + 5x}{(1 - x)(5x - 7)}\, dx =$

   (A) $3 \ln|1 - x| - 100 \ln|5x - 7| + C$

   (B) $3 \ln|1 - x| - 4 \ln|5x - 7| + C$

   (C) $-3 \ln|1 - x| - 4 \ln|5x - 7| + C$

   (D) $-3 \ln|1 - x| - 20 \ln|5x - 7| + C$

30. If $F(x) = \displaystyle\int_0^{2x} \dfrac{1}{1 - t^3}\, dt$, then $F'(x) =$

   (A) $\dfrac{1}{1 - x^3}$

   (B) $\dfrac{2}{1 - 2x^3}$

   (C) $\dfrac{1}{1 - 8x^3}$

   (D) $\dfrac{2}{1 - 8x^3}$

STOP   End of Part A, Section I

Part B

TIME: 45 MINUTES

Some questions in this part of the examination require the use of a graphing calculator. There are 15 questions in Part B, for which 45 minutes are allowed.

DIRECTIONS: Choose the best answer for each question. If the exact numerical value of the correct answer is not listed as a choice, select the choice that is closest to the exact numerical answer.

31. The series

$$(x - 1) - \frac{(x - 1)^2}{2!} + \frac{(x - 1)^3}{3!} - \frac{(x - 1)^4}{4!} + \cdots$$

converges

(A) for all real x
(B) if $0 \leq x < 2$
(C) if $0 < x \leq 2$
(D) only if $x = 1$

32. If $f(x)$ is continuous at the point where $x = a$, which of the following statements may be false?

(A) $\lim\limits_{x \to a} f(x) = f(a)$
(B) $f'(a)$ exists
(C) $f(a)$ is defined
(D) $\lim\limits_{x \to a^-} f(x) = \lim\limits_{x \to a^+} f(x)$

33. A Maclaurin polynomial is to be used to approximate $y = \sin x$ on the interval $-\pi \leq x \leq \pi$. What is the least number of terms needed to guarantee no error greater than 0.1?

(A) 3
(B) 4
(C) 5
(D) 6

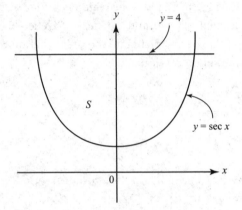

34. The region S in the figure above is bounded by $y = \sec x$ and $y = 4$. What is the volume of the solid formed when S is rotated about the x-axis?

(A) 11.385
(B) 23.781
(C) 53.126
(D) 108.177

x	$f'(x)$
2	1.4
2.5	1.5
3	2.3

35. Values of $f'(x)$ are given in the table above. Using Euler's method with a step size of 0.5, approximate $f(3)$, if $f(2) = 6$.

(A) 7.90
(B) 7.45
(C) 4.55
(D) 4.10

GO ON TO THE NEXT PAGE

36. If $x = 2t - 1$ and $y = 3 - 4t^2$, then $\dfrac{dy}{dx}$ is

    (A) $4t$

    (B) $-4t$

    (C) $-\dfrac{1}{4t}$

    (D) $-8t$

37. For a function, g, it is known that $g(3) = -2$, $g'(3) = 5$, $g''(3) = 4$, and $g'''(3) = 9$. The function has derivatives of all orders. Find the third-degree Taylor polynomial for g about $x = 3$, and use it to approximate $g(3.2)$.

    (A) -0.768

    (B) -0.896

    (C) -0.908

    (D) -0.920

38. The coefficient of x^3 in the Taylor series of $\ln(1 - x)$ about $x = 0$ (the Maclaurin series) is

    (A) $-\dfrac{2}{3}$

    (B) $-\dfrac{1}{3}$

    (C) $-\dfrac{1}{6}$

    (D) $\dfrac{1}{3}$

39. The rate at which a rumor spreads across a campus of college students is given by $\dfrac{dP}{dt} = 0.16(1200 - P)$, where $P(t)$ represents the number of students who have heard the rumor after t days. If 200 students heard the rumor today ($t = 0$), how many will have heard it by midnight the day after tomorrow ($t = 2$)?

    (A) 320

    (B) 474

    (C) 726

    (D) 1,015

40. Given function f, defined by $f(x) = \displaystyle\int_{2}^{x} \ln(t^2 + 1)\,dt$. Find the average rate of change of f on the interval $[-1,1]$.

    (A) -1.433

    (B) 0

    (C) 0.264

    (D) 0.693

41. A 26-foot ladder leans against a building so that its foot moves away from the building at the rate of 3 feet per second. When the foot of the ladder is 10 feet from the building, the top is moving down at the rate of r feet per second, where r is

    (A) 0.80

    (B) 1.25

    (C) 7.20

    (D) 12.50

$f(3) = 3$	$g(3) = -3$	$h(3) = 3$
$f'(3) = 2$	$g'(3) = -4$	$h'(3) = 4$
$f''(3) = 1/2$	$g''(3) = 8$	$h''(3) = 2$

42. The functions $f(x)$, $g(x)$, and $h(x)$ have derivatives of all orders. Listed above are values for the functions and their first and second derivatives at $x = 3$. Find $\displaystyle\lim_{x \to 3} \frac{\left(f(x)\right)^2 - 3h(x)}{g(x) + h(x)}$.

    (A) $-\dfrac{1}{5}$

    (B) $\dfrac{1}{2}$

    (C) 1

    (D) nonexistent

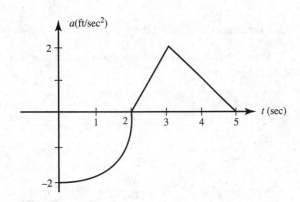

43. The graph above shows an object's acceleration (in ft/sec²). It consists of a quarter-circle and two line segments. If the object was at rest at $t = 5$ seconds, what was its initial velocity?

   (A) -2 ft/sec
   (B) $3 - \pi$ ft/sec
   (C) $\pi - 3$ ft/sec
   (D) $\pi + 3$ ft/sec

44. Water is leaking from a tank at the rate of $R(t) = 5 \arctan\left(\dfrac{t}{5}\right)$ gallons per hour, where t is the number of hours since the leak began. How many gallons will leak out during the first day?

   (A) 7
   (B) 12
   (C) 24
   (D) 124

45. The first-quadrant area inside the rose $r = 3 \sin 2\theta$ is approximately

   (A) 1.5
   (B) 1.767
   (C) 3
   (D) 3.534

End of Section I

Section II

Part A

TIME: 30 MINUTES

2 PROBLEMS

> A graphing calculator is required for some of these problems. See instructions on page 8.

1. When a faulty seam opened at the bottom of an elevated hopper, grain began leaking out onto the ground. After a while, a worker spotted the growing pile below and began making repairs. The following table shows how fast the grain was leaking (in cubic feet per minute) at various times during the 20 minutes it took to repair the hopper.

t (min)	0	4	5	7	10	12	18	20
$L(t)$ (ft^3/min)	4	7	9	8	6	5	2	0

   (a) Estimate $L'(15)$ using the data in the table. Show the computations that lead to your answer. Using correct units, explain the meaning of $L'(15)$ in the context of the problem.

(b) The falling grain forms a conical pile that the worker estimates to be 5 times as far across as it is deep. The pile was 3 feet deep when the repairs had been half-completed. How fast was the depth increasing then?

   NOTE: The volume of a cone with height h and radius r is given by $V = \frac{1}{3}\pi r^2 h$.

(c) Use a trapezoidal sum with seven subintervals as indicated in the table to approximate $\int_0^{20} L(t)\,dt$. Using correct units, explain the meaning of $\int_0^{20} L(t)\,dt$ in the context of the problem.

2. A particle is moving in the plane with position $\left(x(t),y(t)\right)$ at time t. It is known that $\dfrac{dx}{dt} = -2t$ and $\dfrac{dy}{dt} = e^t$. The position at time $t = 0$ is $x(0) = 4$ and $y(0) = 3$.

   (a) Find the speed of the particle at time $t = 2$, and find the acceleration vector at time $t = 2$.

   (b) Find the slope of the tangent line to the path of the particle at $t = 2$.

   (c) Find the position of the particle at $t = 2$.

   (d) Find the total distance traveled by the particle on the interval $0 \le t \le 2$.

STOP   End of Part A, Section II

Part B

TIME: 60 MINUTES

4 PROBLEMS

> No calculator is allowed for any of these problems.
>
> If you finish Part B before time has expired, you may return to work on Part A, but you may not use a calculator.

3. The graph of function f consists of the semicircle and line segment shown in the figure below.

   Define the area function $A(x) = \int_0^x f(t)\,dt$ for $0 \le x \le 18$.

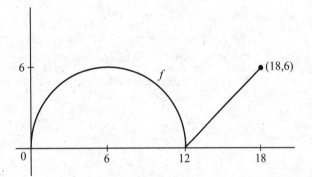

   (a) Find $A(6)$ and $A(18)$.
   (b) What is the average value of f on the interval $0 \le x \le 18$?
   (c) Write an equation of the line tangent to the graph of A at $x = 6$. Use the tangent line to estimate $A(7)$.
   (d) Give the coordinates of any points of inflection on the graph of A. Justify your answer.

4. Let f be the function satisfying the differential equation $\dfrac{dy}{dx} = 2x(y^2 + 1)$ and passing through $(0,-1)$.

   (a) Sketch the slope field for this differential equation at the points shown.

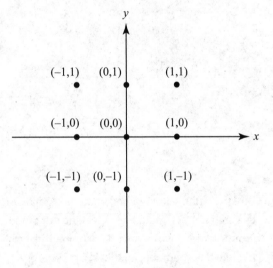

   (b) Use Euler's method with a step size of 0.5 to estimate $f(1)$.
   (c) Solve the differential equation, expressing f as a function of x.

5. The graph above represents the curve C, given by

$$f(x) = \frac{6}{\sqrt[3]{2x+5}} \text{ for } -2 \leq x \leq 11.$$

(a) Let R represent the region between C and the x-axis. Find the area of R.

(b) Set up, but do not solve, an equation to find the value of k such that the line $x = k$ divides R into two regions of equal area.

(c) Set up, but do not solve, an integral for the volume of the solid generated when R is rotated around the x-axis.

6. The function p is given by the series

$$p(x) = 2 + 2(x - 2) + 2(x - 2)^2 + \cdots + 2(x - 2)^n + \cdots$$
$$= \sum_{n=0}^{\infty} 2(x - 2)^n$$

(a) Find the interval of convergence for p. Justify your answer.

(b) The series that defines p is the Taylor series about $x = 2$. Find the sum of the series for p.

(c) Let $q(x) = \int_{2}^{x} p(t)dt$. Find $q\left(\frac{3}{2}\right)$, if it exists, or explain why it cannot be determined.

(d) Let r be defined as $r(x) = p(x^3 + 2)$. Find the first three terms and the general term for the Taylor series for r centered at $x = 0$, and find $r\left(-\frac{1}{2}\right)$.

STOP _____ End of Test _____

ANSWER KEY
Diagnostic Test Calculus BC

SECTION I

Part A

1. **A** (2)	9. **D** (10)	17. **D** (4)	25. **A** (6)
2. **D** (3)	10. **C** (3)	18. **D** (8)	26. **B** (9)
3. **A** (4)	11. **D** (3)	19. **C** (9)	27. **C** (7)
4. **C** (6)	12. **A** (5)	20. **A** (6)	28. **B** (2)
5. **D** (3)	13. **D** (4)	21. **C** (4)	29. **B** (5)
6. **C** (4)	14. **C** (9)	22. **D** (9)	30. **D** (6)
7. **B** (3)	15. **B** (7)	23. **A** (3)	
8. **B** (5)	16. **B** (3)	24. **B** (6)	

Part B

31. **A** (10)	35. **B** (9)	39. **B** (9)	43. **C** (6)
32. **B** (2)	36. **B** (3)	40. **C** (4)	44. **D** (8)
33. **B** (10)	37. **C** (6)	41. **B** (4)	45. **D** (7
34. **D** (7)	38. **B** (10)	42. **B** (3)	

NOTE: Chapters that review and offer additional practice for each topic are specified in parentheses.

Answer Explanations

The explanations for questions not given below can be found in the answer explanation section for the Calculus AB Diagnostic Test on pages 28–38. Some questions on Diagnostic Tests AB and BC are identical.

Section I Multiple-Choice

Part A

1. **(A)** $x'(t) = 2te^{t^2+2}$, $y'(t) = 2(t^3 + 2) \cdot 3t^2$. The velocity vector is $\langle x'(1), y'(1) \rangle = \langle 2e^3, 18 \rangle$.

7. **(B)** The volume is given by $\pi \int_0^\infty e^{-2x} dx$, an improper integral.

$$\lim_{k \to \infty} \pi \int_0^k e^{-2x} dx = \lim_{x \to \infty} \left(-\frac{1}{2} \pi \int_0^k e^{-2x} (-2dx) \right)$$

$$= \lim_{x \to \infty} \left(-\frac{1}{2} \pi e^{-2x} \Big|_0^k \right)$$

$$= \lim_{k \to \infty} \left(-\frac{1}{2} \pi \left(e^{-2k} - e^0 \right) \right)$$

$$= \frac{\pi}{2}$$

8. **(B)** The Maclaurin series for $\sin(x) = x - \frac{x^3}{3!} + \frac{x^5}{5!} - \frac{x^7}{7!} + \cdots$. To get the series for $\sin(x^2)$, substitute x^2 into the series for $\sin(x)$: $\sin(x^2) = x^2 - \frac{x^6}{3!} + \frac{x^{10}}{5!} - \frac{x^{14}}{7!} + \cdots$. Finally, by multiplying all terms by x^2, you can get the series for $x^2 \sin(x^2)$: $x^2 \sin(x^2) = x^4 - \frac{x^8}{3!} + \frac{x^{12}}{5!} - \frac{x^{16}}{7!} + \cdots$.

9. **(D)** $\lim_{n \to \infty} \frac{n}{5n+1} = \frac{1}{5} \neq 0$. Series fails the nth Term Test.

10. **(C)** $\int_1^\infty \frac{12}{(x+8)^{3/2}} dx = \lim_{b \to \infty} \int_1^b \frac{12}{(x+8)^{3/2}} dx = \lim_{b \to \infty} \left(12 \cdot \frac{(x+8)^{-1/2}}{-1/2} \Big|_1^b \right)$

$$= -24 \lim_{b \to \infty} \left(\frac{1}{(b+8)^{1/2}} - \frac{1}{3} \right) = -24 \left(0 - \frac{1}{3} \right) = 8$$

12. **(A)** Use the Parts Formula with $u = x$ and $dv = e^x dx$. Then $du = dx$ and $v = e^x$, and

$$\left(xe^x - \int e^x dx \right) \Big|_0^1 = (xe^x - e^x) \Big|_0^1 = (e - e) - (0 - 1)$$

15. **(B)** The arc length is given by the integral $\int_0^1 \sqrt{1 + (3x^{1/2})^2} \ dx$, which is

$$\frac{1}{9} \int_0^1 (1 + 9x)^{1/2} (9dx) = \frac{1}{9} \cdot \frac{2}{3} \cdot (1 + 9x)^{3/2} \Big|_0^1 = \frac{2}{27} (10^{3/2} - 1)$$

18. **(D)** $\mathbf{v}(t) = \langle \sec^2 t, \cos t \rangle$, so $\mathbf{v}\left(\frac{\pi}{3} \right) = \left\langle 2^2, 2 \cdot \frac{1}{2} \right\rangle$ and $|\mathbf{v}| = \sqrt{4^2 + 1^2}$.

19. **(C)** The initial point given (x_0, y_0) is given to be $(0,2)$; use this initial point with step size $\Delta x = 0.5$.

Note: $f'(x_n, y_n) = \frac{dy}{dx} \Big|_{(x_n, y_n)}$

$x_1 = 0.5$; $y_1 = y_0 + f'(0,2)(0.5) = 2 + (0)(0.5) = 2$

$x_2 = 1$; $y_2 = y_1 + f'(0.5, 2)(0.5) = 2 + \left(\frac{10 \cdot 0.5}{0.5 + 2} \right)(0.5) = 2 + (2)(0.5) = 3$

21. **(C)** The ratio test can be used to find the open interval of convergence.

$$\lim_{n\to\infty}\left|\frac{a_{n+1}}{a_n}\right| = \lim_{n\to\infty}\left|\frac{(x-2)^{n+1}}{(n+1)\cdot 3^{n+1}} \cdot \frac{n\cdot 3^n}{(x-2)^n}\right| = \lim_{n\to\infty}\left|\frac{n}{n+1} \cdot \frac{x-2}{3}\right| = \left|\frac{x-2}{3}\right|$$

To be convergent, the limit must be less than 1: $\left|\dfrac{x-2}{3}\right| < 1 \Rightarrow |x-2| < 3 \Rightarrow$ $-3 < x - 2 < 3 \Rightarrow -1 < x < 5$

Now, check the endpoints of the interval.

At $x = -1$, the series is $\displaystyle\sum_{n=1}^{\infty}\frac{(-3)^n}{n\cdot 3^n} = \sum_{n=1}^{\infty}\frac{(-1)^n \cdot 3^n}{n \cdot 3^n} = \sum_{n=1}^{\infty}\frac{(-1)^n}{n}$, which converges because it is the alternating harmonic series.

At $x = 5$, the series is $\displaystyle\sum_{n=1}^{\infty}\frac{3^n}{n\cdot 3^n} = \sum_{n=1}^{\infty}\frac{1}{n}$, which diverges because it is the harmonic series. Therefore, the interval of convergence is $-1 \le x < 5$.

24. **(B)** Since function f is increasing on the interval [2,6], rectangles based on left endpoints of the subintervals will all lie completely below the curve and thus have smaller areas than any of the other sums or the definite integral. See pages 252–253.

28. **(B)** The function is in the form of the sum of a convergent infinite geometric series, $\dfrac{a}{1-r}$. To be convergent, the absolute value of the common ratio, r, must be less than 1. For this function, the common ratio is $r = 4x^2$. To find the radius of convergence, solve the inequality $|4x^2| < 1 \Rightarrow x^2 < \dfrac{1}{4} \Rightarrow |x| < \dfrac{1}{2}$. Therefore, the radius of convergence is $\dfrac{1}{2}$.

29. **(B)** For this integral, use the method of partial fraction decomposition.

$$\frac{1+5x}{(1-x)(5x-7)} = \frac{A}{1-x} + \frac{B}{5x-7} \Rightarrow 1+5x = A(5x-7) + B(1-x)$$

Set $x = 1 \Rightarrow 1 + 5(1) = A(5(1)-7) + B(1-1) \Rightarrow 6 = -2A \Rightarrow A = -3$

Set $x = \dfrac{7}{5} \Rightarrow 1 + 5\left(\dfrac{7}{5}\right) = A\left(5\left(\dfrac{7}{5}\right)-7\right) + B\left(1-\dfrac{7}{5}\right) \Rightarrow 8 = -\dfrac{2}{5}B \Rightarrow B = -20$

Therefore, $\displaystyle\int \frac{1+5x}{(1-x)(5x-7)}\, dx = \int\left(\frac{-3}{1-x} + \frac{-20}{5x-7}\right)dx = \frac{-3\ln|1-x|}{-1} + \frac{-20\ln|5x-7|}{5} + C$

Part B

31. **(A)** Use the Ratio Test, page 407:

$$\lim_{n\to\infty} \left| \frac{(x-1)^{n+1}}{(n+1)!} \cdot \frac{n!}{(x-1)^n} \right| = \lim_{n\to\infty} \frac{1}{n+1}|x-1|$$

which equals zero if $x \neq 1$. The series also converges if $x = 1$ (each term equals 0).

32. **(B)** The absolute-value function $f(x) = |x|$ is continuous at $x = 0$, but $f'(0)$ does not exist.

33. **(B)** The Maclaurin series is

$$\sin x = x - \frac{x^3}{3!} + \frac{x^5}{5!} - \frac{x^7}{7!} + \frac{x^9}{9!} - \cdots$$

When an alternating series satisfies the Alternating Series Test, the sum is approximated by using a finite number of terms, and the error is less than the first term omitted. On the interval $-\pi \leq x \leq \pi$, the maximum error (numerically) occurs when $x = \pi$. Since

$$\frac{\pi^7}{7!} < 0.6 \text{ and } \frac{\pi^9}{9!} < 0.09$$

four terms will suffice to assure no error greater than 0.1.

34. **(D)** S is the region bounded by $y = \sec x$, the y-axis, and $y = 4$.

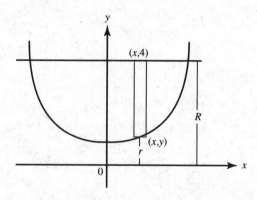

We revolve region S about the x-axis. Using washers, $\Delta V = \pi(R^2 - r^2)\,\Delta x$ (here $R = 4$ and $r = \sec x$). Symmetry allows us to double the volume generated by the first-quadrant portion of S. We find the upper limit of integration by solving $\sec x = 4$ and store the result ($x = 1.31811607$) as A. Then

$$V = 2\pi \int_0^A (16 - \sec^2 x)\,dx = 108.177$$

35. **(B)** $f(2) = 6 \Rightarrow f(2.5) = f(2) + f'(2)(0.5) = 6 + (1.4)(0.5) = 6.7$

$f(2.5) = 6.7 \Rightarrow f(3) = f(2.5) + f'(2.5)(0.5) = 6.7 + (1.5)(0.5) = 7.45$

36. **(B)** $\dfrac{dx}{dt} = 2$ and $\dfrac{dy}{dt} = -8t$. Hence $\dfrac{dy}{dx} = \dfrac{dy/dt}{dx/dt} = \dfrac{-8t}{2}$.

37. **(C)** $T_3(x) = g(3) + g'(3) \cdot (x-3) + \dfrac{g''(3)}{2!} \cdot (x-3)^2 + \dfrac{g'''(3)}{3!}(x-3)^3$

$\qquad = -2 + 5 \cdot (x-3) + \dfrac{4}{2!} \cdot (x-3)^2 + \dfrac{9}{3!}(x-3)^3$

$\qquad = -2 + 5 \cdot (x-3) + 2 \cdot (x-3)^2 + \dfrac{3}{2}(x-3)^3$

$\quad T_3(3.2) = -2 + 5 \cdot (3.2-3) + 2 \cdot (3.2-3)^2 + \dfrac{3}{2}(3.2-3)^3 = -0.908$

38. **(B)** The power series for $\ln(1-x)$, if $x < 1$, is $-x - \dfrac{x^2}{2} - \dfrac{x^3}{3} - \cdots$

39. **(B)** Solve by separation of variables; then

$$\frac{dP}{1200 - P} = 0.16 dt$$

$$-\ln(1200 - P) = 0.16t + C$$

$$1200 - P = ce^{-0.16t}$$

Use $P(0) = 200$; then $c = 1000$, so $P(x) = 1200 - 1000e^{-0.16t}$. Now $P(2) = 473.85$.

40. **(C)** Average rate of change:

$$\frac{f(1) - f(-1)}{1 - (-1)} = \frac{\displaystyle\int_2^1 \ln(t^2 + 1)\, dt - \int_2^{-1} \ln(t^2 + 1)\, dt}{2} = \frac{\displaystyle\int_{-1}^1 \ln(t^2 + 1)\, dt}{2} = 0.264$$

45. **(D)** The first-quadrant area is $\dfrac{1}{2}\displaystyle\int_0^{\pi/2} (3\sin 2\theta)^2\, d\theta \approx 3.534$.

Section II Free-Response

Part A

BC 1. See solution for AB/BC 1 on page 35.

BC 2. (a) Speed $= \sqrt{(x'(2))^2 + (y'(2))^2} = \sqrt{(-4)^2 + (e^2)^2} = 8.402$;
Acceleration $= \langle x''(2), y''(2) \rangle = \langle -2, e^2 \rangle = \langle -2, 7.389 \rangle$

(b) Slope $= \dfrac{y'(2)}{x'(2)} = \dfrac{e^2}{-4} = -1.847$

(c) $x(2) = x(0) + \displaystyle\int_0^2 x'(t)dt = 4 + \int_0^2 (-2t)dt = 0$

$y(2) = y(0) + \displaystyle\int_0^2 y'(t)dt = 3 + \int_0^2 e^t \, dt = 9.389$

(d) Total distance $= \displaystyle\int_0^2 \sqrt{(x'(t))^2 + (y'(t))^2} \; dt = 7.566$

(Review Chapters 4, 6, and 8)

Part B

BC 3.   See solution for AB/BC 3 on page 36.

BC 4.   (a)   Using the differential equation, evaluate the derivative at each point, then sketch a short segment having that slope. For example, at $(-1,-1)$, $\dfrac{dy}{dx} = 2(-1)\big((-1)^2 + 1\big) = -4$; draw a segment at $(-1,-1)$ that decreases steeply. Repeat this process at each of the other points. The result is shown below.

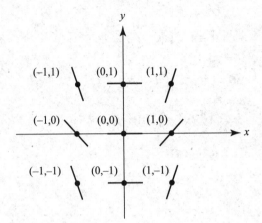

(b)   At $(0,-1)$, $\dfrac{dy}{dx} = 2(0)\big((-1)^2 + 1\big) = 0$. For $\Delta x = 0.5$ and $\dfrac{\Delta y}{\Delta x} = 0$, $\Delta y = 0$, so move to $(0 + 0.5, -1 + 0) = (0.5, -1)$.

At $(0.5, -1)$, $\dfrac{dy}{dx} = 2(0.5)\big((-1)^2 + 1\big) = 2$. Thus, for $\Delta x = 0.5$ and $\dfrac{\Delta y}{\Delta x} = 2$, $\Delta y = 1$.

Move to $(0.5 + 0.5, -1 + 1) = (1, 0)$, then $f(1) \approx 0$.

(c)   The differential equation $\dfrac{dy}{dx} = 2x(y^2 + 1)$ is separable:

$$\int \frac{dy}{y^2 + 1} = \int 2x\,dx$$

$$\arctan(y) = x^2 + c$$

$$y = \tan(x^2 + c)$$

It is given that f passes through $(0,-1)$, so $-1 = \tan(0^2 + c)$ and $c = -\dfrac{\pi}{4}$.

The solution is $f(x) = \tan\left(x^2 - \dfrac{\pi}{4}\right)$.

(Review Chapter 9)

BC 5.   See solution for AB/BC 5 on page 37.

BC 6. (a) The series is a geometric series with common ratio $(x - 2)$. The series converges if $|x - 2| < 1$, so the interval of convergence is $1 < x < 3$.

NOTE: Because this is geometric, there is no need to check endpoints.

(b) Geometric series with first term 2 and common ratio $(x - 2)$:

$$p(x) = \frac{2}{1 - (x - 2)} = \frac{2}{3 - x} \text{ for } 1 < x < 3$$

(c) $q\left(\dfrac{3}{2}\right) = \displaystyle\int_{2}^{3/2} \frac{2}{3 - x}\, dx = -\int_{3/2}^{2} \frac{2}{3 - x}\, dx = \left(2 \ln|3 - x|\right)\Big|_{3/2}^{2} = -2 \ln\left(\frac{3}{2}\right)$

(d) $r(x) = p(x^3 + 2) = 2 + (x^3 + 2 - 2) + 2(x^3 + 2 - 2)^2 + \cdots$

$$r(x) = 2 + 2x^3 + 2x^6 + \cdots + 2x^{2n} + \cdots$$

$$r\left(-\frac{1}{2}\right) = p\left(\left(-\frac{1}{2}\right)^3 + 2\right) = p\left(\frac{15}{8}\right) = \frac{2}{3 - \dfrac{15}{8}} = \frac{16}{9}$$

(Review Chapter 10)

Topical Review and Practice

1

Functions

Learning Objectives

In this chapter, you will review precalculus topics. Although these topics are not directly tested on the AP exam, reviewing them will reinforce some basic principles:

→ General properties of functions: domain, range, composition, inverse
→ Special functions: absolute value, greatest integer; polynomial, rational, trigonometric, exponential, and logarithmic

And some BC topics:

→ Parametrically defined curves
→ Polar curves

A. Definitions

A1. A *function f* is a correspondence that associates with each element a of a set called the *domain* one and only one element b of a set called the *range*. We write

$$f(a) = b$$

to indicate that b is the value of f at a. The elements in the domain are called *inputs*, and those in the range are called *outputs*.

A function is often represented by an equation, a graph, or a table.

A vertical line cuts the graph of a function in at most one point.

Function
Domain
Range

❯ Example 1

The domain of $f(x) = x^2 - 2$ is the set of all real numbers; its range is the set of all reals greater than or equal to -2. Note that

$$f(0) = 0^2 - 2 = -2 \qquad f(-1) = (-1)^2 - 2 = -1$$
$$f(\sqrt{3}) = (\sqrt{3})^2 - 2 = 1 \qquad f(c) = c^2 - 2$$
$$f(x + h) - f(x) = \left[(x + h)^2 - 2\right] - \left[x^2 - 2\right]$$
$$= x^2 + 2hx + h^2 - 2 - x^2 + 2 = 2hx + h^2$$

❯ Example 2

Find the domains of: (a) $f(x) = \dfrac{4}{x - 1}$ (b) $g(x) = \dfrac{x}{x^2 - 9}$ (c) $h(x) = \dfrac{\sqrt{4 - x}}{x}$

✓ Solutions

(a) The domain of $f(x) = \dfrac{4}{x - 1}$ is the set of all reals except $x = 1$ (which we shorten to "$x \neq 1$").

(b)   The domain of $g(x) = \dfrac{x}{x^2 - 9}$ is $x \neq 3, -3$.

(c)   The domain of $h(x) = \dfrac{\sqrt{4 - x}}{x}$ is $x \leq 4$, $x \neq 0$ (which is a short way of writing $\{x \mid x$ is real, $x < 0$ or $0 < x \leq 4\}$).

A2. Two functions f and g with the same domain may be combined to yield their sum and difference: $f(x) + g(x)$ and $f(x) - g(x)$, also written as $(f + g)(x)$ and $(f - g)(x)$, respectively; or their product and quotient: $f(x)g(x)$ and $f(x)/g(x)$, also written as $(fg)(x)$ and $(f/g)(x)$, respectively. The quotient is defined for all x in the shared domain except those values for which $g(x)$, the denominator, equals zero.

> **Example 3**

If $f(x) = x^2 - 4x$ and $g(x) = x + 1$, then find $\dfrac{f(x)}{g(x)}$ and $\dfrac{g(x)}{f(x)}$.

✓ **Solutions**

$\dfrac{f(x)}{g(x)} = \dfrac{x^2 - 4x}{x + 1}$ and has domain $x \neq -1$

$\dfrac{g(x)}{f(x)} = \dfrac{x + 1}{x^2 - 4x} = \dfrac{x + 1}{x(x - 4)}$ and has domain $x \neq 0, 4$

Composition

A3. The *composition* (or *composite*) of f with g, written as $f(g(x))$ and read as "f of g of x," is the function obtained by replacing x wherever it occurs in $f(x)$ by $g(x)$. We also write $(f \circ g)(x)$ for $f(g(x))$. The domain of $(f \circ g)(x)$ is the set of all x in the domain of g for which $g(x)$ is in the domain of f.

> **Example 4A**

If $f(x) = 2x - 1$ and $g(x) = x^2$, then does $f(g(x)) = g(f(x))$?

✓ **Solution**

$f(g(x)) = 2(x^2) - 1 = 2x^2 - 1$

$g(f(x)) = (2x - 1)^2 = 4x^2 - 4x + 1$

In general, $f(g(x)) \neq g(f(x))$.

> **Example 4B**

If $f(x) = 4x^2 - 1$ and $g(x) = \sqrt{x}$, find $f(g(x))$ and $g(f(x))$.

✓ **Solutions**

$f(g(x)) = 4x - 1 \quad (x \geq 0); \quad g(f(x)) = \sqrt{4x^2 - 1} \quad \left(|x| \geq \dfrac{1}{2}\right)$

Symmetry

A4. A function f is *odd* if, for all x in the domain of f, $f(-x) = -f(x)$. A function f is *even* if, for all x in the domain of f, $f(-x) = f(x)$.

The graph of an odd function is symmetric about the origin; the graph of an even function is symmetric about the y-axis.

> ## Example 5

The graphs of $f(x) = \frac{1}{2}x^3$ and $g(x) = 3x^2 - 1$ are shown in Figure 1.1; $f(x)$ is odd, $g(x)$ even.

Figure 1.1

A5. If a function f yields a single output for each input and also yields a single input for every output, then f is said to be *one-to-one*. Geometrically, this means that any horizontal line cuts the graph of f in at most one point. The function sketched at the left in Figure 1.1 is one-to-one; the function sketched at the right is not. A function that is increasing (or decreasing) on an interval I is one-to-one on that interval (see pages 163–164 for definitions of increasing and decreasing functions). Graphs that are strictly increasing (or decreasing) on their domain are one-to-one functions.

A6. If f is one-to-one with domain X and range Y, then there is a function f^{-1}, with domain Y and range X, such that

$$f^{-1}(y_0) = x_0 \quad \text{if and only if} \quad f(x_0) = y_0$$

Inverse

The function f^{-1} is the *inverse* of f. It can be shown that f^{-1} is also one-to-one and that its inverse is f. The graphs of a function and its inverse are symmetric with respect to the line $y = x$.

> **To find the inverse of $y = f(x)$, interchange x and y, then solve for y.**

> ## Example 6

Find the inverse of the one-to-one function $f(x) = x^3 - 1$.

> ## ✓ Solution

Interchange x and y: $x = y^3 - 1$
Solve for y: $y = \sqrt[3]{x + 1} = f^{-1}(x)$

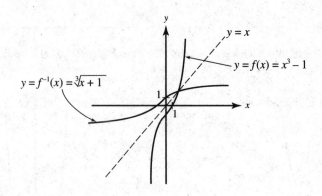

Figure 1.2

Note that the graphs of f and f^{-1} in Figure 1.2 are mirror images, with the line $y = x$ as the mirror.

Zeros

A7. The zeros of a function f are the values of x for which $f(x) = 0$; they are the x-intercepts of the graph of $y = f(x)$.

> **Example 7**

Find zeros of $f(x) = x^4 - 2x^2$.

✓ **Solution**

The zeros are the x's for which $x^4 - 2x^2 = 0$. The function has three zeros, since $x^4 - 2x^2 = x^2(x^2 - 2)$ equals zero if $x = 0$, $+\sqrt{2}$, or $-\sqrt{2}$.

B. Special Functions

The *absolute-value* function $f(x) = |x|$ and the *greatest-integer* function $g(x) = [x]$ are sketched in Figure 1.3.

Absolute value

Greatest integer

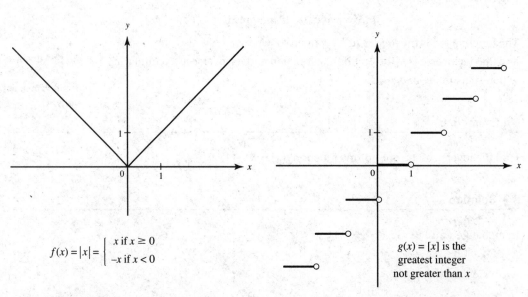

$$f(x) = |x| = \begin{cases} x \text{ if } x \geq 0 \\ -x \text{ if } x < 0 \end{cases}$$

$g(x) = [x]$ is the greatest integer not greater than x

Absolute-value function

Greatest-integer function

Figure 1.3

> Example 8

A function f is defined on the interval $[-2,2]$ and has the graph shown in Figure 1.4.

(a)   Sketch the graph of $y = |f(x)|$.

(b)   Sketch the graph of $y = f(|x|)$.

(c)   Sketch the graph of $y = -f(x)$.

(d)   Sketch the graph of $y = f(-x)$.

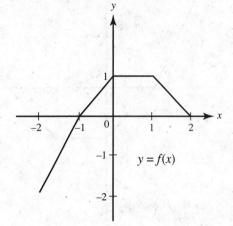

Figure 1.4

✓ Solutions

The graphs are shown in Figures 1.4a through 1.4d.

Figure 1.4a   **Figure 1.4b**

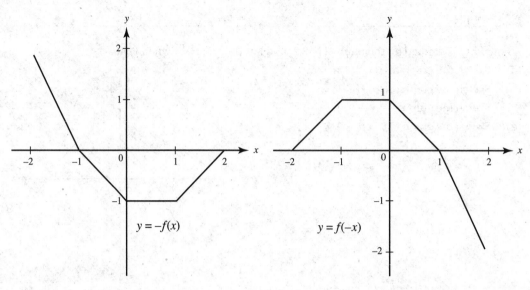

Figure 1.4c   **Figure 1.4d**

Note that graph (c) of $y = -f(x)$ is the reflection of $y = f(x)$ in the x-axis, whereas graph (d) of $y = f(-x)$ is the reflection of $y = f(x)$ in the y-axis. How do the graphs of $|f(x)|$ and $f(|x|)$ compare with the graph of $f(x)$?

❯ Example 9

Let $f(x) = x^3 - 3x^2 + 2$. Graph the following functions on your calculator in the window $[-3,3] \times [-3,3]$: (a) $y = f(x)$ (b) $y = |f(x)|$ (c) $y = f(|x|)$

✔ Solutions

(a) $y = f(x)$
      See Figure 1.5a.

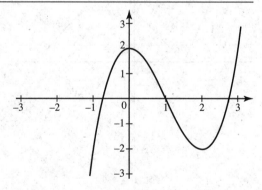

Figure 1.5a

(b) $y = |f(x)|$
      See Figure 1.5b.

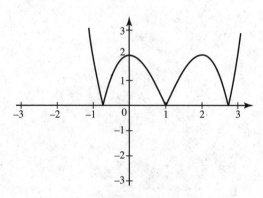

Figure 1.5b

(c) $y = f(|x|)$
See Figure 1.5c.

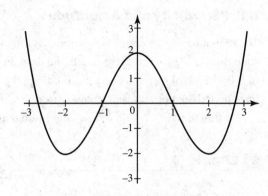

Figure 1.5c

Note how the graphs for (b) and (c) compare with the graph for (a).

C. Polynomial and Other Rational Functions

C1. Polynomial Functions

A *polynomial function* is of the form

$$f(x) = a_n x^n + a_{n-1} x^{n-1} + a_{n-2} x^{n-2} + \cdots + a_1 x + a_0$$

Polynomial functions

where n is a positive integer or zero, and the *coefficients*, $a_0, a_1, a_2, \ldots a_n$, are constants. If $a_n \neq 0$, the degree of the polynomial is n.

A *linear* function, $f(x) = mx + b$, is of the first degree; its graph is a straight line with slope m, the constant rate of change of $f(x)$ (or y) with respect to x, and b is the line's y-intercept.

A *quadratic* function, $f(x) = ax^2 + bx + c$, has degree 2; its graph is a parabola that opens up if $a > 0$, down if $a < 0$, and whose axis is the line $x = -\dfrac{b}{2a}$.

A *cubic* function, $f(x) = a_3 x^3 + a_2 x^2 + a_1 x + a_0$, has degree 3; calculus enables us to sketch its graph easily; and so on. The domain of every polynomial is the set of all reals.

C2. Rational Functions

A *rational function* is of the form

$$f(x) = \frac{P(x)}{Q(x)}$$

Rational functions

where $P(x)$ and $Q(x)$ are polynomials. The domain of f is the set of all reals for which $Q(x) \neq 0$.

D. Trigonometric Functions

The fundamental trigonometric identities, graphs, and reduction formulas are given in the Appendix.

Trigonometric functions

D1. Periodicity and Amplitude

The trigonometric functions are periodic. A function f is *periodic* if there is a positive number p such that $f(x + p) = f(x)$ for each x in the domain of f. The smallest such p is called the *period* of f. The graph of f repeats every p units along the x-axis. The functions $\sin x$, $\cos x$, $\csc x$, and $\sec x$ have period 2π; $\tan x$ and $\cot x$ have period π.

The function $f(x) = A \sin bx$ has amplitude A and period $\dfrac{2\pi}{b}$; $g(x) = \tan cx$ has period $\dfrac{\pi}{c}$.

> ### Example 10

Consider the function $f(x) = \dfrac{1}{k}\cos(kx)$.

(a)   For what value of k does f have period 2?

(b)   What is the amplitude of f for this k?

✓ **Solutions**

(a)   Function f has period $\dfrac{2\pi}{k}$; since this must equal 2, we solve the equation $\dfrac{2\pi}{k} = 2$, getting $k = \pi$.

(b)   It follows that the amplitude of f that equals $\dfrac{1}{k}$ has a value of $\dfrac{1}{\pi}$.

> ### Example 11

Consider the function $f(x) = 3 - \sin\dfrac{\pi x}{3}$.

Find (a) the period and (b) the maximum value of f.

(c)   What is the smallest positive x for which f is a maximum?

(d)   Sketch the graph.

✓ **Solutions**

(a)   The period of f is $2\pi \div \dfrac{\pi}{3}$, or 6.

(b)   Since the maximum value of $-\sin x$ is $-(-1)$ or $+1$, the maximum value of f is $3 + 1$ or 4.

(c) $-\left(\sin\dfrac{\pi x}{3}\right)$ equals $+1$ when $\sin\dfrac{\pi x}{3} = -1$, that is, when $\dfrac{\pi x}{3} = \dfrac{3\pi}{2}$. Solving yields $x = \dfrac{9}{2}$.

(d)   We graph $y = 3 - \sin\dfrac{\pi x}{3}$ in $[-5,8] \times [0,5]$ in Figure 1.6:

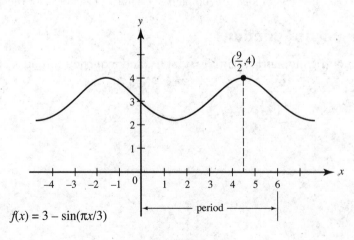

$f(x) = 3 - \sin(\pi x/3)$

Figure 1.6

D2. Inverses

We obtain *inverses* of the trigonometric functions by limiting the domains of the latter so each trigonometric function is one-to-one over its restricted domain. For example, we restrict

$$\sin x \text{ to } \quad -\frac{\pi}{2} \le x \le \frac{\pi}{2}$$

$$\cos x \text{ to } \quad 0 \le x \le \pi$$

$$\tan x \text{ to } \quad -\frac{\pi}{2} < x < \frac{\pi}{2}$$

Inverse trig functions

The graphs of $f(x) = \sin x$ on $\left[-\frac{\pi}{2}, \frac{\pi}{2}\right]$ and of its inverse $f^{-1}(x) = \sin^{-1} x$ are shown in Figure 1.7. The inverse trigonometric function $\sin^{-1} x$ is also commonly denoted by arcsin x, which denotes *the* angle whose sine is x. The graph of $\sin^{-1} x$ is, of course, the reflection of the graph of $\sin x$ in the line $y = x$.

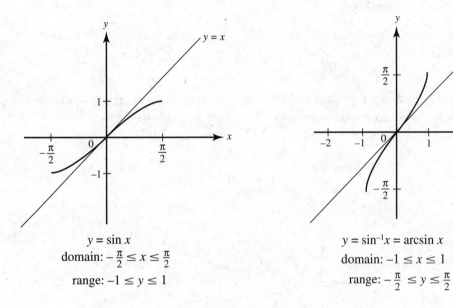

$$y = \sin x$$
domain: $-\frac{\pi}{2} \le x \le \frac{\pi}{2}$
range: $-1 \le y \le 1$

$$y = \sin^{-1} x = \arcsin x$$
domain: $-1 \le x \le 1$
range: $-\frac{\pi}{2} \le y \le \frac{\pi}{2}$

Figure 1.7

Also, for other inverse trigonometric functions,

$y = \cos^{-1} x$ (or arccos x) has domain $-1 \le x \le 1$ and range $0 \le y \le \pi$

$y = \tan^{-1} x$ (or arctan x) has domain the set of reals and range $-\frac{\pi}{2} < y < \frac{\pi}{2}$

Note also that

$$\sec^{-1}(x) = \cos^{-1}\left(\frac{1}{x}\right) \qquad \csc^{-1}(x) = \sin^{-1}\left(\frac{1}{x}\right) \qquad \text{and} \qquad \cot^{-1}(x) = \frac{\pi}{2} - \tan^{-1}(x)$$

E. Exponential and Logarithmic Functions

E1. Exponential Functions

Exponential functions

The following laws of exponents hold for all rational m and n, provided that $a > 0$, $a \neq 1$:

$$a^0 = 1 \qquad a^1 = a \qquad a^m \cdot a^n = a^{m+n} \qquad a^m \div a^n = a^{m-n}$$

$$(a^m)^n = a^{mn} \qquad a^{-m} = \frac{1}{a^m}$$

The exponential function $f(x) = a^x$ ($a > 0$, $a \neq 1$) is thus defined for all real x; its range is the set of positive reals. The graph of $y = a^x$, when $a = 2$, is shown in Figure 1.8.

Of special interest and importance in calculus is the exponential function $f(x) = e^x$, where e is an irrational number whose decimal approximation to five decimal places is 2.71828. We define e on page 99.

E2. Logarithmic Functions

Log functions

Since $f(x) = a^x$ is one-to-one, it has an inverse, $f^{-1}(x) = \log_a x$, called the *logarithmic* function with base a. We note that

$$y = \log_a x \qquad \text{if and only if} \qquad a^y = x$$

The domain of $\log_a x$ is the set of positive reals; its range is the set of all reals. It follows that the graphs of the pair of mutually inverse functions $y = 2^x$ and $y = \log_2 x$ are symmetric to the line $y = x$, as can be seen in Figure 1.8.

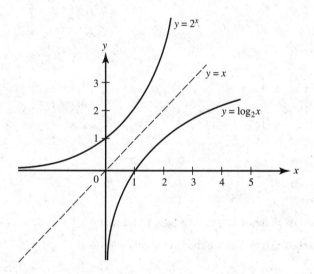

Figure 1.8

The logarithmic function $\log_a x$ ($a > 0$, $a \neq 1$) has the following properties:

$$\log_a 1 = 0 \qquad \log_a a = 1 \qquad \log_a mn = \log_a m + \log_a n$$

$$\log_a \frac{m}{n} = \log_a m - \log_a n \qquad \log_a x^m = m \log_a x$$

The logarithmic base e is so important and convenient in calculus that we use a special symbol:

$$\log_e x = \ln x$$

Logarithms with base *e* are called *natural* logarithms. The domain of ln *x* is the set of positive reals; its range is the set of all reals. The graphs of the mutually inverse functions ln *x* and e^x are given in the Appendix.

F. Parametrically Defined Functions

BC ONLY

If the *x*- and *y*-coordinates of a point on a graph are given as functions *f* and *g* of a third variable, say *t*, then

$$x = f(t) \qquad y = g(t)$$

are called *parametric equations* and *t* is called the *parameter*. When *t* represents time, as it often does, then we can view the curve as that followed by a moving particle as the time varies.

Parametric equations

> ### Example 12

Find the Cartesian equation of, and sketch, the curve defined by the parametric equations

$$x = 4 \sin t \qquad y = 5 \cos t \qquad (0 \le t \le 2\pi)$$

✓ Solution

We can eliminate the parameter *t* as follows:

$$\sin t = \frac{x}{4} \qquad \cos t = \frac{y}{5}$$

Since $\sin^2 t + \cos^2 t = 1$, we have

$$\left(\frac{x}{4}\right)^2 + \left(\frac{y}{5}\right)^2 = 1 \text{ or } \frac{x^2}{16} + \frac{y^2}{25} = 1$$

The curve is the ellipse shown in Figure 1.9.

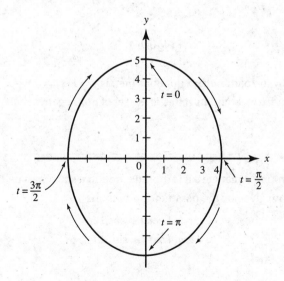

Figure 1.9

Note that as *t* increases from 0 to 2π, a particle moving in accordance with the given parametric equations starts at point (0,5) (when *t* = 0) and travels in a clockwise direction along the ellipse, returning to (0,5) when *t* = 2π.

> **Example 13**

Given the pair of parametric equations,

$$x = 1 - t \qquad y = \sqrt{t} \qquad (t \geq 0)$$

write an equation of the curve in terms of x and y, and sketch the graph.

✓ **Solution**

We can eliminate t by squaring the second equation and substituting for t in the first; then we have

$$y^2 = t \qquad \text{and} \qquad x = 1 - y^2$$

We see the graph of the equation $x = 1 - y^2$ on the left in Figure 1.10. At the right we see only the upper part of this graph, the part defined by the parametric equations for which t and y are both restricted to nonnegative numbers.

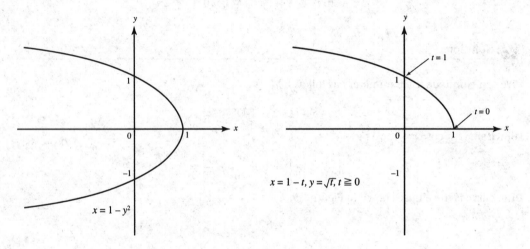

Figure 1.10

The function defined by the parametric equations here is $y = F(x) = \sqrt{1 - x}$, whose graph is at the right above; its domain is $x \leq 1$, and its range is the set of nonnegative reals.

> **Example 14**

A satellite is in orbit around a planet that is orbiting around a star. The satellite makes 12 orbits each year. Graph its path given by the parametric equations

$$x = 4 \cos t + \cos 12t$$
$$y = 4 \sin t + \sin 12t$$

✓ **Solution**

Figure 1.11 shows the graph of the satellite's path using the calculator's parametric mode for $0 \le t \le 2\pi$.

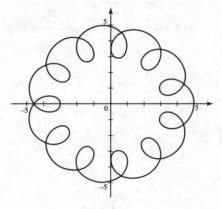

Figure 1.11

> **Example 15**

Graph $x = y^2 - 6y + 8$.

✓ **Solution**

We encounter a difficulty here. The calculator is constructed to graph y as a function of x: it accomplishes this by scanning horizontally across the window and plotting points in varying vertical positions. Ideally, we want the calculator to scan *down* the window and plot points at appropriate horizontal positions. But it won't do that.

One alternative is to interchange variables, entering x as Y_1 and y as X, thus entering $Y_1 = X^2 - 6X + 8$. But then, during all subsequent processing we must remember that we have made this interchange.

Less risky and more satisfying is to switch to parametric mode: Enter $x = t^2 - 6t + 8$ and $y = t$. Then graph these equations in $[-10,10] \times [-10,10]$, for t in $[-10,10]$. See Figure 1.12.

Figure 1.12

> **Example 16**

Let $f(x) = x^3 + x$; graph $f^{-1}(x)$.

✓ **Solution**

Recalling that f^{-1} interchanges x and y, we use parametric mode to graph

$$f: x = t, y = t^3 + t$$

$$\text{and } f^{-1}: x = t^3 + t, y = t$$

Figure 1.13 shows both $f(x)$ and $f^{-1}(x)$.

Figure 1.13

Parametric equations give rise to vector functions, which will be discussed in connection with motion along a curve in Chapter 4.

G. Polar Functions

Polar coordinates of the form (r,θ) identify the location of a point by specifying θ, an angle of rotation from the positive x-axis, and r, a distance from the origin, as shown in Figure 1.14.

Polar functions

A **polar function** defines a curve with an equation of the form $r = f(\theta)$. Some common polar functions include:

Figure 1.14

Spiral	Rose	Cardioid	Limacon
$r = \theta$	$r = 4\sin(2\theta)$	$r = 2(1 + \sin\theta)$	$r = 1 + 3\cos\theta$

> **Example 17**

Consider the polar function $r = 2 + 4 \sin \theta$.

(a)   For what values of θ in the interval $[0, 2\pi]$ does the curve pass through the origin?

(b)   For what value of θ in the interval $[0, \pi/2]$ does the curve intersect the circle $r = 3$?

✓ **Solutions**

(a)   At the origin $r = 0$, so we want $2 + 4 \sin \theta = 0$. Solving for θ yields $\sin \theta = -\frac{1}{2}$, which occurs at $\theta = \frac{7\pi}{6}$ and $\theta = \frac{11\pi}{6}$.

(b)   The curves $r = 2 + 4 \sin \theta$ and $r = 3$ intersect when $2 + 4 \sin \theta = 3$, or $\sin \theta = \frac{1}{4}$. From the calculator we find $\theta = \arcsin \frac{1}{4} \approx 0.253$. See Figure 1.15.

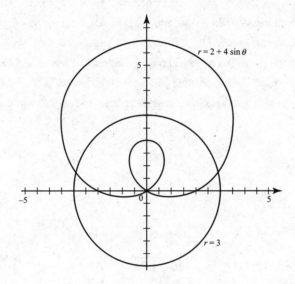

Figure 1.15

A polar function may also be expressed parametrically:

$$x = r \cos \theta \qquad y = r \sin \theta$$

In this form, the curve $r = 2 + 4 \sin \theta$ from Example 17 would be defined by:

$$x(\theta) = (2 + 4 \sin \theta) \cos \theta \qquad y(\theta) = (2 + 4 \sin \theta) \sin \theta$$

BC ONLY

> **Example 18**

Find the (x,y) coordinates of the point on $r = 1 + \cos \theta$ where $\theta = \frac{\pi}{3}$.

✓ **Solution**

At $\theta = \frac{\pi}{3}$, $r = 1 + \cos \frac{\pi}{3} = 1 + \frac{1}{2} = \frac{3}{2}$.

Since $x = r \cos \theta = \frac{3}{2} \cos \frac{\pi}{3} = \frac{3}{2} \cdot \frac{1}{2} = \frac{3}{4}$ and $y = r \sin \theta = \frac{3}{2} \sin \frac{\pi}{3} = \frac{3}{2} \cdot \frac{\sqrt{3}}{2} = \frac{3\sqrt{3}}{4}$, the point is $\left(\frac{3}{4}, \frac{3\sqrt{3}}{4} \right)$.

CHAPTER SUMMARY

This chapter reviewed some important precalculus topics. These topics are not directly tested on the AP exam; rather, they represent basic principles important in calculus. These include finding the domain, range, and inverse of a function and understanding the properties of polynomial and rational functions, trigonometric and inverse trig functions, and exponential and logarithmic functions.

For BC students, this chapter also reviewed parametrically defined functions and polar curves.

Practice Exercises

PART A. NO CALCULATOR—DIRECTIONS: Answer these questions *without* using your calculator.

A1. If $f(x) = x^3 - 2x - 1$, then $f(-2) =$

 (A) -13
 (B) -5
 (C) -1
 (D) 7

A2. The domain of $f(x) = \dfrac{x-1}{x^2+1}$ is

 (A) all $x \neq 1$
 (B) all $x \neq 1, -1$
 (C) all $x \neq -1$
 (D) all reals

A3. The domain of $g(x) = \dfrac{\sqrt{x-2}}{x^2-x}$ is

 (A) all $x \neq 0, 1$
 (B) $x \leq 2, x \neq 0, 1$
 (C) $x \geq 2$
 (D) $x > 2$

A4. If $f(x) = x^3 - 3x^2 - 2x + 5$ and $g(x) = 2$, then $g(f(x)) =$

 (A) $2x^3 - 6x^2 - 2x + 10$
 (B) $2x^2 - 6x + 1$
 (C) -3
 (D) 2

A5. If $f(x) = x^3 - 3x^2 - 2x + 5$ and $g(x) = 2$, then $f(g(x)) =$

 (A) $2x^3 - 6x^2 - 2x + 10$
 (B) $2x^2 - 6x + 1$
 (C) -3
 (D) 2

A6. If $f(x) = x^3 + Ax^2 + Bx - 3$ and if $f(1) = 4$ and $f(-1) = -6$, what is the value of $2A + B$?

 (A) 12
 (B) 8
 (C) 0
 (D) -2

A7. Which of the following equations has a graph that is symmetric with respect to the origin?

 (A) $y = \dfrac{x-1}{x}$
 (B) $y = 2x^4 + 1$
 (C) $y = x^3 + 2x$
 (D) $y = x^3 + 2$

A8. Let g be a function defined for all reals. Which of the following conditions is not sufficient to guarantee that g has an inverse function?

 (A) $g(x) = ax + b, a \neq 0$
 (B) g is strictly decreasing
 (C) g is symmetric to the origin
 (D) g is one-to-one

A9. Let $y = f(x) = \sin(\arctan x)$. Then the range of f is

 (A) $\{y \mid -1 < y < 1\}$
 (B) $\{y \mid -1 \leq y \leq 1\}$
 (C) $\left\{y \mid -\dfrac{\pi}{2} < y < \dfrac{\pi}{2}\right\}$
 (D) $\left\{y \mid -\dfrac{\pi}{2} \leq y \leq \dfrac{\pi}{2}\right\}$

A10. Let $g(x) = |\cos x - 1|$. The maximum value attained by g on the closed interval $[0, 2\pi]$ is for x equal to

 (A) 0
 (B) $\dfrac{\pi}{2}$
 (C) 2
 (D) π

A11. Which of the following functions is not odd?

 (A) $f(x) = \sin x$
 (B) $f(x) = \sin 2x$
 (C) $f(x) = x^3 + 1$
 (D) $f(x) = \dfrac{x}{x^2+1}$

A12. If the solutions to the equation $f(x) = 0$ are $x = 1, -2$, then $f(2x) = 0$ at $x =$

(A) $\frac{1}{2}$ and -1

(B) $-\frac{1}{2}$ and 1

(C) 2 and -4

(D) -2 and 4

A13. The set of zeros of $f(x) = x^3 + 4x^2 + 4x$ is

(A) $\{-2\}$

(B) $\{0, -2\}$

(C) $\{0, 2\}$

(D) $\{2\}$

A14. The values of x for which the graphs of $y = x + 2$ and $y^2 = 4x$ intersect are

(A) -2 and 2

(B) -2

(C) 2

(D) no intersection

A15. The function whose graph is a reflection in the y-axis of the graph of $f(x) = 1 - 3^x$ is

(A) $g(x) = 1 - 3^{-x}$

(B) $g(x) = 3^x - 1$

(C) $g(x) = \log_3 (x - 1)$

(D) $g(x) = \log_3 (1 - x)$

A16. Let $f(x)$ have an inverse function $g(x)$. Then $f(g(x)) =$

(A) 1

(B) x

(C) $\frac{1}{x}$

(D) $f(x) \cdot g(x)$

A17. The function $f(x) = 2x^3 + x - 5$ has exactly one real zero. It is between

(A) -1 and 0

(B) 0 and 1

(C) 1 and 2

(D) 2 and 3

A18. The period of $f(x) = \sin\left(\frac{2\pi}{3}x\right)$ is

(A) $\frac{2}{3}$

(B) $\frac{3}{2}$

(C) 3

(D) 6

A19. The range of $y = f(x) = \ln (\cos x)$ is

(A) $\{y \mid -\infty < y \le 0\}$

(B) $\{y \mid 0 < y \le 1\}$

(C) $\{y \mid -1 < y < 1\}$

(D) $\{y \mid 0 \le y \le 1\}$

A20. If $\log_b (3^b) = \frac{b}{2}$, then $b =$

(A) $\frac{1}{9}$

(B) $\frac{1}{2}$

(C) 3

(D) 9

A21. Let f^{-1} be the inverse function of $f(x) = x^3 + 2$. Then $f^{-1}(x) =$

(A) $\dfrac{1}{x^3 - 2}$

(B) $(x - 2)^3$

(C) $\sqrt[3]{x + 2}$

(D) $\sqrt[3]{x - 2}$

A22. The set of x-intercepts of the graph of $f(x) = x^3 - 2x^2 - x + 2$ is

(A) $\{-1, 1\}$

(B) $\{1, 2\}$

(C) $\{-1, 1, 2\}$

(D) $\{-1, -2, 2\}$

A23. If the domain of f is restricted to the open interval $\left(-\frac{\pi}{2}, \frac{\pi}{2}\right)$, then the range of $f(x) = e^{\tan x}$ is

(A) the set of all reals

(B) the set of positive reals

(C) the set of nonnegative reals

(D) $\{y \mid 0 < y \le 1\}$

A24. Which of the following is a reflection of the graph of $y = f(x)$ in the x-axis?

(A) $y = -f(x)$
(B) $y = f(-x)$
(C) $y = f(|x|)$
(D) $y = -f(-x)$

A25. The smallest positive x for which the function $f(x) = \sin\left(\dfrac{x}{3}\right) - 1$ is a maximum is

(A) $\dfrac{\pi}{2}$
(B) π
(C) $\dfrac{3\pi}{2}$
(D) 3π

A26. $\tan\left(\arccos\left(-\dfrac{\sqrt{2}}{2}\right)\right) =$

(A) -1
(B) $-\dfrac{\sqrt{3}}{3}$
(C) $\dfrac{\sqrt{3}}{3}$
(D) 1

A27. If $f^{-1}(x)$ is the inverse of $f(x) = 2e^{-x}$, then $f^{-1}(x) =$

(A) $\ln\left(\dfrac{2}{x}\right)$
(B) $\ln\left(\dfrac{x}{2}\right)$
(C) $\left(\dfrac{1}{2}\right)\ln x$
(D) $\sqrt{\ln x}$

A28. Which of the following functions does not have an inverse function?

(A) $y = \sin x \left(-\dfrac{\pi}{2} \le x \le \dfrac{\pi}{2}\right)$
(B) $y = x^3 + 2$
(C) $y = \dfrac{x}{x^2 + 1}$
(D) $y = \ln(x - 2)$ (where $x > 2$)

A29. Suppose that $f(x) = \ln x$ for all positive x and $g(x) = 9 - x^2$ for all real x. The domain of $f(g(x))$ is

(A) $\{x \mid x \le 3\}$
(B) $\{x \mid |x| > 3\}$
(C) $\{x \mid |x| < 3\}$
(D) $\{x \mid 0 < x < 3\}$

A30. Suppose that $f(x) = \ln x$ for all positive x and $g(x) = 9 - x^2$ for all real x. The range of $y = f(g(x))$ is

(A) $\{y \mid y > 0\}$
(B) $\{y \mid 0 < y \le \ln 9\}$
(C) $\{y \mid y \le \ln 9\}$
(D) $\{y \mid y < 0\}$

BC ONLY

A31. The curve defined parametrically by $x(t) = t^2 + 3$ and $y(t) = t^2 + 4$ is part of a(n)

(A) line
(B) circle
(C) parabola
(D) ellipse

A32. Which equation includes the curve defined parametrically by $x(t) = \cos^2(t)$ and $y(t) = 2\sin(t)$?

(A) $x^2 + y^2 = 4$
(B) $4x^2 + y^2 = 4$
(C) $4x + y^2 = 4$
(D) $x + 4y^2 = 1$

A33. Find the smallest value of θ in the interval $[0, 2\pi]$ for which the rose $r = 2\cos(5\theta)$ passes through the origin.

(A) 0
(B) $\dfrac{\pi}{20}$
(C) $\dfrac{\pi}{10}$
(D) $\dfrac{\pi}{5}$

A34. For what value of θ in the interval $[0, \pi]$ do the polar curves $r = 3$ and $r = 2 + 2\cos\theta$ intersect?

(A) $\dfrac{\pi}{6}$
(B) $\dfrac{\pi}{4}$
(C) $\dfrac{\pi}{3}$
(D) $\dfrac{2\pi}{3}$

PART B. CALCULATOR ACTIVE—DIRECTIONS: Some of the following questions require the use of a graphing calculator.

B1. The graph of the function $f(x) = 2e^{\sin(x)} - 3$ crosses the x-axis once in the interval $[0,1]$. What is the x-coordinate of this x-intercept?

(A) 0.209
(B) 0.417
(C) 0.552
(D) 0.891

B2. Find the x-intercept of the graph of $f(x) = \sqrt{\sin(x) + 1} + x^3 - 3e^{\cos(x)} + 6$ on the portion of the graph where $f(x)$ is decreasing.

(A) -1.334
(B) -0.065
(C) -0.801
(D) 0.472

B3. You are given the function $f(x) = e^{x^5 - 2x^3 + 2} - 7$ on the closed interval $[-2,2]$. Find all intervals where $f(x) < 0$.

(A) $(-2,-1.421)$ and $(0.305,1.407)$
(B) $(-1.421,0.305)$ only
(C) $(-2,-1.421)$ only
(D) $(0.305,1.407)$ only

B4. You are given the function $f(x) = (4 - 2x - 2x^2)\cos(3x - 4)$ on the closed interval $[-3,2]$. How many times does $f(x)$ cross the x-axis in the interval?

(A) four
(B) five
(C) six
(D) seven

B5. On the interval $[0,2\pi]$ there is one point on the curve $r = \theta - 2\cos\theta$ whose x-coordinate is 2. Find the y-coordinate there.

BC ONLY

(A) -4.594
(B) -3.764
(C) 1.979
(D) 5.201

B6. If $f(x) = (1 + e^x)$ then the domain of $f^{-1}(x)$ is

(A) $(-\infty,\infty)$
(B) $(0,\infty)$
(C) $(1,\infty)$
(D) $(2,\infty)$

Answer Key

A1. **B**	A11. **C**	A21. **D**	A31. **A**
A2. **D**	A12. **A**	A22. **C**	A32. **C**
A3. **C**	A13. **B**	A23. **B**	A33. **C**
A4. **D**	A14. **D**	A24. **A**	A34. **C**
A5. **C**	A15. **A**	A25. **C**	B1. **B**
A6. **B**	A16. **B**	A26. **A**	B2. **C**
A7. **C**	A17. **C**	A27. **A**	B3. **A**
A8. **C**	A18. **C**	A28. **C**	B4. **D**
A9. **A**	A19. **A**	A29. **C**	B5. **B**
A10. **D**	A20. **D**	A30. **C**	B6. **C**

Answer Explanations

A1. **(B)** $f(-2) = (-2)^3 - 2(-2) - 1 = -5$.

A2. **(D)** The denominator, $x^2 + 1$, is never 0.

A3. **(C)** Since $x - 2$ may not be negative, $x \geq 2$. The denominator equals 0 at $x = 0$ and $x = 1$, but these values are not in the interval $x \geq 2$.

A4. **(D)** Since $g(x) = 2$, g is a constant function. Thus, for all $f(x)$, $g(f(x)) = 2$.

A5. **(C)** $f(g(x)) = f(2) = -3$.

A6. **(B)** Solve the pair of equations
$$\begin{cases} 4 = 1 + A + B - 3 \\ -6 = -1 + A - B - 3 \end{cases}$$
Add to get A; substitute in either equation to get B. $A = 2$ and $B = 4$.

A7. **(C)** The graph of $f(x)$ is symmetrical to the origin if $f(-x) = -f(x)$. In (C),
$$f(-x) = (-x)^3 + 2(-x) = -x^3 - 2x = -(x^3 + 2x) = -f(x).$$

A8. **(C)** For g to have an inverse function it must be one-to-one. Note, on page 329, that although the graph of $y = xe^{-x^2}$ is symmetric to the origin, it is not one-to-one.

A9. **(A)** Note that $-\dfrac{\pi}{2} < \arctan x < \dfrac{\pi}{2}$; the sine function varies from -1 to 1 as the argument varies from $-\dfrac{\pi}{2}$ to $\dfrac{\pi}{2}$.

A10. **(D)** The maximum value of g is 2, attained when $\cos x = -1$. On $[0, 2\pi]$, $\cos x = -1$ for $x = \pi$.

A11. **(C)** f is odd if $f(-x) = -f(x)$. In (C), $f(-x) = (-x)^3 + 1 = -x^3 + 1 \neq -f(x)$.

A12. **(A)** Since $f(q) = 0$ if $q = 1$ or $q = -2$, $f(2x) = 0$ if $2x$, a replacement for q, equals 1 or -2.

A13. **(B)** $f(x) = x(x^2 + 4x + 4) = x(x + 2)^2$; $f(x) = 0$ for $x = 0$ and $x = -2$.

A14. **(D)** Solving simultaneously yields $(x + 2)^2 = 4x$; $x^2 + 4x + 4 = 4x$; $x^2 + 4 = 0$. There are no real solutions.

A15. **(A)** The reflection of $y = f(x)$ in the y-axis is $y = f(-x)$.

A16. **(B)** If g is the inverse of f, then f is the inverse of g. This implies that the function f assigns to each value $g(x)$ the number x.

A17. **(C)** Since f is continuous (see page 99), then if f is negative at a and positive at b, f must equal 0 at some intermediate point. Since $f(1) = -2$ and $f(2) = 13$, this point is between 1 and 2.

A18. **(C)** The function $\sin bx$ has period $\dfrac{2\pi}{b}$. Then $2\pi \div \dfrac{2\pi}{3} = 3$.

A19. **(A)** Since $\ln q$ is defined only if $q > 0$, the domain of $\ln \cos x$ is the set of x for which $\cos x > 0$, that is, when $0 < \cos x \leq 1$. Thus $-\infty < \ln \cos x \leq 0$.

A20. **(D)** $\log_b 3^b = \dfrac{b}{2}$ implies $b \log_b 3 = \dfrac{b}{2}$. Then $\log_b 3 = \dfrac{1}{2}$ and $3 = b^{1/2}$. So $3^2 = b$.

A21. **(D)** Interchange x and y: $x = y^3 + 2$.

Solve for y: $y^3 = x - 2$, so $y = \sqrt[3]{x - 2} = f^{-1}(x)$.

A22. **(C)** Since $f(1) = 0$, $x - 1$ is a factor of f. Since $f(x)$ divided by $x - 1$ yields $x^2 - x - 2$, $f(x) = (x - 1)(x + 1)(x - 2)$; the roots are $x = 1, -1$, and 2.

A23. **(B)** If $-\frac{\pi}{2} < x < \frac{\pi}{2}$, then $-\infty < \tan x < \infty$ and $0 < e^{\tan x} < \infty$.

A24. **(A)** The reflection of $f(x)$ in the x-axis is $-f(x)$.

A25. **(C)** $f(x)$ attains its maximum when $\sin\left(\frac{x}{3}\right)$ does. The maximum value of the sine function is 1; the smallest positive occurrence is at $\frac{\pi}{2}$. Set $\frac{x}{3}$ equal to $\frac{\pi}{2}$.

A26. **(A)** $\arccos\left(\frac{-\sqrt{2}}{2}\right) = \frac{3\pi}{4}$; $\tan\left(\frac{3\pi}{4}\right) = -1$.

A27. **(A)** Interchange x and y: $x = 2e^{-y}$

Solve for y: $e^{-y} = \frac{x}{2}$, so $-y = \ln\frac{x}{2}$ and $y = -\ln\frac{x}{2}$

Thus $f^{-1}(x) = \ln\frac{2}{x}$.

A28. **(C)** The function in (C) is not one-to-one since, for each y between $-\frac{1}{2}$ and $\frac{1}{2}$ (except 0), there are two x's in the domain.

A29. **(C)** The domain of the ln function is the set of positive reals. The function $g(x) > 0$ if $x^2 < 9$.

A30. **(C)** Since the domain of $f(g)$ is $(-3,3)$, $\ln(9 - x^2)$ takes on every real value less than or equal to $\ln 9$.

A31. **(A)** Substituting $t^2 = x - 3$ in $y(t) = t^2 + 4$ yields $y = x + 1$.

A32. **(C)** Using the identity $\cos^2(t) + \sin^2(t) = 1$, $x + \left(\frac{y}{2}\right)^2 = 1$.

A33. **(C)** $2 \cos 5\theta = 0$ when $5\theta = \frac{\pi}{2}$.

A34. **(C)** If $2 + 2\cos\theta = 3$, then $\cos\theta = \frac{1}{2}$.

B1. **(B)** The graph of the function $f(x) = 2e^{\sin(x)} - 3$ is shown below. The graph crosses the x-axis at $x = 0.417$ in the interval $[0,1]$.

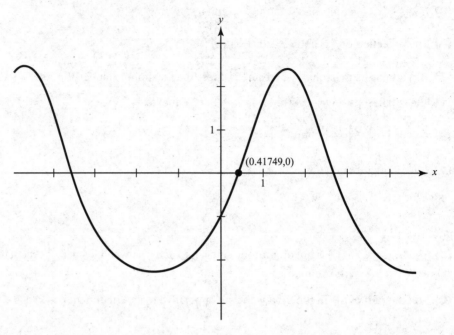

B2. **(C)** The graph of the function $f(x) = \sqrt{\sin(x) + 1} + x^3 - 3e^{\cos(x)} + 6$ is shown below. There is only one root on an interval where the graph of $f(x)$ is decreasing, and that is at $x = -0.801$.

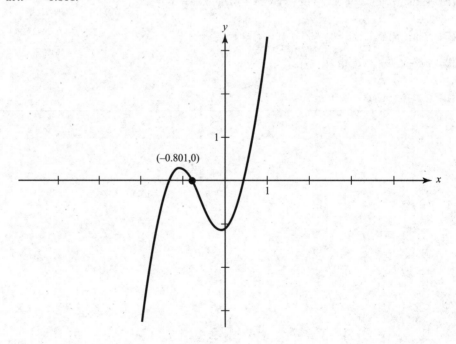

B3. **(A)** The graph of the function $f(x) = e^{x^5 - 2x^3 + 2} - 7$ is shown below. The graph of $f(x)$ is below the x-axis on the intervals $(-2, -1.421)$ and $(0.305, 1.407)$.

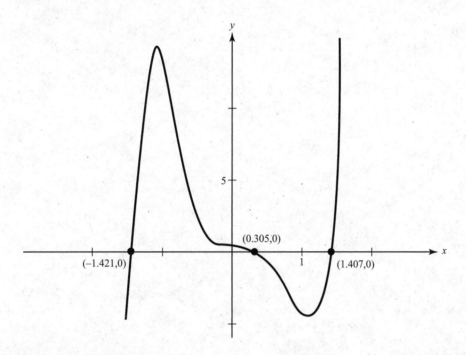

B4. **(D)** The graph of the function $f(x) = (4 - 2x - 2x^2)\cos(3x - 4)$ is shown below. The graph of $f(x)$ crosses the x-axis seven times.

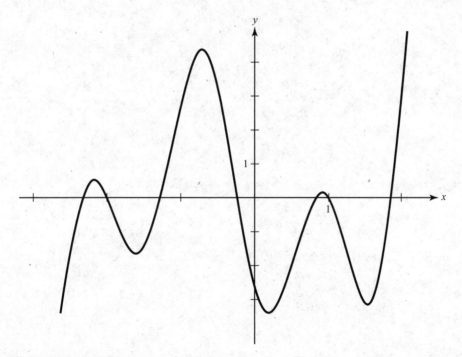

B5. **(B)** For polar functions, $x = r \cos \theta$. Solving $(\theta - 2 \cos \theta)\cos \theta = 2$ yields $\theta \approx 5.201$, and thus $y = r \sin \theta = (5.201 - 2 \cos 5.201)\sin 5.201$.

B6. **(C)** The inverse of $y = 1 + e^x$ is $x = 1 + e^y$ or $y = \ln(x - 1)$; $(x - 1)$ must be positive.

2
Limits and Continuity

Learning Objectives

In this chapter, you will review:

→ General properties of limits
→ How to find limits using algebraic expressions, tables, and graphs
→ Horizontal and vertical asymptotes
→ Continuity
→ Removable, jump, and infinite discontinuities
→ Some important theorems, including the Squeeze (Sandwich) Theorem, the Extreme Value Theorem, and the Intermediate Value Theorem

A. Definitions and Examples

The number L is the *limit of the function* $f(x)$ as x approaches c if, as the values of x get arbitrarily close (but not equal) to c, the values of $f(x)$ approach (or equal) L. We write

Limit

$$\lim_{x \to c} f(x) = L$$

In order for $\lim_{x \to c} f(x)$ to exist, the values of f must tend to the same number L as x approaches c from either the left or the right. We write

$$\lim_{x \to c^-} f(x)$$

for the *left-hand* limit of f at c (as x approaches c through values *less* than c), and

One-sided limits

$$\lim_{x \to c^+} f(x)$$

for the *right-hand* limit of f at c (as x approaches c through values *greater* than c).

> ### Example 1

The greatest-integer function $g(x) = [x]$, shown in Figure 2.1, has different left-hand and right-hand limits at *every* integer. For example,

$$\lim_{x \to 1^-} [x] = 0 \quad \text{but} \quad \lim_{x \to 1^+} [x] = 1$$

This function, therefore, does not have a limit at $x = 1$ or, by the same reasoning, at any other integer.

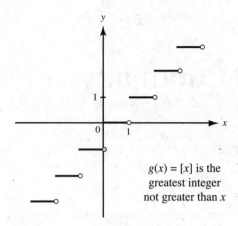

Figure 2.1

However, $[x]$ does have a limit at every nonintegral real number. For example,

$$\lim_{x \to 0.6}[x] = 0 \qquad \lim_{x \to 0.99}[x] = 0 \qquad \lim_{x \to 2.01}[x] = 2 \qquad \lim_{x \to 2.95}[x] = 2$$

$$\lim_{x \to -3.1}[x] = -4 \qquad \lim_{x \to -2.9}[x] = -3 \qquad \lim_{x \to -0.9}[x] = -1 \qquad \lim_{x \to -0.01}[x] = -1$$

> ## ❯ Example 2

Suppose the function $y = f(x)$, graphed in Figure 2.2, is defined as follows:

$$f(x) = \begin{cases} x+1 & (-2 < x < 0) \\ 2 & (x = 0) \\ -x & (0 < x < 2) \\ 0 & (x = 2) \\ x-4 & (2 < x \le 4) \end{cases}$$

Determine whether limits of f, if any, exist at

(a) $x = -2$ (b) $x = 0$

(c) $x = 2$ (d) $x = 4$

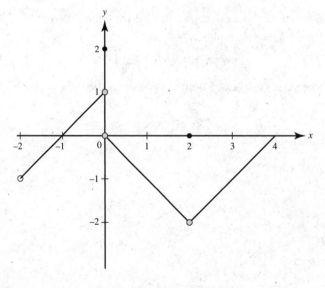

Figure 2.2

✓ Solutions

(a) $\lim\limits_{x \to 2^+} f(x) = -1$, so the right-hand limit exists at $x = -2$, even though f is not defined at $x = -2$.

(b) $\lim\limits_{x \to 0} f(x)$ does not exist. Although f is defined at $x = 0$ ($f(0) = 2$), we observe that $\lim\limits_{x \to 0^-} f(x) = 1$, whereas $\lim\limits_{x \to 0^+} f(x) = 0$. For the limit to exist at a point, the left-hand and right-hand limits must be the same.

(c) $\lim\limits_{x \to 2} f(x) = -2$. This limit exists because $\lim\limits_{x \to 2^-} f(x) = \lim\limits_{x \to 2^+} f(x) = -2$. Indeed, the limit exists at $x = 2$ even though it is different from the value of f at 2 ($f(2) = 0$).

(d) $\lim\limits_{x \to 4^-} f(x) = 0$, so the left-hand limit exists at $x = 4$.

❯ Example 3

Prove that $\lim\limits_{x \to 0} |x| = 0$.

✓ Solution

The graph of $|x|$ is shown in Figure 2.3.

We examine both left- and right-hand limits of the absolute-value function as $x \to 0$. Since

$$|x| = \begin{cases} -x \text{ if } x < 0 \\ x \text{ if } x > 0 \end{cases}$$

it follows that $\lim\limits_{x \to 0^-} |x| = \lim\limits_{x \to 0^-} (-x) = 0$ and $\lim\limits_{x \to 0^+} |x| = \lim\limits_{x \to 0^+} x = 0$.

Since the left-hand and right-hand limits both equal 0, $\lim\limits_{x \to 0} |x| = 0$.

Note that $\lim\limits_{x \to c} |x| = c$ if $c > 0$ but equals $-c$ if $c < 0$.

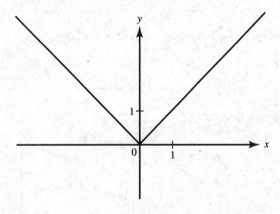

Figure 2.3

Definition

The function $f(x)$ is said to *become infinite* (positively or negatively) as x approaches c if $f(x)$ can be made arbitrarily large (positively or negatively) by taking x sufficiently close to c. We write

$$\lim\limits_{x \to c} f(x) = +\infty \ (\text{or} \lim\limits_{x \to c} f(x) = -\infty)$$

Since, for the limit to exist, it must be a finite number, neither of the preceding limits exists.

This definition can be extended to include x approaching c from the left or from the right. The following examples illustrate these definitions.

❯ Example 4

Describe the behavior of $f(x) = \dfrac{1}{x}$ near $x = 0$ using limits.

✓ Solution

The graph (Figure 2.4) shows that:

$$\lim_{x \to 0^-} \frac{1}{x} = -\infty$$

$$\lim_{x \to 0^+} \frac{1}{x} = +\infty$$

$$\lim_{x \to 0} \frac{1}{x} \text{ does not exist}$$

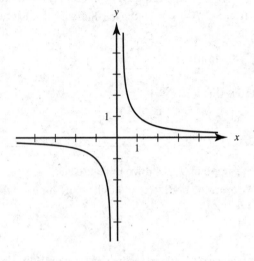

Figure 2.4

❯ Example 5

Describe the behavior of $g(x) = \dfrac{1}{(x-1)^2}$ near $x = 1$ using limits.

✓ Solution

The graph (Figure 2.5) shows that:

$$\lim_{x \to 1^-} g(x) = \lim_{x \to 1^+} g(x) = \infty$$

$$\lim_{x \to 1^-} g(x) = \infty$$

$$\lim_{x \to 1^+} g(x) = \infty$$

$$\lim_{x \to 1} g(x) = \infty$$

Figure 2.5

NOTE: Using $+\infty$ or $-\infty$ to indicate a limit is describing the behavior of the function and not actually a limit. Remember that none of the limits in Examples 4 and 5 exist!

Definition

We write

$$\lim_{x \to \infty} f(x) = L \ (\text{or} \lim_{x \to -\infty} f(x) = L)$$

if the difference between $f(x)$ and L can be made arbitrarily small by making x sufficiently large positively (or negatively).

In Examples 4 and 5, note that $\lim_{x \to \infty} f(x) = \lim_{x \to -\infty} f(x) = 0$ and $\lim_{x \to \infty} g(x) = \lim_{x \to -\infty} g(x) = 0$.

❯ Example 6

From the graph of $h(x) = 1 + \dfrac{3}{x - 2} = \dfrac{x + 1}{x - 2}$ (Figure 2.6), describe the behavior of h using limits.

✓ Solution

$$\lim_{x \to -\infty} h(x) = \lim_{x \to +\infty} h(x) = 1$$

$$\lim_{x \to 2^-} h(x) = -\infty$$

$$\lim_{x \to 2^+} h(x) = +\infty$$

Figure 2.6

Definition

The theorems that follow in Section C of this chapter confirm the conjectures made about limits of functions from their graphs.

Finally, if the function $f(x)$ becomes infinite as x becomes infinite, then one or more of the following may hold:

$$\lim_{x \to +\infty} f(x) = +\infty \text{ or } -\infty \quad \text{or} \quad \lim_{x \to -\infty} f(x) = +\infty \text{ or } -\infty$$

End Behavior of Polynomials

Every polynomial whose degree is greater than or equal to 1 becomes infinite as x does. It becomes positively or negatively infinite, depending only on the sign of the leading coefficient and the degree of the polynomial.

> **Example 7**

For each function given below, describe $\lim\limits_{x \to +\infty}$ and $\lim\limits_{x \to -\infty}$.

(a) $f(x) = x^3 - 3x^2 + 7x + 2$

(b) $g(x) = -4x^4 + 1{,}000{,}000x^3 + 100$

(c) $h(x) = -5x^3 + 3x^2 - 4\pi + 8$

(d) $k(x) = \pi - 0.001x$

✓ **Solutions**

(a) $\lim\limits_{x \to +\infty} f(x) = +\infty$ $\lim\limits_{x \to -\infty} f(x) = -\infty$

(b) $\lim\limits_{x \to +\infty} g(x) = -\infty$ $\lim\limits_{x \to -\infty} g(x) = -\infty$

(c) $\lim\limits_{x \to +\infty} h(x) = -\infty$ $\lim\limits_{x \to -\infty} h(x) = +\infty$

(d) $\lim\limits_{x \to +\infty} k(x) = -\infty$ $\lim\limits_{x \to -\infty} k(x) = +\infty$

It's easy to write rules for the behavior of a polynomial as x becomes infinite!

B. Asymptotes

Horizontal asymptote

The line $y = b$ is a *horizontal asymptote* of the graph of $y = f(x)$ if

$$\lim_{x \to \infty} f(x) = b \quad \text{or} \quad \lim_{x \to -\infty} f(x) = b$$

The graph of $f(x) = \dfrac{1}{x}$ (Figure 2.4) on page 92 has the x-axis ($y = 0$) as the horizontal asymptote.

So does the graph of $g(x) = \dfrac{1}{(x-1)^2}$ (Figure 2.5) on page 92.

The graph of $h(x) = \dfrac{x+1}{x-2}$ (Figure 2.6) on page 93 has

the line $y = 1$ as the horizontal asymptote.

Note, unlike vertical asymptotes, horizontal asymp-

totes can be crossed. The graph of $p(x) = \dfrac{3x^2 - 8x + 7}{x^2 - 4x + 5}$

has a horizontal asymptote at $y = 3$, as shown

to the right.

The line $x = a$ is a *vertical asymptote* of the graph of $y = f(x)$ if one or more of the following holds:

$$\lim_{x \to a^-} f(x) = +\infty \quad \text{or} \quad \lim_{x \to a^-} f(x) = -\infty$$

or

$$\lim_{x \to a^+} f(x) = +\infty \quad \text{or} \quad \lim_{x \to a^+} f(x) = -\infty$$

The graph of $f(x) = \dfrac{1}{x}$ (Figure 2.4) has $x = 0$ (the y-axis) as the vertical asymptote.

The graph of $g(x) = \dfrac{1}{(x-1)^2}$ (Figure 2.5) has $x = 1$ as the vertical asymptote.

The graph of $h(x) = \dfrac{x+1}{x-2}$ (Figure 2.6) has the line $x = 2$ as the vertical asymptote.

> **Example 8**

From the graph of $k(x) = \dfrac{2x-4}{x-3}$ in Figure 2.7, describe the asymptotes of k using limits.

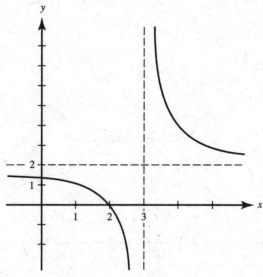

Figure 2.7

✓ **Solution**

We see that $y = 2$ is a horizontal asymptote since

$$\lim_{x \to +\infty} k(x) = \lim_{x \to -\infty} k(x) = 2$$

Also, $x = 3$ is a vertical asymptote; the graph shows that

$$\lim_{x \to 3^-} k(x) = -\infty \quad \text{and} \quad \lim_{x \to 3^+} k(x) = +\infty$$

C. Theorems on Limits

If B, D, c, and k are real numbers and if the limits of functions f and g exist at $x = c$ such that $\lim\limits_{x \to c} f(x) = B$ and $\lim\limits_{x \to c} g(x) = D$, then:

(1) The Constant Rule. $\lim\limits_{x \to c} k = k$

(2) The Constant Multiple Rule. $\lim\limits_{x \to c} k \cdot f(x) = k \cdot \lim\limits_{x \to c} f(x) = k \cdot B$

(3) The Sum Rule. $\lim\limits_{x \to c}(f(x) + g(x)) = \lim\limits_{x \to c} f(x) + \lim\limits_{x \to c} g(x) = B + D$

(4) The Difference Rule. $\lim\limits_{x \to c}(f(x) - g(x)) = \lim\limits_{x \to c} f(x) - \lim\limits_{x \to c} g(x) = B - D$

(5) The Product Rule. $\lim\limits_{x \to c}(f(x) \cdot g(x)) = \lim\limits_{x \to c} f(x) \cdot \lim\limits_{x \to c} g(x) = B \cdot D$

(6) The Quotient Rule. $\lim\limits_{x \to c}\left(\dfrac{f(x)}{g(x)}\right) = \dfrac{\lim\limits_{x \to c} f(x)}{\lim\limits_{x \to c} g(x)} = \dfrac{B}{D}$, provided that $D \neq 0$

(7) The Composition Rule. If the limit of $g(x)$ at $x = c$ exists, such that $\lim\limits_{x \to c} g(x) = D$, and $f(x)$ is continuous at $x = D$, such that $\lim\limits_{x \to D} f(x) = f(D)$, then $\lim\limits_{x \to c}\big(f(g(x))\big) = f\big(\lim\limits_{x \to c} g(x)\big) = f(D)$.

Sandwich (Squeeze) Theorem

(8) The Squeeze (Sandwich) Theorem. If $f(x) \leq g(x) \leq h(x)$ and if $\lim\limits_{x \to c} f(x) = \lim\limits_{x \to c} h(x) = L$, then $\lim\limits_{x \to c} g(x) = L$.

Figure 2.8 illustrates this theorem.

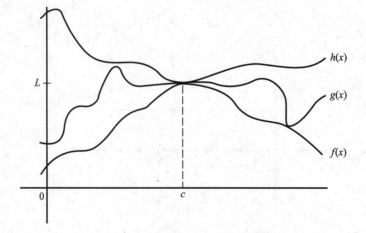

Figure 2.8

Squeezing function g between functions f and h forces g to have the same limit L at $x = c$ as do f and h.

> **Example 9**

$$\lim_{x \to 2}\left(5x^2 - 3x + 1\right) = 5\lim_{x \to 2} x^2 - 3\lim_{x \to 2} x + \lim_{x \to 2} 1$$
$$= 5 \cdot 4 \quad - 3 \cdot 2 \quad + 1$$
$$= 15$$

> **Example 10**

$$\lim_{x \to 0}(x \cos 2x) = \lim_{x \to 0} x \cdot \lim_{x \to 0}(\cos 2x)$$
$$= 0 \quad \cdot 1$$
$$= 0$$

> **Example 11**

$$\lim_{x \to -1}\frac{3x^2 - 2x - 1}{x^2 + 1} = \lim_{x \to -1}\left(3x^2 - 2x - 1\right) \div \lim_{x \to -1}\left(x^2 + 1\right)$$
$$= (3 + 2 - 1) \qquad \div (1 + 1)$$
$$= 2$$

> **Example 12**

$$\lim_{x \to 3}\frac{x^2 - 9}{x - 3} = \lim_{x \to 3}\frac{(x - 3)(x + 3)}{x - 3} = \lim_{x \to 3}(x + 3) = 6$$

since, by the definition of $\lim_{x \to c} f(x)$ in Section A, x must be different from 3 as $x \to 3$, the factor $x - 3$ may be removed *before* taking the limit.

> **Example 13**

$$\lim_{x \to -2}\frac{x^3 + 8}{x^2 - 4} = \lim_{x \to -2}\frac{(x + 2)(x^2 - 2x + 4)}{(x + 2)(x - 2)} = \lim_{x \to -2}\frac{x^2 - 2x + 4}{x - 2} = \frac{4 + 4 + 4}{-4} = -3$$

> **Example 14**

$\lim_{x \to 0}\dfrac{x}{x^3} = \lim_{x \to 0}\dfrac{1}{x^2} = \infty$. As $x \to 0$, the numerator approaches 1 while the denominator approaches 0; the limit does *not* exist.

> **Example 15**

$$\lim_{x \to 1}\frac{x^2 - 9}{x^2 - 1} = \lim_{x \to 1}1 = 1$$

> **Example 16**

$$\lim_{\Delta x \to 0}\frac{(3 + \Delta x)^2 - 3^2}{\Delta x} = \lim_{\Delta x \to 0}\frac{6\Delta x + \Delta x^2}{\Delta x} = \lim_{\Delta x \to 0} 6 + \Delta x = 6$$

> **Example 17**

$$\lim_{h \to 0}\frac{1}{h}\left(\frac{1}{2 + h} - \frac{1}{2}\right) = \lim_{h \to 0}\frac{2 - (2 + h)}{2h(2 + h)} = \lim_{h \to 0}\frac{-h}{2h(2 + h)}$$
$$= \lim_{h \to 0} -\frac{1}{2(2 + h)} = -\frac{1}{4}$$

D. Limit of a Quotient of Polynomials

To find $\lim\limits_{x\to\infty}\dfrac{P(x)}{Q(x)}$, where $P(x)$ and $Q(x)$ are polynomials in x, we can divide both the numerator and denominator by the highest power of x that occurs and use the fact that $\lim\limits_{x\to\infty}\dfrac{1}{x}=0$.

> **Example 18**

$$\lim_{x\to\infty}\frac{3-x}{4+x+x^2}=\lim_{x\to\infty}\frac{\dfrac{3}{x^2}-\dfrac{1}{x}}{\dfrac{4}{x^2}+\dfrac{1}{x}+1}=\frac{0-0}{0+0+1}=0$$

> **Example 19**

$$\lim_{x\to\infty}\frac{4x^4+5x+1}{37x^3-9}=\lim_{x\to\infty}\frac{4+\dfrac{5}{x^3}+\dfrac{1}{x^4}}{\dfrac{37}{x}-\dfrac{9}{x^4}}=\infty\ (\text{no limit})$$

> **Example 20**

$$\lim_{x\to\infty}\frac{x^3-4x^2+7}{3-6x-2x^3}=\lim_{x\to\infty}\frac{1-\dfrac{4}{x}+\dfrac{7}{x^3}}{\dfrac{3}{x^3}-\dfrac{6}{x^2}-2}=\frac{1-0+0}{0-0-2}=-\frac{1}{2}$$

The Rational Function Theorem

We see from Examples 18, 19, and 20 that: if the degree of $P(x)$ is less than that of $Q(x)$, then $\lim\limits_{x\to\infty}\dfrac{P(x)}{Q(x)}=0$; if the degree of $P(x)$ is higher than that of $Q(x)$, then $\lim\limits_{x\to\infty}\dfrac{P(x)}{Q(x)}=\infty$ or $-\infty$ (i.e., does not exist); and if the degrees of $P(x)$ and $Q(x)$ are the same, then $\lim\limits_{x\to\infty}\dfrac{P(x)}{Q(x)}=\dfrac{a_n}{b_n}$, where a_n and b_n are the coefficients of the highest powers of x in $P(x)$ and $Q(x)$, respectively.

This theorem holds also when we replace "$x\to\infty$" by "$x\to-\infty$."

Note also that:

(i) when $\lim\limits_{x\to\pm\infty}\dfrac{P(x)}{Q(x)}=0$, then $y=0$ is a horizontal asymptote of the graph of $y=\dfrac{P(x)}{Q(x)}$

(ii) when $\lim\limits_{x\to\pm\infty}\dfrac{P(x)}{Q(x)}=+\infty$ or $-\infty$, then the graph of $y=\dfrac{P(x)}{Q(x)}$ has no horizontal asymptotes

(iii) when $\lim\limits_{x\to\pm\infty}\dfrac{P(x)}{Q(x)}=\dfrac{a_n}{b_n}$, then $y=\dfrac{a_n}{b_n}$ is a horizontal asymptote of the graph of $y=\dfrac{P(x)}{Q(x)}$

> **Example 21**

$$\lim_{x\to\infty}\frac{100x^2-19}{x^3+5x^2+2}=0 \qquad \lim_{x\to-\infty}\frac{x^3-5}{1+x^2}=-\infty\ (\text{no limit}) \qquad \lim_{x\to\infty}\frac{x-4}{13+5x}=\frac{1}{5}$$

$$\lim_{x\to-\infty}\frac{4+x^2-3x^3}{x+7x^3}=-\frac{3}{7} \qquad \lim_{x\to\infty}\frac{x^3+1}{2-x^2}=-\infty\ (\text{no limit})$$

E. Other Basic Limits

E1. The basic trigonometric limit is:

$$\lim_{\theta \to 0} \frac{\sin \theta}{\theta} = 1 \text{ if } \theta \text{ is measured in radians}$$

> **Example 22**

Prove that $\lim\limits_{x \to \infty} \dfrac{\sin x}{x} = 0$.

✓ **Solution**

Since, for all x, $-1 \leq \sin x \leq 1$, it follows that, if $x > 0$, then $-\dfrac{1}{x} \leq \dfrac{\sin x}{x} \leq \dfrac{1}{x}$. But as $x \to \infty$, $-\dfrac{1}{x}$ and $\dfrac{1}{x}$ both approach 0; therefore, by the Squeeze Theorem, $\dfrac{\sin x}{x}$ must also approach 0. To obtain graphical confirmation of this fact, and of the additional fact that $\lim\limits_{x \to -\infty} \dfrac{\sin x}{x}$ also equals 0, graph

$$y_1 = \frac{\sin x}{x}, y_2 = \frac{1}{x}, \text{ and } y_3 = -\frac{1}{x}$$

in $[-4\pi, 4\pi] \times [-1,1]$. Observe, as $x \to \pm\infty$, that y_2 and y_3 approach 0 and that y_1 is squeezed between them.

> **Example 23**

Find $\lim\limits_{x \to 0} \dfrac{\sin 3x}{x}$.

✓ **Solution**

$$\lim_{x \to 0} \frac{\sin 3x}{x} = \lim_{x \to 0} \frac{3 \sin 3x}{3x} = 3 \lim_{x \to 0} \frac{\sin 3x}{3x} = 3 \cdot 1 = 3$$

E2. The number e can be defined as follows:

$$e = \lim_{n \to \infty} \left(1 + \frac{1}{n} \right)^n$$

Limit definition of e

The value of e can be approximated on a graphing calculator to a large number of decimal places by evaluating

$$y_1 = \left(1 + \frac{1}{x} \right)^x$$

for large values of x.

F. Continuity

If a function is continuous over an interval, we can draw its graph without lifting pencil from paper. The graph has no holes, breaks, or jumps on the interval.

Conceptually, if $f(x)$ is continuous at a point $x = c$, then the closer x is to c, the closer $f(x)$ gets to $f(c)$. This is made precise by the following definition:

Definition

Continuous

The function $y = f(x)$ is continuous at $x = c$ if

(1) $f(c)$ exists (that is, c is in the domain of f)

(2) $\lim\limits_{x \to c} f(x)$ exists

(3) $\lim\limits_{x \to c} f(x) = f(c)$

A function is continuous over the closed interval $[a,b]$ if it is continuous at each x such that $a \le x \le b$.

A function that is not continuous at $x = c$ is said to be discontinuous at that point. We then call $x = c$ a *point of discontinuity*.

Continuous Functions

Polynomials are continuous everywhere—namely, at every real number.

Rational functions, $\dfrac{P(x)}{Q(x)}$, are continuous at each point in their domain—that is, except where $Q(x) = 0$. The function $f(x) = \dfrac{1}{x}$, for example, is continuous except at $x = 0$, where f is not defined.

The absolute-value function $f(x) = |x|$ (sketched in Figure 2.3, page 91) is continuous everywhere.

The trigonometric, inverse trigonometric, exponential, and logarithmic functions are continuous at each point in their domains.

Functions of the type $\sqrt[n]{x}$ (where n is a positive integer ≥ 2) are continuous at each x for which $\sqrt[n]{x}$ is defined.

The greatest-integer function $f(x) = [x]$ (Figure 2.1, page 90) is discontinuous at each integer since it does not have a limit at any integer.

Kinds of Discontinuities

In Example 2, page 90, $y = f(x)$ is defined as follows:

$$f(x) = \begin{cases} x + 1 & (-2 < x < 0) \\ 2 & (x = 0) \\ -x & (0 < x < 2) \\ 0 & (x = 2) \\ x - 4 & (2 < x \le 4) \end{cases}$$

The graph of f is shown at the right.

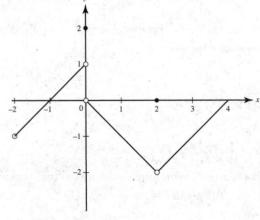

We observe that f is not continuous at $x = -2$, $x = 0$, or $x = 2$.

At $x = -2$, f is not defined.

At $x = 0$, f is defined; in fact, $f(0) = 2$. However, since $\lim\limits_{x \to 0^-} f(x) = 1$ and $\lim\limits_{x \to 0^+} f(x) = 0$, $\lim\limits_{x \to 0} f(x)$ does not exist. Where the left- and right-hand limits exist but are different, the function has a *jump discontinuity*. The greatest-integer (or step) function, $y = [x]$, has a jump discontinuity at every integer. (See page 90.)

Jump
discontinuity

At $x = 2$, f is defined; in fact, $f(2) = 0$. Also, $\lim\limits_{x \to 2} f(x) = -2$; the limit exists.

However, $\lim\limits_{x \to 2} f(x) \neq f(2)$. This discontinuity is called *removable*. If we were to redefine the function at $x = 2$ to be $f(2) = -2$, the new function would no longer have a discontinuity there. We cannot, however, "remove" a jump discontinuity by any redefinition whatsoever.

Removable
discontinuity

Whenever the graph of a function $f(x)$ has the line $x = a$ as a vertical asymptote, then $f(x)$ becomes positively or negatively infinite as $x \to a^+$ or as $x \to a^-$. The function is then said to have an *infinite discontinuity*. See, for example, Figure 2.4 (page 92) for $f(x) = \frac{1}{x}$, Figure 2.5 (page 92) for $g(x) = \frac{1}{(x-1)^2}$, or Figure 2.7 (page 95) for $k(x) = \frac{2x - 4}{x - 3}$. Each of these functions exhibits an infinite discontinuity.

Infinite
discontinuity

› Example 24

$f(x) = \dfrac{x - 1}{x^2 + x} = \dfrac{x - 1}{x(x + 1)}$ is not continuous at $x = 0$ or $= -1$ since the function is not defined for either of these numbers. Note also that neither $\lim\limits_{x \to 0} f(x)$ nor $\lim\limits_{x \to -1} f(x)$ exists.

› Example 25

Discuss the continuity of f, as graphed in Figure 2.9.

✓ Solution

$f(x)$ is continuous on $[(0,1), (1,3), \text{ and } (3,5)]$. The discontinuity at $x = 1$ is removable; the one at $x = 3$ is not. (Note that f is continuous from the right at $x = 0$ and from the left at $x = 5$.)

Figure 2.9

In Examples 26 through 31, we determine whether the functions are continuous at the points specified.

❯ Example 26

Is $f(x) = \frac{1}{2}x^4 - \sqrt{3}\,x^2 + 7$ continuous at $x = -1$?

✓ Solution

Since f is a polynomial, it is continuous everywhere, including, of course, at $x = -1$.

❯ Example 27

Is $g(x) = \dfrac{1}{x-3}$ continuous (a) at $x = 3$; (b) at $x = 0$?

✓ Solution

This function is continuous except where the denominator equals 0 (where g has an infinite discontinuity). It is not continuous at $x = 3$ but is continuous at $x = 0$.

❯ Example 28

Is $h(x) = \begin{cases} \dfrac{4}{x-2} & \text{if } x \neq 2 \\ 1 & \text{if } x = 2 \end{cases}$ continuous

(a)   at $x = 2$; (b) at $x = 3$?

Figure 2.10

✓ Solutions

(a) $h(x)$ has an infinite discontinuity at $x = 2$; this discontinuity is not removable.

(b) $h(x)$ is continuous at $x = 3$ and at every other point different from 2.
      (See Figure 2.10)

> **Example 29**

Is $k(x) = \dfrac{x^2 - 4}{x - 2}$ $(x \neq 2)$ continuous at $x = 2$?

✓ **Solution**

Note that $k(x) = x + 2$ for all $x \neq 2$. The function is continuous everywhere except at $x = 2$, where k is not defined. The discontinuity at 2 is removable. If we redefine $f(2)$ to equal 4, the new function will be continuous everywhere. (See Figure 2.11)

Figure 2.11

> **Example 30**

Is $f(x) = \begin{cases} x^2 + 2 & x \leq 1 \\ 4 & x > 1 \end{cases}$ continuous at $x = 1$?

✓ **Solution**

$f(x)$ is not continuous at $x = 1$ since $\lim\limits_{x \to 1^-} f(x) = 3 \neq \lim\limits_{x \to 1^+} f(x) = 4$. This function has a jump discontinuity at $x = 1$ (which cannot be removed). (See Figure 2.12)

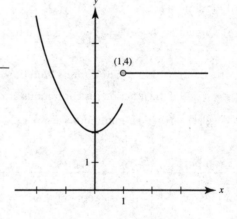

Figure 2.12

> **Example 31**

Is $g(x) = \begin{cases} x^2 & x \neq 2 \\ 1 & x = 2 \end{cases}$ continuous at $x = 2$?

✓ **Solution**

$g(x)$ is not continuous at $x = 2$ since $\lim\limits_{x \to 2} g(x) = 4 \neq g(2) = 1$. This discontinuity can be removed by redefining $g(2)$ to equal 4. (See Figure 2.13)

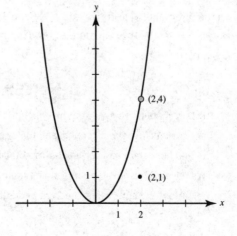

Figure 2.13

Theorems on Continuous Functions

Extreme Value Theorem

(1) **The Extreme Value Theorem.** If f is continuous on the closed interval $[a,b]$, then f attains a minimum value and a maximum value somewhere in that interval.

Intermediate Value Theorem

(2) **The Intermediate Value Theorem.** If f is continuous on the closed interval $[a,b]$ and M is a number such that $f(a) \leq M \leq f(b)$, then there is at least one number, c, in the interval $[a,b]$ such that $f(c) = M$.

Note an important special case of the Intermediate Value Theorem:

If f is continuous on the closed interval $[a,b]$ and $f(a)$ and $f(b)$ have opposite signs, then f has a zero in that interval (there is a value, c, in $[a,b]$ where $f(c) = 0$).

(3) **The Continuous Functions Theorem.** If functions f and g are both continuous at $x = c$, then so are the following functions:

(a) Constant Multiples: $k \cdot f(x)$ for any real number k
(b) Sums: $f(x) + g(x)$
(c) Differences: $f(x) - g(x)$
(d) Products: $f(x) \cdot g(x)$
(e) Quotients: $\dfrac{f(x)}{g(x)}$ provided that $g(c) \neq 0$

(4) **The Composition of Continuous Functions Theorem.** If the function g is continuous at $x = c$, and if the function f is continuous at $x = g(c)$, then the composite function $(f \circ g)(x) = f(g(x))$ is continuous at $x = c$.

> **Example 32**

Show that $f(x) = \dfrac{x^2 - 5}{x + 1}$ has a root between $x = 2$ and $x = 3$.

> **Solution**

The rational function f is discontinuous only at $x = -1$, $f(2) = -\frac{1}{3}$, and $f(3) = 1$. Since f is continuous on the interval $[2,3]$ and $f(2)$ and $f(3)$ have opposite signs, there is a value, c, in the interval where $f(c) = 0$, by the Intermediate Value Theorem.

CHAPTER SUMMARY

In this chapter, we reviewed the concept of a limit. We practiced finding limits using algebraic expressions, graphs, and the Squeeze (Sandwich) Theorem. We used limits to find horizontal and vertical asymptotes and to assess the continuity of a function. We reviewed removable, jump, and infinite discontinuities. We also looked at the very important Extreme Value Theorem and Intermediate Value Theorem.

Practice Exercises

PART A. NO CALCULATOR—DIRECTIONS: Answer these questions *without* using your calculator.

A1. $\lim\limits_{x \to 2} \dfrac{x^2 - 4}{x^2 + 4}$ is

(A) 1

(B) 0

(C) $-\dfrac{1}{2}$

(D) ∞

A2. $\lim\limits_{x \to \infty} \dfrac{4 - x^2}{x^2 - 1}$ is

(A) 1

(B) 0

(C) -1

(D) ∞

A3. $\lim\limits_{x \to 3} \dfrac{x - 3}{x^2 - 2x - 3}$ is

(A) 0

(B) 1

(C) $\dfrac{1}{4}$

(D) ∞

A4. $\lim\limits_{x \to 0} \dfrac{x}{x}$ is

(A) 1

(B) 0

(C) -1

(D) nonexistent

A5. $\lim\limits_{x \to 2} \dfrac{x^3 - 8}{x^2 - 4}$ is

(A) 1

(B) 3

(C) 6

(D) ∞

A6. $\lim\limits_{x \to \infty} \dfrac{4 - x^2}{4x^2 - x - 2}$ is

(A) -2

(B) $-\dfrac{1}{4}$

(C) 1

(D) 2

A7. $\lim\limits_{x \to -\infty} \dfrac{5x^3 + 27}{20x^2 + 10x + 9}$ is

(A) $-\infty$

(B) -1

(C) 3

(D) ∞

A8. $\lim\limits_{x \to \infty} \dfrac{3x^2 + 27}{x^3 - 27}$ is

(A) 0

(B) 1

(C) 3

(D) ∞

A9. $\lim\limits_{x \to \infty} \dfrac{2^{-x}}{2^x}$ is

(A) -1

(B) 1

(C) 0

(D) ∞

A10. $\lim\limits_{x \to -\infty} \dfrac{2^{-x}}{2^x}$ is

(A) -1

(B) 1

(C) 0

(D) ∞

A11. $\lim\limits_{x \to 0} \dfrac{\sin 5x}{x}$

(A) $= \dfrac{1}{5}$

(B) $= 1$

(C) $= 5$

(D) does not exist

A12. $\lim\limits_{x \to 0} \dfrac{\sin 2x}{3x}$

(A) $= \dfrac{2}{3}$

(B) $= 1$

(C) $= \dfrac{3}{2}$

(D) does not exist

A13. The graph of $y = \arctan x$ has

    (A) vertical asymptotes at $x = 0$ and $x = \pi$

    (B) horizontal asymptotes at $y = \pm \dfrac{\pi}{2}$

    (C) horizontal asymptotes at $y = 0$ and $y = \pi$

    (D) vertical asymptotes at $x = \pm \dfrac{\pi}{2}$

A14. The graph of $y = \dfrac{x^2 - 9}{3x - 9}$ has

    (A) a vertical asymptote at $x = 3$

    (B) a horizontal asymptote at $y = \dfrac{1}{3}$

    (C) a removable discontinuity at $x = 3$

    (D) an infinite discontinuity at $x = 3$

A15. $\displaystyle\lim_{x \to 0} \dfrac{\sin x}{x^2 + 3x}$ is

    (A) 1

    (B) $\dfrac{1}{3}$

    (C) 3

    (D) ∞

A16. $\displaystyle\lim_{x \to 0} \sin \dfrac{1}{x}$ is

    (A) ∞

    (B) 1

    (C) -1

    (D) nonexistent

A17. Which statement is true about the curve
$$y = \dfrac{2x^2 + 4}{2 + 7x - 4x^2}?$$

    (A) The line $x = -\dfrac{1}{4}$ is a vertical asymptote.

    (B) The line $x = 1$ is a vertical asymptote.

    (C) The line $y = -\dfrac{1}{4}$ is a horizontal asymptote.

    (D) The line $y = 2$ is a horizontal asymptote.

A18. $\displaystyle\lim_{x \to \infty} \dfrac{2x^2 + 1}{(2 - x)(2 + x)}$ is

    (A) -2

    (B) 1

    (C) 2

    (D) nonexistent

A19. $\displaystyle\lim_{x \to 0} \dfrac{|x|}{x}$ is

    (A) 0

    (B) 1

    (C) -1

    (D) nonexistent

A20. $\displaystyle\lim_{x \to \infty} x \sin \dfrac{1}{x}$ is   **CHALLENGE**

    (A) 0

    (B) -1

    (C) 1

    (D) nonexistent

A21. $\displaystyle\lim_{x \to \pi} \dfrac{\sin(\pi - x)}{(\pi - x)}$ is

    (A) 1

    (B) 0

    (C) π

    (D) nonexistent

A22. Let $f(x) = \begin{cases} \dfrac{x^2 - 1}{x - 1} & \text{if } x \neq 1 \\ 4 & \text{if } x = 1 \end{cases}$

Which of the following statements is (are) true?

    I. $\displaystyle\lim_{x \to 1} f(x)$ exists

    II. $f(1)$ exists

    III. f is continuous at $x = 1$

    (A) I only

    (B) II only

    (C) I and II only

    (D) I, II, and III

A23. If $\begin{cases} f(x) = \dfrac{x^2 - x}{2x} & \text{for } x \neq 0 \\ f(0) = k \end{cases}$

and if f is continuous at $x = 0$, then $k =$

    (A) -1

    (B) $-\dfrac{1}{2}$

    (C) $\dfrac{1}{2}$

    (D) 1

A24. Suppose $\begin{cases} f(x) = \dfrac{3x(x - 1)}{x^2 - 3x + 2} & \text{for } x \neq 1, 2 \\ f(2) = -3 \\ f(2) = 4 \end{cases}$

Then $f(x)$ is continuous

    (A) except at $x = 1$

    (B) except at $x = 2$

    (C) except at $x = 1$ or 2

    (D) at each real number

A25. The graph of $f(x) = \dfrac{4}{x^2 - 1}$ has

  (A) one vertical asymptote, at $x = 1$
  (B) the y-axis as its vertical asymptote
  (C) the x-axis as its horizontal asymptote and $x = \pm 1$ as its vertical asymptotes
  (D) two vertical asymptotes, at $x = \pm 1$, but no horizontal asymptote

A26. The graph of $y = \dfrac{2x^2 + 2x + 3}{4x^2 - 4x}$ has

  (A) a horizontal asymptote at $y = \dfrac{1}{2}$ but no vertical asymptote
  (B) no horizontal asymptote but two vertical asymptotes, at $x = 0$ and $x = 1$
  (C) a horizontal asymptote at $y = \dfrac{1}{2}$ and two vertical asymptotes, at $x = 0$ and $x = 1$
  (D) a horizontal asymptote at $y = \dfrac{1}{2}$ and two vertical asymptotes, at $x = \pm 1$

A27. Let $f(x) = \begin{cases} \dfrac{x^2 + x}{x} & \text{if } x \neq 0 \\ 1 & \text{if } x = 0 \end{cases}$

Which of the following statements is (are) true?

  I. $f(0)$ exists
  II. $\lim\limits_{x \to 0} f(x)$ exists
  III. f is continuous at $x = 0$

  (A) I only
  (B) II only
  (C) I and II only
  (D) I, II, and III

PART B. CALCULATOR ACTIVE—DIRECTIONS: Some of the following questions require the use of a graphing calculator.

B1. If $[x]$ denotes the greatest integer not greater than x, then $\lim\limits_{x \to 1/2} [x]$ is

  (A) $\dfrac{1}{2}$
  (B) 1
  (C) 0
  (D) nonexistent

B2. If $[x]$ denotes the greatest integer not greater than x, then $\lim\limits_{x \to -2} [x]$ is

  (A) -3
  (B) -2
  (C) -1
  (D) nonexistent

B3. $\lim\limits_{x \to \infty} \sin x$

  (A) is -1
  (B) is infinity
  (C) is zero
  (D) does not exist

B4. The function $f(x) = \begin{cases} \dfrac{x^2}{x} & (x \neq 0) \\ 0 & (x = 0) \end{cases}$

  (A) is continuous everywhere
  (B) is continuous except at $x = 0$
  (C) has a removable discontinuity at $x = 0$
  (D) has an infinite discontinuity at $x = 0$

Questions B5–B9 are based on the function f shown in the graph and defined below:

$$f(x) = \begin{cases} 1 - x & (-1 \le x < 0) \\ 2x^2 - 2 & (0 \le x \le 1) \\ -x + 2 & (1 < x < 2) \\ 1 & (x = 2) \\ 2x - 4 & (2 < x \le 3) \end{cases}$$

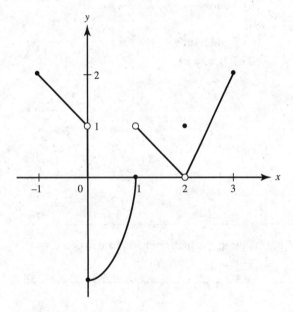

B5. $\lim\limits_{x \to 2} f(x)$

(A) equals 0
(B) equals 1
(C) equals 2
(D) does not exist

B6. The function f is defined on $[-1,3]$

(A) if $x \ne 0$
(B) if $x \ne 1$
(C) if $x \ne 2$
(D) at each x in $[-1,3]$

B7. The function f has a removable discontinuity at

(A) $x = 0$
(B) $x = 1$
(C) $x = 2$
(D) $x = 3$

B8. On which of the following intervals is f continuous?

(A) $-1 \le x \le 0$
(B) $0 < x < 1$
(C) $1 \le x \le 2$
(D) $2 \le x \le 3$

B9. The function f has a jump discontinuity at

(A) $x = -1$
(B) $x = 1$
(C) $x = 2$
(D) $x = 3$

B10. $\lim\limits_{x \to 0} \sqrt{3 + \arctan \frac{1}{x}}$ **CHALLENGE**

(A) $= \sqrt{3 - \frac{\pi}{2}}$

(B) $= \sqrt{3 + \frac{\pi}{2}}$

(C) $= \infty$

(D) does not exist

B11. Suppose $\lim\limits_{x \to -3^-} f(x) = -1$, $\lim\limits_{x \to -3^+} f(x) = -1$, and $f(-3)$ is not defined. Which of the following statements is (are) true?

   I. $\lim\limits_{x \to -3} f(x) = -1$

   II. f is continuous everywhere except at $x = -3$

   III. f has a removable discontinuity at $x = -3$

(A) I only
(B) III only
(C) I and III only
(D) I, II, and III

B12. If $y = \dfrac{1}{2 + 10^{1/x}}$, then $\lim\limits_{x \to 0} y$ **CHALLENGE**

(A) $= 0$

(B) $= \dfrac{1}{12}$

(C) $= \dfrac{1}{2}$

(D) does not exist

Questions B13–B15 are based on the function f shown in the graph.

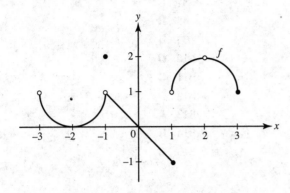

B13. For what value(s) of a is it true that $\lim\limits_{x \to a} f(x)$ exists and $f(a)$ exists, but $\lim\limits_{x \to a} f(x) \neq f(a)$? It is possible that $a =$

(A) −1 only

(B) 2 only

(C) −1 or 1 only

(D) −1 or 2 only

B14. $\lim\limits_{x \to a} f(x)$ does not exist for $a =$

(A) −1 only

(B) 1 only

(C) 2 only

(D) 1 and 2 only

B15. Which of the following statements about limits at $x = 1$ is (are) true?

  I. $\lim\limits_{x \to 1^-} f(x)$ exists

  II. $\lim\limits_{x \to 1^+} f(x)$ exists

  III. $\lim\limits_{x \to 1} f(x)$ exists

(A) I only

(B) II only

(C) I and II only

(D) I, II, and III

Answer Key

A1. **B**	A12. **A**	A23. **B**	B7. **C**
A2. **C**	A13. **B**	A24. **B**	B8. **B**
A3. **C**	A14. **C**	A25. **C**	B9. **B**
A4. **A**	A15. **B**	A26. **C**	B10. **D**
A5. **B**	A16. **D**	A27. **D**	B11. **C**
A6. **B**	A17. **A**	B1. **C**	B12. **D**
A7. **A**	A18. **A**	B2. **D**	B13. **A**
A8. **A**	A19. **D**	B3. **D**	B14. **B**
A9. **C**	A20. **C**	B4. **A**	B15. **C**
A10. **D**	A21. **A**	B5. **A**	
A11. **C**	A22. **C**	B6. **D**	

Answer Explanations

A1. **(B)** The limit as $x \to 2$ is $0 \div 8$.

A2. **(C)** Use the Rational Function Theorem (page 98). The degrees of $P(x)$ and $Q(x)$ are the same.

A3. **(C)** Remove the common factor $x - 3$ from the numerator and denominator.

A4. **(A)** The fraction equals 1 for all nonzero x.

A5. **(B)** Note that $\dfrac{x^3 - 8}{x^2 - 4} = \dfrac{(x - 2)(x^2 + 2x + 4)}{(x - 2)(x + 2)}$.

A6. **(B)** Use the Rational Function Theorem. The degree of the numerator and the degree of the denominator are equal.

A7. **(A)** Use the Rational Function Theorem. The degree of the numerator is greater than the degree of the denominator.

A8. **(A)** Use the Rational Function Theorem. The degree of the numerator is less than the degree of the denominator.

A9. **(C)** The fraction is equivalent to $\dfrac{1}{2^{2x}}$; the denominator approaches ∞.

A10. **(D)** Since $\dfrac{2^{-x}}{2^x} = 2^{-2x}$, therefore, as $x \to -\infty$, the fraction $\to +\infty$.

A11. **(C)** $\lim\limits_{x \to 0} \dfrac{\sin 5x}{x} = \lim\limits_{x \to 0} \dfrac{\sin 5x}{x} \cdot \dfrac{5}{5} = 5 \lim\limits_{x \to 0} \dfrac{\sin 5x}{5x} = 5$

A12. **(A)** $\lim\limits_{x \to 0} \dfrac{\sin 2x}{3x} = \dfrac{1}{3} \lim\limits_{x \to 0} \dfrac{\sin 2x}{x} \cdot \dfrac{2}{2} = \dfrac{2}{3} \lim\limits_{x \to 0} \dfrac{\sin 2x}{2x} = \dfrac{2}{3}$

A13. **(B)** Because the graph of $y = \tan x$ has vertical asymptotes at $x = \pm\dfrac{\pi}{2}$, the graph of the inverse function $y = \arctan x$ has horizontal asymptotes at $y = \pm\dfrac{\pi}{2}$.

A14. **(C)** Since $\dfrac{x^2 - 9}{3x - 9} = \dfrac{(x - 3)(x + 3)}{3(x - 3)} = \dfrac{x + 3}{3}$ (provided $x \neq 3$), y can be defined to be equal to 2 at $x = 3$, removing the discontinuity at that point.

A15. **(B)** Note that $\dfrac{\sin x}{x^2 + 3x} = \dfrac{\sin x}{x(x + 3)} = \dfrac{\sin x}{x} \cdot \dfrac{1}{x + 3} \to 1 \cdot \dfrac{1}{3}$.

A16. **(D)** As $x \to 0$, $\dfrac{1}{x}$ takes on varying finite values as it increases. Since the sine function repeats, $\sin \dfrac{1}{x}$ oscillates, taking on, infinitely many times, each value between -1 and 1. The calculator graph of $Y_1 = \sin(1/X)$ exhibits this oscillating discontinuity at $x = 0$.

A17. **(A)** Note that, since $y = \dfrac{2x^2 + 4}{(2 - x)(1 + 4x)}$, both $x = 2$ and $x = -\dfrac{1}{4}$ are vertical asymptotes. Also, $y = -\dfrac{1}{2}$ is a horizontal asymptote.

A18. **(A)** $\dfrac{2x^2 + 1}{(2 - x)(2 + x)} = \dfrac{2x^2 + 1}{4 - x^2}$. Use the Rational Function Theorem (page 98).

A19. **(D)** Since $|x| = x$ if $x > 0$ but equals $-x$ if $x < 0$, $\lim\limits_{x \to 0^+} \dfrac{|x|}{x} = \lim\limits_{x \to 0^+} \dfrac{x}{x} = 1$ while $\lim\limits_{x \to 0^-} \dfrac{|x|}{x} = \lim\limits_{x \to 0^-} \dfrac{-x}{x} = -1$.

A20. **(C)** Note that $x \sin \dfrac{1}{x}$ can be rewritten as $\dfrac{\sin\frac{1}{x}}{\frac{1}{x}}$ and that, as $x \to -\infty, \dfrac{1}{x} \to 0$.

A21. **(A)** As $x \to \pi, (\pi - x) \to 0$.

A22. **(C)** Since $f(x) = x + 1$ if $x \neq 1$, $\lim\limits_{x\to1} f(x)$ exists (and is equal to 2).

A23. **(B)** $f(x) = \dfrac{x(x-1)}{2x} = \dfrac{(x-1)}{2}$, for all $x \neq 0$. For f to be continuous at $x = 0$, $\lim\limits_{x\to0} f(x)$ must equal $f(0)$. $\lim\limits_{x\to0} f(x) = -\dfrac{1}{2}$.

A24. **(B)** Only $x = 1$ and $x = 2$ need be checked. Since $f(x) = \dfrac{3x}{x-2}$ for $x \neq 1, 2$, and $\lim\limits_{x\to1} f(x) = -3 = f(1)$, f is continuous at $x = 1$. Since $\lim\limits_{x\to2} f(x)$ does not exist, f is not continuous at $x = 2$.

A25. **(C)** As $x \to \pm\infty, y = f(x) \to 0$, so the x-axis is a horizontal asymptote. Also, as $x \to \pm1$, $y \to \infty$, so $x = \pm1$ are vertical asymptotes.

A26. **(C)** As $x \to \infty, y \to \dfrac{1}{2}$; the denominator (but not the numerator) of y equals 0 at $x = 0$ and at $x = 1$.

A27. **(D)** The function is defined at 0 to be 1, which is also $\lim\limits_{x\to0} \dfrac{x^2 + x}{x} = \lim\limits_{x\to0}(x + 1)$.

B1. **(C)** See Figure 2.1 on page 90.

B2. **(D)** Note, from Figure 2.1, that $\lim\limits_{x\to-2^-} [x] = -3$ but $\lim\limits_{x\to-2^+} [x] = -2$.

B3. **(D)** As $x \to \infty$, the function $\sin x$ oscillates between -1 and 1; hence the limit does not exist.

B4. **(A)** Note that $\dfrac{x^2}{x} = x$ if $x \neq 0$ and that $\lim\limits_{x\to0} f = 0$.

B5. **(A)** $\lim\limits_{x\to2^-} f(x) = \lim\limits_{x\to2^+} f(x) = 0$.

B6. **(D)** Verify that f is defined at $x = 0, 1, 2$, and 3 (as well as at all other points in $[-1,3]$).

B7. **(C)** Note that $\lim\limits_{x\to2^-} f(x) = \lim\limits_{x\to2^+} f(x) = 0$. However, $f(2) = 1$. Redefining $f(2)$ as 0 removes the discontinuity.

B8. **(B)** The function is not continuous at $x = 0, 1,$ or 2.

B9. **(B)** $\lim\limits_{x\to1^-} f(x) = 0 \neq \lim\limits_{x\to1^+} f(x) = 1$.

B10. **(D)** As $x \to 0^-$, $\arctan \dfrac{1}{x} \to -\dfrac{\pi}{2}$, so $y \to \sqrt{3 - \dfrac{\pi}{2}}$. As $x \to 0^+$, $y \to \sqrt{3 - \dfrac{\pi}{2}}$. The graph has a jump discontinuity at $x = 0$. (Verify with a calculator.)

B11. **(C)** No information is given about the domain of f except in the neighborhood of $x = -3$.

B12. **(D)** As $x \to 0^+$, $10^{1/x} \to \infty$ and therefore $y \to 0$. As $x \to 0^-$, $\dfrac{1}{x} \to -\infty$, so $10^{1/x} \to 0$ and therefore $y \to \dfrac{1}{2}$. Because the two one-sided limits are not equal, the limit does not exist. (Verify with a calculator.)

B13. **(A)** $\lim\limits_{x\to-1} f(x) = 1$, but $f(-1) = 2$. The limit does not exist at $a = 1$ and $f(2)$ does not exist.

B14. **(B)** $\lim\limits_{x\to-1} f(x) = 1$ and $\lim\limits_{x\to2} f(x) = 2$.

B15. **(C)** $\lim\limits_{x\to1^-} f(x) = -1$ and $\lim\limits_{x\to1^+} f(x) = 1$, but since these two limits are not the same, $\lim\limits_{x\to1} f(x)$ does not exist.

3
Differentiation

Learning Objectives

In this chapter, you will review:

→ Derivatives as instantaneous rates of change
→ Estimating derivatives using graphs and tables
→ Derivatives of basic functions
→ The Product, Quotient, and Chain Rules
→ Implicit differentiation
→ Derivatives of inverse functions
→ Mean Value Theorem
→ L'Hospital's Rule for evaluating limits of indeterminate forms $\left(\text{only } \frac{0}{0} \text{ and } \frac{\infty}{\infty}\right)$

In addition, BC Calculus students will review:

→ Derivatives of parametrically defined functions

A. Definition of Derivative

At any x in the domain of the function $y = f(x)$, the *derivative* is defined as

Derivative

$$\lim_{\Delta x \to 0} \frac{f(x + \Delta x) - f(x)}{\Delta x} \text{ or } \lim_{\Delta x \to 0} \frac{\Delta y}{\Delta x} \tag{1}$$

The function is said to be *differentiable* at every x for which this limit exists, and its derivative may be denoted by $f'(x)$, y', $\frac{dy}{dx}$, or $D_x y$. Frequently Δx is replaced by h or some other symbol.

Differentiable

The derivative of $y = f(x)$ at $x = a$, denoted by $f'(a)$ or $y'(a)$, may be defined as follows:

$$f'(a) = \lim_{h \to 0} \frac{f(a + h) - f(a)}{h} \tag{2}$$

The fraction $\dfrac{f(a + h) - f(a)}{h}$ is called the *difference quotient for f at a* and represents the *average rate of change of f from a to a + h*. Geometrically, it is the slope of the secant PQ to the curve $y = f(x)$ through the points $P(a, f(a))$ and $Q(a + h, f(a + h))$. The limit, $f'(a)$, of the difference quotient is the (*instantaneous*) *rate of change of f* at point a. Geometrically (see Figure 3.1a), the derivative $f'(a)$ is the limit of the slope of secant PQ as Q approaches P—that is, as h approaches zero. This limit is the *slope of the curve at P*. The *tangent to the curve at P* is the line through P with this slope.

Difference quotient

Average rate of change

Instantaneous rate of change

Slope of a curve

Figure 3.1a

In Figure 3.1a, PQ is the secant line through $(a, f(a))$ and $(a + h, f(a + h))$. The average rate of change from a to $a + h$ equals $\dfrac{RQ}{PR}$, which is the slope of secant PQ.

PT is the tangent to the curve at P. As h approaches zero, point Q approaches point P along the curve, PQ approaches PT, and the slope of PQ approaches the slope of PT, which equals $f'(a)$.

If we replace $(a + h)$ by x, in (2) above, so that $h = x - a$, we get the equivalent expression

$$f'(a) = \lim_{x \to a} \frac{f(x) - f(a)}{x - a} \tag{3}$$

See Figure 3.1b.

Figure 3.1b

The second derivative, denoted by $f''(x)$ or $\dfrac{d^2y}{dx^2}$ or y'', is the (first) derivative of $f'(x)$.

Also, $f''(a)$ is the second derivative of $f(x)$ at $x = a$.

B. Formulas

The formulas in this section for finding derivatives are so important that familiarity with them is essential. If a and n are constants and u and v are differentiable functions of x, then:

$$\frac{da}{dx} = 0 \tag{1}$$

$$\frac{d}{dx}au = a\frac{du}{dx} \tag{2}$$

$$\frac{d}{dx}u^n = nu^{n-1}\frac{du}{dx} \quad \text{(the Power Rule)}; \frac{d}{dx}x^n = nx^{n-1} \tag{3}$$

$$\frac{d}{dx}(u+v) = \frac{d}{dx}u + \frac{d}{dx}v; \ \frac{d}{dx}(u-v) = \frac{d}{dx}u - \frac{d}{dx}v \tag{4}$$

$$\frac{d}{dx}(uv) = u\frac{dv}{dx} + v\frac{du}{dx} \quad \text{(the Product Rule)} \tag{5}$$ **Product Rule**

$$\frac{d}{dx}\left(\frac{u}{v}\right) = \frac{v\dfrac{du}{dx} - u\dfrac{dv}{dx}}{v^2} \quad (v \neq 0) \quad \text{(the Quotient Rule)} \tag{6}$$ **Quotient Rule**

$$\frac{d}{dx}\sin u = \cos u\frac{du}{dx} \tag{7}$$

$$\frac{d}{dx}\cos u = -\sin u\frac{du}{dx} \tag{8}$$

$$\frac{d}{dx}\tan u = \sec^2 u\frac{du}{dx} \tag{9}$$

$$\frac{d}{dx}\cot u = -\csc^2 u\frac{du}{dx} \tag{10}$$

$$\frac{d}{dx}\sec u = \sec u \tan u\frac{du}{dx} \tag{11}$$

$$\frac{d}{dx}\csc u = -\csc u \cot u\frac{du}{dx} \tag{12}$$

$$\frac{d}{dx}\ln u = \frac{1}{u}\frac{du}{dx} \tag{13}$$

$$\frac{d}{dx}e^u = e^u\frac{du}{dx} \tag{14}$$

$$\frac{d}{dx}a^u = a^u \ln a\frac{du}{dx} \tag{15}$$

$$\frac{d}{dx}\sin^{-1} u = \frac{d}{dx}\arcsin u = \frac{1}{\sqrt{1-u^2}}\frac{du}{dx} \quad (-1 < u < 1) \tag{16}$$

$$\frac{d}{dx}\cos^{-1} u = \frac{d}{dx}\arccos u = -\frac{1}{\sqrt{1-u^2}}\frac{du}{dx} \quad (-1 < u < 1) \tag{17}$$

$$\frac{d}{dx}\tan^{-1} u = \frac{d}{dx}\arctan u = \frac{1}{1+u^2}\frac{du}{dx} \tag{18}$$

$$\frac{d}{dx}\cot^{-1} u = \frac{d}{dx}\text{arccot } u = -\frac{1}{1+u^2}\frac{du}{dx} \tag{19}$$

$$\frac{d}{dx}\sec^{-1} u = \frac{d}{dx}\text{arcsec } u = \frac{1}{|u|\sqrt{u^2-1}}\frac{du}{dx} \quad \left(|u| > 1\right) \tag{20}$$

$$\frac{d}{dx}\csc^{-1} u = \frac{d}{dx}\text{arccsc } u = -\frac{1}{|u|\sqrt{u^2-1}}\frac{du}{dx} \quad \left(|u| > 1\right) \tag{21}$$

C. The Chain Rule: The Derivative of a Composite Function

Formula (3) on page 115 says that

$$\frac{d}{dx}u^n = nu^{n-1}\frac{du}{dx}$$

This formula is an application of the *Chain Rule*. For example, if we use formula (3) to find the derivative of $(x^2 - x + 2)^4$, we get

$$\frac{d}{dx}\left(x^2 - x + 2\right)^4 = 4\left(x^2 - x + 2\right)^3 \cdot (2x - 1)$$

In this last equation, if we let $y = \left(x^2 - x + 2\right)^4$ and let $u = x^2 - x + 2$, then $y = u^4$. The preceding derivative now suggests one form of the Chain Rule:

$$\frac{dy}{dx} = \frac{dy}{du}\cdot\frac{du}{dx} = 4u^3 \cdot \frac{du}{dx} = 4\left(x^2 - x + 2\right)^3 \cdot (2x - 1)$$

as before. Formula (3) on page 115 gives the general case where $y = u^n$ and u is a differentiable function of x.

Now suppose we think of y as the composite function $f(g(x))$, where $y = f(u)$ and $u = g(x)$ are differentiable functions. Then

Chain Rule

$$(f(g(x)))' = f'(g(x)) \cdot g'(x)$$
$$= f'(u) \cdot g'(x)$$
$$= \frac{dy}{du}\cdot\frac{du}{dx}$$

as we obtained above. The Chain Rule tells us how to differentiate the composite function: "Find the derivative of the 'outside' function first, then multiply by the derivative of the 'inside' one."

For example:

$$\frac{d}{dx}\left(x^3 + 1\right)^{10} = 10\left(x^3 + 1\right)^9 \cdot 3x^2 = 30x^2\left(x^3 + 1\right)^9$$

$$\frac{d}{dx}\sqrt{7x - 2} = \frac{d}{dx}(7x - 2)^{1/2} = \frac{1}{2}(7x - 2)^{-1/2}\cdot 7$$

$$\frac{d}{dx}\frac{3}{(2 - 4x^2)^4} = \frac{d}{dx}3\left(2 - 4x^2\right)^{-4} = 3\cdot(-4)\left(2 - 4x^2\right)^{-5}\cdot(-8x)$$

$$\frac{d}{dx}\sin\!\left(\frac{\pi}{2} - x\right) = \cos\!\left(\frac{\pi}{2} - x\right)\cdot(-1)$$

$$\frac{d}{dx}\cos^3 2x = \frac{d}{dx}(\cos 2x)^3 = 3(\cos 2x)^2 \cdot (-\sin 2x \cdot 2)$$

Many of the formulas listed in Section B and most of the illustrative examples that follow use the Chain Rule. Often the Chain Rule is used more than once in finding a derivative.

Note that the algebraic simplifications that follow are included only for completeness.

> **Example 1**

If $y = 4x^3 - 5x + 7$, find $y'(1)$ and $y''(1)$.

✓ **Solution**

$$y' = \frac{dy}{dx} = 12x^2 - 5 \qquad \text{and} \qquad y'' = \frac{d^2y}{dx^2} = 24x$$

Then $y'(1) = 12 \cdot 1^2 - 5 = 7$ and $y''(1) = 24 \cdot 1 = 24$.

> **Example 2**

If $f(x) = (3x + 2)^5$, find $f'(x)$.

Handwritten: $f'(x) = 5(3x+2)^4 (3)$

$\boxed{15(3x+2)^4}$

✓ **Solution**

$f'(x) = 5(3x + 2)^4 \cdot 3 = 15(3x + 2)^4$

> **Example 3**

If $y = \sqrt{3 - x - x^2}$, find $\dfrac{dy}{dx}$.

Handwritten: $(3-x-x^2)^{1/2}$ $\frac{1}{2}(3-x-x^2)^{-1/2}(-1-2x)$

$-\frac{1}{2}(3-x-x^2)^{-1/2}$ $\dfrac{-1-2x}{2\sqrt{3-x-x^2}}$

✓ **Solution**

$y = (3 - x - x^2)^{1/2}$, so $\dfrac{dy}{dx} = \dfrac{1}{2}(3 - x - x^2)^{-1/2}(-1 - 2x)$

$= -\dfrac{1 + 2x}{2\sqrt{3 - x - x^2}}$

> **Example 4**

If $y = \dfrac{5}{\sqrt{(1 - x^2)^3}}$, find $\dfrac{dy}{dx}$.

Handwritten: $5\left[\left((1-x^2)^3\right)^{-1/2}\right] \Rightarrow 5(1-x^2)^{-3/2}$

$-\frac{15}{2}(1-x^2)^{-5/2}(-2x)$ $15x(1-x^2)^{-5/2}$

$\dfrac{15x}{\sqrt{(1-x^2)^5}}$

or

$\dfrac{15x}{(1-x^2)^{3/2}}$

✓ **Solution**

$y = 5(1 - x^2)^{-3/2}$, so $\dfrac{dy}{dx} = \dfrac{-15}{2}(1 - x^2)^{-5/2}(-2x)$

$= \dfrac{15x}{(1 - x^2)^{5/2}}$

Handwritten: $(1-t)(1-t)$ $1-t-t+t^2$ $1-2t+t^2$

> **Example 5**

If $s(t) = (t^2 + 1)(1 - t)^2$, find $s'(t)$.

Handwritten: ↑ requires chain rule

Handwritten: $(t^2+1)\cdot 2(1-t)(-1) + (1-t)^2\cdot 2t$

$t^2+1 \cdot 2(1-t)$

$-1+2t-t^2$ $2-4t^2+2t^3$

✓ **Solution**

$s'(t) = (t^2 + 1) \cdot 2(1 - t)(-1) + (1 - t)^2 \cdot 2t$ (Product Rule)

$= 2(1 - t)(-1 + t - 2t^2)$

> **Example 6**

If $f(t) = e^{2t}\sin 3t$, find $f'(0)$.

Handwritten: $e^{2t}(\cos 3t)(3) + \sin 3t(e^{2t})(2)$

$e^{2t} 3\cos 3t + 2\sin 3t \, e^{2t}$

$e^{2t}(3\cos 3t + 2\sin 3t)$

$1(\cos) = 1(3\cdot 1 + 2\cdot 0) = 3$

✓ **Solution**

$f'(t) = e^{2t}(\cos 3t \cdot 3) + \sin 3t(e^{2t} \cdot 2)$ (Product Rule)

$= e^{2t}(3\cos 3t + 2\sin 3t)$

Then, $f'(0) = 1(3 \cdot 1 + 2 \cdot 0) = 3$.

Handwritten at top:

$$\frac{(1-2v^2) \cdot (2) - 2v(-4v)}{(1-2v^2)} = \frac{2+4v^2+8v}{(1-2v^2)}$$

> **Example 7**

If $f(v) = \dfrac{2v}{1-2v^2}$, find $f'(v)$.

✓ **Solution**

Handwritten: $2-4v^2+8v^2 \Rightarrow \dfrac{2+4v^2}{(1-2v^2)^2}$

$$f'(v) = \frac{(1-2v^2) \cdot 2 - 2v(-4v)}{(1-2v^2)^2} = \frac{2+4v^2}{(1-2v^2)^2} \qquad \text{(Quotient Rule)}$$

Note that neither $f(v)$ nor $f'(v)$ exists where the denominator equals zero, namely, where $1 - 2v^2 = 0$ or where v equals $\pm\dfrac{\sqrt{2}}{2}$.

> **Example 8**

If $f(x) = \dfrac{\sin x}{x^2}$, $x \neq 0$, find $f'(x)$.

✓ **Solution**

$$f'(x) = \frac{x^2 \cos x - \sin x \cdot 2x}{x^4} = \frac{x \cos x - 2 \sin x}{x^3}$$

> **Example 9**

If $y = \tan(2x^2 + 1)$, find y'.

✓ **Solution**

$$y' = 4x \sec^2(2x^2 + 1)$$

> **Example 10**

If $x = \cos^3(1-3\theta)$, find $\dfrac{dx}{d\theta}$.

✓ **Solution**

$$\frac{dx}{d\theta} = -3 \cos^2(1-3\theta) \sin(1-3\theta)(-3)$$
$$= 9 \cos^2(1-3\theta) \sin(1-3\theta)$$

> **Example 11**

If $y = e^{(\sin x) + 1}$, find $\dfrac{dy}{dx}$.

✓ **Solution**

$$\frac{dy}{dx} = \cos x \cdot e^{(\sin x) + 1}$$

> **Example 12**

If $y = (x+1) \ln^2 (x+1)$, find $\dfrac{dy}{dx}$.

✓ **Solution**

$\dfrac{dy}{dx} = (x+1) \dfrac{2 \ln (x+1)}{x+1} + \ln^2 (x+1)$ (Product and Chain Rules)

$\qquad = 2 \ln (x+1) + \ln^2 (x+1)$

> **Example 13**

If $g(x) = \left(1 + \sin^2 3x\right)^4$, find $g'\left(\dfrac{\pi}{2}\right)$.

✓ **Solution**

$g'(x) = 4\left(1 + \sin^2 3x\right)^3 (2 \sin 3x \cos 3x) \cdot (3)$

$\qquad = 24\left(1 + \sin^2 3x\right)^3 (\sin 3x \cos 3x)$

Then $g'\left(\dfrac{\pi}{2}\right) = 24\left(1 + (-1)^2\right)^3 (-1 \cdot 0) = 24 \cdot 8 \cdot 0 = 0$.

> **Example 14**

If $y = \sin^{-1} x + x\sqrt{1 - x^2}$, find y'.

✓ **Solution**

$y' = \dfrac{1}{\sqrt{1 - x^2}} + \dfrac{x(-2x)}{2\sqrt{1 - x^2}} + \sqrt{1 - x^2}$

$\quad = \dfrac{1 - x^2 + 1 - x^2}{\sqrt{1 - x^2}} = 2\sqrt{1 - x^2}$

> **Example 15**

If $u = \ln \sqrt{v^2 + 2v - 1}$, find $\dfrac{du}{dv}$.

✓ **Solution**

$u = \dfrac{1}{2} \ln \left(v^2 + 2v - 1\right)$, so

$\dfrac{du}{dv} = \dfrac{1}{2} \dfrac{2v + 2}{v^2 + 2v - 1} = \dfrac{v + 1}{v^2 + 2v - 1}$

> **Example 16**

If $s = e^{-t}(\sin t - \cos t)$, find s'.

✓ **Solution**

$s' = e^{-t}(\cos t + \sin t) + (\sin t - \cos t)(-e^{-t})$
$\quad = e^{-t}(2 \cos t) = 2e^{-t}\cos t$

> **Example 17**

Let $y = 2u^3 - 4u^2 + 5u - 3$ and $u = x^2 - x$. Find $\dfrac{dy}{dx}$.

✓ **Solution**

$\dfrac{dy}{dx} = \left(6u^2 - 8u + 5\right)(2x - 1)$
$\quad = \left[6\left(x^2 - x\right)^2 - 8\left(x^2 - x\right) + 5\right](2x - 1)$

> **Example 18**

If $y = \sin(ax + b)$, with a and b constants, find $\dfrac{dy}{dx}$.

✓ **Solution**

$\dfrac{dy}{dx} = \left[\cos(ax + b)\right] \cdot a = a \cos(ax + b)$

> **Example 19**

If $f(x) = ae^{kx}$ (with a and k constants), find f' and f''.

✓ **Solution**

$f'(x) = kae^{kx} \qquad \text{and} \qquad f'' = k^2 ae^{kx}$

> **Example 20**

If $y = \ln(kx)$, where k is a constant, find $\dfrac{dy}{dx}$.

✓ **Solution**

We can use both formula (13), page 115, and the Chain Rule to get

$$\frac{dy}{dx} = \frac{1}{kx} \cdot k = \frac{1}{x}$$

Alternatively, we can rewrite the given function using a property of logarithms:
$\ln(kx) = \ln k + \ln x$. Then,

$$\frac{dy}{dx} = 0 + \frac{1}{x} = \frac{1}{x}$$

> **Example 21**

Given $f(u) = u^2 - u$ and $u = g(x) = x^3 - 5$ and $F(x) = f(g(x))$, evaluate $F'(2)$.

✓ **Solution**

$F'(2) = f'\big(g(2)\big)g'(2) = f'(3) \cdot (12) = 5 \cdot 12 = 60$
Now, since $g'(x) = 3x^2$, $g'(2) = 12$, and since $f'(u) = 2u - 1$, $f'(3) = 5$.
Of course, we get exactly the same answer as follows.
Since $F(x) = \big(x^3 - 5\big)^2 - \big(x^3 - 5\big)$
$$F'(x) = 2\big(x^3 - 5\big) \cdot 3x^2 - 3x^2$$
$$F'(2) = 2 \cdot (3) \cdot 12 - 12 = 60$$

D. Differentiability and Continuity

If a function f has a derivative at $x = c$, then f is continuous at $x = c$.

This statement is an immediate consequence of the definition of the derivative of $f'(c)$ in the form

$$f'(c) = \lim_{x \to c} \frac{f(x) - f(c)}{x - c}$$

If $f'(c)$ exists, then it follows that $\lim\limits_{x \to c} f(x) = f(c)$, which guarantees that f is continuous at $x = c$.

If f is differentiable at c, its graph cannot have a hole or jump at c, nor can $x = c$ be a vertical asymptote of the graph. The tangent to the graph of f cannot be vertical at $x = c$; there cannot be a corner or cusp at $x = c$.

Each of the "prohibitions" in the preceding paragraph (each "cannot") tells how a function may fail to have a derivative at c. These cases are illustrated in Figures 3.2 (a) through (f).

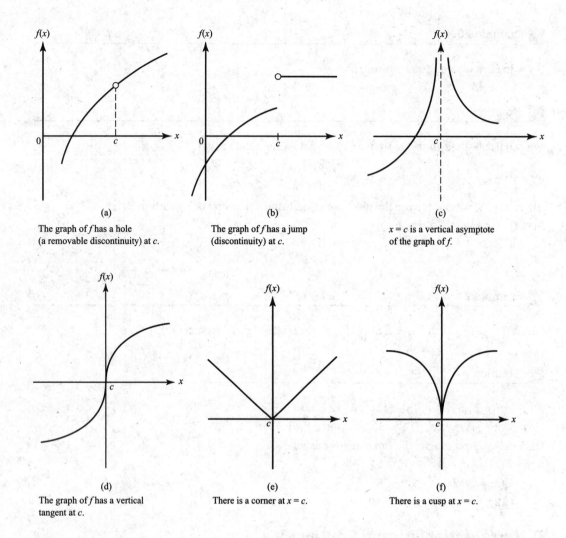

(a)

The graph of f has a hole (a removable discontinuity) at c.

(b)

The graph of f has a jump (discontinuity) at c.

(c)

$x = c$ is a vertical asymptote of the graph of f.

(d)

The graph of f has a vertical tangent at c.

(e)

There is a corner at $x = c$.

(f)

There is a cusp at $x = c$.

Figure 3.2

The graph in (e) is for the absolute function, $f(x) = |x|$. Since $f'(x) = -1$ for all negative x but $f'(x) = +1$ for all positive x, $f'(0)$ does not exist.

We may conclude from the preceding discussion that, although differentiability implies continuity, the converse is false. The functions in (d), (e), and (f) in Figure 3.2 are all continuous at $x = 0$, but not one of them is differentiable at the origin.

E. Estimating a Derivative

E1. Numerically

> **Example 22**

The table shown gives the temperatures of a polar bear on a very cold arctic day ($t =$ minutes; $T =$ degrees Fahrenheit):

t	0	1	2	3	4	5	6	7	8
T	98	94.95	93.06	91.90	91.17	90.73	90.45	90.28	90.17

Our task is to estimate the derivative of T numerically at various times. One possible graph of $T(t)$ is sketched in Figure 3.3, but this assumes the curve is smooth. In estimating derivatives, we shall use only the data from the table.

Figure 3.3

Using the difference quotient $\dfrac{T(t+h) - T(t)}{h}$ with h equal to 1, we see that

$$T'(0) \approx \frac{T(1) - T(0)}{1} = -3.05°/\text{min}$$

Also,

$$T'(1) \approx \frac{T(2) - T(1)}{1} = -1.89°/\text{min}$$

$$T'(2) \approx \frac{T(3) - T(2)}{1} = -1.16°/\text{min}$$

$$T'(3) \approx \frac{T(4) - T(3)}{1} = -0.73°/\text{min}$$

and so on.

The following table shows the *approximate* values of $T'(t)$ obtained from the difference quotients above:

t	0	1	2	3	4	5	6	7
$T'(t)$	−3.05	−1.89	−1.16	−0.73	−0.47	−0.28	−0.17	−0.11

Note that the entries for $T'(t)$ also represent the approximate slopes of the T curve at times 0.5, 1.5, 2.5, and so on.

From a Symmetric Difference Quotient

In Example 22 we approximated a derivative numerically from a table of values. We can also estimate $f'(a)$ numerically using the *symmetric difference quotient*, which is defined as follows:

Symmetric difference quotient

$$f'(a) \approx \frac{f(a+h) - f(a-h)}{2h}$$

Note that the symmetric difference quotient is equal to

$$\frac{1}{2}\left[\frac{f(a+h) - f(a)}{h} + \frac{f(a) - f(a-h)}{h}\right]$$

We see that it is just the average of two difference quotients. Many calculators use the symmetric difference quotient in finding derivatives.

> ### Example 23

For the function $f(x) = x^4$, approximate $f'(1)$ using the symmetric difference quotient with $h = 0.01$.

✓ Solution

$$f'(1) \approx \frac{(1.01)^4 - (0.99)^4}{2(0.01)} = 4.0004$$

The exact value of $f'(1)$, of course, is 4.

The use of the symmetric difference quotient is particularly convenient when, as is often the case, obtaining a derivative precisely (with formulas) is cumbersome and an approximation is all that is needed for practical purposes.

A word of caution is in order. Sometimes a wrong result is obtained using the symmetric difference quotient. On page 122 we noted that $f(x) = |x|$ does not have a derivative at $x = 0$ since $f'(x) = -1$ for all $x < 0$ but $f'(x) = 1$ for all $x > 0$. Our calculator (which uses the symmetric difference quotient) tells us (incorrectly!) that $f'(0) = 0$. Note that, if $f(x) = |x|$, the symmetric difference quotient gives 0 for $f'(0)$ for every $h \neq 0$. If, for example, $h = 0.01$, then we get

$$f'(0) \approx \frac{|0.01| - |-0.01|}{0.02} = \frac{0}{0.02} = 0$$

which, as previously noted, is incorrect. The graph of the derivative of $f(x) = |x|$, which we see in Figure 3.4, shows that $f'(0)$ does not exist.

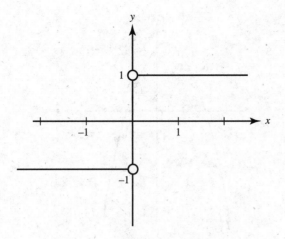

Figure 3.4

E2. Graphically

If we have the graph of a function $f(x)$, we can use it to graph $f'(x)$. We accomplish this by estimating the slope of the graph of $f(x)$ at enough points to assure a smooth curve for $f'(x)$. In Figure 3.5 we see the graph of $y = f(x)$. Below it is a table of the approximate slopes estimated from the graph.

Figure 3.5

x	-3	-2.5	-2	-1.5	-1	0	0.5	1	1.5	2	2.5
$f'(x)$	-6	-3	-0.5	1	2	2	1.5	0.5	-2	-4	-7

Figure 3.6 was obtained by plotting the points from the table of slopes above and drawing a smooth curve through these points. The result is the graph of $y = f'(x)$.

Figure 3.6

From the graphs above we can make the following observations:

(1) At the points where the slope of f (in Figure 3.5) equals 0, the graph of f' (Figure 3.6) has x-intercepts: approximately $x = -1.8$ and $x = 1.1$. We've drawn horizontal broken lines at these points on the curve in Figure 3.5.

(2) On open intervals where the derivative is positive, the graph of f increases. On open intervals where the derivative is negative, the graph of f decreases. We see here that f decreases for $x < -1.8$ (approximately) and for $x > 1.1$ (approximately), and that f increases for $-1.8 < x < 1.1$ (approximately). In Chapter 4 we discuss other behaviors of f that are reflected in the graph of f'.

BC ONLY

F. Derivatives of Parametrically Defined Functions

Parametric equations were defined on page 73.

If $x = f(t)$ and $y = g(t)$ are differentiable functions of t, then

Parametric differentiation

$$\frac{dy}{dx} = \frac{\dfrac{dy}{dt}}{\dfrac{dx}{dt}} \quad \text{and} \quad \frac{d^2y}{dx^2} = \frac{d}{dx}\left(\frac{dy}{dx}\right) = \frac{\dfrac{d}{dt}\left(\dfrac{dy}{dx}\right)}{\dfrac{dx}{dt}}$$

> **Example 24**

If $x = 2\sin\theta$ and $y = \cos 2\theta$, find $\dfrac{dy}{dx}$ and $\dfrac{d^2y}{dx^2}$.

✓ **Solution**

$$\frac{dy}{dx} = \frac{\dfrac{dy}{d\theta}}{\dfrac{dx}{d\theta}} = \frac{-2\sin 2\theta}{2\cos\theta} = -\frac{2\sin\theta\cos\theta}{\cos\theta} = -2\sin\theta$$

Also,

$$\frac{d^2y}{dx^2} = \frac{\dfrac{d}{d\theta}\left(\dfrac{dy}{dx}\right)}{\dfrac{dx}{d\theta}} = \frac{-2\cos\theta}{2\cos\theta} = -1$$

> **Example 25**

Find an equation of the tangent to the curve in Example 24 for $\theta = \frac{\pi}{6}$.

✓ **Solution**

When $\theta = \frac{\pi}{6}$, the slope of the tangent, $\frac{dy}{dx}$, equals $-2 \sin\left(\frac{\pi}{6}\right) = -1$. Since
$x = 2 \sin\left(\frac{\pi}{6}\right) = 1$ and $y = \cos\left(2 \cdot \frac{\pi}{6}\right) = \cos\frac{\pi}{3} = \frac{1}{2}$, the equation is

$$y - \frac{1}{2} = -1(x-1) \qquad \text{or} \qquad y = -x + \frac{3}{2}$$

> **Example 26**

Suppose two objects are moving in a plane during the time interval $0 \le t \le 4$. Their positions at time t are described by the parametric equations

$$x_1 = 2t \qquad y_1 = 4t - t^2 \qquad \text{and} \qquad x_2 = t+1 \qquad y_2 = 4 - t$$

(a)   Find all collision points. Justify your answer.
(b)   Use a calculator to help you sketch the paths of the objects, indicating the direction in which each object travels.

✓ **Solutions**

(a)   Equating x_1 and x_2 yields $t = 1$. When $t = 1$, both y_1 and y_2 equal 3. So $t = 1$ yields a *true* collision point (not just an intersection point) at (2,3). (An *intersection point* is any point that is on both curves but not necessarily at the same time.)

(b)   Using parametric mode, we graph both curves with t in $[0,4]$, in the window $[0,8] \times [0,4]$ as shown in Figure 3.7.

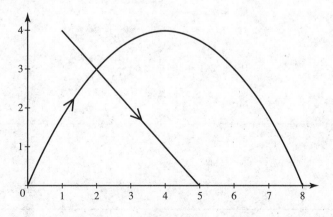

Figure 3.7

We've inserted arrows to indicate the direction of motion.

Note that if your calculator can draw the curves in simultaneous graphing mode, you can watch the objects as they move, seeing that they do indeed pass through the intersection point at the same time.

G. Implicit Differentiation

When a functional relationship between x and y is defined by an equation of the form $F(x,y) = 0$, we say that the equation defines y *implicitly* as a function of x. Some examples are $x^2 + y^2 - 9 = 0$, $y^2 - 4x = 0$, and $\cos(xy) = y^2 - 5$ (which can be written as $\cos(xy) - y^2 + 5 = 0$). Sometimes two (or more) explicit functions are defined by $F(x,y) = 0$. For example, $x^2 + y^2 - 9 = 0$ defines the two functions $y_1 = +\sqrt{9 - x^2}$ and $y_2 = -\sqrt{9 - x^2}$, the upper and lower halves, respectively, of the circle centered at the origin with radius 3. Each function is differentiable except at the points where $x = 3$ and $x = -3$.

Implicit differentiation is the technique we use to find a derivative when y is not defined explicitly in terms of x but is differentiable.

In the following examples, we differentiate both sides with respect to x, using appropriate formulas, and then solve for $\dfrac{dy}{dx}$.

> ### Example 27

If $x^2 + y^2 - 9 = 0$, then

$$2x + 2y\frac{dy}{dx} = 0 \quad \text{and} \quad \frac{dy}{dx} = -\frac{x}{y}$$

Note that the derivative above holds for every point on the circle and exists for all y different from 0 (where the tangents to the circle are vertical).

> ### Example 28

If $x^2 - 2xy + 3y^2 = 2$, find $\dfrac{dy}{dx}$.

✓ **Solution**

$$2x - 2\left(x\frac{dy}{dx} + y \cdot 1\right) + 6y\frac{dy}{dx} = 0$$

$$\frac{dy}{dx}(6y - 2x) = 2y - 2x, \text{ so } \frac{dy}{dx} = \frac{y - x}{3y - x}$$

> ### Example 29

If $x \sin y = \cos(x + y)$, find $\dfrac{dy}{dx}$.

✓ **Solution**

$$x\cos y\frac{dy}{dx} + \sin y = -\sin(x + y)\left(1 + \frac{dy}{dx}\right)$$

$$\frac{dy}{dx} = -\frac{\sin y + \sin(x + y)}{x\cos y + \sin(x + y)}$$

> **Example 30**

Find $\dfrac{dy}{dx}$ and $\dfrac{d^2y}{dx^2}$ using implicit differentiation on the equation $x^2 + y^2 = 1$.

✓ **Solution**

$$2x + 2y\frac{dy}{dx} = 0 \qquad \rightarrow \qquad \frac{dy}{dx} = -\frac{x}{y} \tag{1}$$

Then,

$$\frac{d^2y}{dx^2} = -\frac{y \cdot 1 - x\left(\dfrac{dy}{dx}\right)}{y^2} = -\frac{y - x\left(-\dfrac{x}{y}\right)}{y^2} \tag{2}$$

$$= -\frac{y^2 + x^2}{y^3} = -\frac{1}{y^3} \tag{3}$$

where we substituted for $\dfrac{dy}{dx}$ from (1) in (2), then used the given equation to simplify in (3).

> **Example 31**

Using implicit differentiation, verify the formula for the derivative of the inverse sine function, $y = \sin^{-1} x = \arcsin x$, with domain $[-1,1]$ and range $\left[-\dfrac{\pi}{2}, \dfrac{\pi}{2}\right]$.

✓ **Solution**

$$y = \sin^{-1} x \qquad \leftrightarrow \qquad x = \sin y$$

Now we differentiate with respect to x:

$$1 = \cos y \frac{dy}{dx}$$

$$\frac{dy}{dx} = \frac{1}{\cos y} = \frac{1}{+\sqrt{1 - \sin^2 y}} = \frac{1}{\sqrt{1 - x^2}}$$

where we chose the positive sign for $\cos y$ since $\cos y$ is nonnegative if $-\dfrac{\pi}{2} < y < \dfrac{\pi}{2}$. Note that this derivative exists only if $-1 < x < 1$.

> **Example 32**

Where is the tangent to the curve $4x^2 + 9y^2 = 36$ vertical?

✓ **Solution**

We differentiate the equation implicitly to get $\dfrac{dy}{dx}$: $8x + 18y\dfrac{dy}{dx} = 0$, so $\dfrac{dy}{dx} = -\dfrac{4x}{9y}$. Since the tangent line to a curve is vertical when $\dfrac{dx}{dy} = 0$, we conclude that $-\dfrac{9y}{4x}$ must equal zero; that is, y must equal zero with $x \neq 0$. When we substitute $y = 0$ in the original equation, we get $x = \pm 3$. The points $(\pm 3, 0)$ are the ends of the major axis of the ellipse, where the tangents are indeed vertical.

H. Derivative of the Inverse of a Function

Suppose f and g are inverse functions. What is the relationship between their derivatives? Recall that the graphs of inverse functions are the reflections of each other in the line $y = x$ and that at corresponding points their x- and y-coordinates are interchanged.

Figure 3.8 shows a function f passing through point (a,b) and the line tangent to f at that point. The slope of the curve there, $f'(a)$, is represented by the ratio of the legs of the triangle, $\dfrac{dy}{dx}$. When this figure is reflected across the line $y = x$, we obtain the graph of f^{-1}, passing through point (b,a), with the horizontal and vertical sides of the slope triangle interchanged. Note that the slope of the line tangent to the graph of f^{-1} at $x = b$ is represented by $\dfrac{dx}{dy}$, the reciprocal of the slope of f at $x = a$. We have, therefore,

Derivative of inverse function

$$\left(f^{-1}\right)'(b) = \frac{1}{f'(a)} \qquad \text{or} \qquad \left(f^{-1}\right)'(x) = \frac{1}{f'\left(f^{-1}(x)\right)}$$

Simply put, the derivative of the inverse of a function at a point is the *reciprocal* of the derivative of the function *at the corresponding point.*

Figure 3.8

> **Example 33**

If $f(3) = 8$ and $f'(3) = 5$, what do we know about f^{-1}?

✓ **Solution**

Since f passes through the point $(3,8)$, f^{-1} must pass through the point $(8,3)$. Furthermore, since the graph of f has slope 5 at $(3,8)$, the graph of f^{-1} must have slope $\dfrac{1}{5}$ at $(8,3)$.

> **Example 34**

A function *f* and its derivative take on the values shown in the table. If *g* is the inverse of *f*, find $g'(6)$.

x	$f(x)$	$f'(x)$
2	6	$\frac{1}{3}$
6	8	$\frac{3}{2}$

✓ **Solution**

To find the slope of *g* at the point where $x = 6$, we must look at the point on *f* where $y = 6$, namely, $(2,6)$. Since $f'(2) = \frac{1}{3}$, $g'(6) = 3$.

> **Example 35**

Let $y = f(x) = x^3 + x - 2$, and let *g* be the inverse function. Evaluate $g'(0)$.

✓ **Solution**

Since $f'(x) = 3x^2 + 1$, $g'(y) = \frac{1}{3x^2 + 1}$. To find *x* when $y = 0$, we must solve the equation $x^3 + x - 2 = 0$. Note by inspection that $x = 1$, so

$$g'(0) = \frac{1}{3(1)^2 + 1} = \frac{1}{4}$$

I. The Mean Value Theorem

If the function $f(x)$ is continuous at each point on the closed interval $a \le x \le b$ and has a derivative at each point on the open interval $a < x < b$, then there is at least one number *c*, $a < c < b$, such that $\frac{f(b) - f(a)}{b - a} = f'(c)$. This important theorem, which relates average rate of change and instantaneous rate of change, is illustrated in Figure 3.9. For the function sketched in the figure there are two numbers, c_1 and c_2, between *a* and *b* where the slope of the curve equals the slope of the chord *PQ* (i.e., where the tangent to the curve is parallel to the secant line).

Mean Value Theorem

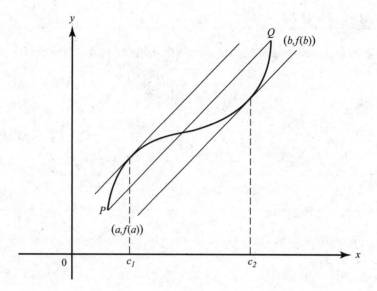

Figure 3.9

We will often refer to the Mean Value Theorem by its initials, MVT.

Rolle's Theorem

If, in addition to the hypotheses of the MVT, it is given that $f(a) = f(b) = k$, then there is a number, c, between a and b such that $f'(c) = 0$. This special case of the MVT is called *Rolle's Theorem*, as seen in Figure 3.10 for $k = 0$. *NOTE:* Rolle's Theorem is an important special case of MVT; however, it will not be tested by name on the AP exam.

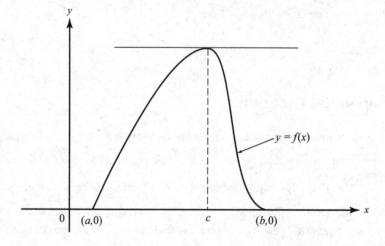

Figure 3.10

The Mean Value Theorem is one of the most useful laws when properly applied.

> **Example 36**

You left home one morning and drove to a cousin's house 300 miles away, arriving 6 hours later. What does the Mean Value Theorem say about your speed along the way?

✓ **Solution**

Your journey was continuous, with an average speed (the average rate of change of distance traveled) given by

$$\frac{\Delta \text{distance}}{\Delta \text{time}} = \frac{300 \text{ miles}}{6 \text{ hours}} = 50 \text{ mph}$$

Furthermore, the derivative (your instantaneous speed) existed everywhere along your trip. The MVT, then, guarantees that at least at one point your instantaneous speed was equal to your average speed for the entire 6-hour interval. Hence, your car's speedometer must have read exactly 50 mph at least once on your way to your cousin's house.

> **Example 37**

Demonstrate Rolle's Theorem using $f(x) = x \sin x$ on the interval $[0,\pi]$.

✓ **Solution**

First, we check that the conditions of Rolle's Theorem are met:

(1) $f(x) = x \sin x$ is continuous on $(0,\pi)$ and exists for all x in $[0,\pi]$.
(2) $f'(x) = x \cos x + \sin x$ exists for all x in $(0,\pi)$.
(3) $f(0) = 0 \sin 0 = 0$ and $f(\pi) = \pi \sin \pi = 0$.

Hence there must be a point, $x = c$, in the interval $0 < x < \pi$ where $f'(c) = 0$. Using the calculator to solve $x \cos x + \sin x = 0$, we find $c = 2.029$ (to three decimal places). As predicted by Rolle's Theorem, $0 \leq c \leq \pi$.

Note that this result indicates that at $x = c$ the line tangent to f is horizontal. The MVT (here as Rolle's Theorem) tells us that any function that is continuous and differentiable must have at least one turning point between any two roots.

J. Indeterminate Forms and L'Hospital's Rule

Limits of the following forms are called *indeterminate*:

$$\frac{0}{0} \text{ or } \frac{\infty}{\infty}, \, 0 \cdot \infty, \, \infty - \infty, \, 0^0, \, 1^\infty, \, \infty^0$$

Although all indeterminate forms are listed above, only the indeterminate forms $\frac{0}{0}$ and $\frac{\infty}{\infty}$ are tested on both the AP Calculus AB and AP Calculus BC exams.

To find the limit of an indeterminate form of the type $\frac{0}{0}$ or $\frac{\infty}{\infty}$, we apply L'Hospital's Rule, which involves taking derivatives of the functions in the numerator and denominator. In the following, a is a finite number. The rule has several parts:

(a) If $\lim\limits_{x \to a} f(x) = \lim\limits_{x \to a} g(x) = 0$ and if $\lim\limits_{x \to a} \dfrac{f'(x)}{g'(x)}$ exists[*], then

$$\lim_{x \to a} \frac{f(x)}{g(x)} = \lim_{x \to a} \frac{f'(x)}{g'(x)}$$

if $\lim\limits_{x \to a} \dfrac{f'(x)}{g'(x)}$ does not exist, then L'Hospital's Rule cannot be applied.

---

[*]The limit can be finite or infinite ($+\infty$ or $-\infty$).

(b) If $\lim\limits_{x \to a} f(x) = \lim\limits_{x \to a} g(x) = \infty$, the same consequences follow as in case (a). The rules in

(a) and (b) both hold for one-sided limits.

(c) If $\lim\limits_{x \to \infty} f(x) = \lim\limits_{x \to \infty} g(x) = 0$ and if $\lim\limits_{x \to \infty} \dfrac{f'(x)}{g'(x)}$ exists, then

$$\lim_{x \to \infty} \frac{f(x)}{g(x)} = \lim_{x \to \infty} \frac{f'(x)}{g'(x)}$$

if $\lim\limits_{x \to \infty} \dfrac{f'(x)}{g'(x)}$ does not exist, then L'Hospital's Rule cannot be applied. (Here the

notation "$x \to \infty$" represents either "$x \to +\infty$" or "$x \to -\infty$.")

(d) If $\lim\limits_{x \to \infty} f(x) = \lim\limits_{x \to \infty} g(x) = \infty$, the same consequences follow as in case (c).

In applying any of the above rules, if we obtain $\dfrac{0}{0}$ or $\dfrac{\infty}{\infty}$ again, we can apply the rule once

more, repeating the process until the form we obtain is no longer indeterminate.

❯ Example 38

$\lim\limits_{x \to 3} \dfrac{x^2 - 9}{x - 3}$ is of type $\dfrac{0}{0}$ and thus equals $\lim\limits_{x \to 3} \dfrac{2x}{1} = 6$.

(Compare with Example 12, page 97.)

❯ Example 39

$\lim\limits_{x \to 0} \dfrac{\tan x}{x}$ is of type $\dfrac{0}{0}$ and therefore equals $\lim\limits_{x \to 0} \dfrac{\sec^2 x}{1} = 1$.

❯ Example 40

$\lim\limits_{x \to -2} \dfrac{x^3 + 8}{x^2 - 4}$ (Example 13, page 97) is of type $\dfrac{0}{0}$ and thus equals $\lim\limits_{x \to -2} \dfrac{3x^2}{2x} = -3$, as before. Note

that $\lim\limits_{x \to -2} \dfrac{3x^2}{2x}$ is *not* the limit of an indeterminate form!

❯ Example 41

$\lim\limits_{h \to 0} \dfrac{e^h - 1}{h}$ is of type $\dfrac{0}{0}$ and therefore equals $\lim\limits_{h \to 0} \dfrac{e^h}{1} = 1$.

❯ Example 42

$\lim\limits_{x \to \infty} \dfrac{x^3 - 4x^2 + 7}{3 - 6x - 2x^3}$ (Example 20, page 98) is of type $\dfrac{\infty}{\infty}$ so that it equals $\lim\limits_{x \to \infty} \dfrac{3x^2 - 8x}{-6 - 6x^2}$, which is

again of type $\dfrac{\infty}{\infty}$. Apply L'Hospital's Rule twice more:

$$\lim_{x \to \infty} \frac{6x - 8}{-12x} = \lim_{x \to \infty} \frac{6}{-12} = -\frac{1}{2}$$

For this problem, it is easier and faster to apply the Rational Function Theorem!

> **Example 43**

Find $\lim\limits_{x\to\infty} \dfrac{\ln x}{x}$.

✓ **Solution**

$\lim\limits_{x\to\infty} \dfrac{\ln x}{x}$ is of type $\dfrac{\infty}{\infty}$ and equals $\lim\limits_{x\to\infty} \dfrac{1/x}{1} = 0$

> **Example 44**

Find $\lim\limits_{x\to2} \dfrac{x^3 + 8}{x^2 + 4}$.

✓ **Solution**

$$\lim_{x\to2} \frac{x^3 + 8}{x^2 + 4} = \frac{16}{8} = 2$$

BEWARE: L'Hospital's Rule applies only to indeterminate forms $\dfrac{0}{0}$ and $\dfrac{\infty}{\infty}$. Trying to use it in other situations leads to incorrect results, like this:

$$\lim_{x\to2} \frac{x^3 + 8}{x^2 + 4} = \lim_{x\to2} \frac{3x^2}{2x} = 3 \ (\text{WRONG!})$$

For more practice, redo the Practice Exercises on pages 105–109, applying L'Hospital's Rule wherever possible.

NOTE: Below is a description of how to determine the limit for other indeterminate forms by transforming them into $\dfrac{0}{0}$ or $\dfrac{\infty}{\infty}$, but only limits originally of the form $\dfrac{0}{0}$ and $\dfrac{\infty}{\infty}$ will be tested on the AP Calculus exam as given in Examples 38–44. Examples 45, 46, and 47 are presented to complete the discussion of indeterminate forms and L'Hospital's Rule, but questions similar to those in Examples 45–47 will not appear on the AP Calculus exam.

L'Hospital's Rule can also be applied to indeterminate forms of the types $0 \cdot \infty$ and $\infty - \infty$, if the forms can be transformed to either $\dfrac{0}{0}$ or $\dfrac{\infty}{\infty}$.

> **Example 45**

Find $\lim\limits_{x\to\infty} x \sin \dfrac{1}{x}$.

✓ **Solution**

$\lim\limits_{x\to\infty} x \sin \dfrac{1}{x}$ is of the type $\infty \cdot 0$. Since $x \sin \dfrac{1}{x} = \dfrac{\sin \dfrac{1}{x}}{\dfrac{1}{x}}$ and, as $x \to \infty$, the latter is the indeterminate form $\dfrac{0}{0}$, we see that

$$\lim_{x\to\infty} x \sin \frac{1}{x} = \lim_{x\to\infty} \frac{-\dfrac{1}{x^2}\cos \dfrac{1}{x}}{-\dfrac{1}{x^2}} = \lim_{x\to\infty} \cos \frac{1}{x} = 1$$

(Note the easier solution $\lim\limits_{x\to\infty} x \sin \dfrac{1}{x} = \lim\limits_{x\to\infty} \dfrac{\sin \dfrac{1}{x}}{\dfrac{1}{x}} = 1$)

Other indeterminate forms, such as 0^0, 1^∞, and ∞^0, may be resolved by taking the natural logarithm and then applying L'Hospital's Rule.

> **Example 46**

Find $\lim\limits_{x \to 0} (1 + x)^{1/x}$.

> ✓ **Solution**

$\lim\limits_{x \to 0}(1 + x)^{1/x}$ is of type 1^∞. Let $y = (1 + x)^{1/x}$, so that $\ln y = \dfrac{1}{x} \ln (1 + x)$.

Then $\lim\limits_{x \to 0} \ln y = \lim\limits_{x \to 0} \dfrac{\ln (1 + x)}{x}$, which is of type $\dfrac{0}{0}$. Thus,

$$\lim\limits_{x \to 0} \ln y = \lim\limits_{x \to 0} \frac{\dfrac{1}{1 + x}}{1} = \frac{1}{1} = 1$$

and since $\lim\limits_{x \to 0} \ln y = 1$, $\lim\limits_{x \to 0} y = e^1 = e$.

> **Example 47**

Find $\lim\limits_{x \to \infty} x^{1/x}$.

> ✓ **Solution**

$\lim\limits_{x \to \infty} x^{1/x}$ is of type ∞^0. Let $y = x^{1/x}$, so that $\ln y = \dfrac{1}{x} \ln x = \dfrac{\ln x}{x}$ (which, as $x \to \infty$, is of type $\dfrac{\infty}{\infty}$).

Then $\lim\limits_{x \to \infty} \ln y = \dfrac{\dfrac{1}{x}}{1} = 0$, and $\lim\limits_{x \to \infty} y = e^0 = 1$.

K. Recognizing a Given Limit as a Derivative

It is often extremely useful to evaluate a limit by recognizing that it is merely an expression for the definition of the derivative of a specific function (often at a specific point). The relevant definition is the limit of the difference quotient:

$$f'(c) = \lim\limits_{h \to 0} \frac{f(c + h) - f(c)}{h}$$

> **Example 48**

Find $\lim\limits_{h \to 0} \dfrac{(2 + h)^4 - 2^4}{h}$.

> ✓ **Solution**

$\lim\limits_{h \to 0} \dfrac{(2 + h)^4 - 2^4}{h}$ is the derivative of $f(x) = x^4$ at the point where $x = 2$. Since $f'(x) = 4x^3$, the value of the given limit is $f'(2) = 4(2^3) = 32$.

> **Example 49**

Find $\lim\limits_{h \to 0} \dfrac{\sqrt{9+h}-3}{h}$.

✓ **Solution**

$\lim\limits_{h \to 0} \dfrac{\sqrt{9+h}-3}{h} = f'(9)$, where $f(x) = \sqrt{x}$. The value of the limit is $\dfrac{1}{2}x^{-1/2}$ when $x = 9$, or $\dfrac{1}{6}$.

> **Example 50**

Find $\lim\limits_{h \to 0} \dfrac{1}{h}\left(\dfrac{1}{2+h} - \dfrac{1}{2}\right)$.

✓ **Solution**

$\lim\limits_{h \to 0} \dfrac{1}{h}\left(\dfrac{1}{2+h} - \dfrac{1}{2}\right) = f'(2)$, where $f(x) = \dfrac{1}{x}$.

Verify that $f'(2) = -\dfrac{1}{4}$ and compare with Example 17, page 97.

> **Example 51**

Find $\lim\limits_{h \to 0} \dfrac{e^h - 1}{h}$.

✓ **Solution**

$\lim\limits_{h \to 0} \dfrac{e^h - 1}{h} = f'(0)$, where $f(x) = e^x$. The limit has value e^0 or 1 (see also Example 41, on page 134).

> **Example 52**

Find $\lim\limits_{x \to 0} \dfrac{\sin x}{x}$.

✓ **Solution**

$\lim\limits_{x \to 0} \dfrac{\sin x}{x}$ is $f'(0)$, where $f(x) = \sin x$, because we can write

$$f'(0) = \lim_{x \to 0} \frac{\sin(0 + x) - \sin 0}{x} = \lim_{x \to 0} \frac{\sin x}{x}$$

The answer is 1 since $f'(x) = \cos x$ and $f'(0) = \cos 0 = 1$. Of course, we already know that the given limit is the basic trigonometric limit with value 1. Also, L'Hospital's Rule yields 1 as the answer immediately.

CHAPTER SUMMARY

In this chapter, we reviewed differentiation. We defined the derivative as the instantaneous rate of change of a function and looked at estimating derivatives using tables and graphs. We reviewed the formulas for derivatives of basic functions, as well as the Product, Quotient, and Chain Rules. We looked at derivatives of implicitly defined functions and inverse functions and reviewed two important theorems: Rolle's Theorem and the Mean Value Theorem.

For BC Calculus students, we reviewed derivatives of parametrically defined functions and the use of L'Hospital's Rule for evaluating limits of indeterminate forms.

Practice Exercises

> **PART A. NO CALCULATOR—DIRECTIONS:** Answer these questions *without* using your calculator.

In each of Questions A1–A20, a function is given. Choose the alternative that is the derivative, $\frac{dy}{dx}$, of the function.

A1. $y = x^5 \tan x$

   (A) $5x^4 \tan x$
   (B) $5x^4 \sec^2 x$
   (C) $5x^4 + \sec^2 x$
   (D) $5x^4 \tan x + x^5 \sec^2 x$

A2. $y = \frac{2-x}{3x+1}$

   (A) $-\frac{7}{(3x+1)^2}$
   (B) $\frac{5-6x}{(3x+1)^2}$
   (C) $-\frac{9}{(3x+1)^2}$
   (D) $\frac{7}{(3x+1)^2}$

A3. $y = \sqrt{3-2x}$

   (A) $\frac{1}{2\sqrt{3-2x}}$
   (B) $-\frac{1}{\sqrt{3-2x}}$
   (C) $-\frac{4}{3}(3-2x)^{3/2}$
   (D) $\frac{2}{3}(3-2x)^{3/2}$

A4. $y = \frac{2}{(5x+1)^3}$

   (A) $-\frac{30}{(5x+1)^2}$
   (B) $\frac{-30}{(5x+1)^4}$
   (C) $\frac{-6}{(5x+1)^4}$
   (D) $\frac{6}{(5x+1)^2}$

A5. $y = 3x^{2/3} - 4x^{1/2} - 2$

   (A) $2x^{1/3} - 2x^{-1/2}$
   (B) $3x^{-1/3} - 2x^{-1/2}$
   (C) $\frac{9}{5}x^{5/3} - \frac{8}{3}x^{3/2}$
   (D) $2x^{-1/3} - 2x^{-1/2}$

A6. $y = 2\sqrt{x} - \frac{1}{2\sqrt{x}}$

   (A) $x + \frac{1}{x\sqrt{x}}$
   (B) $\frac{4x-1}{4x\sqrt{x}}$
   (C) $\frac{1}{\sqrt{x}} + \frac{1}{4x\sqrt{x}}$
   (D) $\frac{4}{\sqrt{x}} + \frac{1}{x\sqrt{x}}$

A7. $y = \sqrt{x^2 + 2x - 1}$

   (A) $\frac{x+1}{y}$
   (B) $4y(x+1)$
   (C) $\frac{1}{2\sqrt{x^2+2x-1}}$
   (D) $-\frac{x+1}{(x^2+2x-1)^{3/2}}$

A8. $y = \frac{x^2}{\cos x}$

   (A) $-\frac{2x}{\sin x}$
   (B) $\frac{2x\cos x - x^2\sin x}{\cos^2 x}$
   (C) $\frac{2x\cos x + x^2\sin x}{\cos^2 x}$
   (D) $\frac{2x\cos x - x^2\sin x}{\sin^2 x}$

A9. $y = \ln\left(\frac{e^x}{e^x - 1}\right)$

   (A) $x - \frac{e^x}{e^x - 1}$
   (B) $\frac{1}{e^x - 1}$
   (C) $-\frac{1}{e^x - 1}$
   (D) $\frac{e^x - 2}{e^x - 1}$

A10. $y = \tan^{-1}\frac{x}{2}$

(A) $\dfrac{4}{4 + x^2}$

(B) $\dfrac{1}{2\sqrt{4 - x^2}}$

(C) $\dfrac{2}{\sqrt{4 - x^2}}$

(D) $\dfrac{2}{x^2 + 4}$

A11. $y = \ln(\sec x + \tan x)$

(A) $\sec x$

(B) $\dfrac{1}{\sec x}$

(C) $\dfrac{1}{\sec x + \tan x}$

(D) $-\dfrac{1}{\sec x + \tan x}$

A12. $y = \dfrac{e^x - e^{-x}}{e^x + e^{-x}}$

(A) 0

(B) 1

(C) $\dfrac{4}{(e^x + e^{-x})^2}$

(D) $\dfrac{1}{e^{2x} + e^{-2x}}$

A13. $y = \ln\left(\sqrt{x^2 + 1}\right)$

(A) $\dfrac{1}{\sqrt{x^2 + 1}}$

(B) $\dfrac{2x}{\sqrt{x^2 + 1}}$

(C) $\dfrac{1}{2(x^2 + 1)}$

(D) $\dfrac{x}{x^2 + 1}$

A14. $y = \sin\left(\frac{1}{x}\right)$

(A) $\cos\left(\frac{1}{x}\right)$

(B) $\cos\left(-\frac{1}{x^2}\right)$

(C) $-\frac{1}{x^2}\cos\left(\frac{1}{x}\right)$

(D) $-\frac{1}{x^2}\sin\left(\frac{1}{x}\right) + \frac{1}{x}\cos\left(\frac{1}{x}\right)$

A15. $y = \dfrac{1}{2\sin 2x}$

(A) $-\csc 2x \cot 2x$

(B) $\dfrac{1}{4\cos 2x}$

(C) $-4\csc 2x \cot 2x$

(D) $-\csc^2 2x$

A16. $y = e^{-x}\cos 2x$

(A) $-e^{-x}(\cos 2x + 2\sin 2x)$

(B) $e^{-x}(\sin 2x - \cos 2x)$

(C) $2e^{-x}\sin 2x$

(D) $-e^{-x}(\cos 2x + \sin 2x)$

A17. $y = \sec^2(x)$

(A) $2\sec x$

(B) $2\sec x \tan x$

(C) $2\sec^2 x \tan x$

(D) $\tan x$

A18. $y = x(\ln x)^3$

(A) $\dfrac{3(\ln x)^2}{x}$

(B) $3(\ln x)^2$

(C) $-3x(\ln x)^2 + (\ln x)^3$

(D) $(\ln x)^3 + 3(\ln x)^2$

A19. $y = \dfrac{1 + x^2}{1 - x^2}$

(A) $-\dfrac{4x}{(1 - x^2)^2}$

(B) $\dfrac{4x}{(1 - x^2)^2}$

(C) $-\dfrac{4x^3}{(1 - x^2)^2}$

(D) $\dfrac{2x}{1 - x^2}$

A20. $y = \sin^{-1}x - \sqrt{1 - x^2}$

(A) $\dfrac{1}{2\sqrt{1 - x^2}}$

(B) $\dfrac{2}{\sqrt{1 - x^2}}$

(C) $\dfrac{1 + x}{\sqrt{1 - x^2}}$

(D) $\dfrac{x^2}{\sqrt{1 - x^2}}$

In each of Questions A21–A24, *y* is a differentiable function of *x*. Choose the alternative that is the derivative $\frac{dy}{dx}$.

A21. $x^3 - y^3 = 1$

    (A) x

    (B) $3x^2$

    (C) $\frac{x^2}{y^2}$

    (D) $\frac{3x^2 - 1}{y^2}$

A22. $x + \cos(x + y) = 0$

    (A) $\csc(x + y) - 1$

    (B) $\csc(x + y)$

    (C) $\frac{x}{\sin(x + y)}$

    (D) $\frac{1 - \sin x}{\sin y}$

A23. $\sin x - \cos y - 2 = 0$

    (A) $-\cot x$

    (B) $\frac{\cos x}{\sin y}$

    (C) $-\csc y \cos x$

    (D) $\frac{2 - \cos x}{\sin y}$

A24. $3x^2 - 2xy + 5y^2 = 1$

    (A) $\frac{3x + y}{x - 5y}$

    (B) $\frac{y - 3x}{5y - x}$

    (C) $3x + 5y$

    (D) $\frac{3x + 4y}{x}$

BC ONLY

A25. If $x = t^2 + 1$ and $y = 2t^3$, then $\frac{dy}{dx} =$

    (A) $3t$

    (B) $6t^2$

    (C) $\frac{6t^2}{(t^2 + 1)^2}$

    (D) $\frac{2t^4 + 6t^2}{(t^2 + 1)^2}$

A26. If $f(x) = x^4 - 4x^3 + 4x^2 - 1$, then the set of values of *x* for which the derivative equals zero is

    (A) {1,2}

    (B) {0,−1,−2}

    (C) {−1,2}

    (D) {0,1,2}

A27. If $16\sqrt{x}$, then $f''(4)$ is equal to

    (A) -16

    (B) -4

    (C) -2

    (D) $-\frac{1}{2}$

A28. If $f(x) = \ln(x^3)$, then $f''(3)$ is

    (A) $-\frac{1}{3}$

    (B) -1

    (C) -3

    (D) 1

A29. If a point moves on the curve $x^2 + y^2 = 25$, then, at $(0,5)$, $\frac{d^2 y}{dx^2}$ is

    (A) 0

    (B) $\frac{1}{5}$

    (C) -5

    (D) $-\frac{1}{5}$

A30. If $x = t^2 - 1$ and $y = t^4 - 2t^3$, then, when $t = 1$, $\frac{d^2 y}{dx^2}$ is

    (A) 1

    (B) -1

    (C) 3

    (D) $\frac{1}{2}$

BC ONLY

A31. If $f(x) = 5^x$ and $5^{1.002} \approx 5.016$, which is closest to $f'(1)$?

    (A) 0.016

    (B) 5.0

    (C) 8.0

    (D) 32.0

A32. If $y = e^x(x-1)$, then $y''(0)$ equals

(A) -2

(B) -1

(C) 1

(D) e

BC ONLY

A33. If $x = e^\theta \cos\theta$ and $y = e^\theta \sin\theta$, then, when $\theta = \dfrac{\pi}{2}, \dfrac{dy}{dx}$ is

(A) 1

(B) 0

(C) $e^{\pi/2}$

(D) -1

A34. Given $x = \cos t$ and $y = \cos 2t$, if $\sin(t) \neq 0$, then $\dfrac{d^2y}{dx^2}$ is

(A) $4\cos t$

(B) 4

(C) -4

(D) $-4\cot t$

A35. $\lim\limits_{h\to 0} \dfrac{(1+h)^6 - 1}{h}$ is

(A) 0

(B) 1

(C) 6

(D) nonexistent

A36. $\lim\limits_{h\to 0} \dfrac{\sqrt[3]{8+h} - 2}{h}$ is

(A) 0

(B) $\dfrac{1}{12}$

(C) 1

(D) 192

A37. $\lim\limits_{h\to 0} \dfrac{\ln(e+h) - 1}{h}$ is

(A) 0

(B) $\dfrac{1}{e}$

(C) 1

(D) e

A38. $\lim\limits_{x\to 0} \dfrac{\cos x - 1}{x}$ is

(A) -1

(B) 0

(C) 1

(D) ∞

A39. If $f(x) = \begin{cases} \dfrac{4x^2 - 4}{x - 1}, & x \neq 1 \\ 4, & x = 1 \end{cases}$, which of the following statements is(are) true?

I. $\lim\limits_{x\to 1} f(x)$ exists

II. f is continuous at $x = 1$

III. f is differentiable at $x = 1$

(A) none

(B) I only

(C) I and II only

(D) I, II, and III

A40. If $g(x) = \begin{cases} x^2, & x \leq 3 \\ 6x - 9, & x > 3 \end{cases}$, which of the following statements is (are) true?

I. $\lim\limits_{x\to 3} g(x)$ exists

II. g is continuous at $x = 3$

III. g is differentiable at $x = 3$

(A) I only

(B) II only

(C) I and II only

(D) I, II, and III

A41. The function $f(x) = x^{2/3}$ on $[-8,8]$ does not satisfy the conditions of the Mean Value Theorem because

(A) $f(0)$ is not defined

(B) $f(x)$ is not continuous on $[-8,8]$

(C) $f(x)$ is not defined for $x < 0$

(D) $f'(0)$ does not exist

A42. If $f(x) = 2x^3 - 6x$, at what point on the interval $0 \leq x \leq \sqrt{3}$, if any, is the tangent to the curve parallel to the secant line on that interval?

(A) 1

(B) $\sqrt{2}$

(C) 0

(D) nowhere

A43. If h is the inverse function of f and if $f(x) = \dfrac{1}{x}$, then $h'(3) =$

(A) -9

(B) $-\dfrac{1}{9}$

(C) $\dfrac{1}{9}$

(D) 9

A44. $\lim\limits_{x\to\infty}\dfrac{e^x}{x^{50}}$ equals

(A) 0

(B) 1

(C) $\dfrac{1}{50!}$

(D) ∞

A45. If $\sin(xy)=x$, then $\dfrac{dy}{dx}=$

(A) $\dfrac{\sec(xy)}{x}$

(B) $\dfrac{\sec(xy)-y}{x}$

(C) $\dfrac{1+\sec(xy)}{x}$

(D) $\sec(xy)-1$

A46. $\lim\limits_{x\to 0}\dfrac{\sin 2x}{x}$ is

(A) 1

(B) 2

(C) $\dfrac{1}{2}$

(D) 0

A47. $\lim\limits_{x\to 0}\dfrac{\sin 3x}{\sin 4x}$ is

(A) 1

(B) $\dfrac{4}{3}$

(C) $\dfrac{3}{4}$

(D) 0

A48. $\lim\limits_{x\to 0}\dfrac{1-\cos x}{x}$ is

(A) 0

(B) 1

(C) 2

(D) ∞

A49. $\lim\limits_{x\to 0}\dfrac{\tan \pi x}{x}$ is

(A) 0

(B) 1

(C) π

(D) ∞

CHALLENGE A50. $\lim\limits_{x\to\infty} x^2 \sin\dfrac{1}{x}$

(A) is 1

(B) is 0

(C) is ∞

(D) oscillates between -1 and 1

A51. The graph in the xy-plane represented by **BC ONLY**
$x=3+2\sin t$ and $y=2\cos t-1$, for
$-\pi \le t \le \pi$, is

(A) a semicircle

(B) a circle

(C) an ellipse

(D) half of an ellipse

A52. $\lim\limits_{x\to 0}\dfrac{\sec x-\cos x}{x^2}$

(A) $=0$

(B) $=\dfrac{1}{2}$

(C) $=1$

(D) $=2$

In each of Questions A53–A56, a pair of equations **BC ONLY**
that represent a curve parametrically is given.
Choose the alternative that is the derivative $\dfrac{dy}{dx}$.

A53. $x=t-\sin t$ and $y=1-\cos t$

(A) $\dfrac{\sin t}{1-\cos t}$

(B) $\dfrac{1-\cos t}{\sin t}$

(C) $\dfrac{\sin t}{\cos t-1}$

(D) $\dfrac{1-x}{y}$

A54. $x=\cos^3\theta$ and $y=\sin^3\theta$

(A) $-\cot\theta$

(B) $\cot\theta$

(C) $-\tan\theta$

(D) $-\tan^2\theta$

A55. $x=1-e^{-t}$ and $y=t+e^{-t}$

(A) $\dfrac{e^{-1}}{1-e^{-t}}$

(B) e^t+1

(C) e^t-e^{-2t}

(D) e^t-1

A56. $x=\dfrac{1}{1-t}$ and $y=1-\ln(1-t)$ $(t<1)$

(A) $\dfrac{1}{1-t}$

(B) $t-1$

(C) $\dfrac{1}{x}$

(D) $\dfrac{(1-t)^2}{t}$

A57. The "left half" of the parabola defined by $y = x^2 - 8x + 10$ for $x \leq 4$ is a one-to-one function; therefore, its inverse is also a function. Call that inverse g. Find $g'(3)$.

(A) $-\dfrac{1}{2}$

(B) $-\dfrac{1}{6}$

(C) $\dfrac{1}{6}$

(D) $\dfrac{1}{2}$

A58. If $f(u) = \sin u$ and $u = g(x) = x^2 - 9$, then $(f \circ g)'(3)$ equals

(A) 0

(B) 1

(C) 3

(D) 6

A59. If $f(x) = \dfrac{x}{(x-1)^2}$, then the set of x's for which $f'(x)$ exists is

(A) all reals

(B) all reals except $x = 1$ and $x = -1$

(C) all reals except $x = -1$

(D) all reals except $x = 1$

A60. If $f(x) = \dfrac{1}{x^2 + 1}$ and $g(x) = \sqrt{x}$, then the derivative of $f(g(x))$ is

(A) $-(x+1)^{-2}$

(B) $\dfrac{-2x}{(x^2+1)^2}$

(C) $\dfrac{1}{(x+1)^2}$

(D) $\dfrac{1}{2\sqrt{x}\,(x+1)}$

A61. Suppose $y = f(x) = 2x^3 - 3x$. If $h(x)$ is the inverse function of f, then $h'(-1) =$

(A) -1

(B) $\dfrac{1}{3}$

(C) 1

(D) 3

A62. If $f(x) = x^3 - 3x^2 + 8x + 5$ and $g(x) = f^{-1}(x)$, then $g'(5) =$

(A) 8

(B) $\dfrac{1}{8}$

(C) $\dfrac{1}{53}$

(D) 53

PART B. CALCULATOR ACTIVE—DIRECTIONS: Some of the following questions require the use of a graphing calculator.

In Questions B1–B8, differentiable functions f and g have the values shown in the table.

x	f	f'	g	g'
0	2	1	5	-4
1	3	2	3	-3
2	5	3	1	-2
3	10	4	0	-1

B1. If $A(x) = f(x) + 2g(x)$, then $A'(3) =$

(A) -2

(B) 2

(C) 7

(D) 8

B2. If $B(x) = f(x) \cdot g(x)$, then $B'(2) =$

(A) -20

(B) -7

(C) -6

(D) 13

B3. If $D(x) = \dfrac{1}{g(x)}$, then $D'(1) =$

(A) $-\dfrac{1}{3}$

(B) $-\dfrac{1}{9}$

(C) $\dfrac{1}{9}$

(D) $\dfrac{1}{3}$

B4. If $H(x) = \sqrt{f(x)}$, then $H'(3) =$

(A) $\frac{1}{4}$

(B) $\frac{1}{2\sqrt{10}}$

(C) $\frac{2}{\sqrt{10}}$

(D) $4\sqrt{10}$

B5. If $K(x) = \dfrac{f(x)}{g(x)}$, then $K'(0) =$

(A) $-\frac{13}{25}$

(B) $-\frac{3}{25}$

(C) $\frac{13}{25}$

(D) $\frac{13}{16}$

B6. If $M(x) = f(g(x))$, then $M'(1) =$

(A) -12

(B) -6

(C) 6

(D) 12

B7. If $P(x) = f(x^3)$, then $P'(1) =$

(A) 2

(B) 6

(C) 8

(D) 12

B8. If $S(x) = f^{-1}(x)$, then $S'(3) =$

(A) -2

(B) $\frac{1}{4}$

(C) $\frac{1}{2}$

(D) 2

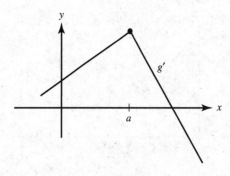

B9. The graph of g' is shown here. Which of the following statements is (are) true of g?

  I. g is continuous at $x = a$

 II. g is differentiable at $x = a$

III. g is increasing in an interval containing $x = a$

(A) I only

(B) III only

(C) I and III only

(D) I, II, and III

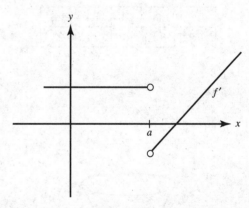

B10. A function f has the derivative shown. Which of the following statements must be false?

(A) f is continuous at $x = a$

(B) f has a vertical asymptote at $x = a$

(C) f has a jump discontinuity at $x = a$

(D) f has a removable discontinuity at $x = a$

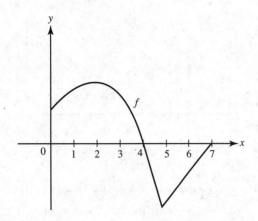

B11. The function f whose graph is shown has $f' = 0$ at $x =$

(A) 2 only
(B) 2 and 5
(C) 4 and 7
(D) 2, 4, and 7

B12. A differentiable function f has the values shown. Estimate $f'(1.5)$.

x	1.0	1.3	1.4	1.6
$f(x)$	8	10	14	22

(A) 8
(B) 18
(C) 40
(D) 80

B13. Water is poured into a conical reservoir at a constant rate. If $h(t)$ is the rate of change of the depth of the water, then h is

(A) linear and increasing
(B) linear and decreasing
(C) nonlinear and increasing
(D) nonlinear and decreasing

Use the figure to answer Questions B14–B16. The graph of f consists of two line segments and a semi-circle of radius 2.

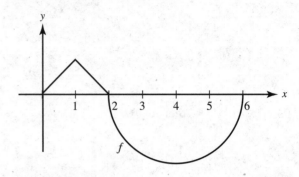

B14. $f'(x) = 0$ for $x =$

(A) 1 only
(B) 2 only
(C) 4 only
(D) 1 and 4

B15. For $0 < x < 6$, $f'(x)$ does not exist for $x =$

(A) 1 only
(B) 2 only
(C) 1 and 2
(D) 2 and 6

B16. $f'(5) =$

(A) $\frac{1}{2}$
(B) $\frac{1}{\sqrt{3}}$
(C) 2
(D) $\sqrt{3}$

B17. At how many points on the interval $[-5,5]$ is a tangent to $y = x + \cos x$ parallel to the secant line on $[-5,5]$?

(A) none
(B) 1
(C) 2
(D) 3

B18. From the values of f shown, estimate $f'(2)$.

x	1.92	1.93	1.95	1.98	2.00
$f(x)$	6.00	5.00	4.40	4.10	4.00

(A) -0.10

(B) -0.20

(C) -5

(D) -10

B19. Using the values shown in the table for Question B18, estimate $(f^{-1})'(4)$.

(A) -0.2

(B) -0.1

(C) -5

(D) -10

B20. The table below shows some points on a function f that is both continuous and differentiable on the closed interval $[2,10]$.

x	2	4	6	8	10
$f(x)$	30	25	20	25	30

Which must be true?

(A) $f'(6) = 0$

(B) $f'(8) > 0$

(C) The maximum value of f on the interval $[2,10]$ is 30.

(D) For some value of x on the interval $[2,10]$ $f'(x) = 0$.

B21. If f is differentiable and difference quotients overestimate the slope of f at $x = a$ for all $h > 0$, which must be true?

(A) $f'(x) \geq 0$ on $[a,h]$

(B) $f'(x) \leq 0$ on $[a,h]$

(C) $f''(x) \geq 0$ on $[a,h]$

(D) $f''(x) \leq 0$ on $[a,h]$

B22. If $f(a) = f(b) = 0$ and $f(x)$ is continuous on $[a,b]$, then

(A) $f(x)$ must be identically zero

(B) $f(x)$ may be different from zero for all x on $[a,b]$

(C) there exists at least one number c, $a < c < b$, such that $f'(c) = 0$

(D) $f(x)$ must exist for every x on (a,b)

B23. Suppose $f'(1) = 2, f'(1) = 3$, and $f'(2) = 4$. Then $(f^{-1})'(2)$

(A) equals $-\frac{1}{3}$

(B) equals $-\frac{1}{4}$

(C) equals $\frac{1}{4}$

(D) equals $\frac{1}{3}$

B24. Suppose $\lim\limits_{x \to 0} \dfrac{g(x) - g(0)}{x} = 1$. It follows necessarily that

(A) g is not defined at $x = 0$

(B) g is not continuous at $x = 0$

(C) the limit of $g(x)$ as x approaches 0 equals 1

(D) $g'(0) = 1$

Use this graph of $y = f(x)$ for Questions B25 and B26.

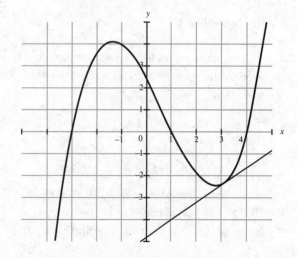

B25. $f'(3)$ is most closely approximated by

(A) 0.3

(B) 0.8

(C) 1.5

(D) 1.8

B26. The rate of change of $f(x)$ is least at $x \approx$

(A) -3

(B) -1.3

(C) 0.7

(D) 2.7

Use the following definition of the *symmetric difference quotient* as an approximation for $f'(x_0)$ for Questions B27–B29. For small values of h,

$$f'(x_0) \approx \frac{f(x_0 + h) - f(x_0 - h)}{2h}$$

B27. For $f(x) = 5^x$, what is the estimate of $f'(2)$ obtained by using the symmetric difference quotient with $h = 0.03$?

(A) 40.25

(B) 40.252

(C) 41.818

(D) 80.503

B28. To how many places is the symmetric difference quotient accurate when it is used to approximate $f'(0)$ for $f(x) = 4^x$ and $h = 0.08$?

(A) 1

(B) 2

(C) 3

(D) more than 3

B29. To how many places is $f'(x_0)$ accurate when it is used to approximate $f'(0)$ for $f(x) = 4^x$ and $h = 0.001$?

(A) 1

(B) 2

(C) 3

(D) more than 3

B30. The value of $f'(0)$ obtained using the symmetric difference quotient with $f(x) = |x|$ and $h = 0.001$ is

(A) -1

(B) 0

(C) 1

(D) indeterminate

B31. If $\frac{d}{dx} f(x) = g(x)$ and $h(x) = \sin x$, then $\frac{d}{dx} f(h(x))$ equals

(A) $g(\sin x)$

(B) $\cos x \cdot g(x)$

(C) $\cos x \cdot g(\sin x)$

(D) $\sin x \cdot g(\sin x)$

B32. Let $f(x) = 3^x - x^3$. The tangent to the curve is parallel to the secant through $(0,1)$ and $(3,0)$ for $x =$

(A) 1.244 only

(B) 2.727 only

(C) 1.244 and 2.727 only

(D) no such value of x exists

Questions B33–B37 are based on the following graph of $f(x)$, sketched on $-6 \leq x \leq 7$. Assume the horizontal and vertical grid lines are equally spaced at unit intervals.

B33. On the interval $1 < x < 2$, $f(x)$ equals

(A) $-x - 2$

(B) $-x - 3$

(C) $-x - 4$

(D) $-x + 2$

B34. Over which of the following intervals does $f'(x)$ equal zero?

I. $(-6, -3)$

II. $(-3, -1)$

III. $(2, 5)$

(A) II only

(B) I and II only

(C) I and III only

(D) II and III only

B35. How many points of discontinuity does $f'(x)$ have on the interval $-6 < x < 7$?

(A) 2
(B) 3
(C) 4
(D) 5

B36. For $-6 < x < -3$, $f'(x)$ equals

(A) $-\frac{3}{2}$
(B) -1
(C) 1
(D) $\frac{3}{2}$

B37. Which of the following statements about the graph of $f'(x)$ is false?

(A) $f'(x)$ consists of six horizontal segments
(B) $f'(x)$ has four jump discontinuities
(C) $f'(x)$ is discontinuous at each x in the set $\{-3,-1,1,2,5\}$
(D) On the interval $-1 < x < 1$, $f'(x) = -3$

B38. The table gives the values of a function f that is differentiable on the interval $[0,1]$:

x	0.10	0.20	0.26	0.40	0.52	0.60
$f(x)$	0.171	0.288	0.357	0.384	0.375	0.336

According to this table, the best approximation of $f'(0.10)$ is

(A) 0.12
(B) 1.08
(C) 1.17
(D) 1.77

B39. At how many points on the interval $[a,b]$ does the function graphed satisfy the Mean Value Theorem?

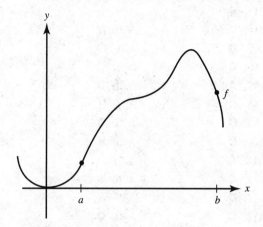

(A) none
(B) 1
(C) 2
(D) 3

Answer Key

A1. **D**	A27. **D**	A53. **A**	B17. **D**
A2. **A**	A28. **A**	A54. **C**	B18. **C**
A3. **B**	A29. **D**	A55. **D**	B19. **A**
A4. **B**	A30. **D**	A56. **C**	B20. **D**
A5. **D**	A31. **C**	A57. **B**	B21. **C**
A6. **C**	A32. **C**	A58. **D**	B22. **B**
A7. **A**	A33. **D**	A59. **D**	B23. **D**
A8. **C**	A34. **B**	A60. **A**	B24. **D**
A9. **C**	A35. **C**	A61. **B**	B25. **B**
A10. **D**	A36. **B**	A62. **B**	B26. **C**
A11. **A**	A37. **B**	B1. **B**	B27. **B**
A12. **C**	A38. **B**	B2. **B**	B28. **B**
A13. **D**	A39. **B**	B3. **D**	B29. **D**
A14. **C**	A40. **D**	B4. **C**	B30. **B**
A15. **A**	A41. **D**	B5. **C**	B31. **C**
A16. **A**	A42. **A**	B6. **A**	B32. **C**
A17. **C**	A43. **B**	B7. **B**	B33. **A**
A18. **D**	A44. **D**	B8. **C**	B34. **D**
A19. **B**	A45. **B**	B9. **D**	B35. **D**
A20. **C**	A46. **B**	B10. **B**	B36. **D**
A21. **C**	A47. **C**	B11. **A**	B37. **B**
A22. **A**	A48. **A**	B12. **C**	B38. **C**
A23. **C**	A49. **C**	B13. **D**	B39. **D**
A24. **B**	A50. **C**	B14. **C**	
A25. **A**	A51. **B**	B15. **C**	
A26. **D**	A52. **C**	B16. **B**	

Answer Explanations

Many of the explanations provided include intermediate steps that would normally be reached on the way to a final algebraically simplified result. You may not need to reach the final answer.

NOTE: The formulas or rules cited in parentheses in the explanations are given on page 115.

A1. **(D)** By the Product Rule, (5),

$$y' = x^5(\tan x)' + (x^5)'(\tan x)$$

A2. **(A)** By the Quotient Rule, (6),

$$y' = \frac{(3x+1)(-1) - (2-x)(3)}{(3x+1)^2} = -\frac{7}{(3x+1)^2}$$

A3. **(B)** Since $y = (3 - 2x)^{1/2}$, by the Power Rule, (3),

$$y' = \frac{1}{2}(3 - 3x)^{-1/2} \cdot (-2) = -\frac{1}{\sqrt{3-2x}}$$

A4. **(B)** Since $y = 2(5x+1)^{-3}$, $y' = -6(5x+1)^{-4}$, (5).

A5. **(D)** $y' = 3\left(\frac{2}{3}\right)x^{-1/3} - 4\left(\frac{1}{2}\right)x^{-1/2}$

A6. **(C)** Rewrite: $y = 2x^{1/2} - \frac{1}{2}x^{-1/2}$; so $y' = x^{-1/2} + \frac{1}{4}x^{-3/2}$

A7. **(A)** Rewrite: $y = (x^2 + 2x - 1)^{1/2}$; then $y' = \frac{1}{2}(x^2 + 2x - 1)^{-1/2}(2x + 2)$

A8. **(C)** Use the Quotient Rule:

$$y' = \frac{2x \cos x - x^2(-\sin x)}{\cos^2 x}$$

A9. **(C)** Since

$$y = \ln e^x - \ln(e^x - 1)$$
$$= x - \ln(e^x - 1)$$

then,

$$y' = 1 - \frac{e^x}{e^x - 1} = \frac{e^x - 1 - e^x}{e^x - 1} = -\frac{1}{e^x - 1}$$

A10. **(D)** Use formula (18): $y' = \dfrac{\frac{1}{2}}{1 + \frac{x^2}{4}}$

A11. **(A)** Use formulas (13), (11), and (9):

$$y' = \frac{\sec x \tan x + \sec^2 x}{\sec x + \tan x} = \frac{\sec x(\tan x + \sec x)}{\sec x + \tan x}$$

A12. **(C)** By the Quotient Rule,

$$y' = \frac{(e^x + e^{-x})(e^x + e^{-x}) - (e^x - e^{-x})(e^x - e^{-x})}{(e^x + e^{-x})^2}$$

$$= \frac{(e^{2x} + 2 + e^{-2x}) - (e^{2x} + 2 + e^{-2x})}{(e^x + e^{-x})^2} = \frac{4}{(e^x + e^{-x})^2}$$

A13. **(D)** Since $y = \frac{1}{2}\ln(x^2+1)$,

$$y' = \frac{1}{2} \cdot \frac{2x}{x^2+1}$$

A14. **(C)** $y' = \sin'\left(\frac{1}{x}\right) \cdot \left(\frac{1}{x}\right)' = \cos\left(\frac{1}{x}\right) \cdot \left(-\frac{1}{x^2}\right)$

A15. **(A)** Since $y = \frac{1}{2}\csc 2x$, $y' = \frac{1}{2}(-\csc 2x \cot 2x \cdot 2)$.

A16. **(A)** $y' = e^{-x}(-2\sin 2x) + \cos 2x(-e^{-x})$

A17. **(C)** $y' = (2\sec x)(\sec x \tan x)$

A18. **(D)** $y' = (\ln x)^3 + x \cdot 3(\ln x)^2 \cdot \frac{1}{x} = (\ln x)^3 + 3(\ln x)^2$

A19. **(B)** $y' = \dfrac{(1-x^2)(2x) - (1+x^2)(-2x)}{(1-x^2)^2}$

A20. **(C)** $y' = \dfrac{1}{\sqrt{1-x^2}} - \dfrac{1 \cdot (-2x)}{2\sqrt{1-x^2}}$

A21. **(C)** Let y' be $\dfrac{dy}{dx}$; then $3x^2 - 3y^2y' = 0$; $y' = \dfrac{-3x^2}{-3y^2}$

A22. **(A)** $1 - \sin(x+y)(1+y') = 0$; $\dfrac{1 - \sin(x+y)}{\sin(x+y)} = y'$

A23. **(C)** $\cos x + \sin y \cdot y' = 0$; $y' = -\dfrac{\cos x}{\sin y}$

A24. **(B)** $6x - 2(xy' + y) + 10yy' = 0$; $y'(10y - 2x) = 2y - 6x$

A25. **(A)** $\dfrac{dy}{dx} = \dfrac{dy/dt}{dx/dt} = \dfrac{6t^2}{2t}$

A26. **(D)** $f'(x) = 4x^3 - 12x^2 + 8x = 4x(x-1)(x-2)$

A27. **(D)** $f'(x) = 8x^{-1/2}$; $f''(x) = -4x^{-3/2} = -\dfrac{4}{x^{3/2}}$; $f''(4) = -\dfrac{4}{8}$

A28. **(A)** $f(x) = 3\ln x$; $f'(x) = \dfrac{3}{x}$; $f''(x) = \dfrac{-3}{x^2}$. Replace x by 3.

A29. **(D)** $2x + 2yy' = 0$; $y' = -\dfrac{x}{y}$; $y'' = -\dfrac{y - xy'}{y^2}$. At $(0,5)$, $y'' = \dfrac{5-0}{25}$

A30. **(D)** $\dfrac{dy}{dx} = \dfrac{4t^3 - 6t^2}{2t} = 2t^2 - 3t$ $(t \neq 0)$; $\dfrac{d^2y}{dx^2} = \dfrac{4t-3}{2t}$. Replace t by 1.

A31. **(C)** $f'(1) \approx \dfrac{5^{1.002} - 5^1}{0.002} = \dfrac{5.016 - 5}{0.002}$

A32. **(C)** $y' = e^x \cdot 1 + e^x(x-1) = xe^x$
$y'' = xe^x + e^x$ and $y''(0) = 0 \cdot 1 + 1 = 1$

A33. **(D)** When simplified, $\dfrac{dy}{dx} = \dfrac{\cos\theta + \sin\theta}{\cos\theta - \sin\theta}$.

A34. **(B)** Since (if $\sin(t) \neq 0$)

$$\frac{dy}{dt} = -2\sin 2t = -4\sin t \cos t \text{ and } \frac{dx}{dt} = -\sin t$$

then $\dfrac{dt}{dx} = 4\cos t$. Thus:

$$\frac{d^2y}{dx^2} = -\frac{4\sin t}{-\sin t}$$

NOTE: Since each of the limits in Questions A35–A39 yields an indeterminate form of the type $\frac{0}{0}$, we can apply L'Hospital's Rule in each case, getting identical answers.

A35. **(C)** The given limit is the derivative of $f(x) = x^6$ at $x = 1$.

A36. **(B)** The given limit is the definition for $f'(8)$, where $f(x) = \sqrt[3]{x}$

$$f'(x) = \frac{1}{3\,x^{2/3}}$$

A37. **(B)** The given limit is $f'(e)$, where $f(x) = \ln x$.

A38. **(B)** The given limit is the derivative of $f(x) = \cos x$ at $x = 0$; $f'(x) = -\sin x$.

A39. **(B)** $\lim\limits_{x \to 1} \dfrac{4x^2 - 4}{x - 1} = \lim\limits_{x \to 1} \dfrac{4(x + 1)(x - 1)}{(x - 1)} = \lim\limits_{x \to 1} 4(x + 1) = 8$, but $f(1) = 4$.

Thus f is discontinuous at $x = 1$, so it cannot be differentiable.

A40. **(D)** $\lim\limits_{x \to 3^-} x^2 = \lim\limits_{x \to 3^+} (6x - 9) = 9$, so the limit exists. Because $g(3) = 9$, g is continuous at $x = 3$.

Since $g'(x) = \begin{cases} 2x, x < 3 \\ 6, x > 3 \end{cases}$, so $g'(3) = 6$.

A41. **(D)** Since $f'(x) = \dfrac{2}{3\,x^{1/3}}$, $f'(0)$ is not defined; $f'(x)$ must be defined on $(-8,8)$.

A42. **(A)** Note that $f(0) = f(\sqrt 3) = 0$ and that $f'(x)$ exists on the given interval. By the MVT, there is a number, c, in the interval such that $f'(c) = 0$. If $c = 1$, then $6c^2 - 6 = 0$.
(-1 is not in the interval.)

A43. **(B)** Since the inverse, h, of $f(x) = \frac{1}{x}$ is $h(x) = \frac{1}{x}$, then $h'(x) = -\dfrac{1}{x^2}$. Replace x by 3.

A44. **(D)** After 50(!) applications of L'Hospital's Rule, we get $\lim\limits_{x \to \infty} \dfrac{e^x}{50!}$, which "equals" ∞.

A perfunctory examination of the limit, however, shows immediately that the answer is ∞. In fact, $\lim\limits_{x \to \infty} \dfrac{e^x}{x^n}$ for any positive integer n, no matter how large, is ∞.

A45. **(B)** $\cos(xy)(xy' + y) = 1$; $x \cos(xy)y' = 1 - y \cos(xy)$

$$y' = \frac{1 - y \cos(xy)}{x \cos(xy)}$$

NOTE: In Questions A46–A50, the limits are all indeterminate forms of the type $\frac{0}{0}$. We have therefore applied L'Hospital's Rule in each one. The indeterminacy can also be resolved by introducing $\frac{\sin a}{a}$, which approaches 1 as a approaches 0. The latter technique is presented in square brackets.

A46. **(B)** $\lim\limits_{x \to 0} \dfrac{\sin 2x}{x} = \lim\limits_{x \to 0} \dfrac{2 \cos 2x}{1} = \dfrac{2 \cdot 1}{1} = 2$

$\left[\text{Using } \sin 2x = 2 \sin x \cos x \text{ yields } \lim\limits_{x \to 0} 2\left(\dfrac{\sin x}{x}\right)\cos x = 2 \cdot 1 \cdot 1 = 2\right]$

A47. **(C)** $\lim\limits_{x \to 0} \dfrac{\sin 3x}{\sin 4x} = \lim\limits_{x \to 0} \dfrac{3 \cos 3x}{4 \cos 4x} = \dfrac{3 \cdot 1}{4 \cdot 1} = \dfrac{3}{4}$

$\left[\text{We rewrite } \dfrac{\sin 3x}{\sin 4x} \text{ as } \dfrac{\sin 3x}{3x} \cdot \dfrac{4x}{\sin 4x} \cdot \dfrac{3}{4}. \text{ As } x \to 0, \text{ so do } 3x \text{ and } 4x; \text{ the fraction approaches } 1 \cdot 1 \cdot \dfrac{3}{4}.\right]$

A48. **(A)** $\lim\limits_{x \to 0} \dfrac{1 - \cos x}{x} = \lim\limits_{x \to 0} \dfrac{\sin x}{1} = 0$

$\left[\text{We can replace } 1 - \cos x \text{ by } 2 \sin^2 \dfrac{x}{2}, \text{ getting}\right.$

$$\lim\limits_{x \to 0} \frac{2 \sin^2 \dfrac{x}{2}}{x} = \lim\limits_{x \to 0} \frac{\sin^2 \dfrac{x}{2}}{\dfrac{x}{2}} = \lim\limits_{x \to 0} \sin \dfrac{x}{2}\left(\frac{\sin \dfrac{x}{2}}{\dfrac{x}{2}}\right) = 0 \cdot 1 \Big]$$

A49. **(C)** $\lim\limits_{x \to 0} \dfrac{\tan \pi x}{x} = \lim\limits_{x \to 0} \dfrac{(\sec^2 \pi x) \cdot \pi}{1} = 1 \cdot \pi = \pi$

$\left[\dfrac{\tan \pi x}{x} = \dfrac{\sin \pi x}{x \cos \pi x} = \pi \cdot \dfrac{\sin \pi x}{\pi x} \cdot \dfrac{1}{\cos x} \right.$; as x (or πx) approaches 0, the original fraction

approaches $\left. \pi \cdot 1 \cdot \dfrac{1}{1} = \pi \right]$

A50. **(C)** The limit is easiest to obtain here if we rewrite:

$$\lim\limits_{x \to \infty} x^2 \sin \dfrac{1}{x} = \lim\limits_{x \to \infty} x \dfrac{\sin(1/x)}{(1/x)} = \infty \cdot 1 = \infty$$

A51. **(B)** Since $x - 3 = 2 \sin t$ and $y + 1 = 2 \cos t$

$$(x - 3)^2 + (y + 1)^2 = 4$$

This is the equation of a circle with center at $(3, -1)$ and radius 2. In the domain given, $-\pi \le t \le \pi$, the entire circle is traced by a particle moving counterclockwise, starting from and returning to $(3, -3)$.

A52. **(C)** Use L'Hospital's Rule; then

$$\lim\limits_{x \to 0} \dfrac{\sec x - \cos x}{x^2} = \lim\limits_{x \to 0} \dfrac{\sec x \tan x + \sin x}{2x}$$

$$= \lim\limits_{x \to 0} \dfrac{\sec^3 x + \sec x \tan^2 x + \cos x}{2} = \dfrac{1 + 1 \cdot 0 + 1}{2} = 1$$

This can be done without using L'Hospital's Rule as follows:

$$\lim\limits_{x \to 0} \dfrac{\sec x - \cos x}{x^2} = \lim\limits_{x \to 0} \dfrac{\dfrac{1}{\cos x} - \cos x}{x^2} = \lim\limits_{x \to 0} \dfrac{1 - \cos^2 x}{x^2 \cos x}$$

$$= \lim\limits_{x \to 0} \dfrac{\sin^2 x}{x^2 \cos x} = \lim\limits_{x \to 0} \dfrac{\sin x}{x} \cdot \dfrac{\sin x}{x} \cdot \dfrac{1}{\cos x} = 1 \cdot 1 \cdot 1 = 1$$

A53. **(A)** $\dfrac{dy}{dx} = \dfrac{\dfrac{dy}{dt}}{\dfrac{dx}{dt}} = \dfrac{\sin t}{1 - \cos t}$

A54. **(C)** $\dfrac{dy}{dx} = \dfrac{\dfrac{dy}{d\theta}}{\dfrac{dx}{d\theta}} = \dfrac{3 \sin^2 \theta \cos \theta}{-3 \cos^2 \sin \theta}$

A55. **(D)** $\dfrac{dy}{dx} = \dfrac{\dfrac{dy}{dt}}{\dfrac{dx}{dt}} = \dfrac{1 - e^t}{e^{-t}} = e^t - 1$

A56. **(C)** Since $\dfrac{dy}{dx} = \dfrac{1}{1 - t}$ and $\dfrac{dy}{dt} = \dfrac{1}{(1 - t)^2}$, then

$$\dfrac{dy}{dx} = 1 - t = \dfrac{1}{x}$$

A57. **(B)** When $x = 3$ on g^{-1}, $y = 3$ on the original half-parabola. $3 = x^2 - 8x + 10$ at $x = 1$ (and at $x = 7$, but that value is not in the given domain).

$g'(3) = \dfrac{1}{y'(1)} = \dfrac{1}{2x - 8}\Big|_{x=1} = -\dfrac{1}{6}.$

A58. **(D)** $(f \circ g)'$ at $x = 3$ equals $f'(g(3)) \cdot g'(3)$ equals $\cos u$ (at $u = 0$) times $2x$ (at $x = 3$) $= 1 \cdot 6 = 6$.

A59. **(D)** Here $f'(x)$ equals $\dfrac{-x - 1}{(x - 1)^3}.$

A60. **(A)** Note that $f(g(x)) = \dfrac{1}{x + 1}.$

A61. **(B)** Since $f'(x) = 6x^2 - 3$, $h'(x) = \dfrac{1}{6x^2 - 3}$; also, $f(x)$, or $2x^3 - 3x$, equals -1, by observation, for $x = 1$. So $h'(-1)$ or $\dfrac{1}{6x^2 - 3}$ (when $x = 1$) equals $\dfrac{1}{6-3} = \dfrac{1}{3}$.

A62. **(B)** Since $f(0) = 5$, $g'(5) = \dfrac{1}{f'(0)} = \dfrac{1}{3x^2 - 6x + 8}\Big|_{x=0} = \dfrac{1}{8}$.

B1. **(B)** $(f + 2g)'(3) = f'(3) + 2g'(3) = 4 + 2(-1)$

B2. **(B)** $(f \cdot g)'(2) = f(2) \cdot g'(2) + g(2) \cdot f'(2) = 5(-2) + 1(3)$

B3. **(D)** $\left(\dfrac{1}{g}\right)'(1) = -1 \cdot \dfrac{1}{[g(1)]^2} \cdot g'(1) = -1 \cdot \dfrac{1}{3^2}(-3)$

B4. **(C)** $\left(\sqrt{f}\right)'(3) = \dfrac{1}{2}[f(3)]^{-\frac{1}{2}} \cdot f'(3) = \dfrac{1}{2}\left(10^{\frac{-1}{2}}\right) \cdot 4$

B5. **(C)** $\left(\dfrac{f}{g}\right)'(0) = \dfrac{g(0) \cdot f'(0) - f(0) \cdot g'(0)}{[g(0)]^2} = \dfrac{5(1) - 2(-4)}{5^2}$

B6. **(A)** $M'(1) = f'(g(1)) \cdot g'(1) = f'(3)g'(1) = 4(-3)$

B7. **(B)** $[f(x^3)]' = f'(x^3) \cdot 3x^2$, so $P'(1) = f'(1^3) \cdot 3 \cdot 1^2 = 2 \cdot 3$

B8. **(C)** $f(S(x)) = x$ implies that $f'(S(x)) \cdot S'(x) = 1$, so
$$S'(3) = \dfrac{1}{f'(S(3))} = \dfrac{1}{f'(f^{-1}(3))} = \dfrac{1}{f'(1)}$$

B9. **(D)** Since $g'(a)$ exists, g is differentiable and thus continuous; $g'(a) > 0$.

B10. **(B)** Near a vertical asymptote the slopes must approach $\pm\infty$.

B11. **(A)** There is only one horizontal tangent.

B12. **(C)** Use the symmetric difference quotient; then
$$f'(1.5) \approx \dfrac{f(1.6) - f(1.4)}{1.6 - 1.4} = \dfrac{8}{0.2}$$

B13. **(D)** Since the water level rises more slowly as the cone fills, the rate of depth change is decreasing, as in (B) and (D). However, at every instant the portion of the cone containing water is similar to the entire cone; the volume is proportional to the cube of the depth of the water. The rate of change of depth (the derivative) is therefore not linear, as in (B).

B14. **(C)** The only horizontal tangent is at $x = 4$. Note that $f'(1)$ does not exist.

B15. **(C)** The graph has corners at $x = 1$ and $x = 2$.

B16. **(B)** Consider triangle ABC: $AB = 1$; radius $AC = 2$; thus, $BC = \sqrt{3}$ and AC has $m = -\sqrt{3}$. The tangent line is perpendicular to the radius.

B17. **(D)** The graph of $y = x + \cos x$ is shown in window $[-5,5] \times [-6,6]$. The average rate of change is represented by the slope of secant segment $\overline{AB}$. There appear to be 3 points at which tangent lines are parallel to $\overline{AB}$.

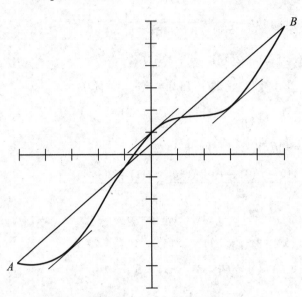

B18. **(C)** $f'(2) \approx \dfrac{f(2) - f(1.98)}{2 - 1.98} = \dfrac{4.00 - 4.10}{0.02}$

B19. **(A)** Since an estimate of the answer for Question B18 is $f'(2) \approx -5$, then
$(f^{-1})'(4) = \dfrac{1}{f'(2)} \approx \dfrac{1}{-5} = -0.2$.

B20. **(D)** f satisfies Rolle's Theorem on $[2,10]$.

B21. **(C)** The diagrams show secant lines (whose slope is the difference quotient) with greater slopes than the tangent line. In both cases, f is concave upward.

B22. **(B)** Sketch the graph of $f(x) = 1 - |x|$; note that $f(-1) = f(1) = 0$ and that f is continuous on $[-1,1]$. Only (B) holds.

B23. **(D)** $(f^{-1})'(2) = \dfrac{1}{f'(1)} = \dfrac{1}{3}$

B24. **(D)** The given limit is the derivative of $g(x)$ at $x = 0$.

B25. **(B)** The tangent line appears to contain the points $(0, -4.8)$ and $(3, -2.4)$. Its slope is approximately $\dfrac{-2.4 - (-4.8)}{3 - 0} = \dfrac{2.4}{3} = 0.8$.

B26. **(C)** $f'(x)$ is least at the point of inflection of the curve, at about 0.7.

B27. **(B)** $\dfrac{5^{2.03} - 5^{1.97}}{2.03 - 1.97} \approx 40.25158$

B28. **(B)** By calculator, $f'(0) = 1.386294805$ and $\dfrac{4^{0.08} - 4^{-0.08}}{0.16} = 1.3891\ldots.$

B29. **(D)** Now $\dfrac{4^{0.001} - 4^{-0.001}}{0.002} = 1.386294805.$

B30. **(B)** Note that any line determined by two points equidistant from the origin will necessarily be horizontal.

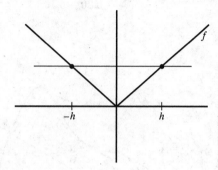

Therefore, the symmetric difference quotient yields:

$$\frac{f(0 + 0.001) - f(0 - 0.001)}{2(0.001)} = \frac{0.001 - 0.001}{0.002} = 0$$

B31. **(C)** Note that $\dfrac{d}{dx} f(h(x)) = f'(h(x)) \cdot h'(x) = g(h(x)) \cdot h'(x) = g(\sin x) \cdot \cos x.$

B32. **(C)** Since $f(x) = 3^x - x^3$, then $f'(x) = 3^x \ln 3 - 3x^2$. Furthermore, f is continuous on $[0,3]$ and f' is differentiable on $(0,3)$, so the MVT applies. We therefore seek c such that $f'(c) = \dfrac{f(3) - f(0)}{3} = -\dfrac{1}{3}$. Solving $3^x \ln 3 - 3x^2 = -\dfrac{1}{3}$ with a calculator, we find that c may be either 1.244 or 2.727. These values are the x-coordinates of points on the graph of $f(x)$ at which the tangents are parallel to the secant through points $(0,1)$ and $(3,0)$ on the curve.

B33. **(A)** The line segment passes through $(1,-3)$ and $(2,-4)$.

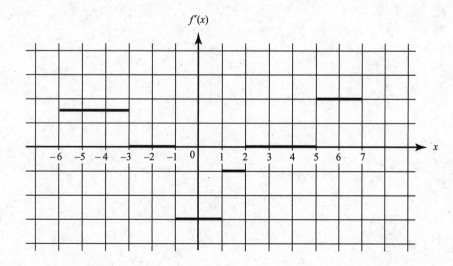

Use the graph of $f'(x)$, shown above, for Questions B34–B37.

B34. **(D)** $f'(x) = 0$ when the slope of $f(x)$ is 0—that is, when the graph of f is a horizontal segment.

B35. **(D)** The graph of $f'(x)$ jumps at each corner of the graph of $f(x)$, namely, at x equal to -3, -1, 1, 2, and 5.

B36. **(D)** On the interval $(-6, -3)$, $f(x) = \frac{3}{2}(x + 5)$.

B37. **(B)** Verify that all choices but (B) are true. The graph of $f'(x)$ has five (not four) jump discontinuities.

B38. **(C)** The best approximation to $f'(0.10)$ is $\dfrac{f(0.20) - f(0.10)}{0.20 - 0.10}$.

B39. **(D)**

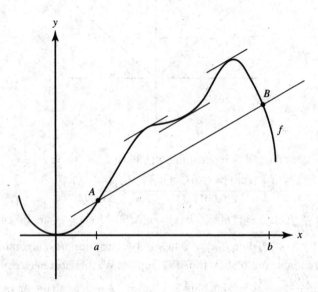

The average rate of change is represented by the slope of secant segment $\overline{AB}$. There appear to be 3 points at which the tangent lines are parallel to $\overline{AB}$.

4

Applications of
Differential Calculus

Learning Objectives

In this chapter, you will review how to use derivatives to:

→ Find slopes of curves and equations of tangent lines
→ Find a function's maxima, minima, and points of inflection
→ Describe where the graph of a function is increasing, decreasing, concave upward, and concave downward
→ Analyze motion along a line
→ Create local linear approximations
→ Work with related rates

In addition, BC Calculus students will review how to:

→ Find the slope of parametric and polar curves
→ Use vectors to analyze motion along parametrically defined curves

A. Slope; Critical Points

If the derivative of $y = f(x)$ exists at $P(x_1, y_1)$, then the *slope* of the curve at P (which is defined to be the slope of the tangent to the curve at P) is $f'(x_1)$, the derivative of $f(x)$ at $x = x_1$.

Slope of a curve

Any c in the domain of f such that either $f'(c) = 0$ or $f'(c)$ is undefined is called a *critical point* or *critical value* of f. If f has a derivative everywhere, we find the critical points by solving the equation $f'(x) = 0$.

Critical point

> **Example 1**

For $f(x) = 4x^3 - 6x^2 - 8$, what are the critical points?

✓ **Solution**

$$f'(x) = 12x^2 - 12x = 12x(x - 1)$$

which equals zero if x is 0 or 1. Thus, 0 and 1 are critical points.

> **Example 2**

Find any critical points of $f(x) = 3x^3 + 2x$.

✓ **Solution**

$$f'(x) = 9x^2 + 2$$

Since $f'(x)$ never equals zero (indeed, it is always positive), f has no critical values.

> **Example 3**

Find any critical points of $f(x) = (x - 1)^{1/3}$.

✓ **Solution**

$$f'(x) = \frac{1}{3(x-1)^{2/3}}$$

Although f' is never zero, $x = 1$ is a critical value of f because f' does not exist at $x = 1$.

Average and Instantaneous Rates of Change

Both average and instantaneous rates of change were defined in Chapter 3. If as x varies from a to $a + h$, the function f varies from $f(a)$ to $f(a + h)$, then we know that the difference quotient

$$\frac{f(a + h) - f(a)}{h}$$

is the average rate of change of f over the interval from a to $a + h$.

Thus, the *average velocity* of a moving object over some time interval is the change in distance divided by the change in time. The *average rate of growth* of a colony of fruit flies over some interval of time is the change in size of the colony divided by the time elapsed. The *average rate of change* in the profit of a company on some gadget with respect to production is the change in profit divided by the change in the number of gadgets produced.

The (instantaneous) rate of change of f at a, or the derivative of f at a, is the limit of the average rate of change as $h \to 0$:

$$f'(a) = \lim_{h \to 0} \frac{f(a + h) - f(a)}{h}$$

On the graph of $y = f(x)$, the rate at which the y-coordinate changes with respect to the x-coordinate is $f'(x)$, the slope of the curve. The rate at which $s(t)$, the distance traveled by a particle in t seconds, changes with respect to time is $s'(t)$, the velocity of the particle; the rate at which a manufacturer's profit $P(x)$ changes relative to the production level x is $P'(x)$.

> **Example 4**

Let $G = 400(15 - t)^2$ be the number of gallons of water in a cistern t minutes after an outlet pipe is opened. Find the average rate of drainage during the first 5 minutes and the rate at which the water is running out at the end of 5 minutes.

✓ **Solution**

The average rate of change during the first 5 minutes equals

$$\frac{G(5) - G(0)}{5} = \frac{400 \cdot 100 - 400 \cdot 225}{5} = -10{,}000 \text{ gal/min}$$

The average rate of drainage during the first 5 minutes is 10,000 gal/min.

The instantaneous rate of change at $t = 5$ is $G'(5)$. Since

$$G'(t) = -800(15 - t)$$

$G'(5) = -800(10) = -8000$ gal/min. Thus the rate of drainage at the end of 5 minutes is 8000 gal/min.

B. Tangents to a Curve

An *equation of the tangent* to the curve $y = f(x)$ at point $P(x_1, y_1)$ is

$$y - y_1 = f'(x_1)(x - x_1)$$

Tangent to
a curve

If the tangent to a curve is horizontal at a point, then the derivative at the point is 0. If the tangent is vertical at a point, then the derivative does not exist at the point.

Tangents to Parametrically Defined Curves

BC ONLY

If the curve is defined parametrically, say in terms of t (as in Chapter 1, page 73), then we obtain the slope at any point from the parametric equations. We then evaluate the slope and the x- and y-coordinates by replacing t by the value specified in the question (see Example 9, page 162).

❯ **Example 5**

Find an equation of the tangent to the curve of $f(x) = x^3 - 3x^2$ at the point $(1, -2)$.

✓ **Solution**

Since $f'(x) = 3x^2 - 6x$ and $f'(1) = -3$, an equation of the tangent is

$$y + 2 = -3(x - 1) \qquad \text{or} \qquad y + 3x = 1$$

❯ **Example 6**

Find an equation of the tangent to $x^2y - x = y^3 - 8$ at the point where $x = 0$.

✓ **Solution**

Here we differentiate implicitly to get $\dfrac{dy}{dx} = \dfrac{1 - 2xy}{x^2 - 3y^2}$.

Since $y = 2$ when $x = 0$ and the slope at this point is $\dfrac{1-0}{0-12} = -\dfrac{1}{12}$, an equation of the tangent is

$$y - 2 = -\frac{1}{12}x \quad \text{or} \quad 12y + x = 24$$

> **Example 7**

Find the coordinates of any point on the curve of $y^2 - 4xy = x^2 + 5$ for which the tangent is horizontal.

✓ **Solution**

Since $\dfrac{dy}{dx} = \dfrac{x + 2y}{y - 2x}$ and the tangent is horizontal when $\dfrac{dy}{dx} = 0$, then $x = -2y$. If we substitute this in the equation of the curve, we get

$$y^2 - 4y(-2y) = 4y^2 + 5$$
$$5y^2 = 5$$

Thus $y = \pm 1$. The points, then, are $(2,-1)$ and $(-2,1)$.

> **Example 8**

Find the x-coordinate of any point on the curve of $y = \sin^2(x + 1)$ for which the tangent is parallel to the line $3x - 3y - 5 = 0$.

✓ **Solution**

Since $\dfrac{dy}{dx} = 2\sin(x + 1)\cos(x + 1) = \sin 2(x + 1)$ and since the given line has slope 1, we seek x such that $\sin 2(x + 1) = 1$. Then

$$2(x + 1) = \frac{\pi}{2} + 2n\pi \quad (n \text{ is an integer})$$

or

$$x + 1 = \frac{\pi}{4} + n\pi \quad \text{and} \quad x = \frac{\pi}{4} + n\pi - 1$$

BC ONLY

> **Example 9**

Find an equation of the tangent to $F(t) = \left(\cos t, 2\sin^2 t\right)$ at the point where $t = \dfrac{\pi}{3}$.

✓ **Solution**

Since $\dfrac{dx}{dt} = -\sin t$ and $\dfrac{dy}{dt} = 4\sin t \cos t$, we see that

$$\frac{dy}{dx} = \frac{4\sin t \cos t}{-\sin t} = -4\cos t$$

At $t = \frac{\pi}{3}$, $x = \frac{1}{2}$, $y = 2\left(\frac{\sqrt{3}}{2}\right)^2 = \frac{3}{2}$, and $\frac{dy}{dx} = -2$. An equation of the tangent is

$$y - \frac{3}{2} = -2\left(x - \frac{1}{2}\right) \qquad \text{or} \qquad 4x + 2y = 5$$

C. Increasing and Decreasing Functions

Case I. Functions with Continuous Derivatives

A function $y = f(x)$ is *increasing* on an interval if, for all $x = a$ and $x = b$ in the interval such that $a < b$, $f(a) < f(b)$. Likewise, a function $y = f(x)$ is *decreasing* on an interval if, for all $x = a$ and $x = b$ in the interval such that $a < b$, $f(a) > f(b)$.

Increasing
Decreasing

Analyzing the signs of the first derivative, $f'(x)$, will allow us to determine intervals over which $f(x)$ is increasing or decreasing, that is, over which the graph is rising or falling. For intervals where $f(x)$ is increasing, $f'(x) \geq 0$, and for intervals where $f(x)$ is decreasing, $f'(x) \leq 0$.

> **Example 10**

For what values of x is $f(x) = x^4 - 4x^3$ increasing and for what values is it decreasing?

✓ Solution

$$f'(x) = 4x^3 - 12x^2 = 4x^2(x - 3)$$

With critical values at $x = 0$ and $x = 3$, we analyze the signs of f' in three intervals:

Based on the sign chart, we can see that $f'(x) \leq 0$ on $(-\infty, 3]$ and $f'(x) \geq 0$ on $[3, \infty)$. Therefore, $f(x)$ is increasing for $x \geq 3$, and $f(x)$ is decreasing for $x \leq 3$.

Note that $x = 0$ is included in the interval where f is decreasing even though $f'(0) = 0$. The graph is given as Figure 4.5 on page 169, and it is clear that indeed the graph is decreasing for $x \leq 3$.

Also, note that $x = 3$ is included in both the increasing and decreasing intervals. Given the definition of increasing and decreasing, it is valid that an x-value can be in both an increasing and decreasing interval, provided it is the endpoint of each as it is in this example. Looking at the graph of $f(x)$, at $x = 3$, the graph transitions from decreasing to increasing. Thus, $f(3)$ is the lowest value on the interval $x \leq 3$ (i.e., it is less than all function values in that decreasing interval). Also, $f(3)$ is the lowest value on the interval $x \geq 3$ (i.e., it is less than all function values in that increasing interval). Those two properties of $f(3)$ allow $x = 3$ to be included in both intervals.

Case II. Functions Whose Derivatives Have Discontinuities

Here we proceed as in Case I but also consider intervals bounded by any points of discontinuity of f or f'.

> **Example 11**

For what values of x is $f(x) = \dfrac{1}{x+1}$ increasing and for what values is it decreasing?

✓ **Solution**

$$f'(x) = -\frac{1}{(x+1)^2}$$

We note that neither f nor f' is defined at $x = -1$; furthermore, $f'(x)$ never equals zero. We need therefore examine only the signs of $f'(x)$ when $x < -1$ and when $x > -1$.

When $x < -1$, $f'(x) < 0$; when $x > -1$, $f'(x) < 0$. Therefore, f decreases on both intervals. The curve is a hyperbola whose center is at the point $(-1,0)$.

D. Maximum, Minimum, Concavity, and Inflection Points: Definitions

Local (relative) max/min

The graph of $y = f(x)$ has a *local maximum* (or *relative maximum*) at $x = c$ if $f(c) \geq f(x)$ for all x in the immediate neighborhood of $x = c$, that is from a little less than c to a little more than c. The graph of $y = f(x)$ has a *local minimum* (or *relative minimum*) at $x = c$ if $f(c) \leq f(x)$ for all x in the immediate neighborhood of $x = c$. If the graph has a local maximum at $x = c$, then the graph changes from increasing to decreasing at $x = c$. If the graph has a local minimum at $x = c$, then the graph changes from decreasing to increasing at $x = c$. If a function is differentiable on the closed interval $[a,b]$ and has a local maximum or minimum at $x = c$ ($a < c < b$), then $f'(c) = 0$. The converse of this statement is not true.

If $f(c)$ is either a local maximum or a local minimum, then $f(c)$ is called a *local extreme value* or *local extremum*. (The plural of *extremum* is *extrema*.)

Global (absolute) max/min

The *global maximum* (or *absolute maximum*) of a function on a closed interval $[a,b]$ occurs at $x = c$ if $f(c) \geq f(x)$ for all x on $[a,b]$. The *global minimum* (or *absolute minimum*) of a function on a closed interval $[a,b]$ occurs at $x = c$ if $f(c) \leq f(x)$ for all x on $[a,b]$. Note that this allows for the maximum and/or the minimum to occur at an endpoint.

Concavity

A graph is *concave upward* on an interval (a,b) if the graph lies above the tangent lines at each point in the interval (a,b). A graph is *concave downward* on an interval (a,b) if the graph lies below the tangent lines at each point in the interval (a,b). If $y'' > 0$ at every point in the interval (a,b), the graph is concave up. If $y'' < 0$ at every point in the interval (a,b), the graph is concave down. In Figure 4.1, the curves sketched in (a) and (b) are concave downward, while in (c) and (d) they are concave upward.

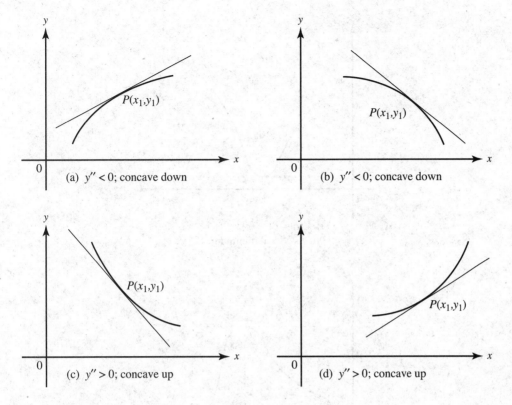

(a) $y'' < 0$; concave down

(b) $y'' < 0$; concave down

(c) $y'' > 0$; concave up

(d) $y'' > 0$; concave up

Figure 4.1

A *point of inflection* is a point where the curve changes its concavity from upward to downward or from downward to upward. See Section I, page 177, for a table relating a function and its derivatives. It tells how to graph the derivatives of f, given the graph of f. On pages 179 and 253 we graph f, given the graph of f'.

Point of inflection

E. Maximum, Minimum, and Inflection Points: Curve Sketching

Case I. Functions That Are Everywhere Differentiable

The following procedure is suggested to determine any maximum, minimum, or inflection point of a curve and to sketch the curve.

1. Find y' and y''.
2. Find all critical points of y, that is, all x for which $y' = 0$. At each of these x's the tangent to the curve is horizontal.
3. Let c be a number for which y' is 0; investigate the sign of y'' at c. If $y''(c) > 0$, then c yields a local minimum; if $y''(c) < 0$, then c yields a local maximum. This procedure is known as the *Second Derivative Test* (for extrema). See Figure 4.2. If $y''(c) = 0$, the Second Derivative Test fails and we must use the test in step (4), which follows.

Second Derivative Test

(a) $y'(c) = 0;\ y''(c) > 0;$
c yields a local minimum.

(b) $y'(c) = 0;\ y''(c) < 0;$
c yields a local maximum.

Figure 4.2

4. If $y'(c) = 0$ and $y''(c) = 0$, investigate the signs of y' as x increases through c. If
$y'(x) > 0$ for x's (just) less than c but $y'(x) < 0$ for x's (just) greater than c, then the situation is that indicated in (a) of Figure 4.3, where the tangent lines have been sketched as x increases through c; here c yields a local maximum. If the situation is reversed and the sign of y' changes from $-$ to $+$ as x increases through c, then c yields a local minimum. In Figure 4.3, (b) shows this case. The schematic sign pattern of y', $+\ 0\ -$ or $-\ 0\ +$, describes each situation completely. If y' does not change sign as x increases through c, then c yields neither a local maximum nor a local minimum. Two examples of this appear in (c) and (d) of Figure 4.3.

(a) y' changes from $+$ to 0 to $-$;
   c yields a maximum.

(b) y' changes from $-$ to 0 to $+$;
   c yields a minimum.

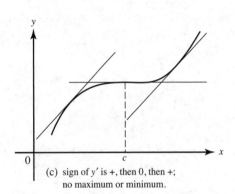

(c) sign of y' is $+$, then 0, then $+$;
   no maximum or minimum.

(d) sign of y' is $-$, then 0, then $-$;
   no maximum or minimum.

Figure 4.3

5. Find all x's for which $y'' = 0$; these are x-values of possible points of inflection. If c is such an x and the sign of y'' changes (from $+$ to $-$ or from $-$ to $+$) as x increases through c, then c is the x-coordinate of a point of inflection. If the signs do not change, then c does not yield a point of inflection.

The crucial points found as indicated in (1) through (5) above should be plotted along with the intercepts. Care should be exercised to ensure that the tangent to the curve is horizontal whenever $\dfrac{dy}{dx} = 0$ and that the curve has the proper concavity.

> **Example 12**

Find any maximum, minimum, or inflection points on the graph of $f(x) = x^3 - 5x^2 + 3x + 6$, and sketch the curve.

✓ **Solution**

For the steps listed previously:

1. Here $f'(x) = 3x^2 - 10x + 3$ and $f''(x) = 6x - 10$.

2. $f'(x) = (3x - 1)(x - 3)$, which is zero when $x = \dfrac{1}{3}$ or 3.

3. Since $f'\left(\dfrac{1}{3}\right) = 0$ and $f''\left(\dfrac{1}{3}\right) < 0$, we know that the point $\left(\dfrac{1}{3}, f\left(\dfrac{1}{3}\right)\right)$ is a local maximum; since $f'(3) = 0$ and $f''(3) > 0$, the point $\left(3, f(3)\right)$ is a local minimum. Thus, $\left(\dfrac{1}{3}, \dfrac{175}{27}\right)$ is a local maximum and $(3, -3)$ a local minimum.

4. This step is unnecessary for this problem.

5. $f''(x) = 0$ when $x = \dfrac{5}{3}$, and f'' changes from negative to positive as x increases through $\dfrac{5}{3}$, so the graph of f has an inflection point. See Figure 4.4.

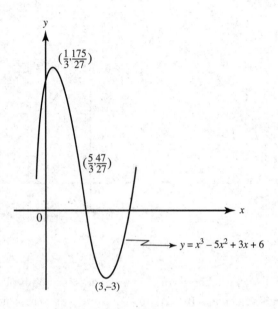

Figure 4.4

Verify the graph and information obtained above on your graphing calculator.

> **Example 13**

Sketch the graph of $f(x) = x^4 - 4x^3$.

✓ **Solution**

1. $f'(x) = 4x^3 - 12x^2$ and $f''(x) = 12x^2 - 24x$.

2. $f'(x) = 4x^2(x - 3)$, which is zero when $x = 0$ or $x = 3$.

3. Since $f''(x) = 12x(x - 2)$ and $f''(3) > 0$ with $f'(3) = 0$, the point $(3, -27)$ is a local minimum. Since $f''(0) = 0$, the Second Derivative Test fails to tell us whether $x = 0$ yields a local maximum or a local minimum.

4. Since $f'(x)$ does not change sign as x increases through 0, the point $(0, 0)$ yields neither a local maximum nor a local minimum.

5. $f''(x) = 0$ when x is 0 or 2; f'' changes signs as x increases through 0 ($+$ to $-$) and also as x increases through 2 ($-$ to $+$). Thus both $(0, 0)$ and $(2, -16)$ are inflection points of the curve.

The curve is sketched in Figure 4.5 on page 169.

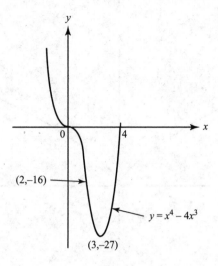

Figure 4.5

Verify the preceding on your calculator.

Case II. Functions Whose Derivatives May Not Exist Everywhere

If there are values of x for which a first or second derivative does not exist, we consider those values separately, recalling that a local maximum or minimum point is one of transition between intervals of rise and fall and that an inflection point is one of transition between intervals of upward and downward concavity.

> **Example 14**
_____

Sketch the graph of $y = x^{2/3}$.

✓ **Solution**
_____

$$\frac{dy}{dx} = \frac{2}{3x^{1/3}} \quad \text{and} \quad \frac{d^2y}{dx^2} = -\frac{2}{9x^{4/3}}$$

Neither derivative is zero anywhere; both derivatives fail to exist when $x = 0$. As x increases through 0, $\frac{dy}{dx}$ changes from $-$ to $+$; (0,0) is therefore a minimum. Note that the tangent is vertical at the origin, and that since $\frac{d^2y}{dx^2}$ is negative everywhere except at 0, the curve is everywhere concave down. See Figure 4.6.

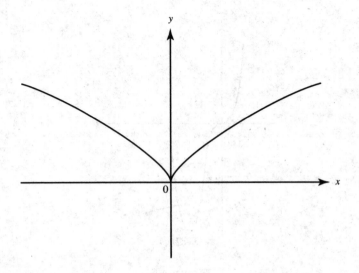

Figure 4.6

> **Example 15**

Sketch the graph of $y = x^{1/3}$.

✓ **Solution**

$$\frac{dy}{dx} = \frac{1}{3x^{2/3}} \quad \text{and} \quad \frac{d^2y}{dx^2} = -\frac{2}{9x^{5/3}}$$

As in Example 14, neither derivative ever equals zero and both fail to exist when $x = 0$. Here, however, as x increases through 0, $\dfrac{dy}{dx}$ does not change sign.

Since $\dfrac{dy}{dx}$ is positive for all x except 0, the curve rises for all x and can have neither maximum nor minimum points. The tangent is again vertical at the origin. Note here that $\dfrac{d^2y}{dx^2}$ does change sign (from + to −) as x increases through 0, so that (0,0) is a point of inflection of the curve. See Figure 4.7.

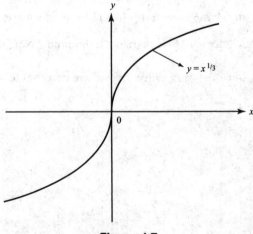

$y = x^{1/3}$

Figure 4.7

Verify the graph on your calculator.

F. Global Maximum or Minimum

Case I. Differentiable Functions

If a function f is differentiable on a closed interval $a \leq x \leq b$, then f is also continuous on the closed interval $[a,b]$ and we know from the Extreme Value Theorem (page 104) that f attains both a (global) maximum and a (global) minimum on $[a,b]$. To find these, we solve the equation $f'(x) = 0$ for critical points on the interval $[a,b]$, then evaluate f at each of those and also at $x = a$ and $x = b$. The largest value of f obtained is the global max, and the smallest the global min. This procedure is called the *Closed Interval Test*, or the *Candidates Test*.

Closed Interval Test (Candidates Test)

> **Example 16**

Find the global max and global min of f on (a) $-2 \leq x \leq 3$, and (b) $0 \leq x \leq 3$, if $f(x) = 2x^3 - 3x^2 - 12x$.

✓ **Solutions**

(a) $f'(x) = 6x^2 - 6x - 12 = 6(x + 1)(x - 2)$, which equals zero if $x = -1$ or 2. Since $f(-2) = -4, f(-1) = 7, f(2) = -20$, and $f(3) = -9$, the global max of f occurs at $x = -1$ and equals 7, and the global min of f occurs at $x = 2$ and equals -20.

(b) Only the critical value 2 lies in $[0,3]$. We now evaluate f at 0, 2, and 3. Since $f(0) = 0$, $f(2) = -20$, and $f(3) = -9$, the global max of f equals 0 and the global min equals -20.

Case II. Functions That Are Not Everywhere Differentiable

We proceed as for Case I but now evaluate f also at each point in a given interval for which f is defined but for which f' does not exist.

> **Example 17**

The absolute-value function $f(x) = |x|$ is defined for all real x, but $f'(x)$ does not exist at $x = 0$. Since $f'(x) = -1$ if $x < 0$, but $f'(x) = 1$ if $x > 0$, we see that f has a global min at $x = 0$.

> **Example 18**

The function $f(x) = \frac{1}{x}$ has neither a global max nor a global min on *any* interval that contains zero (see Figure 2.4, page 92). However, it does attain both a global max and a global min on every closed interval that does not contain zero. For instance, on $[2,5]$ the global max of f is $\frac{1}{2}$, the global min $\frac{1}{5}$.

G. Further Aids in Sketching

It is often very helpful to investigate one or more of the following before sketching the graph of a function or of an equation:

1. Intercepts. Set $x = 0$ and $y = 0$ to find any y- and x-intercepts, respectively.
2. Symmetry. Let the point (x,y) satisfy an equation. Then its graph is symmetric about

> the x-axis if $(x,-y)$ also satisfies the equation
> the y-axis if $(-x,y)$ also satisfies the equation
> the origin if $(-x,-y)$ also satisfies the equation

3. Asymptotes. The line $y = b$ is a horizontal asymptote of the graph of a function f if either $\lim\limits_{x \to \infty} f(x) = b$ or $\lim\limits_{x \to -\infty} f(x) = b$. If $f(x) = \dfrac{P(x)}{Q(x)}$, inspect the degrees of $P(x)$ and $Q(x)$, then use the Rational Function Theorem, page 98. The line $x = c$ is a vertical asymptote of the rational function $\dfrac{P(x)}{Q(x)}$ if $Q(c) = 0$ but $P(c) \neq 0$.

4. Points of discontinuity. Identify points not in the domain of a function, particularly where the denominator equals zero.

> ## Example 19

Sketch the graph of $y = \dfrac{2x + 1}{x - 1}$.

✓ Solution

If $x = 0$, then $y = -1$. Also, $y = 0$ when the numerator equals zero, which is when $x = -\dfrac{1}{2}$. A check shows that the graph does not possess any of the symmetries described previously. Since $y \to 2$ as $x \to \pm\infty$, $y = 2$ is a horizontal asymptote; also, $x = 1$ is a vertical asymptote. The function is defined for all reals except $x = 1$; the latter is the only point of discontinuity.

We find derivatives: $y' = -\dfrac{3}{(x - 1)^2}$ and $y'' = \dfrac{6}{(x - 1)^3}$

From y' we see that the function decreases everywhere (except at $x = 1$), and from y'' that the curve is concave down if $x < 1$, up if $x > 1$. See Figure 4.8.

Figure 4.8

Verify the preceding on your calculator using $[-4, 4] \times [-4, 8]$.

> **Example 20**

Describe any symmetries of the graphs of

(a) $3y^2 + x = 2$ (b) $y = x + \dfrac{1}{x}$ (c) $x^2 - 3y^2 = 27$

✓ **Solutions**

(a)   Suppose point (x,y) is on this graph. Then so is point $(x,-y)$ since $3(-y)^2 + x = 2$ is equivalent to $3y^2 + x = 2$. Then (a) is symmetric about the x-axis.

(b)   Note that point $(-x,-y)$ satisfies the equation if point (x,y) does:

$$(-y) = (-x) + \frac{1}{(-x)} \leftrightarrow y = x + \frac{1}{x}$$

Therefore the graph of this function is symmetric about the origin.

(c)   This graph is symmetric about the x-axis, the y-axis, and the origin. It is easy to see that, if point (x,y) satisfies the equation, so do points $(x,-y)$, $(-x,y)$, and $(-x,-y)$.

H. Optimization: Problems Involving Maxima and Minima

The techniques described above can be applied to problems in which a function is to be maximized (or minimized). Often it helps to draw a figure. If y, the quantity to be maximized (or minimized), can be expressed explicitly in terms of x, then the procedure outlined above can be used. If the domain of y is restricted to some closed interval, one should always check the endpoints of this interval so as not to overlook possible extrema. Often, implicit differentiation, sometimes of two or more equations, is indicated.

> **Example 21**

The region in the first quadrant bounded by the curves of $y^2 = x$ and $y = x$ is rotated about the y-axis to form a solid. Find the area of the largest cross section of this solid that is perpendicular to the y-axis.

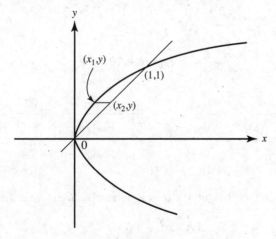

Figure 4.9

✓ **Solution**

See Figure 4.9. The curves intersect at the origin and at $(1,1)$, so $0 < y < 1$. A cross section of the solid is a ring whose area A is the difference between the areas of two circles, one with radius x_2, the other with radius x_1. Thus

$$A = \pi x_2^2 - \pi x_1^2 = \pi(y^2 - y^4); \quad \frac{dA}{dy} = \pi(2y - 4y^3) = 2\pi y\left(1 - 2y^2\right)$$

The only relevant zero of the first derivative is $y = \dfrac{1}{\sqrt{2}}$. There the area A is

$$A = \pi\left(\frac{1}{2} - \frac{1}{4}\right) = \frac{\pi}{4}$$

Note that $\dfrac{d^2A}{dy^2} = \pi\left(2 - 12y^2\right)$ and that this is negative when $y = \dfrac{1}{\sqrt{2}}$, assuring a maximum there. Note further that A equals zero at each endpoint of the interval $[0,1]$ so that $\frac{\pi}{4}$ is the global maximum area.

> **Example 22**

The volume of a cylinder equals V cubic inches, where V is a constant. Find the proportions of the cylinder that minimize the total surface area. See Figure 4.10.

Figure 4.10

✓ **Solution**

We know that the volume is

$$V = \pi r^2 h \tag{1}$$

where r is the radius and h the height. We seek to minimize S, the total surface area, where

$$S = 2\pi r^2 + 2\pi rh \tag{2}$$

Solving (1) for h, we have $h = \dfrac{V}{\pi r^2}$, which we substitute in (2):

$$S = 2\pi r^2 + 2\pi r\frac{V}{\pi r^2} = 2\pi r^2 + \frac{2V}{r} \tag{3}$$

Differentiating (3) with respect to r yields

$$\frac{dS}{dr} = 4\pi r - \frac{2V}{r^2}$$

Now we set $\dfrac{dS}{dr}$ equal to zero to determine the conditions that make S a minimum:

$$4\pi r - \frac{2V}{r^2} = 0$$

$$4\pi r = \frac{2V}{r^2}$$

$$4\pi r = \frac{2(\pi r^2 h)}{r^2}$$

$$2r = h$$

The total surface area of a cylinder of fixed volume is thus a minimum when its height equals its diameter.

(Note that we need not concern ourselves with the possibility that the value of r that renders $\dfrac{dS}{dr}$ equal to zero will produce a maximum surface area rather than a minimum one. With V fixed, we can choose r and h so as to make S as large as we like.)

> Example 23

A charter bus company advertises a trip for a group as follows: At least 20 people must sign up. The cost when 20 participate is $80 per person. The price will drop by $2 per ticket for each member of the traveling group in excess of 20. If the bus can accommodate 28 people, how many participants will maximize the company's revenue?

✓ Solution

Let x denote the number who sign up in excess of 20. Then $0 \le x \le 8$. The total number who agree to participate is $(20 + x)$, and the price per ticket is $(80 - 2x)$ dollars. Then the revenue R, in dollars, is

$$R = (20 + x)(80 - 2x)$$

$$R'(x) = (20 + x)(-2) + (80 - 2x) \cdot 1$$

$$= 40 - 4x$$

$R'(x)$ is zero if $x = 10$. Although $x = 10$ yields maximum R—note that $R''(x) = -4$ and is always negative—this value of x is not within the restricted interval. We therefore evaluate R at the endpoints 0 and 8: $R(0) = 1600$ and $R(8) = 28 \cdot 64 = 1792$; 28 participants will maximize revenue.

> Example 24

A utilities company wants to deliver gas from a source S to a plant P located across a straight river 3 miles wide, then downstream 5 miles, as shown in Figure 4.11. It costs $4 per foot to lay the pipe in the river but only $2 per foot to lay it on land.

(a)   Express the cost of laying the pipe in terms of u.

(b)   How can the pipe be laid most economically?

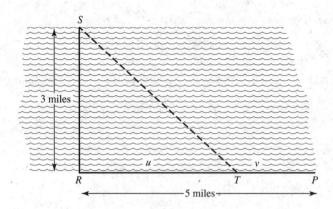

Figure 4.11

✓ Solutions

(a)   Note that the problem "allows" us to (1) lay all of the pipe in the river, along a line from S to P; (2) lay pipe along SR, in the river, then along RP on land; or (3) lay some pipe in the river, say, along ST, and lay the rest on land along TP. When T coincides with P, we have case (1), with $v = 0$; when T coincides with R, we have case (2), with $u = 0$. Case (3) includes both (1) and (2).

In any event, we need to find the lengths of pipe needed (that is, the distances involved); then we must figure out the cost.

In terms of u:

		In the River	On Land
Distances:			
	miles	$ST = \sqrt{9 + u^2}$	$TP = v = 5 - u$
	feet	$ST = 5280\sqrt{9 + u^2}$	$TP = 5280(5 - u)$
Costs (dollars):		$4(5280)\sqrt{9 + u^2}$	$2[5280(5 - u)]$

If $C(u)$ is the total cost,

$$C(u) = 21{,}120\sqrt{9 + u^2} + 10{,}560(5 - u)$$
$$= 10{,}560\left(2\sqrt{9 + u^2} + 5 - u\right)$$

(b)   We now minimize $C(u)$:

$$C'(u) = 10{,}560\left(2 \cdot \frac{1}{2}\frac{2u}{\sqrt{9 + u^2}} - 1\right) = 10{,}560\left(\frac{2u}{\sqrt{9 + u^2}} - 1\right)$$

We now set $C'(u)$ equal to zero and solve for u:

$$\frac{2u^2}{\sqrt{9 + u^2}} - 1 = 0 \quad \rightarrow \quad \frac{2u}{\sqrt{9 + u^2}} = 1 \quad \rightarrow \quad \frac{4u^2}{\sqrt{9 + u^2}} = 1$$

where, in the last step, we squared both sides; then

$$4u^2 = 9 + u^2 \qquad 3u^2 = 9 \qquad u^2 = 3 \qquad u = \sqrt{3}$$

where we discard $u = -\sqrt{3}$ as meaningless for this problem.

The domain of $C(u)$ is [0,5] and C is continuous on [0,5]. Since

$$C(0) = 10{,}560\left(2\sqrt{9} + 5\right) = \$116{,}160$$

$$C(5) = 10{,}560\left(2\sqrt{34}\right) \approx \$123{,}150$$

$$C\left(\sqrt{3}\right) = 10{,}560\left(2\sqrt{12} + 5 - \sqrt{3}\right) = \$107{,}671$$

So $u = \sqrt{3}$ yields minimum cost. Thus, the pipe can be laid most economically if some of it is laid in the river from the source S to a point T that is $\sqrt{3}$ miles toward the plant P from R and the rest is laid along the road from T to P.

I. Relating a Function and Its Derivatives Graphically

The following table shows the characteristics of a function f and their implications for f's derivatives. These are crucial in obtaining one graph from another. The table can be used reading from left to right or from right to left.

Note that the slope at $x = c$ of any graph of a function is equal to the ordinate at c of the derivative of the function.

	f	**f'**	**f''**
ON AN INTERVAL	increasing	≥ 0	
	decreasing	≤ 0	
AT c	local maximum	$x < c$ $\quad$ $x = c$ $\quad$ $x > c$ $+$ $\qquad$ 0 $\qquad$ $-$ (f' is decreasing)	$f''(c) < 0$
	local minimum	$-$ $\qquad$ 0 $\qquad$ $+$ (f' is increasing)	$f''(c) > 0$
	neither local maximum nor local minimum	$+$ $\qquad$ 0 $\qquad$ $+$ $-$ $\qquad$ 0 $\qquad$ $-$ (f' does not change sign)	
AT c	point of inflection	$f'(c)$ is a minimum; f' changes from decreasing to increasing	$x < c$ $\quad$ $x = c$ $\quad$ $x > c$ $-$ $\qquad$ 0 $\qquad$ $+$
		$f'(c)$ is a maximum; f' changes from increasing to decreasing	$+$ $\qquad$ 0 $\qquad$ $-$
ON AN INTERVAL	concave up	f' is increasing	$f'' \geq 0$
	concave down	f' is decreasing	$f'' \leq 0$

If $f'(c)$ does not exist, check the signs of f' as x increases through c: plus-to-minus yields a local maximum; minus-to-plus yields a local minimum; no sign change means no maximum or minimum, but check the possibility of a point of inflection.

AN IMPORTANT NOTE:

Tables and number lines showing sign changes of the function and its derivatives can be very helpful in organizing all of this information. *Note, however, that the AP exam requires that students write sentences that describe the behavior of the function based on the sign of its derivative.*

❯ Example 25A

Given the graph of $f(x)$ shown in Figure 4.12, sketch $f'(x)$.

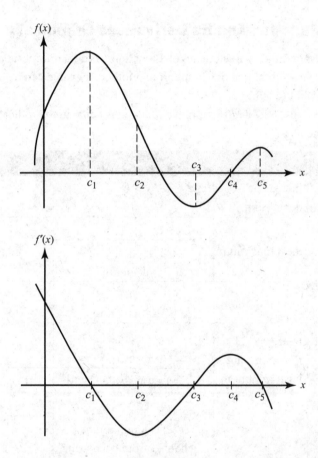

Figure 4.12

Point $x =$	Behavior of f	Behavior of f'
c_1	$f(c_1)$ is a local maximum	$f'(c_1) = 0$; f' changes sign from $+$ to $-$
c_2	c_2 is an inflection point of f; the graph of f changes concavity from down to up	f' changes from decreasing to increasing; $f''(c_2)$ is a local minimum
c_3	$f(c_3)$ is a local minimum	$f'(c_3) = 0$; f' changes sign from $-$ to $+$
c_4	c_4 is an inflection point of f; the graph of f changes concavity from up to down	f' changes from increasing to decreasing; $f''(c_4)$ is a local maximum
c_5	$f(c_5)$ is a local maximum	$f'(c_5) = 0$; f' changes sign from $+$ to $-$

> **Example 25B**

Given the graph of $f'(x)$ shown in Figure 4.13, sketch a possible graph of f.

Figure 4.13

✓ **Solution**

First, we note that $f'(-3)$ and $f'(2)$ are both 0. Thus the graph of f must have horizontal tangents at $x = -3$ and $x = 2$. Since $f'(x) < 0$ for $x < -3$, we see that f must be decreasing there. Below is a complete signs analysis of f', showing what it implies for the behavior of f.

f	dec	-3	inc	2	inc
f'	$-$		$+$		$+$

Because f' changes from negative to positive at $x = -3$, f must have a minimum there, but f has neither a minimum nor a maximum at $x = 2$.

We note next from the graph that f' is increasing for $x < -1$. This means that the derivative of f', f'', must be positive for $x < -1$ and that f is concave upward there. Analyzing the signs of f'' yields the following:

f	conc. upward	-1	conc. down	2	conc. upward
f'	inc		dec		inc
f''	$+$		$-$		$+$

We conclude that the graph of f has two points of inflection, because it changes concavity from upward to downward at $x = -1$ and back to upward at $x = 2$. We use the information obtained to sketch a possible graph of f, shown in Figure 4.14. Note that other graphs are possible; in fact, any vertical translation of this f will do!

Figure 4.14

J. Motion Along a Line

If a particle moves along a line according to the law $s = f(t)$, where s represents the position

Velocity

of the particle P on the line at time t, then the velocity v of P at time t is given by $\dfrac{ds}{dt}$ and its

Acceleration

acceleration a by $\dfrac{dv}{dt}$ or by $\dfrac{d^2s}{dt^2}$. The speed of the particle is $|v|$, the magnitude of v. If the line of

motion is directed positively to the right, then the motion of the particle P is subject to the fol-

Speed

lowing: At any instant,

1. if $v > 0$, then P is moving to the right (its position s is increasing); if $v < 0$, then P is moving to the left (its position s is decreasing)
2. if $a > 0$, then v is increasing; if $a < 0$, then v is decreasing
3. if a and v are both positive or both negative, then (1) and (2) imply that the speed of P is increasing or that P is accelerating; if a and v have opposite signs, then the speed of P is decreasing or P is decelerating
4. if s is a continuous function of t, then P reverses direction whenever v is zero and a is different from zero; note that zero velocity does not necessarily imply a reversal in direction

> **Example 26**

A particle moves along a horizontal line such that its position $s = 2t^3 - 9t^2 + 12t - 4$, for $t \geq 0$.
(a)   Find all t for which the particle is moving to the right.
(b)   Find all t for which the velocity is increasing.
(c)   Find all t for which the speed of the particle is increasing.
(d)   Find the speed when $t = \dfrac{3}{2}$.
(e)   Find the total distance traveled between $t = 0$ and $t = 4$.

✓ Solutions

$$v = \frac{ds}{dt} = 6t^2 - 18t + 12 = 6(t^2 - 3t + 2) = 6(t - 2)(t - 1)$$

and

$$a = \frac{dv}{dt} = \frac{d^2s}{dt^2} = 12t - 18 = 12\left(t - \frac{3}{2}\right)$$

Velocity $v = 0$ at $t = 1$ and $t = 2$, and:

$$
\begin{array}{lll}
\text{if} & t < 1, & \text{then} \quad v > 0 \\
& 1 < t < 2, & \quad v < 0 \\
& t > 2, & \quad v > 0
\end{array}
$$

Acceleration $a = 0$ at $t = \dfrac{3}{2}$, and:

$$
\begin{array}{lll}
\text{if} & t < \dfrac{3}{2}, & \text{then} \quad a < 0 \\
& t > \dfrac{3}{2}, & \quad a > 0
\end{array}
$$

These signs of v and a immediately yield the answers, as follows:
(a)   The particle moves to the right when $t < 1$ or $t > 2$.
(b) v increases when $t > \dfrac{3}{2}$.

(c)   The speed $|v|$ is increasing when v and a are both positive, that is, for $t > 2$, and when v and a are both negative, that is, for $1 < t < \frac{3}{2}$.

(d)   The speed when $t = \frac{3}{2}$ equals $|v| = \left| -\frac{3}{2} \right| = \frac{3}{2}$.

Figure 4.15

(e) P's motion can be indicated as shown in Figure 4.15. P moves to the right if $t < 1$, reverses its direction at $t = 1$, moves to the left when $1 < t < 2$, reverses again at $t = 2$, and continues to the right for all $t > 2$. The position of P at certain times t are shown in the following table:

t:	0	1	2	4
s:	-4	1	0	28

Thus P travels a total of 34 units between times $t = 0$ and $t = 4$.

> **Example 27**

Answer the questions of Example 26 if the law of motion is

$$s = t^4 - 4t^3$$

✓ **Solutions**

Since $v = 4t^3 - 12t^2 = 4t^2(t - 3)$ and $a = 12t^2 - 24t = 12t(t - 2)$, the signs of v and a are as follows:

if	$t < 3,$	then	$v < 0$
	$3 < t,$		$v > 0$
if	$t < 0,$	then	$a > 0$
	$0 < t < 2,$		$a < 0$
	$2 < t,$		$a > 0$

Thus

(a) s increases if $t > 3$.

(b) v increases if $t < 0$ or $t > 2$.

(c)   Since v and a have the same sign if $0 < t < 2$ or if $t > 3$, the speed increases on these intervals.

(d)   The speed when $t = \frac{3}{2}$ equals $|v| = \left| -\frac{27}{2} \right| = \frac{27}{2}$.

Figure 4.16

(e)   The motion is shown in Figure 4.16. The particle moves to the left if $t < 3$ and to the right if $t > 3$, stopping instantaneously when $t = 0$ and $t = 3$, but reversing direction only when $t = 3$. Thus:

$$
\begin{array}{llll}
t\text{:} & 0 & 3 & 4 \\
s\text{:} & 0 & -27 & 0
\end{array}
$$

The particle travels a total of 54 units between $t = 0$ and $t = 4$.

(Compare with Example 13, page 168, where the function $f(x) = x^4 - 4x^3$ is investigated for maximum and minimum values; also see the accompanying Figure 4.5 on page 169.)

BC ONLY

K. Motion Along a Curve: Velocity and Acceleration Vectors

Vector

If a point P moves along a curve defined parametrically by $P(t) = (x(t), y(t))$, where t represents time, then the vector from the origin to P is called the *position vector*, with x as its *horizontal component* and y as its *vertical component*. The set of position vectors for all values of t in the domain common to $x(t)$ and $y(t)$ is called the *vector function*.

Components

A vector may be symbolized either by a boldface letter (**R**) or an italic letter with an arrow written over it ($\vec{R}$). The position vector, then, may be written as $\vec{R}(t) = \langle x, y \rangle$ or as $\mathbf{R} = \langle x, y \rangle$. In print the boldface notation is clearer and will be used in this book; when writing by hand, the arrow notation is simpler.

The *velocity vector* is the derivative of the vector function (the position vector):

Velocity vector

$$
\mathbf{v} = \frac{d\mathbf{R}}{dt} = \left\langle \frac{dx}{dt}, \frac{dy}{dt} \right\rangle \quad \text{or} \quad \vec{v}(t) = \left\langle \frac{dx}{dt}, \frac{dy}{dt} \right\rangle
$$

Alternative notations for $\dfrac{dx}{dt}$ and $\dfrac{dy}{dt}$ are v_x and v_y, respectively; these are the components of **v** in the horizontal and vertical directions, respectively. The slope of **v** is

$$
\frac{\dfrac{dy}{dt}}{\dfrac{dx}{dt}} = \frac{dy}{dx}
$$

which is the slope of the curve; the *magnitude* of **v** is the vector's length:

Magnitude

$$
|\mathbf{v}| = \sqrt{\left(\frac{dx}{dt}\right)^2 + \left(\frac{dy}{dt}\right)^2} = \sqrt{v_x^{\,2} + v_y^{\,2}}
$$

Thus, if the vector **v** is drawn initiating at P, it will be tangent to the curve at P and its magnitude will be the *speed* of the particle at P.

The *acceleration vector* **a** is $\dfrac{d\mathbf{v}}{dt}$ or $\dfrac{d^2\mathbf{R}}{dt^2}$ and can be obtained by a second differentiation of the components of **R**. Thus

$$\mathbf{a} = \left\langle \frac{d^2x}{dt^2}, \frac{d^2y}{dt^2} \right\rangle, \text{ or } \vec{a}(t) = \left\langle \frac{d^2x}{dt^2}, \frac{d^2y}{dt^2} \right\rangle$$

and its magnitude is the vector's length:

$$|\mathbf{a}| = \sqrt{\left(\frac{d^2x}{dt^2}\right)^2 + \left(\frac{d^2y}{dt^2}\right)^2} = \sqrt{a_x^2 + a_y^2}$$

where we have used a_x and a_y for $\dfrac{d^2x}{dt^2}$ and $\dfrac{d^2y}{dt^2}$, respectively.

> **Example 28**

A particle moves according to the equations $x = 3 \cos t$, $y = 2 \sin t$.
(a)   Find a single equation in x and y for the path of the particle and sketch the curve.
(b)   Find the velocity and acceleration vectors at any time t, and show that $\mathbf{a} = -\mathbf{R}$ at all times.
(c)   Find **R**, **v**, and **a** when (1) $t_1 = \frac{\pi}{6}$, (2) $t_2 = \pi$, and draw them on the sketch.
(d)   Find the speed of the particle and the magnitude of its acceleration at each instant in (c).
(e)   When is the speed a maximum? When is the speed a minimum?

✓ **Solutions**

(a)   Since $\dfrac{x^2}{9} = \cos^2 t$ and $\dfrac{y^2}{4} = \sin^2 t$, therefore

$$\frac{x^2}{9} + \frac{y^2}{4} = 1$$

and the particle moves in a counterclockwise direction along an ellipse, starting, when $t = 0$, at $(3,0)$ and returning to this point when $t = 2\pi$. See Figure 4.17 for the sketch of the curve.

(b)   We have

$$\mathbf{R} = \langle 3 \cos t, 2 \sin t \rangle$$
$$\mathbf{v} = \langle -3 \sin t, 2 \cos t \rangle$$
$$\mathbf{a} = \langle -3 \cos t, -2 \sin t \rangle = -\mathbf{R}$$

The acceleration, then, is always directed toward the center of the ellipse.

(c)   At $t_1 = \frac{\pi}{6}$,

$$\mathbf{R}_1 = \left\langle \frac{3\sqrt{3}}{2}, 1 \right\rangle$$
$$\mathbf{v}_1 = \left\langle -\frac{3}{2}, \sqrt{3} \right\rangle$$
$$\mathbf{a}_1 = \left\langle -\frac{3\sqrt{3}}{2}, -1 \right\rangle$$

At $t_2 = \pi$,

$$\mathbf{R}_2 = \langle -3, 0 \rangle$$
$$\mathbf{v}_2 = \langle 0, -2 \rangle$$
$$\mathbf{a}_2 = \langle 3, 0 \rangle$$

The curve, and **v** and **a** at t_1 and t_2, are sketched in Figure 4.17, below.

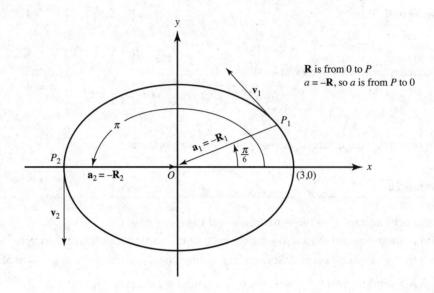

Figure 4.17

(d)   At $t_1 = \frac{\pi}{6}$,   At $t_2 = \pi$,

$$|\mathbf{v}_1| = \sqrt{\frac{9}{4} + 3} = \frac{\sqrt{21}}{2} \qquad\qquad |\mathbf{v}_2| = \sqrt{0 + 4} = 2$$

$$|\mathbf{a}_1| = \sqrt{\frac{27}{4} + 1} = \frac{\sqrt{31}}{2} \qquad\qquad |\mathbf{a}_2| = \sqrt{9 + 0} = 3$$

(e)   For the speed $|\mathbf{v}|$ at any time t

$$|\mathbf{v}| = \sqrt{9\sin^2 t + 4\cos^2 t}$$
$$= \sqrt{4\sin^2 t + 4\cos^2 t + 5\sin^2 t}$$
$$= \sqrt{4 + 5\sin^2 t}$$

We see immediately that the speed is a maximum when $t = \frac{\pi}{2}$ or $\frac{3\pi}{2}$, and a minimum when $t = 0$ or π. The particle goes fastest at the ends of the minor axis and most slowly at the ends of the major axis. Generally, one can determine maximum or minimum speed by finding $\frac{d}{dt}|\mathbf{v}|$, setting it equal to zero, and applying the usual tests to sort out values of t that yield maximum or minimum speeds.

> **Example 29**

A particle moves along the parabola $y = x^2 - x$ with constant speed $\sqrt{10}$. Find **v** at (2,2).

✓ **Solution**

Since

$$v_y = \frac{dy}{dt} = (2x - 1)\frac{dx}{dt} = (2x - 1)v_x \qquad (1)$$

and

$$v_x^2 + v_y^2 = 10 \qquad (2)$$

$$v_x^2 + (2x - 1)^2\, v_x^2 = 10 \qquad (3)$$

Relation (3) holds at all times; specifically, at (2,2), $v_x^2 + 9v_x^2 = 10$ so that $v_x = \pm 1$. From (1), then, we see that $v_y = \pm 3$. Therefore **v** at (2,2) is either $\langle 1,3 \rangle$ or $\langle -1,-3 \rangle$. The former corresponds to counterclockwise motion along the parabola, as shown in Figure 4.18a, and the latter to clockwise motion, as shown in Figure 4.18b.

Figure 4.18a

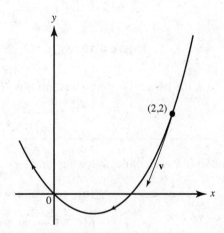

Figure 4.18b

L. Tangent-Line Approximations

Local linear approximation

If $f'(a)$ exists, then the *local linear approximation* of $f(x)$ at a is

$$f(a) + f'(a)(x - a)$$

Since the equation of the tangent line to $y = f(x)$ at $x = a$ is

$$y - f(a) = f'(a)(x - a)$$

Tangent-line approximation

we see that the y-value on the tangent line is an approximation for the actual or true value of f. Local linear approximation is therefore also called *tangent-line approximation.*[†] For values of x close to a, we have

$$f(x) \approx f(a) + f'(a)(x - a) \qquad (1)$$

where $f(a) + f'(a)(x - a)$ is the linear or tangent-line approximation for $f(x)$, and $f'(a)(x - a)$ is the approximate change in f as we move along the curve from a to x. See Figure 4.19.

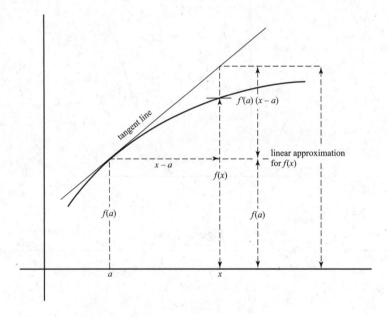

Figure 4.19

In general, the closer x is to a, the better the approximation is to $f(x)$.

> **Example 30**

Find tangent-line approximations for each of the following functions at the values indicated:

(a) $\sin x$ at $a = 0$ (b) $\cos x$ at $a = \dfrac{\pi}{2}$

(c) $2x^3 - 3x$ at $a = 1$ (d) $\sqrt{1 + x}$ at $a = 8$

[†] Local linear approximation is also referred to as "local linearization" or even "best linear approximation" (the latter because it is better than any other linear approximation).

✓ **Solutions**

(a)   At $a = 0$, $\sin x \approx \sin(0) + \cos(0)(x - 0) \approx 0 + 1 \cdot x \approx x$

(b)   At $a = \frac{\pi}{2}$, $\cos x \approx \cos \frac{\pi}{2} - \sin\left(\frac{\pi}{2}\right)\left(x - \frac{\pi}{2}\right) \approx -x + \frac{\pi}{2}$

(c)   At $a = 1$, $2x^3 - 3x \approx -1 + 3(x - 1) \approx 3x - 4$

(d)   At $a = 8$, $\sqrt{1 + x} = \sqrt{1 + 8} + \dfrac{1}{2\sqrt{1 + 8}}(x - 8) = 3 + \dfrac{1}{6}(x - 8)$

> **Example 31**

Using the tangent lines obtained in Example 30 and a calculator, we evaluate each function, then its linear approximation, at the indicated x-values:

Function	(a)	(b)	(c)	(d)
x-value	−0.80	2.00	1.10	5.50
True value	−0.72	−0.42	−0.64	2.55
Approximation	−0.80	−0.43	−0.70	2.58

Example 31 shows how small the errors can be when tangent lines are used for approximations and x is near a.

> **Example 32**

A very useful and important local linearization enables us to approximate $(1 + x)^k$ by $1 + kx$ for k any real number and for x near 0. Equation (1) yields

$$(1 + x)^k \approx (1 + 0)^k + k(1 + x)^{k-1}_{\text{at } 0} \cdot (x - 0)$$

$$\approx 1 + kx \tag{2}$$

Then, near 0, for example,

$$\sqrt{1 + x} \approx 1 + \frac{1}{2}x \qquad \text{and} \qquad (1 + x)^3 \approx 1 + 3x$$

> **Example 33**

Estimate the value of $\dfrac{3}{(1 - x)^2}$ at $x = 0.05$.

✓ **Solution**

Use the line tangent to $f(x) = \dfrac{3}{(1 - x)^2}$ at $x = 0$; $f(0) = 3$.

$f'(x) = \dfrac{6}{(1 - x)^3}$, so $f'(0) = 6$; hence, the line is $y = 6x + 3$.

Our tangent-line approximation, then, is $\dfrac{3}{(1 - x)^2} \approx 6x + 3$.

At $x = 0.05$, we have $f(0.05) \approx 6(0.05) + 3 = 3.3$.

To determine if this is an overestimate or an underestimate, we examine the second derivative.

$f''(x) = \dfrac{18}{(1 - x)^4} > 0$ for $x \neq 1$; therefore, $f(x)$ is always concave up.

Therefore the tangent line at $x = 0$ lies below the graph of $f(x)$ on the interval $[0, 0.05]$. The estimate is on the tangent line, so it must be an underestimate. We can verify this by evaluating $f(0.05)$ or by graphing $f(x)$ and its tangent line at $x = 0$ on a calculator.

M. Related Rates

If several variables that are functions of time t are related by an equation, we can obtain a relation involving their (time) rates of change by differentiating with respect to t.

> **Example 34**

If one leg AB of a right triangle increases at the rate of 2 inches per second while the other leg AC decreases at 3 inches per second, find how fast the hypotenuse is changing when $AB = 6$ feet and $AC = 8$ feet.

Figure 4.20

> **✓ Solution**

See Figure 4.20. Let u, v, and z denote the lengths, respectively, of AB, AC, and BC. We know that $\dfrac{du}{dt} = \dfrac{1}{6}$ (ft/sec) and $\dfrac{dv}{dt} = -\dfrac{1}{4}$. Since (at any time) $z^2 = u^2 + v^2$, then

$$2z\frac{dz}{dt} = 2u\frac{du}{dt} + 2v\frac{dv}{dt} \quad \text{and} \quad \frac{dz}{dt} = \frac{u\dfrac{du}{dt} + v\dfrac{dv}{dt}}{z}$$

At the instant in question, $u = 6$, $v = 8$, and $z = 10$, so

$$\frac{dz}{dt} = \frac{6\left(\dfrac{1}{6}\right) + 8\left(-\dfrac{1}{4}\right)}{10} = -\frac{1}{10} \text{ ft/sec}$$

> **Example 35**

The diameter and height of a paper cup in the shape of a cone are both 4 inches, and water is leaking out at the rate of $\dfrac{1}{2}$ cubic inch per second. Find the rate at which the water level is dropping when the diameter of the surface is 2 inches.

✓ Solution

See Figure 4.21. We know that $\dfrac{dV}{dt} = -\dfrac{1}{2}$ and that $h = 2r$.

Here, $V = \dfrac{1}{3}\pi r^2 h = \dfrac{\pi h^3}{12}$, so

$$\frac{dV}{dt} = \frac{\pi h^2}{4}\frac{dh}{dt} \qquad \text{and} \qquad \frac{dh}{dt} = -\frac{1}{2}\frac{4}{\pi h^2} = -\frac{2}{\pi h^2} \text{ at any time}$$

When the diameter is 2 inches, so is the height, and $\dfrac{dh}{dt} = -\dfrac{1}{2\pi}$. The water level is thus dropping at the rate of $\dfrac{1}{2\pi}$ in./sec.

Figure 4.21

⟩ Example 36

Suppose liquid is flowing into a vessel at a constant rate. The vessel has the shape of a hemisphere capped by a cylinder, as shown in Figure 4.22. Graph $y = h(t)$, the height ($=$ depth) of the liquid at time t, labeling and explaining any salient characteristics of the graph.

Figure 4.22

✓ **Solution**

Liquid flowing in at a constant rate means the change in volume is constant per unit of time. Obviously, the depth of the liquid increases as t does, so $h'(t)$ is positive throughout. To maintain the constant increase in volume per unit of time, when the radius grows, $h'(t)$ must decrease. Thus, the rate of increase of h decreases as h increases from 0 to a (where the cross-sectional area of the vessel is largest). Therefore, since $h'(t)$ decreases, $h''(t) < 0$ from 0 to a and the curve is concave down.

As h increases from a to b, the radius of the vessel (here cylindrical) remains constant, as do the cross-sectional areas. Therefore $h'(t)$ is also constant, implying that $h(t)$ is linear from a to b.

Note that the inflection point at depth a does not exist since $h''(t) < 0$ for all values less than a but is equal to 0 for all depths greater than or equal to a.

BC ONLY

N. Slope of a Polar Curve

We know that, if a smooth curve is given by the parametric equations

$$x = f(t) \quad \text{and} \quad y = g(t)$$

then

$$\frac{dy}{dx} = \frac{g'(t)}{f'(t)} \quad \text{provided that } f'(t) \neq 0$$

To find the slope of a polar curve $r = f(\theta)$, we must first express the curve in parametric form. Since

$$x = r \cos \theta \quad \text{and} \quad y = r \sin \theta$$

therefore,

$$x = f(\theta) \cos \theta \quad \text{and} \quad y = f(\theta) \sin \theta$$

If $f(\theta)$ is differentiable, so are x and y; then

$$\frac{dx}{d\theta} = f'(\theta) \cos \theta - f(\theta) \sin \theta$$

$$\frac{dy}{d\theta} = f'(\theta) \sin \theta + f(\theta) \cos \theta$$

Also, if $\frac{dx}{d\theta} \neq 0$, then

$$\frac{dy}{dx} = \frac{dy/d\theta}{dx/d\theta} = \frac{f'(\theta)\sin\theta + f(\theta)\cos\theta}{f'(\theta)\cos\theta - f(\theta)\sin\theta} = \frac{r'\sin\theta + r\cos\theta}{r'\cos\theta - r\sin\theta}$$

In doing an exercise, it is far easier simply to express the polar equation parametrically, then find dy/dx, rather than to memorize the formula.

> **Example 37**

(a) Find the slope of the cardioid $r = 2(1 + \cos \theta)$ at $\theta = \frac{\pi}{6}$. See Figure 4.23.

(b) Where is the tangent to the curve horizontal?

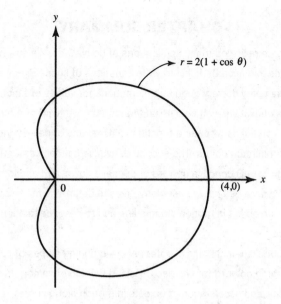

Figure 4.23

✓ **Solutions**

(a) Use $r = 2(1 + \cos \theta)$, $x = r \cos \theta$, $y = r \sin \theta$, and $r' = -2 \sin \theta$; then

$$\frac{dy}{dx} = \frac{dy/d\theta}{dx/d\theta} = \frac{(-2 \sin \theta)\sin \theta + 2(1 + \cos \theta)(\cos \theta)}{(-2 \sin \theta)\cos \theta - 2(1 + \cos \theta)(\sin \theta)}$$

At $\theta = \frac{\pi}{6}$, $\dfrac{dy}{dx} = -1$.

(b) Since the cardioid is symmetric to $\theta = 0$ we need consider only the upper half of the curve for part (b). The tangent is horizontal where $\dfrac{dy}{d\theta} = 0$ (provided $\dfrac{dx}{d\theta} \neq 0$). Since $\dfrac{dy}{d\theta}$ factors into $2(2 \cos \theta - 1)(\cos \theta + 1)$, which equals 0 for $\cos \theta = \frac{1}{2}$ or -1, $\theta = \frac{\pi}{3}$ or π. From part (a), $\dfrac{dx}{d\theta} \neq 0$ at $\frac{\pi}{3}$, but $\dfrac{dx}{d\theta}$ does equal 0 at π. Therefore, the tangent is horizontal only at $\frac{\pi}{3}$ (and, by symmetry, at $\frac{5\pi}{3}$).

It is obvious from Figure 4.23 that $r'(\theta)$ does *not* give the slope of the cardioid. As θ varies from 0 to $\frac{2\pi}{3}$, the slope varies from $-\infty$ to 0 to $+\infty$ (with the tangent rotating counterclockwise), taking on every real value. However, $r'(\theta)$ equals $-2 \sin \theta$, which takes on values only between -2 and 2.

CHAPTER SUMMARY

In this chapter, we reviewed many applications of derivatives. We saw how to find slopes of curves and used that skill to write equations of lines tangent to a curve. Those lines often provide very good approximations for values of functions. We looked at ways derivatives can help us understand the behavior of a function. The first derivative can tell us whether a function is increasing or decreasing and locate maximum and minimum points. The second derivative can tell us whether the graph of the function is concave upward or concave downward and locate points of inflection. We reviewed how to use derivatives to determine the velocity and acceleration of an object in motion along a line and to describe relationships among rates of change.

For BC Calculus students, this chapter reviewed finding slopes of curves defined parametrically or in polar form. We also reviewed the use of vectors to describe the position, velocity, and acceleration of objects in motion along curves.

Practice Exercises

A1. The slope of the curve $y^3 - xy^2 = 4$ at the point where $y = 2$ is

(A) -2

(B) $-\frac{1}{2}$

(C) $\frac{1}{2}$

(D) 2

A2. The slope of the curve $y^2 - xy - 3x = 1$ at the point $(0, -1)$ is

(A) -1

(B) -2

(C) 1

(D) 2

A3. An equation of the tangent to the curve $y = x \sin x$ at the point $\left(\frac{\pi}{2}, \frac{\pi}{2}\right)$ is

(A) $y = x - \pi$

(B) $y = \frac{\pi}{2}$

(C) $y = \pi - x$

(D) $y = x$

A4. The tangent to the curve of $y = xe^{-x}$ is horizontal when x is equal to

(A) 0

(B) 1

(C) -1

(D) $\frac{1}{e}$

A5. The minimum value of the slope of the curve $y = x^5 + x^3 - 2x$ is

(A) 2

(B) 6

(C) -2

(D) -6

A6. An equation of the tangent to the hyperbola $x^2 - y^2 = 12$ at the point $(4,2)$ on the curve is

(A) $x - 2y + 6 = 0$

(B) $y = 2x$

(C) $y = 2x - 6$

(D) $y = \frac{x}{2}$

A7. A tangent to the curve $y^2 - xy + 9 = 0$ is vertical when

(A) $y = 0$

(B) $y = \pm\sqrt{3}$

(C) $y = \frac{1}{2}$

(D) $y = \pm 3$

A8. The volume of a sphere is given by $V = \frac{4}{3}\pi r^3$. Use a tangent line to approximate the increase in volume, in cubic inches, when the radius of a sphere is increased from 3 to 3.1 inches.

(A) $\frac{0.04\pi}{3}$

(B) 0.04π

(C) 1.2π

(D) 3.6π

A9. When $x = 3$, the equation $2x^2 - y^3 = 10$ has the solution $y = 2$. Using a tangent line to the graph of the curve, approximate y when $x = 3.04$.

(A) 1.6

(B) 1.96

(C) 2.04

(D) 2.4

A10. If the side e of a square is increased by 1%, then the area is increased approximately   CHALLENGE

(A) $0.02e$

(B) $0.02e^2$

(C) $0.01e^2$

(D) $0.01e$

CHALLENGE

A11. The edge of a cube has length 10 in., with a possible error of 1%. The possible error, in cubic inches, in the volume of the cube is

(A) 1
(B) 3
(C) 10
(D) 30

A12. The function $f(x) = x^4 - 4x^2$ has

(A) one local minimum and two local maxima
(B) one local minimum and no local maximum
(C) no local minimum and one local maximum
(D) two local minima and one local maximum

A13. The number of inflection points on the graph of $f(x) = x^4 - 4x^2$ is

(A) 0
(B) 1
(C) 2
(D) 3

A14. The maximum value of the function $y = -4\sqrt{2-x}$ is

(A) 0
(B) −4
(C) −2
(D) 2

A15. The total number of local maximum and minimum points of the function whose derivative, for all x, is given by $f'(x) = x(x-3)^2(x+1)^4$ is

(A) 0
(B) 1
(C) 2
(D) 3

A16. For which curve shown below are both f' and f'' negative?

(A)

(B)

(C)

(D)

A17. For which curve shown in Question A16 is f'' positive but f' negative?

(A) Curve (A)
(B) Curve (B)
(C) Curve (C)
(D) Curve (D)

In Questions A18–A21, the position of a particle moving along a horizontal line is given by $s = t^3 - 6t^2 + 12t - 8$.

A18. The object is moving to the right for

   (A) $t < 2$
   (B) all t except $t = 2$
   (C) all t
   (D) $t > 2$

A19. The minimum value of the speed is

   (A) 0
   (B) 1
   (C) 2
   (D) 3

A20. The acceleration is positive

   (A) when $t > 2$
   (B) for all t, $t \neq 2$
   (C) when $t < 2$
   (D) for $1 < t < 2$

A21. The speed of the particle is decreasing for

   (A) $t < 2$
   (B) all t
   (C) $t < 1$ or $t > 2$
   (D) $t > 2$

In Questions A22–A24, a particle moves along a horizontal line and its position at time t is $s = t^4 - 6t^3 + 12t^2 + 3$.

A22. The particle is at rest when t is equal to

   (A) 1 or 2
   (B) 0
   (C) $\dfrac{9}{4}$
   (D) 0, 1, or 2

A23. The velocity, v, is increasing when

   (A) $t > 1$
   (B) $1 < t < 2$
   (C) $t < 2$
   (D) $t < 1$ or $t > 2$

A24. The speed of the particle is increasing for

   (A) $0 < t < 1$ or $t > 2$
   (B) $1 < t < 2$
   (C) $t < 2$
   (D) $t < 0$ or $t > 2$

A25. The displacement from the origin of a particle moving on a line is given by $s = t^4 - 4t^3$. The maximum displacement during the time interval $-2 \leq t \leq 4$ is

   (A) 27
   (B) 3
   (C) 48
   (D) 16

A26. If a particle moves along a line according to the law $s = t^5 + 5t^4$, then the number of times it reverses direction is

   (A) 0
   (B) 1
   (C) 2
   (D) 3

In Questions A27–A30, $R = \left\langle 3\cos\left(\dfrac{\pi}{3}t\right), 2\sin\left(\dfrac{\pi}{3}t\right)\right\rangle$ is the (position) vector $\langle x, y \rangle$ from the origin to a moving point $P(x, y)$ at time t. **BC ONLY**

A27. A single equation in x and y for the path of the point is   **CHALLENGE**

   (A) $x^2 + y^2 = 13$
   (B) $9x^2 + 4y^2 = 36$
   (C) $2x^2 + 3y^2 = 13$
   (D) $4x^2 + 9y^2 = 36$

A28. When $t = 3$, the speed of the particle is

   (A) $\dfrac{2\pi}{3}$
   (B) 2
   (C) 3
   (D) $\dfrac{\sqrt{13}}{3}\pi$

A29. The magnitude of the acceleration when $t = 3$ is

   (A) 2
   (B) $\dfrac{\pi^2}{3}$
   (C) 3
   (D) $\dfrac{2\pi^2}{9}$

BC ONLY

A30. At the point where $t = \frac{1}{2}$, the slope of the curve along which the particle moves is

(A) $-\frac{2\sqrt{3}}{9}$

(B) $-\frac{\sqrt{3}}{2}$

(C) $\frac{2}{\sqrt{3}}$

(D) $-\frac{2\sqrt{3}}{3}$

A31. A balloon is being filled with helium at the rate of 4 ft^3/min. The rate, in square feet per minute, at which the surface area is increasing when the volume is $\frac{32\pi}{3}$ ft^3 is

(A) 4π

(B) 2

(C) 4

(D) 1

A32. A circular conical reservoir, vertex down, has depth 20 ft and radius at the top 10 ft. Water is leaking out so that the surface is falling at the rate of $\frac{1}{2}$ ft/hr. The rate, in cubic feet per hour, at which the water is leaving the reservoir when the water is 8 ft deep is

(A) 4π

(B) 8π

(C) $\frac{1}{4\pi}$

(D) $\frac{1}{8\pi}$

A33. A local minimum value of the function $y = \frac{e^x}{x}$ is

(A) $\frac{1}{e}$

(B) 1

(C) -1

(D) e

CHALLENGE

A34. The area of the largest rectangle that can be drawn with one side along the x-axis and two vertices on the curve of $y = e^{-x^2}$ is

(A) $\sqrt{\frac{2}{e}}$

(B) $\sqrt{2e}$

(C) $\frac{2}{e}$

(D) $\frac{1}{\sqrt{2e}}$

CHALLENGE

A35. A line with a negative slope is drawn through the point (1,2) forming a right triangle with the positive x- and y-axes. The slope of the line forming the triangle of least area is

(A) -1

(B) -2

(C) -3

(D) -4

CHALLENGE

A36. The point(s) on the curve $x^2 - y^2 = 4$ closest to the point (6,0) is (are)

(A) $(2,0)$

(B) $\left(\sqrt{5}, \pm 1\right)$

(C) $\left(3, \pm\sqrt{5}\right)$

(D) $\left(\sqrt{13}, \pm\sqrt{3}\right)$

A37. The sum of the squares of two positive numbers is 200; their minimum product is

(A) 100

(B) 20

(C) 0

(D) there is no minimum

CHALLENGE

A38. The first-quadrant point on the curve $y^2 x = 18$ that is closest to the point (2,0) is

(A) $(2,3)$

(B) $\left(6, \sqrt{3}\right)$

(C) $\left(3, \sqrt{6}\right)$

(D) $(1, 3\sqrt{2})$

A39. If h is a small negative number, then the local linear approximation for $\sqrt[3]{27 + h}$ is

(A) $3 + \frac{h}{27}$

(B) $3 - \frac{h}{27}$

(C) $\frac{h}{27}$

(D) $-\frac{h}{27}$

A40. If $f(x) = xe^{-x}$, then at $x = 0$

(A) f is increasing

(B) f is decreasing

(C) f has a relative maximum

(D) f has a relative minimum

A41. A function f has a derivative for each x such that $|x| < 2$ and has a local minimum at $(2,-5)$. Which statement below must be true?

(A) $f'(2) = 0$
(B) f' exists at $x = 2$
(C) $f'(x) < 0$ if $x < 2$, $f'(x) > 0$ if $x > 2$
(D) none of the preceding is necessarily true

A42. The height of a rectangular box is 10 in. Its length increases at the rate of 2 in./sec; its width decreases at the rate of 4 in./sec. When the length is 8 in. and the width is 6 in., the rate, in cubic inches per second, at which the volume of the box is changing is

(A) 200
(B) 80
(C) −80
(D) −200

A43. The tangent to the curve $x^3 + x^2y + 4y = 1$ at the point $(3,-2)$ has slope

(A) -3
(B) $-\dfrac{27}{13}$
(C) $-\dfrac{11}{9}$
(D) $-\dfrac{15}{13}$

A44. If $f(x) = ax^4 + bx^2$ and $ab > 0$, then

(A) the curve has no horizontal tangents
(B) the curve is concave up for all x
(C) the curve has no inflection point
(D) none of the preceding is necessarily true

A45. A function f is continuous and differentiable on the interval $[0,4]$, where f' is positive but f'' is negative. Which table could represent points on f?

(A)

x	0	1	2	3	4
y	10	12	14	16	18

(B)

x	0	1	2	3	4
y	10	12	15	19	24

(C)

x	0	1	2	3	4
y	10	18	24	28	30

(D)

x	0	1	2	3	4
y	10	14	21	24	26

A46. An equation of the tangent to the curve with parametric equations $x = 2t + 1$, $y = 3 - t^3$ at the point where $t = 1$ is

`BC ONLY`

(A) $2x + 3y = 12$
(B) $3x + 2y = 13$
(C) $6x + y = 20$
(D) $3x - 2y = 5$

A47. Given the function $f(x) = \sqrt[3]{x}$, it is known that $f(64) = 4$. Using the tangent line to the graph of $f(x)$, approximately how much less than 4 is $\sqrt[3]{63}$?

(A) $\dfrac{1}{48}$
(B) $\dfrac{1}{16}$
(C) $\dfrac{1}{3}$
(D) $\dfrac{2}{3}$

A48. The best linear approximation for $f(x) = \tan x$ near $x = \dfrac{\pi}{4}$ is $y =$

(A) $1 + \dfrac{1}{2}\left(x - \dfrac{\pi}{4}\right)$
(B) $1 + \left(x - \dfrac{\pi}{4}\right)$
(C) $1 + \sqrt{2}\left(x - \dfrac{\pi}{4}\right)$
(D) $1 + 2\left(x - \dfrac{\pi}{4}\right)$

A49. Given $f(x) = e^{kx}$, approximate $f(h)$, where h is near zero, using a tangent-line approximation. $f(h) \approx$

(A) k
(B) kh
(C) $1 + k$
(D) $1 + kh$

A50. If $f(x) = cx^2 + dx + e$ for the function shown in the graph, then

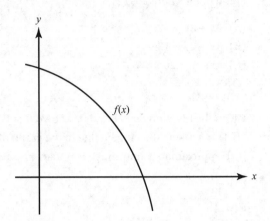

(A) c, d, and e are all positive

(B) $c > 0, d < 0, e > 0$

(C) $c < 0, d > 0, e > 0$

(D) $c < 0, d < 0, e > 0$

A51. Given $f(x) = \log_{10}x$ and $\log_{10}(102) \approx 2.0086$, which is closest to $f'(100)$?

(A) 0.0043

(B) 0.0086

(C) 0.01

(D) 1.0043

PART B. CALCULATOR ACTIVE—DIRECTIONS: Some of the following questions require the use of a graphing calculator.

B1. The point on the curve $y = \sqrt{2x+1}$ at which the tangent is parallel to the line $x - 3y = 6$ is

(A) $(4,3)$

(B) $(0,1)$

(C) $(1,\sqrt{3})$

(D) $(2,\sqrt{5})$

B2. An equation of the tangent to the curve $x^2 = 4y$ at the point on the curve where $x = -2$ is

(A) $x + y - 3 = 0$

(B) $x - y + 3 = 0$

(C) $x + y - 1 = 0$

(D) $x + y + 1 = 0$

B3. The table shows the velocity at time t of an object moving along a line. Estimate the acceleration (in ft/sec^2) at $t = 6$ sec.

t (sec)	0	4	8	10
vel.	18	16	10	0

(A) -6

(B) -1.5

(C) 1.5

(D) 6

Use the graph shown, sketched on [0,7], for Questions B4–B6.

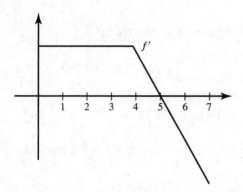

B4. From the graph it follows that

(A) f is discontinuous at $x = 4$
(B) f is decreasing for $4 < x < 7$
(C) $f(5) < f(0)$
(D) $f(2) < f(3)$

B5. Which statement best describes f at $x = 5$?

(A) f has a root.
(B) f has a maximum.
(C) f has a minimum.
(D) The graph of f has a point of inflection.

B6. For which interval is the graph of f concave downward?

(A) (0,4)
(B) (4,5)
(C) (5,7)
(D) (4,7)

Use the graph shown for Questions B7–B13. It shows the velocity of an object moving along a straight line during the time interval $0 \leq t \leq 5$.

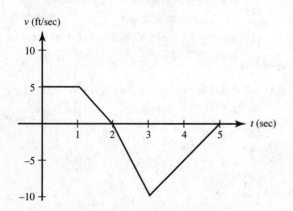

B7. The object attains its maximum speed when $t =$

(A) 0
(B) 1
(C) 2
(D) 3

B8. The speed of the object is increasing during the time interval

(A) (1,2)
(B) (0,2)
(C) (2,3)
(D) (3,5)

B9. The acceleration of the object is positive during the time interval

(A) (1,2)
(B) (0,2)
(C) (2,3)
(D) (3,5)

B10. How many times on $0 < t < 5$ is the object's acceleration undefined?

(A) none
(B) 1
(C) 2
(D) 3

B11. During $2 < t < 3$ the object's acceleration (in ft/sec^2) is

(A) -10
(B) -5
(C) 5
(D) 10

B12. The object is farthest to the right when $t =$

(A) 0
(B) 1
(C) 2
(D) 5

B13. The object's average acceleration (in ft/sec^2) for the interval $0 \leq t \leq 3$ is

(A) -15
(B) -5
(C) -3
(D) -1

B14. The line $y = 3x + k$ is tangent to the curve $y = x^3$ when k is equal to

(A) 1 or −1
(B) 2 or −2
(C) 3 or −3
(D) 4 or −4

B15. The two tangents that can be drawn from the point (3,5) to the parabola $y = x^2$ have slopes

(A) 1 and 5
(B) 0 and 4
(C) 2 and 10
(D) 2 and 4

B16. The table shows the velocity at various times of an object moving along a line. An estimate of its acceleration (in ft/sec^2) at $t = 1$ is

t (sec)	1.0	1.5	2.2	2.5
v (ft/sec)	12.2	13.0	13.4	13.7

(A) 0.8
(B) 1.0
(C) 1.2
(D) 1.6

For Questions B17 and B18, $f'(x) = x \sin x - \cos x$ for $0 < x < 4$.

B17. f has a local maximum when x is approximately

(A) 1.192
(B) 2.289
(C) 3.426
(D) 3.809

B18. The graph of f has a point of inflection when x is approximately

(A) 1.192
(B) 2.289
(C) 3.426
(D) 3.809

In Questions B19–B22, the motion of a particle in a plane is given by the pair of equations $x = 2t$ and $y = 4t - t^2$.

`BC ONLY`

B19. The particle moves along

`CHALLENGE`

(A) an ellipse
(B) a hyperbola
(C) a line
(D) a parabola

B20. The speed of the particle at any time t is

(A) $\sqrt{6 - 2t}$
(B) $2\sqrt{t^2 - 4t + 5}$
(C) $2\sqrt{t^2 - 2t + 5}$
(D) $2\left(|3 - t|\right)$

B21. The minimum speed of the particle is

(A) 0
(B) 1
(C) 2
(D) 4

B22. The acceleration of the particle

(A) depends on t
(B) is always directed upward
(C) is constant both in magnitude and in direction
(D) never exceeds 1 in magnitude

B23. If a particle moves along a curve with constant speed, then

`CHALLENGE`

(A) the magnitude of its acceleration must equal zero
(B) the direction of acceleration must be constant
(C) the curve along which the particle moves must be a straight line
(D) its velocity and acceleration vectors must be perpendicular

B24. A particle is moving on the curve of $y = 2x - \ln x$ so that $\dfrac{dx}{dt} = -2$ at all times t. At the point (1,2), $\dfrac{dy}{dt}$ is

(A) −4
(B) −2
(C) 2
(D) 4

BC ONLY

In Questions B25 and B26, a particle is in motion along the polar curve $r = 6 \cos 2\theta$ such that $\frac{d\theta}{dt} = \frac{1}{3}$ radian/sec when $\theta = \frac{\pi}{6}$.

B25. At $\theta = \frac{\pi}{6}$, find the rate of change (in units per second) of the particle's distance from the origin.

(A) $-6\sqrt{3}$

(B) $-2\sqrt{3}$

(C) $2\sqrt{3}$

(D) $6\sqrt{3}$

B26. At $\theta = \frac{\pi}{6}$, what is the horizontal component of the particle's velocity?

(A) $-\frac{21}{2}$

(B) $-\frac{7}{2}$

(C) -2

(D) $\frac{1}{2}$

Use the graph of f' on [0,5], shown below, for Questions B27 and B28.

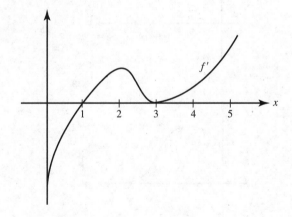

B27. f has a local minimum at $x =$

(A) 0

(B) 1

(C) 2

(D) 3

B28. The graph of f has a point of inflection at $x =$

(A) 1 only

(B) 2 only

(C) 3 only

(D) 2 and 3 only

B29. It follows from the graph of f', shown below, that

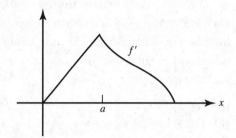

(A) f is not continuous at $x = a$

(B) f is continuous but not differentiable at $x = a$

(C) f has a relative maximum at $x = a$

(D) The graph of f has a point of inflection at $x = a$

B30. A vertical circular cylinder has radius r ft and height h ft. If the height and radius both increase at the constant rate of 2 ft/sec, then the rate, in square feet per second, at which the lateral surface area increases is

(A) $4\pi r$

(B) $2\pi(r + h)$

(C) $4\pi(r + h)$

(D) $4\pi h$

B31. A tangent drawn to the parabola $y = 4 - x^2$ at the point (1,3) forms a right triangle with the coordinate axes. The area of the triangle is

(A) $\frac{5}{4}$

(B) $\frac{5}{2}$

(C) $\frac{25}{2}$

(D) $\frac{25}{4}$

B32. Two cars are traveling along perpendicular roads, car A at 40 mph and car B at 60 mph. At noon, when car A reaches the intersection, car B is 90 miles away, and moving toward the intersection. At 1 P.M. the rate, in miles per hour, at which the distance between the cars is changing is

(A) -68

(B) -4

(C) 4

(D) 68

B33. Two cars are traveling along perpendicular roads, car A at 40 mph and car B at 60 mph. At noon, when car A reaches the intersection, car B is 90 miles away and moving toward the intersection. If t represents the time, in hours, traveled after noon, then the cars are closest together when t is

(A) 1.038
(B) 1.077
(C) 1.5
(D) 1.8

The graph for Questions B34 and B35 shows the velocity of an object moving along a straight line during the time interval $0 \leq t \leq 12$.

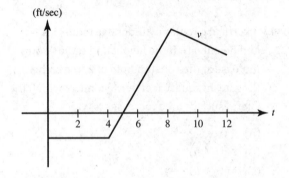

B34. For what t does this object attain its maximum acceleration?

(A) $0 < t < 4$
(B) $4 < t < 8$
(C) $t = 5$
(D) $t = 8$

B35. The object reverses direction at $t =$

(A) 4 only
(B) 5 only
(C) 8 only
(D) 5 and 8

B36. Given f' as graphed, which could be the graph of f?

(A)

(B)

(C)

(D)

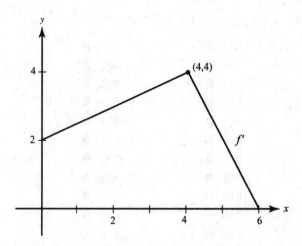

B37. The graph of f' is shown above. If we know that $f(2) = 10$, then the local linearization of f at $x = 2$ is $f(x) \approx$

(A) $\dfrac{x}{2} + 2$

(B) $\dfrac{x}{2} + 9$

(C) $3x - 3$

(D) $3x + 4$

Use the following graph for Questions B38–B40.

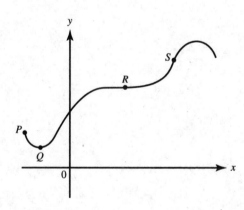

B38. At which labeled point do both $\dfrac{dy}{dx}$ and $\dfrac{d^2y}{dx^2}$ equal zero?

(A) P

(B) Q

(C) R

(D) S

B39. At which labeled point is $\dfrac{dy}{dx}$ positive and $\dfrac{d^2y}{dx^2}$ equal to zero?

(A) P

(B) Q

(C) R

(D) S

B40. At which labeled point is $\dfrac{dy}{dx}$ equal to zero and $\dfrac{d^2y}{dx^2}$ positive?

(A) P

(B) Q

(C) R

(D) S

B41. If $f(6) = 30$ and $f'(x) = \dfrac{x^2}{x+3}$, estimate $f(6.02)$ using the line tangent to f at $x = 6$.

(A) 29.92

(B) 30.02

(C) 30.08

(D) 30.16

B42. A local linear approximation for $f(x) = \sqrt{x^2 + 16}$ near $x = -3$ is

(A) $y = 5 - \dfrac{3}{5}(x - 3)$

(B) $y = 5 + \dfrac{3}{5}(x - 3)$

(C) $y = 5 - \dfrac{3}{5}(x + 3)$

(D) $y = 3 - \dfrac{5}{3}(x - 3)$

Answer Key

A1.	**C**	A20.	**A**	A39.	**A**	B7.	**D**	B26.	**B**
A2.	**A**	A21.	**A**	A40.	**A**	B8.	**C**	B27.	**B**
A3.	**D**	A22.	**B**	A41.	**D**	B9.	**D**	B28.	**D**
A4.	**B**	A23.	**D**	A42.	**D**	B10.	**D**	B29.	**D**
A5.	**C**	A24.	**A**	A43.	**D**	B11.	**A**	B30.	**C**
A6.	**C**	A25.	**C**	A44.	**C**	B12.	**C**	B31.	**D**
A7.	**D**	A26.	**C**	A45.	**C**	B13.	**B**	B32.	**B**
A8.	**D**	A27.	**D**	A46.	**B**	B14.	**B**	B33.	**A**
A9.	**C**	A28.	**A**	A47.	**A**	B15.	**C**	B34.	**B**
A10.	**B**	A29.	**B**	A48.	**D**	B16.	**D**	B35.	**B**
A11.	**D**	A30.	**D**	A49.	**D**	B17.	**C**	B36.	**C**
A12.	**D**	A31.	**C**	A50.	**D**	B18.	**B**	B37.	**D**
A13.	**C**	A32.	**B**	A51.	**A**	B19.	**D**	B38.	**C**
A14.	**A**	A33.	**D**	B1.	**A**	B20.	**B**	B39.	**D**
A15.	**B**	A34.	**A**	B2.	**D**	B21.	**C**	B40.	**B**
A16.	**D**	A35.	**B**	B3.	**B**	B22.	**C**	B41.	**C**
A17.	**A**	A36.	**C**	B4.	**D**	B23.	**D**	B42.	**C**
A18.	**B**	A37.	**D**	B5.	**B**	B24.	**B**		
A19.	**A**	A38.	**C**	B6.	**D**	B25.	**B**		

Answer Explanations

A1. **(C)** Substituting $y = 2$ yields $x = 1$. We find y' implicitly.

$$3y^2 y' - (2xyy' + y^2) = 0; \quad (3y^2 - 2xy)y' - y^2 = 0$$

Replace x by 1 and y by 2; solve for y'.

A2. **(A)** $2yy' - (xy' + y) - 3 = 0$. Replace x by 0 and y by -1; solve for y'.

A3. **(D)** Find the slope of the curve at $x = \frac{\pi}{2}$: $y' = x \cos x + \sin x$. At $x = \frac{\pi}{2}$, $y' = \frac{\pi}{2} \cdot 0 + 1$. An equation is $y - \frac{\pi}{2} = 1\left(x - \frac{\pi}{2}\right)$.

A4. **(B)** Since $y' = e^{-x}(1 - x)$ and $e^{-x} > 0$ for all x, $y' = 0$ when $x = 1$.

A5. **(C)** The slope $y' = 5x^4 + 3x^2 - 2$. Let $g = y'$. Since $g'(x) = 20x^3 + 6x = 2x(10x^2 + 3)$, $g'(x) = 0$ only if $x = 0$. Since $g''(x) = 60x^2 + 6$, g'' is always positive, assuring that $x = 0$ yields the minimum slope. Find y' when $x = 0$.

A6. **(C)** Since $2x - 2yy' = 0$, $y' = \frac{x}{y}$. At $(4,2)$, $y' = 2$. An equation of the tangent at $(4,2)$ is $y - 2 = 2(x - 4)$.

A7. **(D)** Since $y' = \frac{y}{2y - x}$, the tangent is vertical for $x = 2y$. Substitute in the given equation and solve for y.

A8. **(D)** The tangent-line approximation to the volume curve at $r = 3$ goes through the point $(3, 36\pi)$ and has a slope of $\left.\frac{dV}{dr}\right|_{r=3} = 4\pi r^2 \Big|_{r=3} = 36\pi$. So an equation of the tangent line is $y = 36\pi + 36\pi(r - 3)$, and we approximate $V(3.1) \approx y(3.1) = 36\pi + 36\pi(3.1 - 3) = 36\pi + 3.6\pi = 39.6\pi$. This gives us an increase of approximately 3.6π in^3.

A9. **(C)** Differentiating implicitly yields $4x - 3y^2 y' = 0$. So $y' = \frac{4x}{3y^2}$. The tangent-line approximation to the curve at the point $(3,2)$ has a slope of $\left.\frac{dy}{dx}\right|_{(3,2)} = \frac{4(3)}{3(2)^2} = 1$.

So an equation of the tangent line is $y = 2 + 1(x - 3)$, and we approximate $y(3.04) \approx 2 + (3.04 - 3) = 2.04$.

A10. **(B)** We want to approximate the change in area of the square when a side of length e increases by $0.01e$. The answer is

$$A'(e)(0.01e) \quad \text{or} \quad 2e(0.01e)$$

A11. **(D)** Since $V = e^3$, $V' = 3e^2$. Therefore at $e = 10$, the slope of the tangent line is 300. The change in volume is approximately $300(\pm 0.1) = 30$ in.$^{3.}$

A12. **(D)** $f'(x) = 4x^3 - 8x = 4x(x^2 - 2)$. $f' = 0$ if $x = 0$ or $\pm\sqrt{2}$

$f''(x) = 12x^2 - 8$; f'' is positive if $x = \pm\sqrt{2}$, negative if $x = 0$

A13. **(C)** Since $f''(x) = 4(3x^2 - 2)$, it equals 0 if $x = \pm\sqrt{\frac{2}{3}}$. Since f'' changes sign from positive to negative at $x = -\sqrt{\frac{2}{3}}$ and from negative to positive at $x = +\sqrt{\frac{2}{3}}$ both locate inflection points.

A14. **(A)** The domain of y is $\{x \mid x \leq 2\}$. Note that y is negative for each x in the domain except 2, where $y = 0$.

A15. **(B)** $f'(x)$ changes sign (from negative to positive) as x passes through zero only.

A16. **(D)** The graph must be decreasing and concave downward.

A17. **(A)** The graph must be concave upward but decreasing.

A18. **(B)** The object moves to the right when v is positive. Since $v = \dfrac{ds}{dt} = 3(t-2)^2$, $v > 0$ for all $t \neq 2$.

A19. **(A)** The speed $= |v|$. From Question A18, $|v| = v$. The least value of v is 0.

A20. **(A)** The acceleration $a = \dfrac{dv}{dt}$. From Question A18, $a = 6(t-2)$.

A21. **(A)** The speed is decreasing when v and a have opposite signs. The answer is $t < 2$ since for all such t the velocity is positive while the acceleration is negative. For $t > 2$, both v and a are positive.

A22. **(B)** The particle is at rest when $v = 0$; $v = 2t(2t^2 - 9t + 12) = 0$ only if $t = 0$. Note that the discriminant of the quadratic factor ($b^2 - 4ac$) is negative.

A23. **(D)** Since $a = 12(t-1)(t-2)$, we check the signs of a in the intervals $t < 1$, $1 < t < 2$, and $t > 2$. We choose those where $a > 0$.

A24. **(A)** From Questions A22 and A23, we see that $v > 0$ if $t > 0$ and that $a > 0$ if $t < 1$ or $t > 2$. So both v and a are positive if $0 < t < 1$ or $t > 2$. There are no values of t for which both v and a are negative.

A25. **(C)** See the figure below, which shows the motion of the particle during the time interval $-2 \leq t \leq 4$. The particle is at rest when $t = 0$ or 3 but reverses direction only at 3. The endpoints need to be checked here, of course. Indeed, the maximum displacement occurs at one of those, namely, when $t = -2$.

A26. **(C)** Since $v = 5t^3(t+4)$, $v = 0$ when $t = -4$ or 0. Note that v does change sign at each of these times.

A27. **(D)** Since $x = 3 \cos \dfrac{\pi}{3} t$ and $y = 2 \sin \dfrac{\pi}{3} t$, we note that $\left(\dfrac{x}{3}\right)^2 + \left(\dfrac{y}{2}\right)^2 = 1$.

A28. **(A)** Note that $\mathbf{v} = \left\langle -\pi \sin \dfrac{\pi}{3} t, \dfrac{2\pi}{3} \cos \dfrac{\pi}{3} t \right\rangle$. At $t = 3$,

$$|\mathbf{v}| = \sqrt{(-\pi \cdot 0)^2 + \left(\dfrac{2}{3} \cdot -1\right)^2}$$

A29. **(B)** $\mathbf{a} = \left\langle -\dfrac{\pi^2}{3} \cos \dfrac{\pi}{3} t, \dfrac{2\pi^2}{9} \sin \dfrac{\pi}{3} t \right\rangle$. At $t = 3$,

$$|\mathbf{a}| = \sqrt{\left(\dfrac{-\pi^2}{3} \cdot -1\right)^2 + \left(\dfrac{-2\pi^2}{9} \cdot 0\right)^2}$$

A30. **(D)** The slope of the curve is the slope of **v**, namely, $\dfrac{\frac{dy}{dt}}{\frac{dx}{dt}}$. At $t = \dfrac{1}{2}$, the slope is equal to

$$\frac{\frac{2\pi}{3} \cdot \cos\frac{\pi}{3}}{-\pi \cdot \sin\frac{\pi}{6}} = -\frac{2}{3}\cot\frac{\pi}{6}$$

A31. **(C)** Since $V = \dfrac{4}{3}\pi r^3$, $\dfrac{dV}{dt} = 4\pi r^2 \dfrac{dr}{dt}$. Since $\dfrac{dV}{dt} = 4$, $\dfrac{dr}{dt} = \dfrac{1}{\pi r^2}$. When $V = \dfrac{32\pi}{3}$, $r = 2$ and $\dfrac{dr}{dt} = \dfrac{1}{4\pi}$.

$$S = 4\pi r^2; \; \frac{dS}{dt} = 8\pi r \frac{dr}{dt}$$

when $r = 2$, $\dfrac{ds}{dt} = 8\pi(2)\left(\dfrac{1}{4\pi}\right) = 4$.

A32. **(B)** See Figure 4.21 on page 189. Replace the printed measurements of the radius and height by 10 and 20, respectively. We are given here that $r = \dfrac{h}{2}$ and that $\dfrac{dh}{dt} = -\dfrac{1}{2}$. Since $V = \dfrac{1}{3}\pi r^2 h$, we have $V = \dfrac{\pi}{3}\dfrac{h^3}{4}$, so

$$\frac{dV}{dt} = \frac{\pi h^2}{4} \cdot \frac{dh}{dt} = \frac{-\pi h^2}{8}$$

Replace h by 8.

A33. **(D)** $y' = \dfrac{e^x(x-1)}{x^2}$ $(x \neq 0)$. Since $y' = 0$ if $x = 1$ and changes from negative to positive as x increases through 1, $x = 1$ yields a minimum. Evaluate y at $x = 1$.

A34. **(A)** The domain of y is $-\infty < x < \infty$. The graph of y, which is nonnegative, is symmetric to the y-axis. The inscribed rectangle has area $A = 2xe^{-x^2}$.

Thus $A' = \dfrac{2(1 - 2x^2)}{e^{x^2}}$, which is 0 when the positive value of x is $\dfrac{\sqrt{2}}{2}$. This value of x yields maximum area. Evaluate A.

A35. **(B)** See the figure below. If we let m be the slope of the line, then its equation is $y - 2 = m(x - 1)$ with intercepts as indicated in the figure.

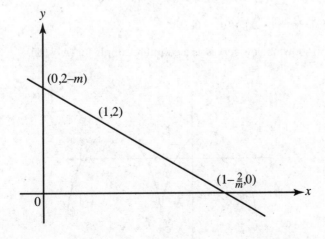

The area A of the triangle is given by

$$A = \frac{1}{2}(2-m)\left(1 - \frac{2}{m}\right) = \frac{1}{2}\left(4 - \frac{4}{m} - m\right)$$

Then $\dfrac{dA}{dm} = \dfrac{1}{2}\left(\dfrac{4}{m^2} - 1\right)$ and equals 0 when $m = \pm 2$; m must be negative.

A36. **(C)** Let $q = (x-6)^2 + y^2$ be the quantity to be minimized. Then

$$q = (x-6)^2 + (x^2 - 4)$$

$q' = 0$ when $x = 3$. Note that it suffices to minimize the square of the distance.

A37. **(D)** Minimize, if possible, xy, where $x^2 + y^2 = 200$ $(x, y > 0)$. The derivative of the product is $\dfrac{2(100 - x^2)}{\sqrt{200 - x^2}}$, which equals 0 for $x = 10$. The derivative is positive to the left of that point and negative to the right, showing that $x = 10$ yields a maximum product. No minimum exists.

A38. **(C)** Minimize $q = (x-2)^2 + \dfrac{18}{x}$. Since

$$q' = 2(x-2) - \frac{18}{x^2} = \frac{2(x^3 - 2x^2 - 9)}{x^2}$$

$q' = 0$ if $x = 3$. Since q' is negative to the left of $x = 3$ and positive to the right, the minimum value of q occurs at $x = 3$.

A39. **(A)** If h is a small negative number, then $27 + h$ is very close to 27. We can use the function $f(x) = \sqrt[3]{x}$ and find the local linear approximation to $f(x)$ at $x = 27$:

$$f(27) = \sqrt[3]{27} = 3$$

$$f'(x) = \frac{1}{3}x^{-2/3} = \frac{1}{3x^{2/3}}$$

$$f'(27) = \frac{1}{3 \cdot 27^{2/3}} = \frac{1}{3 \cdot 9} = \frac{1}{27}$$

This yields $y = 3 + \dfrac{1}{27}(x - 27)$ as the linear approximation. Therefore at $x = 27 + h$, $y = 3 + \dfrac{1}{27}((27 + h) - 27) = 3 + \dfrac{1}{27}h$.

A40. **(A)** Since $f'(x) = e^{-x}(1 - x)$, $f'(0) > 0$.

A41. **(D)** The graph shown below serves as a counterexample for (A)–(C).

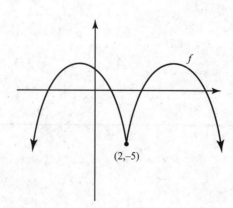

(2,–5)

A42. **(D)** Since $V = 10\ell w$, $V' = 10\left(\ell\,\dfrac{dw}{dt} + w\,\dfrac{d\ell}{dt}\right) = 10(8 \cdot -4 + 6 \cdot 2)$.

A43. **(D)** We differentiate implicitly: $3x^2 + x^2 y' + 2xy + 4y' = 0$. Then $y' = -\dfrac{3x^2 + 2xy}{x^2 + 4}$. At $(3,-2)$, $y' = -\dfrac{27 - 12}{9 + 4} = -\dfrac{15}{13}$.

A44. **(C)** Since $ab > 0$, a and b have the same sign; therefore $f''(x) = 12ax^2 + 2b$ never equals 0. The curve has one horizontal tangent at $x = 0$.

A45. **(C)** Since the first derivative is positive, the function must be increasing. However, the negative second derivative indicates that the rate of increase is slowing down, as seen in table (C).

A46. **(B)** Since $\dfrac{dy}{dx} = -\dfrac{3t^2}{2}$, therefore at $t = 1$, $\dfrac{dy}{dx} = -\dfrac{3}{2}$. Also, $x = 3$ and $y = 2$.

A47. **(A)** Let $f(x) = x^{1/3}$, and find the slope of the tangent line at $(64,4)$. Since $f'(x) = \dfrac{1}{3}x^{-2/3}$, $f'(64) = \dfrac{1}{48}$. If we move one unit to the left of 64, the tangent line will drop approximately $\dfrac{1}{48}$ unit.

A48. **(D)** $f\left(\dfrac{\pi}{4}\right) = \tan\left(\dfrac{\pi}{4}\right) = 1$, $f'(x) = \sec^2 x$, $f'\left(\dfrac{\pi}{4}\right) = \sec^2\left(\dfrac{\pi}{4}\right) = 2$

$\tan x \approx \tan\left(\dfrac{\pi}{4}\right) + \sec^2\left(\dfrac{\pi}{4}\right)\left(x - \dfrac{\pi}{4}\right) = 1 + 2\left(x - \dfrac{\pi}{4}\right)$

A49. **(D)** $f(0) = e^{k \cdot 0} = 1$, $f'(x) = ke^{kx}$, $f'(0) = ke^{k \cdot 0} = k$

$e^{kh} \approx e^{k \cdot 0} + ke^{k \cdot 0}(h - 0) = 1 + kh$

A50. **(D)** Since the curve has a positive y-intercept, $e > 0$. Note that $f'(x) = 2cx + d$ and $f''(x) = 2c$. Since the curve is concave down, $f''(x) < 0$, implying that $c < 0$. Since the curve is decreasing at $x = 0$, $f'(0)$ must be negative, implying, since $f'(0) = d$, that $d < 0$. Therefore $c < 0$, $d < 0$, and $e > 0$.

A51. **(A)** $f'(100) \approx \dfrac{f(102) - f(100)}{102 - 100} = \dfrac{2.0086 - 2}{2}$

B1. **(A)** $y' = \dfrac{1}{\sqrt{2x+1}}$. Solving the equation of the line for y yields $y = \dfrac{1}{3}x - 2$, so to be parallel the slope of the tangent must also be $\dfrac{1}{3}$. If $\dfrac{1}{\sqrt{2x+1}} = \dfrac{1}{3}$, then $2x + 1 = 9$.

B2. **(D)** The slope $y' = \dfrac{2x}{4}$; at the given point $y' = -\dfrac{4}{4} = -1$ and $y = 1$. An equation is therefore

$$y - 1 = -1(x + 2) \quad \text{or} \quad x + y + 1 = 0$$

B3. **(B)** $a \approx \dfrac{\Delta v}{\Delta t} = \dfrac{v(8) - v(4)}{8 - 4} = \dfrac{10 - 16}{4}\,\text{ft/sec}^2$

B4. **(D)** Because $f'(x) > 0$ on the interval $2 < x < 3$, the value of f increases.

B5. **(B)** For $x < 5$, $f' > 0$, so f is increasing; for $x > 5$, f is decreasing.

B6. **(D)** The graph of f being concave downward implies that $f'' < 0$, which implies that f' is decreasing.

B7. **(D)** Speed is the magnitude of velocity; at $t = 3$, speed $= 10$ ft/sec.

B8. **(C)** Speed increases from 0 at $t = 2$ to 10 at $t = 3$; it is constant or decreasing elsewhere.

B9. **(D)** Acceleration is positive when the *velocity* increases.

B10. **(D)** Acceleration is undefined when velocity is not differentiable. Here that occurs at $t = 1, 2, 3$.

B11. **(A)** Acceleration is the derivative of velocity. Since the velocity is linear, its derivative is its slope.

B12. **(C)** Positive velocity implies motion to the right ($t < 2$); negative velocity ($t > 2$) implies motion to the left.

B13. **(B)** The average rate of change of velocity is $\dfrac{v(3) - v(0)}{3 - 0} = \dfrac{-10 - 5}{-3}$ ft/sec^2.

B14. **(B)** The slope of $y = x^3$ is $3x^2$. It is equal to 3 when $x = \pm 1$. At $x = 1$, an equation of the tangent is

$$y - 1 = 3(x - 1) \qquad \text{or} \qquad y = 3x - 2$$

At $x = -1$, an equation is

$$y + 1 = 3(x + 1) \qquad \text{or} \qquad y = 3x + 2$$

B15. **(C)** Let the tangent to the parabola from $(3,5)$ meet the curve at (x_1, y_1). An equation is $y - 5 = 2x_1(x - 3)$. Since the point (x_1, y_1) is on both the tangent and the parabola, we solve simultaneously:

$$y_1 - 5 = 2x_1(x_1 - 3) \qquad \text{and} \qquad y_1 = x_1^2$$

The points of tangency are $(5,25)$ and $(1,1)$. The slopes, which equal $2x_1$, are 10 and 2.

B16. **(D)** $a \approx \dfrac{\Delta v}{\Delta t} = \dfrac{v(1.5) - v(1.0)}{0.5} = \dfrac{13.0 - 12.2}{0.5}$ ft/sec^2.

B17. **(C)** The graph of $f'(x) = x \sin x - \cos x$ is drawn here in the window $[0,4] \times [-3,3]$:

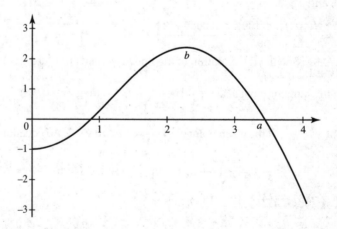

A local maximum exists at $x = a$, where f' changes from positive to negative; use your calculator to approximate a.

B18. **(B)** f'' changes sign when f' changes from increasing to decreasing (or vice versa). Again, use your calculator to approximate the x-coordinate at b.

B19. **(D)** Eliminating t yields the equation $y = -\dfrac{1}{4}x^2 + 2x$.

B20. **(B)** $|\mathbf{v}| = \sqrt{\left(\dfrac{dx}{dt}\right)^2 + \left(\dfrac{dy}{dt}\right)^2} = \sqrt{2^2 + (4 - 2t)^2}$

B21. **(C)** Since $|\mathbf{v}| = 2\sqrt{t^2 - 4t + 5}$, $\dfrac{d|\mathbf{v}|}{dt} = \dfrac{2t - 4}{\sqrt{t^2 - 4t + 5}} = 0$ if $t = 2$. We note that, as t increases through 2, the sign of $|\mathbf{v}|'$ changes from negative to positive, assuring a minimum of $|\mathbf{v}|$ at $t = 2$. Evaluate $|\mathbf{v}|$ at $t = 2$.

B22. **(C)** The direction of $\mathbf{a}$ is $\tan^{-1} \dfrac{\frac{d^2y}{dt^2}}{\frac{d^2x}{dt^2}}$. Since $\dfrac{d^2x}{dt^2} = 0$ and $\dfrac{d^2y}{dt^2} = -2$, the acceleration is always directed downward. Its magnitude, $\sqrt{0^2 + (-2)^2}$, is 2 for all t.

B23. **(D)** Using the notations v_x, v_y, a_x, and a_y, we are given that $|\mathbf{v}| = \sqrt{v_x^2 + v_y^2} = k$, where k is a constant. Then

$$\frac{d|\mathbf{v}|}{dt} = \frac{v_x a_x + v_y a_y}{|\mathbf{v}|} = 0 \quad \text{or} \quad \frac{v_x}{v_y} = -\frac{a_y}{a_x}$$

B24. **(B)** $\dfrac{dy}{dt} = \left(2 - \dfrac{1}{x}\right)\dfrac{dx}{dt} = \left(2 - \dfrac{1}{x}\right)(-2)$

B25. **(B)** The rate of change of the distance from the origin with respect to time is given by $\dfrac{dr}{dt} = -6 \sin 2\theta \cdot 2 \cdot \dfrac{d\theta}{dt}$. At $\theta = \dfrac{\pi}{6}$, $\dfrac{dr}{dt} = -6 \sin\left(\dfrac{\pi}{3}\right) \cdot 2 \cdot \dfrac{1}{3} = -2\sqrt{3}$.

B26. **(B)** In parametric form, $x = r \cos \theta$.

$\dfrac{dx}{dt} = r \cdot (-\sin \theta) \cdot \dfrac{d\theta}{dt} + \cos \theta \cdot \dfrac{dr}{dt}$. At $\theta = \dfrac{\pi}{6}$, $\dfrac{dr}{dt} = -2\sqrt{3}$ (see Answer B25)

and $r\left(\dfrac{\pi}{6}\right) = 3$. Therefore,

$$\frac{dx}{dt} = 3\left(-\sin\left(\frac{\pi}{6}\right)\right) \cdot \left(\frac{1}{3}\right) + \cos\left(\frac{\pi}{6}\right) \cdot (-2\sqrt{3}) = -\frac{7}{2}$$

B27. **(B)** A local minimum exists where f changes from decreasing ($f' < 0$) to increasing ($f' > 0$). Note that f has local maxima at both endpoints, $x = 0$ and $x = 5$.

B28. **(D)** f'' changes sign when f' changes from increasing to decreasing (or vice versa).

B29. **(D)** At $x = a$, f' changes from increasing ($f'' > 0$) to decreasing ($f'' < 0$). Thus f changes from concave upward to concave downward and therefore has a point of inflection at $x = a$. Note that f is differentiable at a (because $f'(a)$ exists) and therefore continuous at a.

B30. **(C)** We know that $\dfrac{dh}{dt} = \dfrac{dr}{dt} = 2$. Since $S = 2\pi rh$, $\dfrac{dS}{dt} = 2\pi\left(r\dfrac{dh}{dt} + h\dfrac{dr}{dt}\right)$.

B31. **(D)** An equation of the tangent is $y = -2x + 5$. Its intercepts are $\dfrac{5}{2}$ and 5.

B32. **(B)** See the figure. At noon, car A is at O, car B at N; the cars are shown t hours after noon.

We know that $\dfrac{dx}{dt} = -60$ and that $\dfrac{dy}{dt} = 40$.

Using $s^2 = x^2 + y^2$, we get

$$\frac{ds}{dt} = \frac{x\dfrac{dx}{dt} + y\dfrac{dy}{dt}}{s} = \frac{-60x + 40y}{s}$$

At 1 P.M., $x = 30$, $y = 40$, and $s = 50$.

B33. **(A)** See the figure from Answer B32. The distance from car A to the intersection is denoted by y on the figure. Since that car is traveling at 40 mph, the distance car A travels in t hours is $40t$. Car A started at point O and is moving away, so $y = 40t$. The distance from car B to the intersection is denoted by x on the figure. Since that car is traveling at 60 mph, the distance car B travels in t hours is $60t$. Car B started 90 miles away from point O, and its distance to point O is decreasing, so $x = 90 - 60t$. Using Pythagorean theorem, the distance between the cars is $s = \sqrt{x^2 + y^2} = \sqrt{(90 - 60t)^2 + (40t)^2}$. Using your calculator, graph this distance function, and find the minimum using the features of the calculator.

B34. **(B)** Maximum acceleration occurs when the derivative (slope) of velocity is greatest.

B35. **(B)** The object changes direction only when velocity changes sign. Velocity changes sign from negative to positive at $t = 5$.

B36. **(C)** At $x = 1$ and 3, $f'(x) = 0$; therefore f has horizontal tangents.

For $x < 1$, $f' > 0$; therefore f is increasing.

For $x > 1$, $f' < 0$, so f is decreasing.

For $x < 2$, f' is decreasing, so $f'' < 0$ and the graph of f is concave downward.

For $x > 2$, f' is increasing, so $f'' > 0$ and the graph of f is concave upward.

B37. **(D)** From the graph, $f'(2) = 3$, and we are told the line passes through $(2,10)$. We therefore have $f(x) \approx 10 + 3(x - 2) = 3x + 4$.

B38. **(C)** Note that $\dfrac{dy}{dx} = 0$ at Q and R. At P and Q, $\dfrac{d^2y}{dx^2} > 0$. At R and S, $\dfrac{d^2y}{dx^2} = 0$.

B39. **(D)** Only at S does the graph both rise and change concavity.

B40. **(B)** Only at Q is the tangent horizontal and the curve concave up.

B41. **(C)** Since $f'(6) = 4$, an equation of the tangent at $(6,30)$ is $y - 30 = 4(x - 6)$. Therefore $f(x) \approx 4x + 6$ and $f(6.02) \approx 30.08$.

B42. **(C)** $f'(x) = \dfrac{x}{\sqrt{x^2 + 16}}$; $f'(-3) = -\dfrac{3}{5}$; $f(-3) = 5 \Rightarrow y = 5 - \dfrac{3}{5}(x + 3)$

5
Antidifferentiation

Learning Objectives

In this chapter, you will review:

→ Indefinite integrals
→ Formulas for antiderivatives of basic functions
→ Techniques for finding antiderivatives (including substitution)

In addition, BC Calculus students will review two important techniques of integration:

→ Integration by partial fractions
→ Integration by parts

A. Antiderivatives

The *antiderivative* or *indefinite integral* of a function $f(x)$ is a function $F(x)$ whose derivative is $f(x)$. Since the derivative of a constant equals zero, the antiderivative of $f(x)$ is not unique; that is, if $F(x)$ is an integral of $f(x)$, then so is $F(x) + C$, where C is any constant. The arbitrary constant C is called the *constant of integration*. The indefinite integral of $f(x)$ is written as $\int f(x)\, dx$; thus

Antiderivative

Indefinite integral

$$\int f(x)\, dx = F(x) + C \quad \text{if } \frac{dF(x)}{dx} = f(x)$$

The function $f(x)$ is called the *integrand*. The Mean Value Theorem can be used to show that, if two functions have the same derivative on an interval, then they differ at most by a constant; that is, if $\dfrac{dF(x)}{dx} = \dfrac{dG(x)}{dx}$, then

$$F(x) - G(x) = C \quad (C \text{ is a constant})$$

B. Basic Formulas

Familiarity with the following fundamental integration formulas is essential.

$$\int k f(x)\, dx = k \int f(x)\, dx \qquad (k \neq 0) \tag{1}$$

$$\int [f(x) + g(x)]\, dx = \int f(x)\, dx + \int g(x)\, dx \tag{2}$$

$$\int u^n\, du = \frac{u^{n+1}}{n+1} + C \qquad (n \neq -1) \tag{3}$$

$$\int \frac{du}{u} = \ln |u| + C \tag{4}$$

$$\int \cos u\, du = \sin u + C \tag{5}$$

$$\int \sin u \, du = -\cos u + C \tag{6}$$

$$\int \tan u \, du = \ln |\sec u| + C \tag{7}$$

$$\text{or } -\ln |\cos u| + C$$

$$\int \cot u \, du = \ln |\sin u| + C \tag{8}$$

$$\text{or } -\ln |\csc u| + C$$

$$\int \sec^2 u \, du = \tan u + C \tag{9}$$

$$\int \csc^2 u \, du = -\cot u + C \tag{10}$$

$$\int \sec u \tan u \, du = \sec u + C \tag{11}$$

$$\int \csc u \cot u \, du = -\csc u + C \tag{12}$$

$$\int e^u \, du = e^u + C \tag{13}$$

$$\int a^u \, du = \frac{a^u}{\ln a} + C \qquad (a > 0, \, a \neq 1) \tag{14}$$

$$\int \frac{du}{\sqrt{1 - u^2}} = \sin^{-1} u + C \tag{15}$$

$$\text{or } \arcsin u + C$$

$$\int \frac{du}{1 + u^2} = \tan^{-1} u + C \tag{16}$$

$$\text{or } \arctan u + C$$

$$\int \frac{du}{u\sqrt{u^2 - 1}} = \sec^{-1} |u| + C \tag{17}$$

$$\text{or } \arcsec |u| + C$$

All the references in the following set of examples are to the preceding basic formulas. In all of these, whenever u is a function of x, we define du to be $u'(x) \, dx$; when u is a function of t, we define du to be $u'(t) \, dt$; and so on.

> **Example 1**

$$\int 5x\,dx = 5\int x\,dx \text{ by (1)}$$

$$= 5\left(\frac{x^2}{2}\right) + C \text{ by (3)}$$

> **Example 2**

$$\int\left(x^4 + \sqrt[3]{x^2} - \frac{2}{x^2} - \frac{1}{3\sqrt[3]{x}}\right)dx = \int\left(x^4 + x^{2/3} - 2x^{-2} - \frac{1}{3}x^{-1/3}\right)dx$$

$$= \int x^4\,dx + \int x^{2/3}\,dx - 2\int x^{-2}\,dx - \frac{1}{3}\int x^{-1/3}\,dx \text{ by (1) and (2)}$$

$$= \frac{x^5}{5} + \frac{x^{5/3}}{\frac{5}{3}} - \frac{2x^{-1}}{-1} - \frac{1}{3}\frac{x^{2/3}}{\frac{2}{3}} + C \text{ by (3)}$$

$$= \frac{x^5}{5} + \frac{3}{5}x^{5/3} + \frac{2}{x} - \frac{1}{2}x^{2/3} + C$$

> **Example 3**

Similarly, $\displaystyle\int(3 - 4x + 2x^3)\,dx = \int 3\,dx - 4\int x\,dx + 2\int x^3\,dx$

$$= 3x - \frac{4x^2}{2} + \frac{2x^4}{4} + C$$

$$= 3x - 2x^2 + \frac{x^4}{2} + C$$

> **Example 4**

$\displaystyle\int 2(1 - 3x)^2\,dx$ is integrated most efficiently by using formula (3) with $u = 1 - 3x$ and $du = u'(x)\,dx = -3\,dx$.

$$2\int(1 - 3x)^2\,dx = \frac{2}{-3}\int(1 - 3x)^2(-3\,dx)$$

$$= -\frac{2}{9}u^3 + C \text{ by (3)}$$

$$= -\frac{2}{9}(1 - 3x)^3 + C$$

> **Example 5**

$\displaystyle\int(2x^3 - 1)^5 \cdot x^2\,dx = \frac{1}{6}\int(2x^3 - 1)^5 \cdot (6x^2\,dx) = \frac{1}{6}\int u^5\,du$, where $u = 2x^3 - 1$ and $du = u'(x)\,dx = 6x^2\,dx$; this, by formula (3), equals

$$\frac{1}{6} \cdot \frac{u^6}{6} + C = \frac{1}{36}(2x^3 - 1)^6 + C$$

> **Example 6**

$\displaystyle\int\sqrt[3]{1 - x}\,dx = \int(1 - x)^{1/3}\,dx = -\int(1 - x)^{1/3}(-1\,dx) = -\int u^{1/3}\,du$, where $u = 1 - x$

and $du = -1\,dx$; this, by formula (3), yields $-\dfrac{u^{4/3}}{\frac{4}{3}} + C = -\dfrac{3}{4}(1 - x)^{4/3} + C.$

› Example 7

$$\int \frac{x}{\sqrt{3-4x^2}}\, dx = \int (3-4x^2)^{-1/2} \cdot (x\, dx) = -\frac{1}{8}\int (3-4x^2)^{-1/2}(-8x\, dx) = -\frac{1}{8}\int u^{-1/2}\, du$$

$$= -\frac{1}{8}\frac{u^{1/2}}{\frac{1}{2}} + C \text{ by (3)} = -\frac{1}{4}\sqrt{3-4x^2} + C$$

› Example 8

$$\int \frac{4x^2}{(x^3-1)^3}\, dx = 4\int (x^3-1)^{-3}\cdot x^2\, dx = \frac{4}{3}\int (x^3-1)^{-3}(3x^2\, dx)$$

$$= \frac{4}{3}\frac{(x^3-1)^{-2}}{-2} + C = -\frac{2}{3}\frac{1}{(x^3-1)^2} + C$$

› Example 9

$$\int \frac{(1+\sqrt{x})^4}{\sqrt{x}}\, dx = \int (1+x^{1/2})^4 \cdot \frac{1}{x^{1/2}}\, dx. \text{ Now let } u = 1+x^{1/2}, \text{ and note that } du = \frac{1}{2}x^{1/2}\, dx;$$

this gives $2\int (1+x^{1/2})^4\left(\frac{1}{2x^{1/2}}\, dx\right) = \frac{2}{5}(1+\sqrt{x})^5 + C.$

› Example 10

$$\int (2-y)^2 \cdot \sqrt{y}\, dy = \int (4-4y+y^2)\cdot y^{1/2}\, dy = \int \left(4y^{1/2} - 4y^{3/2} + y^{5/2}\right) dy$$

$$= 4\cdot\frac{2}{3}y^{3/2} - 4\cdot\frac{2}{5}\cdot y^{5/2} + \frac{2}{7}\cdot y^{7/2} + C \text{ by (2)}$$

$$= \frac{8}{3}y^{3/2} - \frac{8}{5}y^{5/2} + \frac{2}{7}y^{7/2} + C$$

› Example 11

$$\int \frac{x^3-x-4}{2x^2}\, dx = \frac{1}{2}\int\left(x - \frac{1}{x} - \frac{4}{x^2}\right) dx = \frac{1}{2}\left(\frac{x^2}{2} - \ln|x| + \frac{4}{x}\right) + C$$

› Example 12

$$\int \frac{3x-1}{\sqrt[3]{1-2x+3x^2}}\, dx = \int (1-2x+3x^2)^{-1/3}(3x-1)\, dx$$

$$= \frac{1}{2}\int (1-2x+3x^2)^{-1/3}(6x-2)\, dx \text{ or } \frac{1}{2}\cdot\frac{3}{2}(1-2x+3x^2)^{2/3} + C \text{ by (3)}$$

$$= \frac{3}{4}(1-2x+3x^2)^{2/3} + C$$

› Example 13

$$\int \frac{2x^2-4x+3}{(x-1)^2}\, dx = \int \frac{2x^2-4x+3}{x^2-2x+1}\, dx = \int\left(2 + \frac{1}{(x-1)^2}\right) dx = \int 2\, dx + \frac{dx}{(x-1)^2} = 2x - \frac{1}{x-1} + C$$

Example 13 illustrates the following principle:

If the degree of the numerator of a rational function is not less than that of the denominator, divide until a remainder of lower degree is obtained.

> **Example 14**

$$\int \frac{du}{u-3} = \ln|u-3| + C \text{ by (4)}$$

> **Example 15**

$$\int \frac{z\,dz}{1-4z^2} = -\frac{1}{8}\ln|1-4z^2| + C \text{ by (4)}$$

> **Example 16**

$$\int \frac{\cos x}{5+2\sin x}\,dx = \frac{1}{2}\int \frac{2\cos x\,dx}{5+2\sin x} = \frac{1}{2}\ln(5+2\sin x) + C \text{ by (4) with } u = 5+2\sin x. \text{ The}$$
absolute-value sign is not necessary here since $(5+2\sin x) > 0$ for all x.

> **Example 17**

$$\int \frac{e^x}{1-2e^x}\,dx = -\frac{1}{2}\int \frac{-2e^x}{1-2e^x}\,dx = -\frac{1}{2}\ln|1-2e^x| + C \text{ by (4)}$$

> **Example 18**

$$\int \frac{x}{1-x}\,dx = \int \left(-1 + \frac{1}{1-x}\right)dx \text{ (by long division)} = -x - \ln|1-x| + C$$

> **Example 19**

$$\int \sin(1-2y)\,dy = -\frac{1}{2}\int \sin(1-2y)(-2\,dy)$$
$$= -\frac{1}{2}[-\cos(1-2y)] + C \text{ by (6)} = \frac{1}{2}\cos(1-2y) + C$$

> **Example 20**

$$\int \sin^2\frac{x}{2}\cos\frac{x}{2}\,dx = 2\int \left(\sin\frac{x}{2}\right)^2 \cos\frac{x}{2}\,\frac{dx}{2} = \frac{2}{3}\sin^3\frac{x}{2} + C \text{ by (3)}$$

> **Example 21**

$$\int \frac{\sin x}{1+3\cos x}\,dx = -\frac{1}{3}\int \frac{-3\sin x\,dx}{1+3\cos x} = \frac{1}{3}\ln|1+3\cos x| + C$$

> **Example 22**

$$\int e^{\tan y}\sec^2 y\,dy = e^{\tan y} + C \text{ by (13) with } u = \tan y$$

> **Example 23**

$$\int e^x \tan e^x\,dx = \ln|\sec e^x| + C \text{ by (7) with } u = e^x$$

> **Example 24**

$$\int \frac{\cos z}{\sin^2 z}\, dz = \int \csc z \cot z\, dz = -\csc z + C \text{ by (12)}$$

> **Example 25**

$$\int \tan t \sec^2 t\, dt = \frac{\tan^2 t}{2} + C \text{ by (3) with } u = \tan t \text{ and } du = u'(t)\, dt = \sec^2 t\, dt$$

> **Example 26**

(a) $\displaystyle\int \frac{dz}{\sqrt{9-z^2}} = \frac{1}{3}\int \frac{dz}{\sqrt{1-\left(\frac{z}{3}\right)^2}} = 3\cdot\frac{1}{3}\int \frac{\frac{1}{3}\,dz}{\sqrt{1-\left(\frac{z}{3}\right)^2}} = \sin^{-1}\frac{z}{3} + C \text{ by (15) with } u = \frac{z}{3}$

(b) $\displaystyle\int \frac{z\,dz}{\sqrt{9-z^2}} = -\frac{1}{2}\int (9-z^2)^{-1/2}(-2z\,dz) = -\frac{1}{2}\frac{(9-z^2)^{1/2}}{1/2} + C \text{ by (3)} = -\sqrt{9-z^2} + C$

   with $u = 9 - z^2,\ n = -\dfrac{1}{2}$

(c) $\displaystyle\int \frac{z\,dz}{9-z^2} = -\frac{1}{2}\int \frac{(-2z\,dz)}{9-z^2} = -\frac{1}{2}\ln\left|9-z^2\right| + C \text{ by (4) with } u = 9-z^2$

(d) $\displaystyle\int \frac{z\,dz}{(9-z^2)^2} = -\frac{1}{2}\int (9-z^2)^{-2}(-2z\,dz) = \frac{1}{2(9-z^2)} + C \text{ by (3)}$

(e) $\displaystyle\int \frac{dz}{9+z^2} = \frac{1}{9}\int \frac{dz}{1+\left(\frac{z}{3}\right)^2} = 3\cdot\frac{1}{9}\int \frac{\frac{1}{3}dz}{1+\left(\frac{z}{3}\right)^2} = \frac{1}{3}\tan^{-1}\frac{z}{3} + C \text{ by (16) with } u = \frac{z}{3}$

> **Example 27**

$$\int \frac{dx}{\sqrt{x}\,(1+2\sqrt{x})} = \int \frac{x^{-1/2}\,dx}{1+2\sqrt{x}} = \ln\left(1+2\sqrt{x}\right) + C \text{ by (4) with } u = 1 + 2\sqrt{x} \text{ and } du = \frac{dx}{\sqrt{x}}$$

> **Example 28**

$$\int \sin x \cos x\, dx = \frac{1}{2}\sin^2 x + C \text{ by (3) with } u = \sin x;\ OR = -\frac{1}{2}\cos^2 x + C \text{ by (3) with } u = \cos x;$$

$$OR = -\frac{1}{4}\cos 2x + C \text{ by (6), where we use the trigonometric identity } \sin 2x = 2\sin x \cos x$$

> **Example 29**

$$\int \frac{\cos\sqrt{x}}{\sqrt{x}}\, dx = 2\int \cos(x^{1/2})\left(\frac{1}{2}x^{-1/2}\,dx\right) = 2\sin\sqrt{x} + C \text{ by (5) with } u = \sqrt{x}$$

> **Example 30**

$$\int \frac{x\,dx}{x^4+1} = \frac{1}{2}\int \frac{2x\,dx}{1+(x^2)^2} = \frac{1}{2}\tan^{-1}x^2 + C \text{ by (16) with } u = x^2$$

> **Example 31**

$$\int \frac{dx}{x^2 + 4x + 13} = \int \frac{dx}{9 + (x+2)^2} = \frac{1}{9} \int \frac{dx}{1 + \left(\frac{x+2}{3}\right)^2} = 3 \cdot \frac{1}{9} \int \frac{\frac{1}{3}\,dx}{1 + \left(\frac{x+2}{3}\right)^2}$$

$$= \frac{1}{3} \tan^{-1} \frac{x+2}{3} + C \text{ by (16) with } u = \frac{x+2}{3}$$

> **Example 32**

$$\int \frac{dy}{\sqrt{6y - y^2}} = \int \frac{dy}{\sqrt{9 - (y^2 - 6y + 9)}} = \frac{1}{3} \int \frac{dy}{\sqrt{1 - \left(\frac{y-3}{3}\right)^2}} = 3 \cdot \frac{1}{3} \int \frac{\frac{1}{3}\,dy}{\sqrt{1 - \left(\frac{y-3}{3}\right)^2}}$$

$$= \sin^{-1} \frac{y-3}{3} + C \text{ by (15) with } u = \frac{y-3}{3}$$

> **Example 33**

$$\int \frac{e^x}{9 + e^{2x}}\,dx = \frac{1}{9}\int \frac{e^x\,dx}{1 + \left(\frac{e^x}{3}\right)^2} = 3 \cdot \frac{1}{9}\int \frac{\frac{e^x}{3}\,dx}{1 + \left(\frac{e^x}{3}\right)^2} = \frac{1}{3}\tan^{-1}\frac{e^x}{3} + C \text{ by (16) with } u = \frac{e^x}{3}$$

> **Example 34**

$$\int \frac{e^x - e^{-x}}{e^x + e^{-x}}\,dx = \ln\left(e^x + e^{-x}\right) + C \text{ by (4) with } u = e^x + e^{-x}$$

> **Example 35**

$$\int \frac{x+1}{x^2 + 1}\,dx = \int \frac{x\,dx}{x^2 + 1} + \int \frac{dx}{x^2 + 1} = \frac{1}{2}\ln\left(x^2 + 1\right) + \tan^{-1}x + C \text{ by (4) and (16)}$$

> **Example 36**

$$\int \frac{dt}{\sin^2 2t} = \int \csc^2 2t\,dt = \frac{1}{2}\int \csc^2 2t(2\,dt) = -\frac{1}{2}\cot 2t + C \text{ by (10)}$$

> **Example 37**

$$\int \frac{\sin 2x}{1 + \sin^2 x}\,dx = \ln\left(1 + \sin^2 x\right) + C \text{ by (4)}$$

> **Example 38**

$$\int x^2 e^{-x^3}dx = -\frac{1}{3}\int e^{-x^3}\left(-3x^2\,dx\right) = -\frac{1}{3}\,e^{-x^3} + C \text{ by (13) with } u = -x^3$$

> **Example 39**

$$\int \frac{dy}{y\sqrt{1+\ln y}} = \int (1+\ln y)^{-1/2}\left(\frac{1}{y}\,dy\right) = 2\sqrt{1+\ln y} + C \text{ by (3)}$$

†C. Integration by Partial Fractions

The method of partial fractions makes it possible to express a rational function $\frac{f(x)}{g(x)}$ as a sum of simpler fractions. Here $f(x)$ and $g(x)$ are real polynomials in x and it is assumed that $\frac{f(x)}{g(x)}$ is a proper fraction; that is, that $f(x)$ is of lower degree than $g(x)$. If not, we divide $f(x)$ by $g(x)$ to express the given rational function as the sum of a polynomial and a proper rational function. Thus,

$$\frac{x^3 - x^2 - 2}{x(x-1)} = x - \frac{2}{x(x-1)}$$

where the fraction on the right is proper.

Theoretically, every real polynomial can be expressed as a product of (powers of) real linear factors and (powers of) real quadratic factors.†

In the following, the capital letters denote constants to be determined. We consider only nonrepeating linear factors. For each distinct linear factor $(x - a)$ of $g(x)$ we set up one partial fraction of the type $\frac{A}{x-a}$. The techniques for determining the unknown constants are illustrated in the following example.

> **Example 40**

Find $\int \frac{x^2 - x + 4}{x^3 - 3x^2 + 2x}\,dx$.

✓ **Solution**

We factor the denominator and then set

$$\frac{x^2 - x + 4}{x(x-1)(x-2)} = \frac{A}{x} + \frac{B}{x-1} + \frac{C}{x-2} \tag{1}$$

where the constants A, B, and C are to be determined. It follows that

$$x^2 - x + 4 = A(x-1)(x-2) + Bx(x-2) + Cx(x-1) \tag{2}$$

Since the polynomial on the right in (2) is to be identical to the one on the left, we can find the constants by either of the following methods:

Method One. We expand and combine the terms on the right in (2), getting

$$x^2 - x + 4 = (A + B + C)x^2 - (3A + 2B + C)x + 2A$$

We then *equate coefficients of like powers in x* and solve simultaneously. Thus

using the coefficients of x^2, we get	$1 = A + B + C$
using the coefficients of x, we get	$-1 = -(3A + 2B + C)$
using the constant coefficient,	$4 = 2A$

†For the Calculus BC exam, integration by partial fractions is restricted to only linear nonrepeating factors.

These equations yield $A = 2$, $B = -4$, $C = 3$.

METHOD TWO. Although equation (1) is meaningless for $x = 0$, $x = 1$, or $x = 2$, it is still true that equation (2) must hold even for these special values. We see, in (2), that

if $x = 0$,	then $4 = 2A$ and $A = 2$
if $x = 1$,	then $4 = -B$ and $B = -4$
if $x = 2$,	then $6 = 2C$ and $C = 3$

The second method is shorter than the first and more convenient when the denominator of the given fraction can be decomposed into nonrepeating linear factors.

Finally, then, the original integral equals

$$\int \left(\frac{2}{x} - \frac{4}{x-1} + \frac{3}{x-2} \right) dx = 2 \ln |x| - 4 \ln |x - 1| + 3 \ln |x - 2| + C'$$

$$= \ln \frac{x^2 |x - 2|^3}{(x - 1)^4} + C'$$

[The symbol "C'" appears here for the constant of integration because C was used in simplifying the original rational function.]

D. Integration by Parts

The Integration by Parts Formula stems from the equation for the derivative of a product:

$$\frac{d}{dx}(uv) = u\frac{dv}{dx} + v\frac{du}{dx} \text{ or, more conveniently, } d(uv) = u\,dv + v\,du$$

Hence, $u\,dv = d(uv) - v\,du$ and integrating gives us $\int u\,dv = \int d(uv) - \int v\,du$, or

$$\int u\,dv = uv - \int v\,du$$

Integration by Parts Formula

the Integration by Parts Formula. Success in using this important technique depends on being able to separate a given integral into parts u and dv so that (a) dv can be integrated, and (b) $\int v\,du$ is no more difficult to calculate than the original integral.

The following acronym may help to determine which function to designate as u when using Integration by Parts. The acronym is LIPET. Choose u in this order: (1) a **L**ogarithmic function, (2) an **I**nverse trigonometric function, (3) a **P**olynomial function, (4) an **E**xponential function, and (5) a **T**rigonometric function. *NOTE:* Some may advise to swap the last two and always choose Exponential last.

> **Example 41**

Find $\int x \cos x\, dx$.

✓ **Solution**

We let $u = x$ and $dv = \cos x\, dx$. Then $du = dx$ and $v = \sin x$. Thus, the Parts Formula yields

$$\int x \cos x\, dx = x \sin x - \int \sin x\, dx = x \sin x + \cos x + C$$

> **Example 42**

Find $\int x^2 e^x \, dx$.

✓ **Solution**

We let $u = x^2$ and $dv = e^x \, dx$. Then $du = 2x \, dx$ and $v = e^x$, so $\int x^2 e^x \, dx = x^2 e^x - \int 2x \, e^x \, dx$.

We use the Parts Formula again, this time letting $u = x$ and $dv = e^x dx$ so that $du = dx$ and $v = e^x$. Thus,

$$\int x^2 e^x \, dx = x^2 e^x - 2(xe^x - \int e^x \, dx) = x^2 e^x - 2xe^x + 2e^x + C$$

> **Example 43**

Find $I = \int e^x \cos x \, dx$.

✓ **Solution**

To integrate, we can let $u = e^x$ and $dv = \cos x \, dx$; then $du = e^x dx$, $v = \sin x$. Thus,

$$I = e^x \sin x - \int e^x \sin x \, dx$$

To evaluate the integral on the right, again we let $u = e^x$, $dv = \sin x \, dx$, so that $du = e^x dx$ and $v = -\cos x$. Then,

$$I = e^x \sin x - \left(-e^x \cos x + \int e^x \cos x \, dx\right)$$

$$= e^x \sin x + e^x \cos x - I$$

$$2I = e^x (\sin x + \cos x)$$

$$I = \frac{1}{2} e^x (\sin x + \cos x) + C$$

> **Example 44**

Find $\int x^4 \ln x \, dx$.

✓ **Solution**

We let $u = \ln x$ and $dv = x^4 \, dx$. Then, $du = \frac{1}{x} dx$ and $v = \frac{x^5}{5}$. Thus,

$$\int x^4 \ln x \, dx = \frac{x^5}{5} \ln x - \frac{1}{5} \int x^4 \, dx = \frac{x^5}{5} \ln x - \frac{x^5}{25} + C$$

The Tic-Tac-Toe Method[1]

This method of integrating is extremely useful when repeated integration by parts is necessary. To integrate $\int u(x)\,v(x)\,dx$, we construct a table as follows:

Here the column at the left contains the successive derivatives of $u(x)$. The column at the right contains the successive antiderivatives of $v(x)$ (always with $C = 0$); that is, $v_1(x)$ is the antiderivative of $v(x)$, $v_2(x)$ is the antiderivative of $v_1(x)$, and so on. The diagonal arrows join the pairs of factors whose products form the successive terms of the desired integral; above each arrow is the sign of that term. By the tic-tac-toe method,

$$\int u(x)\,v(x)\,dx = u(x)\,v_1(x) - u'(x)\,v_2(x) + u''(x)\,v_3(x) - u'''(x)\,v_4(x) + \dots.$$

> **Example 45**

To integrate $\int x^4 \cos x\,dx$ by the tic-tac-toe method, we let $u(x) = x^4$ and $v(x) = \cos x$, and get the following table:

---

[1]This method was described by K. W. Folley in Vol. 54 (1947) of the *American Mathematical Monthly* and was referred to in the movie *Stand and Deliver*.

The method yields

$$\int x^4 \cos x \, dx = x^4 \sin x - (-4x^3 \cos x) + (-12x^2 \sin x) - 24x \cos x + 24 \sin x + C$$
$$= x^4 \sin x + 4x^3 \cos x - 12x^2 \sin x - 24x \cos x + 24 \sin x + C$$

With the ordinary method we would have had to apply the Parts Formula four times to perform this integration.

E. Applications of Antiderivatives; Differential Equations

The following examples show how we use given conditions to determine constants of integration.

❯ Example 46

Find $f(x)$ if $f'(x) = 3x^2$ and $f(1) = 6$.

$$f(x) = \int 3x^2 \, dx = x^3 + C$$

✓ Solution

Since $f(1) = 6$, $1^3 + C$ must equal 6; so C must equal $6 - 1$ or 5, and $f(x) = x^3 + 5$.

❯ Example 47

Find a curve whose slope at each point (x,y) equals the reciprocal of the x-value if the curve contains the point $(e,-3)$.

✓ Solution

We are given that $\dfrac{dy}{dx} = \dfrac{1}{x}$ and that $y = -3$ when $x = e$. This equation is also solved by integration. Since $\dfrac{dy}{dx} = \dfrac{1}{x}$, $dy = \dfrac{1}{x} \, dx$.

Thus, $y = \ln x + C$. We now use the given condition, by substituting the point $(e,-3)$, to determine C. Since $-3 = \ln e + C$, we have $-3 = 1 + C$, and $C = -4$. Then, the solution of the given equation subject to the given condition is

$$y = \ln x - 4$$

Differential Equations: Motion Problems

Differential equation

An equation involving a derivative is called a *differential equation*. In Examples 46 and 47, we solved two simple differential equations. In each one we were given the derivative of a function and the value of the function at a particular point. The problem of finding the function is called an *initial-value problem,* and the given condition is called the *initial condition.*

In Examples 48 and 49, we use the velocity (or the acceleration) of a particle moving on a line to find the position of the particle. Note especially how the initial conditions are used to evaluate constants of integration.

> **Example 48**

The velocity of a particle moving along a line is given by $v(t) = 4t^3 - 3t^2$ at time t. If the particle is initially at $x = 3$ on the line, find its position when $t = 2$.

> ✓ **Solution**

Since

$$v(t) = \frac{dx}{dt} = 4t^3 - 3t^2$$

$$x = \int (4t^3 - 3t^2)\, dt = t^4 - t^3 + C$$

Since $x(0) = 0^4 - 0^3 + C = 3$, we see that $C = 3$ and that the position function is $x(t) = t^4 - t^3 + 3$. When $t = 2$, we see that

$$x(2) = 2^4 - 2^3 + 3 = 16 - 8 + 3 = 11$$

> **Example 49**

Suppose that $a(t)$, the acceleration of a particle at time t, is given by $a(t) = 4t - 3$, that $v(1) = 6$, and that $f(2) = 5$, where $f(t)$ is the position function.
(a) Find $v(t)$ and $f(t)$.
(b) Find the position of the particle when $t = 1$.

> ✓ **Solutions**

(a)
$$a(t) = v'(t) = \frac{dv}{dt} = 4t - 3$$

$$v = \int (4t - 3)\, dt = 2t^2 - 3t + C_1$$

Using $v(1) = 6$, we get $6 = 2(1)^2 - 3(1) + C_1$, and $C_1 = 7$, from which it follows that $v(t) = 2t^2 - 3t + 7$. Since

$$v(t) = f'(t) = \frac{df}{dt}$$

$$f(t) = \int \left(2t^2 - 3t + 7\right) dt = \frac{2t^3}{3} - \frac{3t^2}{2} + 7t + C_2$$

Using $f(2) = 5$, we get $5 = \frac{2}{3}(2)^3 - \frac{3}{2}(2)^2 + 7(2) + C_2$, $5 = \frac{16}{3} - 6 + 14 + C_2$, so $C_2 = -\frac{25}{3}$. Thus,

$$f(t) = \frac{2}{3} t^3 - \frac{3}{2} t^2 + 7t - \frac{25}{3}$$

(b) When $t = 1, f(1) = \frac{2}{3} - \frac{3}{2} + 7 - \frac{25}{3} = -\frac{13}{6}$.

For more examples of motion along a line, see Chapter 8, "Further Applications of Integration," and Chapter 9, "Differential Equations."

CHAPTER SUMMARY

In this chapter, we reviewed basic skills for finding indefinite integrals. We looked at the antiderivative formulas for all of the basic functions and reviewed techniques for finding antiderivatives of other functions.

We also reviewed the more advanced techniques of integration by partial fractions and integration by parts, both topics only for the BC Calculus course.

Practice Exercises

PART A. NO CALCULATOR—DIRECTIONS: Answer these questions *without* using your calculator.

A1. $\int \left(3x^2 - 2x + 3 \right) dx =$

   (A) $x^3 - x^2 + C$
   (B) $3x^3 - x^2 + 3x + C$
   (C) $x^3 - x^2 + 3x + C$
   (D) $\frac{1}{2} \left(3x^2 - 2x + 3 \right)^2 + C$

A2. $\int \left(x - \frac{1}{2x} \right)^2 dx =$

   (A) $\frac{1}{3} \left(x - \frac{1}{2x} \right)^3 + C$
   (B) $\frac{x^3}{3} - 2x - \frac{1}{4x} + C$
   (C) $\frac{x^3}{3} - x - \frac{4}{x} + C$
   (D) $\frac{x^3}{3} - x - \frac{1}{4x} + C$

A3. $\int \sqrt{4 - 2t} \, dt =$

   (A) $-\frac{1}{3} (4 - 2t)^{3/2} + C$
   (B) $\frac{2}{3} (4 - 2t)^{3/2} + C$
   (C) $-\frac{1}{6} (4 - 2t)^3 + C$
   (D) $\frac{4}{3} (4 - 2t)^{3/2} + C$

A4. $\int (2 - 3x)^5 \, dx =$

   (A) $\frac{1}{6} (2 - 3x)^6 + C$
   (B) $-\frac{1}{2} (2 - 3x)^6 + C$
   (C) $\frac{1}{2} (2 - 3x)^6 + C$
   (D) $-\frac{1}{18} (2 - 3x)^6 + C$

A5. $\int \frac{1 - 3y}{\sqrt{2y - 3y^2}} \, dy =$

   (A) $4\sqrt{2y - 3y^2} + C$
   (B) $\frac{1}{2} \ln \sqrt{2y - 3y^2} + C$
   (C) $\frac{1}{4} \left(2y - 3y^2 \right)^{1/2} + C$
   (D) $\sqrt{2y - 3y^2} + C$

A6. $\int \frac{dx}{3(2x - 1)^2} =$

   (A) $\frac{-3}{2x - 1} + C$
   (B) $\frac{1}{6 - 12x} + C$
   (C) $\frac{6}{2x - 1} + C$
   (D) $\frac{1}{3} \ln |2x - 1| + C$

A7. $\int \frac{2 \, du}{1 + 3u} =$

   (A) $\frac{2}{3} \ln |1 + 3u| + C$
   (B) $-\frac{1}{3(1 + 3u)^2} + C$
   (C) $2 \ln |1 + 3u| + C$
   (D) $\frac{3}{(1 + 3u)^2} + C$

A8. $\int \frac{t}{\sqrt{2t^2 - 1}} \, dt =$

   (A) $\frac{1}{2} \ln \sqrt{2t^2 - 1} + C$
   (B) $8 \sqrt{2t^2 - 1} + C$
   (C) $-\frac{1}{4 \left(2t^2 - 1 \right)} + C$
   (D) $\frac{1}{2} \sqrt{2t^2 - 1} + C$

A9. $\displaystyle\int \cos 3x\, dx =$

(A) $3 \sin 3x + C$

(B) $-\sin 3x + C$

(C) $-\dfrac{1}{3} \sin 3x + C$

(D) $\dfrac{1}{3} \sin 3x + C$

A10. $\displaystyle\int \dfrac{x\, dx}{1 + 4x^2} =$

(A) $\dfrac{1}{8} \ln\left(1 + 4x^2\right) + C$

(B) $\dfrac{1}{4} \sqrt{1 + 4x^2} + C$

(C) $\dfrac{1}{2} \ln\left|1 + 4x^2\right| + C$

(D) $\dfrac{1}{2} \tan^{-1} 2x + C$

A11. $\displaystyle\int \dfrac{dx}{1 + 4x^2} =$

(A) $\tan^{-1}(2x) + C$

(B) $\dfrac{1}{8} \ln(1 + 4x^2) + C$

(C) $\dfrac{1}{2} \tan^{-1}(2x) + C$

(D) $\dfrac{1}{8x} \ln\left|1 + 4x^2\right| + C$

A12. $\displaystyle\int \dfrac{x}{(1 + 4x^2)2}\, dx =$

(A) $\dfrac{1}{8} \ln\left(1 + 4x^2\right)^2 + C$

(B) $-\dfrac{1}{8\left(1 + 4x^2\right)} + C$

(C) $-\dfrac{1}{3\left(1 + 4x^2\right)^3} + C$

(D) $-\dfrac{1}{\left(1 + 4x^2\right)} + C$

A13. $\displaystyle\int \dfrac{x\, dx}{\sqrt{1 + 4x^2}} =$

(A) $\dfrac{1}{8} \sqrt{1 + 4x^2} + C$

(B) $\dfrac{\sqrt{1 + 4x^2}}{4} + C$

(C) $\dfrac{1}{2} \sin^{-1} 2x + C$

(D) $\dfrac{1}{8} \ln \sqrt{1 + 4x^2} + C$

A14. $\displaystyle\int \dfrac{dy}{\sqrt{4 - y^2}} =$

(A) $\dfrac{1}{2} \sin^{-1} \dfrac{y}{2} + C$

(B) $-\sqrt{4 - y^2} + C$

(C) $\sin^{-1} \dfrac{y}{2} + C$

(D) $-\dfrac{1}{2} \ln \sqrt{4 - y^2} + C$

A15. $\displaystyle\int \dfrac{y\, dy}{\sqrt{4 - y^2}} =$

(A) $\dfrac{1}{2} \sin^{-1} \dfrac{y}{2} + C$

(B) $-\sqrt{4 - y^2} + C$

(C) $\sin^{-1} \dfrac{y}{2} + C$

(D) $2\sqrt{4 - y^2} + C$

A16. $\displaystyle\int \dfrac{2x + 1}{2x}\, dx =$

(A) $x + \dfrac{1}{2} \ln |x| + C$

(B) $x + 2 \ln |x| + C$

(C) $x + \ln |2x| + C$

(D) $\dfrac{1}{2}\left(2x - \dfrac{1}{x^2}\right) + C$

A17. $\displaystyle\int \dfrac{(x - 2)^3}{x^2}\, dx =$

(A) $\dfrac{(x - 2)^4}{4x^2} + C$

(B) $\dfrac{x^2}{2} - 6x + 6 \ln |x| - \dfrac{8}{x} + C$

(C) $\dfrac{x^2}{2} - 3x + 6 \ln |x| + \dfrac{4}{x} + C$

(D) $\dfrac{x^2}{2} - 6x + 12 \ln |x| + \dfrac{8}{x} + C$

A18. $\displaystyle\int \left(\sqrt{t} - \dfrac{1}{\sqrt{t}}\right)^2 dt =$

(A) $\dfrac{t^3}{3} - 2t - \dfrac{1}{t} + C$

(B) $\dfrac{t^2}{2} + \ln |t| + C$

(C) $\dfrac{t^2}{2} - 2t + \ln |t| + C$

(D) $\dfrac{t^2}{2} - t - \dfrac{1}{t^2} + C$

A19. $\int \left(4x^{1/3} - 5x^{3/2} - x^{-1/2}\right) dx =$

(A) $3x^{4/3} - 2x^{5/2} - 2x^{1/2} + C$

(B) $3x^{4/3} - 2x^{5/2} + 2x^{1/2} + C$

(C) $6x^{2/3} - 2x^{5/2} - \frac{1}{2}x^2 + C$

(D) $\frac{16}{9}x^{4/3} - \frac{25}{2}x^{5/2} - \frac{1}{2}x^{1/2} + C$

A20. $\int \dfrac{x^3 - x - 1}{x^2} dx =$

(A) $\dfrac{\frac{1}{4}x^4 - \frac{1}{2}x^2 - x}{\frac{1}{3}x^3} + C$

(B) $1 + \dfrac{1}{x^2} + \dfrac{2}{x^3} + C$

(C) $\dfrac{x^2}{2} - \ln|x| - \dfrac{1}{x} + C$

(D) $\dfrac{x^2}{2} - \ln|x| + \dfrac{1}{x} + C$

A21. $\int \dfrac{dy}{\sqrt{y}(1 - \sqrt{y})} =$

(A) $4\sqrt{1 - \sqrt{y}} + C$

(B) $\frac{1}{2}\ln\left|1 - \sqrt{y}\right| + C$

(C) $2\ln\left(1 - \sqrt{y}\right) + C$

(D) $-2\ln\left|1 - \sqrt{y}\right| + C$

A22. $\int \dfrac{u\,du}{\sqrt{4 - 9u^2}} =$

(A) $\frac{1}{3}\sin^{-1}\dfrac{3u}{2} + C$

(B) $-\dfrac{1}{18}\ln\sqrt{4 - 9u^2} + C$

(C) $2\sqrt{4 - 9u^2} + C$

(D) $-\frac{1}{9}\sqrt{4 - 9u^2} + C$

A23. $\int \sin^2 \theta \cos \theta\, d\theta =$

(A) $\dfrac{\sin^2 \theta}{2} + C$

(B) $\dfrac{\sin^3 \theta}{3} + C$

(C) $-\dfrac{\sin^3 \theta}{3} + C$

(D) $-\dfrac{\sin^2 \theta}{2} + C$

A24. $\int \dfrac{\sin\sqrt{x}}{\sqrt{x}} dx =$

(A) $-2\cos^{1/2}x + C$

(B) $-\cos\sqrt{x} + C$

(C) $-2\cos\sqrt{x} + C$

(D) $\frac{1}{2}\cos\sqrt{x} + C$

A25. $\int t\cos\left(4t^2\right)dt =$

(A) $\frac{1}{8}\sin\left(4t^2\right) + C$

(B) $\frac{1}{2}\cos^2(2t) + C$

(C) $-\frac{1}{8}\sin\left(4t^2\right) + C$

(D) $\frac{1}{4}\sin\left(4t^2\right) + C$

A26. $\int \cos^2 2x\, dx =$ **CHALLENGE**

(A) $\dfrac{x}{2} + \dfrac{\sin 4x}{8} + C$

(B) $\dfrac{x}{2} - \dfrac{\sin 4x}{8} + C$

(C) $\dfrac{x}{4} + \dfrac{\sin 4x}{4} + C$

(D) $\dfrac{x}{4} + \dfrac{\sin 4x}{16} + C$

A27. $\int \sin 2\theta\, d\theta =$

(A) $\frac{1}{2}\cos 2\theta + C$

(B) $-2\cos 2\theta + C$

(C) $\cos^2 \theta + C$

(D) $-\frac{1}{2}\cos 2\theta + C$

A28. $\int x\cos x\, dx =$ **BC ONLY**

(A) $x\sin x + C$

(B) $x\sin x + \cos x + C$

(C) $x\sin x - \cos x + C$

(D) $\cos x - x\sin x + C$

A29. $\int \dfrac{du}{\cos^2 3u} =$

(A) $-\dfrac{\sec 3u}{3} + C$

(B) $\tan 3u + C$

(C) $u + \dfrac{\sec 3u}{3} + C$

(D) $\frac{1}{3}\tan 3u + C$

A30. $\displaystyle\int \frac{\cos x \, dx}{\sqrt{1 + \sin x}} =$

(A) $-\dfrac{1}{2}(1 + \sin x)^{1/2} + C$

(B) $\ln \sqrt{1 + \sin x} + C$

(C) $2\sqrt{1 + \sin x} + C$

(D) $\ln |1 + \sin x| + C$

A31. $\displaystyle\int \frac{\cos(\theta - 1) \, d\theta}{\sin^2(\theta - 1)} =$

(A) $2 \ln \sin|\theta - 1| + C$

(B) $-\csc(\theta - 1) + C$

(C) $-\cot(\theta - 1) + C$

(D) $\csc(\theta - 1) + C$

CHALLENGE   A32. $\displaystyle\int \frac{\sin 2x \, dx}{\sqrt{1 + \cos^2 x}} =$

(A) $-2\sqrt{1 + \cos^2 x} + C$

(B) $\dfrac{1}{2} \ln (1 + \cos^2 x) + C$

(C) $\sqrt{1 + \cos^2 x} + C$

(D) $-\ln \sqrt{1 + \cos^2 x} + C$

A33. $\displaystyle\int \sec^{3/2} x \tan x \, dx =$

(A) $\dfrac{2}{5} \sec^{5/2} x + C$

(B) $-\dfrac{2}{3} \cos^{-3/2} x + C$

(C) $\sec^{3/2} x + C$

(D) $\dfrac{2}{3} \sec^{3/2} x + C$

A34. $\displaystyle\int \tan \theta \, d\theta =$

(A) $-\ln |\cos \theta| + C$

(B) $\sec^2 \theta + C$

(C) $\ln |\sin \theta| + C$

(D) $-\ln |\sec \theta| + C$

A35. $\displaystyle\int \frac{dx}{\sin^2 2x} =$

(A) $\dfrac{1}{2} \csc 2x \cot 2x + C$

(B) $-\dfrac{1}{2} \cot 2x + C$

(C) $-\cot x + C$

(D) $-\csc 2x + C$

A36. $\displaystyle\int \frac{\tan^{-1} y}{1 + y^2} \, dy =$

(A) $\left(\tan^{-1} y\right)^2 + C$

(B) $\ln \left(1 + y^2\right) + C$

(C) $\ln \left(\tan^{-1} y\right) + C$

(D) $\dfrac{1}{2}\left(\tan^{-1} y\right)^2 + C$

A37. $\displaystyle\int 2 \sin \theta \cos^2 \theta \, d\theta =$

(A) $-\dfrac{2}{3} \cos^3 \theta + C$

(B) $\dfrac{2}{3} \cos^3 \theta + C$

(C) $\sin^2 \theta \cos \theta + C$

(D) $\cos^3 \theta + C$

A38. $\displaystyle\int \frac{\sin 2t}{1 - \cos 2t} \, dt =$

(A) $\dfrac{2}{(1 - \cos 2t)^2} + C$

(B) $-\ln |1 - \cos 2t| + C$

(C) $\dfrac{1}{2}\ln |1 - \cos 2t| + C$

(D) $2 \ln |1 - \cos 2t| + C$

A39. $\displaystyle\int \cot 2u \, du =$

(A) $\ln |\sin u| + C$

(B) $\dfrac{1}{2} \ln |\sin 2u| + C$

(C) $-\dfrac{1}{2} \csc^2 2u + C$

(D) $2 \ln |\sin 2u| + C$

A40. $\displaystyle\int \frac{e^x}{e^x - 1} \, dx =$

(A) $x + \ln |e^x - 1| + C$

(B) $x - e^x + C$

(C) $x - \dfrac{1}{(e^x - 1)^2} + C$

(D) $\ln |e^x - 1| + C$

A41. $\displaystyle\int \frac{x - 1}{x(x - 2)} \, dx =$

(A) $\dfrac{1}{2} \ln |x| + \ln |x - 2| + C$

(B) $\dfrac{1}{2} \ln \left|\dfrac{x - 2}{x}\right| + C$

(C) $\dfrac{1}{2} \ln |x(x - 2)| + C$

(D) $\ln \left|\dfrac{x}{x - 2}\right| + C$

A42. $\int x e^{x^2}\, dx =$

(A) $\frac{1}{2} e^{x^2} + C$

(B) $2 e^{x^2} + C$

(C) $e^{x^2} + C$

(D) $\frac{1}{2} e^{x^2+1} + C$

A43. $\int \cos\theta\, e^{\sin\theta}\, d\theta =$

(A) $e^{\sin\theta + 1} + C$

(B) $e^{\sin\theta} + C$

(C) $-e^{\sin\theta} + C$

(D) $e^{\cos\theta} + C$

A44. $\int e^{2\theta} \sin e^{2\theta}\, d\theta =$

(A) $\cos e^{2\theta} + C$

(B) $2 e^{4\theta}\left(\cos e^{2\theta} + \sin e^{2\theta}\right) + C$

(C) $-\frac{1}{2} \cos e^{2\theta} + C$

(D) $-2 \cos e^{2\theta} + C$

A45. $\int \frac{e^{\sqrt{x}}}{\sqrt{x}}\, dx =$

(A) $2\sqrt{x}\left(e^{\sqrt{x}} - 1\right) + C$

(B) $2 e^{\sqrt{x}} + C$

(C) $\frac{1}{2} e^{\sqrt{x}} + C$

(D) $\frac{\sqrt{x}}{2} e^{\sqrt{x}} + C$

A46. $\int x e^{-x}\, dx =$

(A) $e^{-x}(1 - x) + C$

(B) $\frac{e^{1-x}}{1-x} + C$

(C) $-e^{-x}(x + 1) + C$

(D) $e^{-x}(x + 1) + C$

A47. $\int x^2 e^x\, dx =$

(A) $e^x\left(x^2 + 2x\right) + C$

(B) $e^x\left(x^2 - 2x - 2\right) + C$

(C) $e^x\left(x^2 - 2x + 2\right) + C$

(D) $e^x(x - 1)^2 + C$

A48. $\int \frac{e^x + e^{-x}}{e^x - e^{-x}}\, dx =$

(A) $x - \ln\left|e^x - e^{-x}\right| + C$

(B) $x + 2\ln\left|e^x - e^{-x}\right| + C$

(C) $\ln\left|e^x - e^{-x}\right| + C$

(D) $\ln\left(e^x + e^{-x}\right) + C$

A49. $\int \frac{e^x}{1 + e^{2x}}\, dx =$

(A) $\tan^{-1} e^x + C$

(B) $\frac{1}{2}\ln\left(1 + e^{2x}\right) + C$

(C) $\ln\left(1 + e^{2x}\right) + C$

(D) $\frac{1}{2}\tan^{-1} e^x + C$

A50. $\int \frac{\ln v\, dv}{v} =$

(A) $\ln\left|\ln v\right| + C$

(B) $\ln\frac{v^2}{2} + C$

(C) $\frac{1}{2}(\ln v)^2 + C$

(D) $\frac{1}{2}\ln v^2 + C$

A51. $\int \frac{\ln\sqrt{x}}{x}\, dx =$

(A) $\frac{(\ln\sqrt{x})^2}{\sqrt{x}} + C$

(B) $\frac{1}{2}\ln\left|\ln x\right| + C$

(C) $\frac{(\ln\sqrt{x})^2}{2} + C$

(D) $\frac{1}{4}(\ln x)^2 + C$

A52. $\int x^3 \ln x\, dx =$

(A) $x^2\left(3\ln x + 1\right) + C$

(B) $\frac{x^4}{16}\left(4\ln x - 1\right) + C$

(C) $\frac{x^4}{4}\left(\ln x - 1\right) + C$

(D) $3x^2\left(\ln x - \frac{1}{2}\right) + C$

A53. $\int \ln\eta\, d\eta =$

(A) $\frac{1}{2}\ln^2\eta + C$

(B) $\eta(\ln\eta - 1) + C$

(C) $\frac{1}{2}\ln\eta^2 + C$

(D) $\eta\ln\eta + \eta + C$

A54. $\int \ln x^3\, dx =$

(A) $\frac{3}{2}(\ln x)^2 + C$

(B) $3x(\ln x - 1) + C$

(C) $3\ln x\,(x-1) + C$

(D) $\frac{3x(\ln x)^2}{2} + C$

A55. $\int \frac{\ln y}{y^2}\, dy =$

(A) $\frac{1}{y}(1 - \ln y) + C$

(B) $\frac{1}{2y}(\ln y)^2 + C$

(C) $-\frac{1}{y}(\ln y + 1) + C$

(D) $\frac{\ln y}{y} - \frac{1}{y} + C$

A56. $\int \frac{dv}{v\ln v} =$

(A) $\frac{1}{\ln v^2} + C$

(B) $-\frac{1}{(\ln v)^2} + C$

(C) $-\ln|\ln v| + C$

(D) $\ln|\ln v| + C$

A57. $\int \frac{y-1}{y+1}\, dy =$

(A) $y - 2\ln|y+1| + C$

(B) $1 - \frac{2}{y+1} + C$

(C) $\ln \frac{|y|}{(y+1)^2} + C$

(D) $1 - 2\ln|y+1| + C$

A58. $\int \frac{dx}{x^2 + 2x + 2} =$

(A) $\ln(x^2 + 2x + 2) + C$

(B) $\ln|x+1| + C$

(C) $\arctan(x+1) + C$

(D) $\frac{1}{\frac{1}{3}x^3 + x^2 + 2x} + C$

A59. $\int \sqrt{x}(\sqrt{x} - 1)dx =$

(A) $2(x^{3/2} - x) + C$

(B) $\frac{x^2}{2} - x + C$

(C) $\frac{1}{2}(\sqrt{x} - 1)^2 + C$

(D) $\frac{1}{2}x^2 - \frac{2}{3}x^{3/2} + C$

A60. $\int e^\theta \cos\theta\, d\theta =$

(A) $e^\theta(\cos\theta - \sin\theta) + C$

(B) $\frac{1}{2}e^\theta(\sin\theta + \cos\theta) + C$

(C) $2e^\theta(\sin\theta + \cos\theta) + C$

(D) $\frac{1}{2}e^\theta(\sin\theta - \cos\theta) + C$

A61. $\int \frac{(1 - \ln t)^2}{t}\, dt =$

(A) $\frac{1}{3}(1 - \ln t)^3 + C$

(B) $\ln t - 2\ln^2 t + \ln^3 t + C$

(C) $\ln t - \ln^2 t + \frac{\ln t^3}{3} + C$

(D) $-\frac{(1 - \ln t)^3}{3} + C$

A62. $\int u\sec^2 u\, du =$

(A) $u\tan u + \ln|\cos u| + C$

(B) $\frac{u^2}{2}\tan u + C$

(C) $\frac{1}{2}\sec u\tan u + C$

(D) $u\tan u - \ln|\sin u| + C$

A63. $\int \frac{2x+1}{4+x^2}\, dx =$

(A) $\ln(x^2 + 4) + C$

(B) $\ln(x^2 + 4) + \tan^{-1}\frac{x}{2} + C$

(C) $\ln(x^2 + 4) + \frac{1}{2}\tan^{-1}\frac{x}{2} + C$

(D) $\ln(x^2 + 4) + \frac{1}{4}\tan^{-1}x + C$

A64. $\int \frac{1-x}{\sqrt{1-x^2}}\, dx =$

(A) $\sqrt{1-x^2} + C$

(B) $\sin^{-1}x + C$

(C) $\sin^{-1}x + \sqrt{1-x^2} + C$

(D) $\sin^{-1}x + \frac{1}{2}\ln\sqrt{1-x^2} + C$

CHALLENGE A65. $\int \dfrac{2x-1}{\sqrt{4x-4x^2}}\,dx =$

(A) $4\ln\sqrt{4x-4x^2}+C$

(B) $\sin^{-1}(1-2x)+C$

(C) $\dfrac{1}{2}\sqrt{4x-4x^2}+C$

(D) $-\dfrac{1}{2}\sqrt{4x-4x^2}+C$

CHALLENGE A66. $\int \dfrac{e^{2x}}{1+e^x}\,dx =$

(A) $\tan^{-1}e^x+C$

(B) $e^x-\ln\left(1+e^x\right)+C$

(C) $e^x-x+\ln|1+e^x|+C$

(D) $e^x+\dfrac{1}{\left(e^x+1\right)^2}+C$

A67. $\int \dfrac{\cos\theta}{1+\sin^2\theta}\,d\theta =$

(A) $\sec\theta\tan\theta+C$

(B) $\ln\left(1+\sin^2\theta\right)+C$

(C) $\tan^{-1}(\sin\theta)+C$

(D) $-\dfrac{1}{\left(1+\sin^2\theta\right)^2}+C$

BC ONLY A68. $\int \arctan x\,dx =$

(A) $x\arctan x-\ln\left(1+x^2\right)+C$

(B) $x\arctan x+\ln\left(1+x^2\right)+C$

(C) $x\arctan x+\dfrac{1}{2}\ln\left(1+x^2\right)+C$

(D) $x\arctan x-\dfrac{1}{2}\ln\left(1+x^2\right)+C$

CHALLENGE A69. $\int \dfrac{dx}{1-e^x} =$

(A) $-\ln|1-e^x|+C$

(B) $x-\ln|1-e^x|+C$

(C) $\dfrac{1}{\left(1-e^x\right)^2}+C$

(D) $e^{-x}\ln|1+e^x|+C$

A70. $\int \dfrac{(2-y)^2}{4\sqrt{y}}\,dy =$

(A) $\dfrac{1}{6}(2-y)^3\sqrt{y}+C$

(B) $2\sqrt{y}-\dfrac{2}{3}y^{3/2}+\dfrac{8}{5}y^{5/2}+C$

(C) $\ln|y|-y+2y^2+C$

(D) $2y^{1/2}-\dfrac{2}{3}y^{3/2}+\dfrac{1}{10}y^{5/2}+C$

A71. $\int e^{2\ln u}\,du =$

(A) $\dfrac{1}{3}e^{u^3}+C$

(B) $e^{u^3/3}+C$

(C) $\dfrac{1}{3}u^3+C$

(D) $\dfrac{2}{u}e^{2\ln u}+C$

A72. $\int \dfrac{dy}{y(1+\ln y^2)} =$

(A) $\dfrac{1}{2}\ln|1+\ln y^2|+C$

(B) $-\dfrac{1}{\left(1+\ln y^2\right)^2}+C$

(C) $\ln|y|+\dfrac{1}{2}\ln|\ln y|+C$

(D) $\tan^{-1}\left(\ln|y|\right)+C$

A73. $\int (\tan\theta-1)^2\,d\theta =$ CHALLENGE

(A) $\sec\theta+\theta+2\ln|\cos\theta|+C$

(B) $\tan\theta+2\ln|\cos\theta|+C$

(C) $\tan\theta-2\sec^2\theta+C$

(D) $\tan\theta-2\ln|\cos\theta|+C$

A74. $\int \dfrac{d\theta}{1+\sin\theta} =$ CHALLENGE

(A) $\sec\theta-\tan\theta+C$

(B) $\ln(1+\sin\theta)+C$

(C) $\ln|\sec\theta+\tan\theta|+C$

(D) $\tan\theta-\sec\theta+C$

A75. A particle starting at rest at $t=0$ moves along a line so that its acceleration at time t is $12t$ ft/sec^2. How much distance does the particle cover during the first 3 seconds?

(A) 32 feet

(B) 48 feet

(C) 54 feet

(D) 108 feet

A76. The equation of the curve whose slope at point (x,y) is $x^2 - 2$ and which contains the point $(1,-3)$ is

(A) $y = \frac{1}{3}x^3 - 2x$

(B) $y = 2x - 1$

(C) $y = \frac{1}{3}x^3 - \frac{10}{3}$

(D) $y = \frac{1}{3}x^3 - 2x - \frac{4}{3}$

A77. A particle moves along a line with acceleration $2 + 6t$ at time t. When $t = 0$, its velocity equals 3 and it is at position $s = 2$. When $t = 1$, it is at position $s =$

(A) 5

(B) 6

(C) 7

(D) 8

A78. Find the acceleration $\left(\text{in ft/sec}^2\right)$ needed to bring a particle moving with a velocity of 75 ft/sec to a stop in 5 seconds.

(A) -3

(B) -6

(C) -15

(D) -25

A79. $\displaystyle\int \frac{x^2}{x^2 - 1} dx =$

BC ONLY

(A) $x + \frac{1}{2} \ln \left| \frac{x-1}{x+1} \right| + C$

(B) $\ln \left| x^2 - 1 \right| + C$

(C) $x + \frac{1}{2} \ln \left| \frac{x+1}{x-1} \right| + C$

(D) $1 + \frac{1}{2} \ln \left| \frac{x+1}{x-1} \right| + C$

Answer Key

A1. **C**	A17. **D**	A33. **D**	A49. **A**	A65. **D**
A2. **D**	A18. **C**	A34. **A**	A50. **C**	A66. **B**
A3. **A**	A19. **A**	A35. **B**	A51. **D**	A67. **C**
A4. **D**	A20. **D**	A36. **D**	A52. **B**	A68. **D**
A5. **D**	A21. **D**	A37. **A**	A53. **B**	A69. **B**
A6. **B**	A22. **D**	A38. **C**	A54. **B**	A70. **D**
A7. **A**	A23. **B**	A39. **B**	A55. **C**	A71. **C**
A8. **D**	A24. **C**	A40. **D**	A56. **D**	A72. **A**
A9. **D**	A25. **A**	A41. **C**	A57. **A**	A73. **B**
A10. **A**	A26. **A**	A42. **A**	A58. **C**	A74. **D**
A11. **C**	A27. **D**	A43. **B**	A59. **D**	A75. **C**
A12. **B**	A28. **B**	A44. **C**	A60. **B**	A76. **D**
A13. **B**	A29. **D**	A45. **B**	A61. **D**	A77. **C**
A14. **C**	A30. **C**	A46. **C**	A62. **A**	A78. **C**
A15. **B**	A31. **B**	A47. **C**	A63. **C**	A79. **A**
A16. **A**	A32. **A**	A48. **C**	A64. **C**	

Answer Explanations

All the references in parentheses below are to the basic integration formulas on pages 213 and 214. In general, if u is a function of x, then $du = u'(x)\,dx$.

A1. **(C)** Use, first, formula (2), then (3), replacing u by x.

A2. **(D)** Hint: Expand. $\int \left(x^2 - 1 + \dfrac{1}{4x^2} \right) dx = \dfrac{x^3}{3} - x - \dfrac{1}{4x} + C$

A3. **(A)** By formula (3), with $u = 4 - 2t$ and $n = \dfrac{1}{2}$,

$$\int \sqrt{4 - 2t}\,dx = -\frac{1}{2} \int \sqrt{4 - 2t} \cdot (-2\,dt) = -\frac{1}{2} \frac{(4 - 2t)^{3/2}}{3/2} + C$$

A4. **(D)** Rewrite: $-\dfrac{1}{3} \int (2 - 3x)^5 (-3\,dx)$

A5. **(D)** Rewrite:

$$\int (2y - 3y^2)^{-1/2}(1 - 3y)\,dy = \frac{1}{2} \int (2y - 3y^2)^{-1/2}(2 - 6y)\,dy$$

Use (3).

A6. **(B)** Rewrite:

$$\frac{1}{3} \int (2x - 1)^{-2}\,dx = \frac{1}{3} \cdot \frac{1}{2} \int (2x - 1)^{-2} \cdot 2\,dx$$

Using (3) yields $-\dfrac{1}{6(2x - 1)} + C.$

A7. **(A)** This is equivalent to $\dfrac{1}{3} \cdot 2 \int \dfrac{3\,du}{1 + 3u}$. Use (4).

A8. **(D)** Rewrite as $\dfrac{1}{4} \int \left(2t^2 - 1 \right)^{-1/2} \cdot 4t\,dt$. Use (3).

A9. **(D)** Use (5) with $u = 3x;\ du = 3\,dx$: $\dfrac{1}{3} \int \cos(3x)(3\,dx)$

A10. **(A)** Use (4). If $u = 1 + 4x^2,\ du = 8x\,dx$: $\dfrac{1}{8} \int \dfrac{8x\,dx}{1 + 4x^2}$

A11. **(C)** Use (16). Let $u = 2x$; then $du = 2\,dx$: $\dfrac{1}{2} \int \dfrac{2\,dx}{1 + (2x)^2}$

A12. **(B)** Rewrite as $\dfrac{1}{8} \int \left(1 + 4x^3 \right)^{-2} \cdot (8x\,dx)$. Use (3) with $n = -2$.

A13. **(B)** Rewrite as $\dfrac{1}{8} \int \left(1 + 4x^2 \right)^{-1/2} \cdot (8x\,dx)$. Use (3) with $n = -\dfrac{1}{2}$.

Note carefully the differences in the integrands in Questions A10–A13.

A14. **(C)** Use (15); rewrite as $\int \dfrac{\frac{1}{2}dy}{\sqrt{1 - \left(\frac{y}{2} \right)^2}}$

A15. **(B)** Rewrite as $-\dfrac{1}{2} \int \left(4 - y^2 \right)^{-1/2} \cdot (-2y\,dy)$. Use (3).

Compare the integrands in Questions A14 and A15, noting the difference.

A16. **(A)** Divide to obtain $\int\left(1+\frac{1}{2}\cdot\frac{1}{x}\right)dx$. Use (2), (3), and (4). Remember that

$\int k\,dx = kx + C$ whenever $k \neq 0$.

A17. **(D)** $\int\frac{(x-2)^3}{x^2}dx = \int\left(x-6+\frac{12}{x}-\frac{8}{x^2}\right)dx = \frac{x^2}{2}-6x+12\ln|x|+\frac{8}{x}+C.$
(Note the Binomial Theorem on page 645 with $n=3$ to expand $(x-2)^3$.)

A18. **(C)** The integral is equivalent to $\int\left(t-2+\frac{1}{t}\right)dt$. Integrate term by term.

A19. **(A)** Integrate term by term. Use (3).

A20. **(D)** Division yields

$$\int\left(x-\frac{1}{x}-\frac{1}{x^2}\right)dx = \int x\,dx - \int\frac{1}{x}dx - \int\frac{1}{x^2}dx$$

A21. **(D)** Use formula (4) with $u = 1-\sqrt{y} = 1-y^{1/2}$. Then $du = -\frac{1}{2\sqrt{y}}dy$. Note that the

integral can be written as $-2\int\frac{1}{(1-\sqrt{y})}\left(\frac{1}{2\sqrt{y}}\right)dy$.

A22. **(D)** Rewrite as $-\frac{1}{18}\int\left(4-9u^2\right)^{-1/2}(-18u\,du)$ and use formula (3).

A23. **(B)** Let $u = \sin\theta$, then $du = \cos\theta\,d\theta$. Rewrite the integral as $\int u^2\,du = \frac{u^3}{3}+C = \frac{\sin^3\theta}{3}+C.$

A24. **(C)** Use formula (6) with $u = \sqrt{x}$; $du = \frac{1}{2\sqrt{x}}dx$; $2\int\sin\left(\sqrt{x}\right)\left(\frac{1}{2\sqrt{x}}dx\right)$

A25. **(A)** Use formula (5) with $u = 4t^2$; $du = 8t\,dt$; $\frac{1}{8}\int\cos\left(4t^2\right)(8t\,dt)$

A26. **(A)** Using the Half-Angle Formula (23) on page 646 with $\alpha = 2x$ yields $\int\left(\frac{1}{2}+\frac{1}{2}\cos 4x\right)dx.$

A27. **(D)** Use formula (6) with $u = 2\theta$; $du = 2\,d\theta$; $\frac{1}{2}\int\sin 2\theta\,(2\,d\theta)$.

A28. **(B)** Integrate by parts (page 221). Let $u = x$ and $dv = \cos x\,dx$. Then $du = dx$ and $v = \sin x$.
The given integral equals $x\sin x - \int\sin x\,dx.$

A29. **(D)** Replace $\frac{1}{\cos^2 3u}$ by $\sec^2 3u$; then use formula (9): $\frac{1}{3}\int\sec^2 3u\,(3\,du)$

A30. **(C)** Rewrite using $u = 1+\sin x$ and $du = \cos x\,dx$ as $\int(1+\sin x)^{-1/2}(\cos x\,dx).$
Use formula (3).

A31. **(B)** The integral is equivalent to $\int\csc(\theta-1)\cot(\theta-1)\,d\theta$. Use formula (12).

A32. **(A)** Replace $\sin 2x$ by $2\sin x\cos x$; then the integral is equivalent to

$$-\int\frac{-2\sin x\cos x}{\sqrt{1+\cos^2 x}}dx = -\int u^{-1/2}\,du$$

where $u = 1+\cos^2 x$ and $du = -2\sin x\cos x\,dx$. Use formula (3).

A33. **(D)** Rewriting in terms of sines and cosines yields

$$\int\frac{\sin x}{\cos^{5/2}x}dx = -\int\cos^{-5/2}x(-\sin x)\,dx = -\left(-\frac{2}{3}\right)\cos^{-3/2}x + C$$

A34. **(A)** Use formula (7).

A35. **(B)** Replace $\dfrac{1}{\sin^2 2x}$ by $\csc^2 2x$ and use formula (10): $\dfrac{1}{2}\displaystyle\int \csc^2 2x\,(2\,dx)$

A36. **(D)** Let $u = \tan^{-1} y$; then integrate $\displaystyle\int u\,du$.

A37. **(A)** Let $u = \cos\theta$, then $du = -\sin\theta\,d\theta$. Rewrite the integral as

$$-2\int u^2\,du = -2\cdot\frac{u^3}{3} + C = -\frac{2}{3}\cos^3\theta + C.$$

A38. **(C)** $\dfrac{1}{2}\displaystyle\int \dfrac{2\sin 2t\,dt}{1 - \cos 2t} = \dfrac{1}{2}\ln|1 - \cos 2t| + C$

A39. **(B)** Rewrite as $\dfrac{1}{2}\displaystyle\int \cot 2u\,(2\,du)$ and use formula (8).

A40. **(D)** Use formula (4) with $u = e^x - 1$; $du = e^x\,dx$

A41. **(C)** Use partial fractions; find A and B such that

$$\frac{x - 1}{x(x - 2)} = \frac{A}{x} + \frac{B}{x - 2}$$

Then $x - 1 = A(x - 2) + Bx$

Set $x = 0$: $-1 = -2A$ and $A = \dfrac{1}{2}$

Set $x = 2$: $1 = 2B$ and $B = \dfrac{1}{2}$

So the given integral equals

$$\int\left(\frac{1}{2x} + \frac{1}{2(x - 2)}\right)dx = \frac{1}{2}\ln|x| + \frac{1}{2}\ln|x - 2| + C$$
$$= \frac{1}{2}\ln|x(x - 2)| + C$$

A42. **(A)** Use formula (13) with $u = x^2$; $du = 2x\,dx$; $\dfrac{1}{2}\displaystyle\int e^{x^2}(2x\,dx)$

A43. **(B)** Use formula (13) with $u = \sin\theta$; $du = \cos\theta\,d\theta$.

A44. **(C)** Use formula (6) with $u = e^{2\theta}$; $du = 2e^{2\theta}\,d\theta$; $\dfrac{1}{2}\displaystyle\int \sin e^{2\theta}\left(2e^{2\theta}\,d\theta\right)$.

A45. **(B)** Use formula (13) with $u = \sqrt{x} = x^{1/2}$; $du = \dfrac{1}{2\sqrt{x}}\,dx$

A46. **(C)** Use the Parts Formula. Let $u = x$, $dv = e^{-x}dx$; $du = dx$, $v = -e^{-x}$. Then,

$$-xe^{-x} + \int e^{-x}\,dx = -xe^{-x} - e^{-x} + C$$

A47. **(C)** See Example 42, page 222.

A48. **(C)** The integral is of the form $\displaystyle\int \dfrac{du}{u}$; use (4).

A49. **(A)** The integral has the form $\displaystyle\int \dfrac{du}{1 + u^2}$. Use formula (16), with $u = e^x$, $du = e^x\,dx$.

A50. **(C)** Let $u = \ln v$; then $du = \dfrac{dv}{v}$. Use formula (3) for $\displaystyle\int \ln v\left(\dfrac{1}{v}\,dv\right)$.

A51. **(D)** Hint: $\ln\sqrt{x} = \dfrac{1}{2}\ln x$; the integral is $\dfrac{1}{2}\displaystyle\int (\ln x)\left(\dfrac{1}{x}\,dx\right)$.

A52. **(B)** Use parts, letting $u = \ln x$ and $dv = x^3\,dx$. Then $du = \dfrac{1}{x}\,dx$ and $v = \dfrac{x^4}{4}$. The integral

equals $\dfrac{x^4}{4}\ln x - \dfrac{1}{4}\displaystyle\int x^3\,dx$.

A53. **(B)** Use parts, letting $u = \ln\eta$ and $dv = dx$. Then $du = \dfrac{1}{\eta}\,d\eta$ and $v = \eta$. The integral equals η

$\ln\eta - \displaystyle\int d\eta$.

A54. **(B)** Rewrite $\ln x^3$ as $3 \ln x$, and use the method from Answer A53.

A55. **(C)** Use parts, letting $u = \ln y$ and $dv = y^{-2}\, dy$. Then $du = \frac{1}{y}\, dy$ and $v = -\frac{1}{y}$. The Parts

Formula yields $\dfrac{-\ln y}{y} + \displaystyle\int \frac{1}{y^2}\, dy.$

A56. **(D)** The integral has the form $\displaystyle\int \frac{du}{u}$, where $u = \ln v$: $\displaystyle\int \frac{\left(\frac{1}{v}\, dv\right)}{\ln v}$

A57. **(A)** By long division, the integrand is equivalent to $1 - \dfrac{2}{y+1}$.

A58. **(C)** $\displaystyle\int \frac{dx}{(x+1)^2 + 1} = \int \frac{dx}{1 + (x+1)^2}$; use formula (16) with $u = x + 1$. *NOTE:* By completing the square, $x^2 + 2x + 2 = (x^2 + 2x + 1) + 1 = (x+1)^2 + 1$.

A59. **(D)** Multiply to get $\displaystyle\int \left(x - \sqrt{x}\right) dx.$

A60. **(B)** See Example 43, page 222. Replace x by θ.

A61. **(D)** The integral equals $-\displaystyle\int (1 - \ln t)^2 \left(-\frac{1}{t}\, dt\right)$; it is equivalent to $-\displaystyle\int u^2\, du$, where $u = 1 - \ln t.$

A62. **(A)** Replace u by x in the given integral to avoid confusion in applying the Parts Formula. To integrate $\displaystyle\int x \sec^2 x\, dx$, let the variable u in the Parts Formula be x, and let dv be $\sec^2 x\, dx$. Then $du = dx$ and $v = \tan x$, so

$$\int x \sec^2 x\, dx = x \tan x - \int \tan x\, dx$$
$$= x \tan x + \ln |\cos x| + C$$

A63. **(C)** The integral is equivalent to $\displaystyle\int \frac{2x}{4 + x^2}\, dx + \int \frac{1}{4 + x^2}\, dx$. Use formula (4) on the first integral and (16) on the second.

A64. **(C)** The integral is equivalent to $\displaystyle\int \frac{1}{\sqrt{1 - x^2}}\, dx - \int \frac{x}{\sqrt{1 - x^2}}\, dx.$

Use formula (15) on the first integral. Rewrite the second integral as

$-\dfrac{1}{2} \displaystyle\int \left(1 - x^2\right)^{-\frac{1}{2}} (-2x)\, dx$, and use (3).

A65. **(D)** Rewrite: $-\dfrac{1}{4} \displaystyle\int \left(4x - 4x^2\right)^{-\frac{1}{2}} (4 - 8x)\, dx$

A66. **(B)** Hint: Divide, getting $\displaystyle\int \left[e^x - \frac{e^x}{1 + e^x}\right] dx.$

A67. **(C)** Letting $u = \sin \theta$ yields the integral $\displaystyle\int \frac{du}{1 + u^2}$. Use formula (16).

A68. **(D)** Use integration by parts, letting $u = \arctan x$ and $dv = dx$. Then

$$du = \frac{dx}{1 + x^2} \quad \text{and} \quad v = x$$

The Parts Formula yields

$x \arctan x - \displaystyle\int \frac{x\, dx}{1 + x^2} \qquad \text{or} \qquad x \arctan x - \frac{1}{2} \ln \left(1 + x^2\right) + C$

A69. **(B)** Hint: Note that

$$\frac{1}{1-e^x} = \frac{1-e^x+e^x}{1-e^x} = 1 + \frac{e^x}{1-e^x}$$

Or multiply the integrand by $\dfrac{e^{-x}}{e^{-x}}$, recognizing that the correct answer is equivalent to $-\ln\left|e^{-x}-1\right|$.

A70. **(D)** Hint: Expand the numerator and divide. Then integrate term by term.

A71. **(C)** Hint: Observe that $e^{2\ln u} = u^2$.

A72. **(A)** Let $u = 1 + \ln y^2 = 1 + 2\ln|y|$; integrate $\dfrac{1}{2}\displaystyle\int \frac{\frac{2}{y}\,dy}{1 + 2\ln|y|}$.

A73. **(B)** Hint: Expand and note that:

$$\int \left(\tan^2\theta - 2\tan\theta + 1\right) d\theta = \int \sec^2\theta\, d\theta - 2\int \tan\theta\, d\theta$$

Use formulas (9) and (7).

A74. **(D)** $\displaystyle\int \frac{1}{1+\sin\theta}\cdot\frac{1-\sin\theta}{1-\sin\theta}\,d\theta = \int \frac{1-\sin\theta}{\cos^2\theta}\,d\theta = \int\left(\sec^2\theta - \sec\theta\tan\theta\right)$

A75. **(C)** Note the initial conditions: when $t = 0$, $v = 0$ and $s = 0$. Integrate twice: $v = 6t^2$ and $s = 2t^3$. Let $t = 3$.

A76. **(D)** Since $y' = x^2 - 2$, $y = \dfrac{1}{3}x^3 - 2x + C$. Replacing x by 1 and y by -3 yields $C = -\dfrac{4}{3}$.

A77. **(C)** When $t = 0$, $v = 3$ and $s = 2$, so

$$v = 2t + 3t^2 + 3 \qquad \text{and} \qquad s = t^2 + t^3 + 3t + 2$$

Let $t = 1$.

A78. **(C)** Let $\dfrac{dv}{dt} = a$; then

$$v = at + C \qquad\qquad (*)$$

Since $v = 75$ when $t = 0$, therefore $C = 75$. Then (*) becomes

$$v = at + 75$$

so

$$0 = at + 75 \qquad \text{and} \qquad a = -15$$

A79. **(A)** Divide to obtain $\displaystyle\int\left(1 + \frac{1}{x^2 - 1}\right) dx$. Use partial fractions to get

$$\frac{1}{x^2 - 1} = \frac{A}{x - 1} + \frac{B}{x + 1} = \frac{\frac{1}{2}}{x - 1} - \frac{\frac{1}{2}}{x + 1}$$

6

Definite Integrals

Learning Objectives

In this chapter, you will review what definite integrals mean and how to evaluate them. You will look at:

→ The all-important Fundamental Theorem of Calculus
→ Other important properties of definite integrals, including the Mean Value Theorem for Integrals
→ Analytic methods for evaluating definite integrals
→ Evaluating definite integrals using tables and graphs
→ Riemann Sums
→ Numerical methods for approximating definite integrals, including left and right rectangular sums, the midpoint rule, and the trapezoidal sum
→ The average value of a function

In addition, BC Calculus students will review how to work with integrals based on parametrically defined functions.

A. Fundamental Theorem of Calculus (FTC); Evaluation of Definite Integrals

If f is continuous on the closed interval $[a,b]$ and $F' = f$, then, according to the Fundamental Theorem of Calculus,

$$\int_a^b f(x)\, dx = F(b) - F(a)$$

Fundamental Theorem of Calculus

Here $\int_a^b f(x)$ is the *definite integral of f from a to b; f(x)* is called the *integrand;* and a and b are called, respectively, the *lower* and *upper limits of integration.*

Definite integral

This important theorem says that if f is the derivative of F then the definite integral of f gives the net change in F as x varies from a to b. It also says that we can evaluate any definite integral for which we can find an antiderivative of a continuous function.

By extension, a definite integral can be evaluated for any function that is bounded and piecewise continuous. Such functions are said to be *integrable.*

Integrable

B. Properties of Definite Integrals

The following theorems about definite integrals are important.

Fundamental Theorem of Calculus

THE FUNDAMENTAL THEOREM OF CALCULUS: $\dfrac{d}{dx}\displaystyle\int_a^x f(t)\,dt = f(x)$ $\hspace{2cm}$ (1)

$$\int_a^b kf(x)\,dx = k\int_a^b f(x)\,dx \hspace{1cm} (k \text{ is a constant}) \hspace{1cm} (2)$$

$$\int_a^a f(x)\,dx = 0 \hspace{5cm} (3)$$

$$\int_a^b f(x)\,dx = -\int_b^a f(x)\,dx \hspace{4cm} (4)$$

$$\int_a^c f(x)\,dx + \int_c^b f(x)\,dx = \int_a^b f(x)\,dx \hspace{1cm} (a < c < b) \hspace{1cm} (5)$$

If f and g are both integrable functions of x on $[a,b]$, then

$$\int_a^b [f(x) \pm g(x)]\,dx = \int_a^b f(x)\,dx \pm \int_a^b g(x)\,dx \hspace{2cm} (6)$$

Mean Value Theorem for Integrals

THE MEAN VALUE THEOREM FOR INTEGRALS: If f is continuous on $[a,b]$ there exists at least one number c, $a < c < b$, such that

$$\int_a^b f(x)\,dx = f(c)(b-a) \hspace{4cm} (7)$$

By the comparison property, if f and g are integrable on $[a,b]$ and if $f(x) \le g(x)$ for all x in $[a,b]$, then

$$\int_a^b f(x)\,dx \le \int_a^b g(x)\,dx \hspace{4cm} (8)$$

The evaluation of a definite integral is illustrated in the following examples. A calculator will be helpful for some numerical calculations.

> **Example 1**

$$\int_{-1}^2 \left(3x^2 - 2x\right)dx = x^3 - x^2 \Big|_{-1}^2 = (8-4) - (-1-1) = 6$$

> **Example 2**

$$\int_1^2 \frac{x^2 + x - 2}{2x^2}\,dx = \frac{1}{2}\int_1^2 \left(1 + \frac{1}{x} - \frac{2}{x^2}\right)dx = \frac{1}{2}\left(x + \ln x + \frac{2}{x}\right)\Big|_1^2$$

$$= \frac{1}{2}[(2 + \ln 2 + 1) - (1 + 2)] = \frac{1}{2}\ln 2 \text{ or } \ln \sqrt{2}$$

> **Example 3**

$$\int_5^8 \frac{dy}{\sqrt{9-y}} = -\int_5^8 (9-y)^{-1/2}\,(-dy) = -2\sqrt{9-y}\,\Big|_5^8 = -2(1-2) = 2$$

> **Example 4**

$$\int_0^1 \frac{x\,dx}{(2-x^2)^3} = -\frac{1}{2}\int_0^1 (2-x^2)^{-3}(-2x\,dx) = -\frac{1}{2}\frac{(2-x^2)^{-2}}{-2}\bigg|_0^1 = \frac{1}{4}\left(1-\frac{1}{4}\right) = \frac{3}{16}$$

> **Example 5**

$$\int_0^3 \frac{dt}{9+t^2} = \frac{1}{3}\tan^{-1}\frac{t}{3}\bigg|_0^3 = \frac{1}{3}\left(\tan^{-1}1 - \tan^{-1}0\right)$$

$$= \frac{1}{3}\left(\frac{\pi}{4}-0\right) = \frac{\pi}{12}$$

> **Example 6**

$$\int_0^1 (3x-2)^3\,dx = \frac{1}{3}\int_0^1 (3x-2)^3(3\,dx)$$

$$= \frac{(3x-2)^4}{12}\bigg|_0^1 = \frac{1}{12}(1-16) = -\frac{5}{4}$$

> **Example 7**

$$\int_0^1 xe^{-x^2}\,dx = -\frac{1}{2}e^{-x^2}\bigg|_0^1 = -\frac{1}{2}\left(\frac{1}{e}-1\right) = \frac{e-1}{2e}$$

> **Example 8**

$$\int_{-\pi/4}^{\pi/4} \cos 2x\,dx = \frac{1}{2}\sin 2x\bigg|_{-\pi/4}^{\pi/4} = \frac{1}{2}(1+1) = 1$$

> **Example 9**

BC ONLY

$$\int_{-1}^1 xe^x\,dx = (xe^x - e^x)\bigg|_{-1}^1 = e - e - \left(-\frac{1}{e}-\frac{1}{e}\right) = \frac{2}{e}\ \text{(by Parts)}$$

> **Example 10**

$$\int_0^{1/2} \frac{dx}{\sqrt{1-x^2}} = \sin^{-1}x\bigg|_0^{1/2} = \frac{\pi}{6}$$

> **Example 11**

BC ONLY

$$\int_0^{e-1} \ln(x+1)\,dx = \left[(x+1)\ln(x+1) - x\right]\bigg|_0^{e-1}\ \text{(Parts Formula)}$$

$$= e\ln e - (e-1) - 0 = 1$$

> **Example 12**

Evaluate $\int_{-1}^{1} \dfrac{dy}{y^2 - 4}$.

✓ **Solution**

We use the method of partial fractions and set

$$\frac{1}{y^2 - 4} = \frac{A}{y+2} + \frac{B}{y-2}$$

Solving for A and B yields $A = -\dfrac{1}{4}$, $B = \dfrac{1}{4}$. Thus,

$$\int_{-1}^{1} \frac{dy}{y^2 - 4} = \frac{1}{4} \ln \left| \frac{y-2}{y+2} \right|_{-1}^{1} = \frac{1}{4}\left(\ln\frac{1}{3} - \ln 3 \right) = -\frac{1}{2}\ln 3$$

> **Example 13**

$$\int_{\pi/3}^{\pi/2} \tan\frac{\theta}{2} \sec^2\frac{\theta}{2}\, d\theta = \frac{2}{2}\tan^2\frac{\theta}{2}\bigg|_{\pi/3}^{\pi/2} = 1 - \frac{1}{3} = \frac{2}{3}$$

> **Example 14**

$$\int_{0}^{\pi/2} \sin^2\frac{1}{2}x\, dx = \int_{0}^{\pi/2}\left(\frac{1}{2} - \frac{\cos x}{2}\right)dx = \frac{x}{2} - \frac{\sin x}{2}\bigg|_{0}^{\pi/2} = \frac{\pi}{4} - \frac{1}{2}$$

> **Example 15**

$$\frac{d}{dx}\int_{-1}^{x} \sqrt{1 + \sin^2 t}\, dt = \sqrt{1 + \sin^2 x} \text{ by theorem (1), page 242}$$

> **Example 16**

$$\frac{d}{dx}\int_{x}^{1} e^{-t^2}\, dt = \frac{d}{dx}\left(-\int_{1}^{x} e^{-t^2}\, dt\right) \text{ by theorem (4), page 242}$$

$$= -\frac{d}{dx}\int_{1}^{x} e^{-t^2}\, dt = -e^{-x^2} \text{ by theorem (1)}$$

> **Example 17**

Given $F(x) = \int_{1}^{x^2} \dfrac{dt}{3+t}$, find $F'(x)$.

✓ **Solution**

$$F'(x) = \frac{d}{dx}\int_1^{x^2} \frac{dt}{3+t}$$

$$= \frac{d}{dx}\int_1^u \frac{dt}{3+t} \qquad (\text{where } u = x^2)$$

$$= \frac{d}{dx}\left(\int_1^u \frac{dt}{3+t}\right)\cdot\frac{du}{dx} \qquad \text{by the Chain Rule}$$

$$= \left(\frac{1}{3+u}\right)(2x) = \frac{2x}{3+x^2}$$

> **Example 18**

If $F(x) = \displaystyle\int_0^{\cos x} \sqrt{1-t^3}\ dt$, find $F'(x)$.

✓ **Solution**

We let $u = \cos x$. Thus

$$\frac{dF}{dx} = \frac{dF}{du}\cdot\frac{du}{dx} = \sqrt{1-u^3}\,(-\sin x) = -\sin x\sqrt{1-\cos^3 x}$$

> **Example 19**

Find $\displaystyle\lim_{h\to 0}\frac{1}{h}\int_x^{x+h} \sqrt{e^t-1}\ dt$.

✓ **Solution**

$$\lim_{h\to 0}\frac{1}{h}\int_x^{x+h} \sqrt{e^t-1}\ dt = \sqrt{e^x-1}$$

Here we have let $f(t) = \sqrt{e^x-1}$ and noted that

$$\lim_{h\to 0}\frac{1}{h}\int_x^{x+h} f(t)dt = \lim_{h\to 0}\frac{F(x+h)-F(x)}{h} \qquad (*)$$

where

$$\frac{dF(x)}{dx} = f(x) = \sqrt{e^x-1}$$

The limit on the right in the starred equation is, by definition, the derivative of $F(x)$, that is, $f(x)$.

> **Example 20**

Reexpress $\displaystyle\int_3^6 x\sqrt{x-2}\ dx$, in terms of u if $u = \sqrt{x-2}$.

✓ **Solution**

When $u = \sqrt{x-2}$, $u^2 = x-2$, and $2u\,du = dx$. The limits of the given integral are values of x. When we write the new integral in terms of the variable u, then the limits, if written, must be the values of u that correspond to the given limits. Thus, when $x = 3$, $u = 1$, and when $x = 6$, $u = 2$. Then

$$\int_3^6 x\sqrt{x-2}\ dx = 2\int_1^2 (u^2+2)u^2 du = 2\int_1^2 \left(u^4 + 2u^2\right)du$$

> **Example 21**

If g' is continuous, find $\lim\limits_{h \to 0} \dfrac{1}{h} \displaystyle\int_c^{c+h} g'(x)\, dx$.

✓ **Solution**

$$\lim_{h \to 0} \frac{1}{h} \int_c^{c+h} g'(x)\, dx = \lim_{h \to 0} \frac{g(c+h) - g(c)}{h} = g'(c)$$

Note that the expanded limit is, by definition, the derivative of $g(x)$ at c.

C. Definition of Definite Integral as the Limit of a Riemann Sum

Most applications of integration are based on the FTC. This theorem provides the tool for evaluating an infinite sum by means of a definite integral. Suppose that a function $f(x)$ is continuous on the closed interval $[a,b]$. Divide the interval into n subintervals of lengths Δx_k (it is not necessary that the widths be of equal length, but the formulation is generally simpler if they are). Choose numbers, one in each subinterval, as follows: x_1 in the first, x_2 in the second, $\ldots$, x_k in the kth, $\ldots$, x_n in the nth. Then, assuming equal width subintervals,

Riemann Sum

$$\lim_{n \to \infty} \sum_{k=1}^{n} f(x_k) \cdot \Delta x_k = \int_a^b f(x)dx, \text{ where } \sum_{k=1}^{n} f(x_k) \cdot \Delta x_k \text{ is called a } \textit{Riemann Sum.}$$

If we choose equal width subintervals on the interval $[a,b]$, then Δx_k is the same for all values of k, and we use $\Delta x = \dfrac{b-a}{n}$ for each width. Given that the set $\{x_0, x_1, x_2, \ldots, x_n\}$ partitions the interval $[a,b]$ into n equal subintervals where $x_0 = a$, $x_n = b$, and each of $x_1, x_2, \ldots, x_n$ are the x-value for the right edge of each subinterval, then $x_k = a + k \cdot \Delta x$ for each k. We can now write the limit as a definite integral: $\lim\limits_{n \to \infty} \displaystyle\sum_{k=1}^{n} f(a + k \cdot \Delta x) \cdot \Delta x = \int_a^b f(x)dx$.

> **Example 22**

Write $\lim\limits_{n \to \infty} \displaystyle\sum_{k=1}^{n} \left(2 + \frac{5k}{n}\right)^2 \cdot \frac{5}{n}$ as a definite integral.

✓ **Solution #1**

Recall that this limit is of the form $\lim\limits_{n \to \infty} \displaystyle\sum_{k=1}^{n} f(a + k \cdot \Delta x) \cdot \Delta x$. If we assume that $x_k = 2 + \dfrac{5k}{n}$, then $a = 2$ and $\Delta x = \dfrac{5}{n} = \dfrac{b-2}{n}$, so $b = 7$; x_k replaced x in $f(x)$ so $f(x) = x^2$. This gives us the definite integral $\displaystyle\int_2^7 x^2 dx$.

✓ **Solution #2**

We could also assume that $x_k = \dfrac{5k}{n}$, then $a = 0$ and $\Delta x = \dfrac{5}{n} = \dfrac{b-0}{n}$, so $b = 5$; x_k replaced x in $f(x)$ so $f(x) = (2 + x)^2$. This gives us the definite integral $\displaystyle\int_0^5 (2 + x)^2\, dx$. Indeed, both definite integrals yield the same value: $\dfrac{335}{3}$.

> **Example 23**

Write $\displaystyle\int_1^4 (x^3 - 2)\, dx$ as the limit of a Riemann Sum.

✓ **Solution**

The endpoints indicate that $a = 1$ and $b = 4$, so $\Delta x = \frac{4-1}{n} = \frac{3}{n}$ and $x_k = 1 + k \cdot \frac{3}{n} = 1 + \frac{3k}{n}$.

Therefore, we can write the definite integral as $\displaystyle\lim_{n \to \infty} \sum_{k=1}^{n} \left(f\left(1 + \frac{3k}{n}\right) \right) \cdot \frac{3}{n} = \lim_{n \to \infty} \sum_{k=1}^{n} \left(\left(1 + \frac{3k}{n}\right)^3 - 2 \right) \cdot \frac{3}{n}$.

D. The Fundamental Theorem Again

Area

If $f(x)$ is nonnegative on $[a,b]$, we see (Figure 6.1) that $f(x_k)\Delta x$ can be regarded as the area of a typical approximating rectangle. Assuming equal width rectangles, as the number of rectangles increases, or, equivalently, as the width Δx of the rectangles approaches zero, the rectangles become an increasingly better fit to the curve. The sum of their areas gets closer and closer to the exact area under the curve. Finally, the area bounded by the x-axis, the curve, and the vertical lines $x = a$ and $x = b$ is given exactly by

$$\lim_{n \to \infty} \sum_{k=1}^{n} f(x_k)\Delta x \qquad \text{and hence by} \qquad \int_a^b f(x)\, dx$$

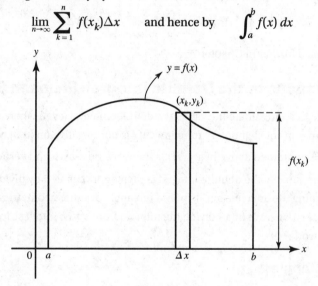

Figure 6.1

What if $f(x)$ is negative? Then *any area above the graph and below the x-axis is counted as negative* (Figure 6.2).

Geometrically, area is always positive, so the shaded area above the curve and below the x-axis equals

$$-\int_a^b f(x)dx$$

where the integral yields a negative number. Note that every product $f(x_k)\Delta x$ in the shaded region is negative since $f(x_k)$ is negative for all x between a and b.

We see from Figure 6.3 that the graph of f crosses the x-axis at c, that area A_1 lies above the x-axis, and that area A_2 lies below the x-axis. Since, by property (5) on page 242,

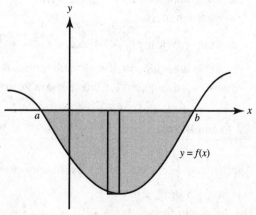

Figure 6.2

$$\int_a^b f(x)\,dx = \int_a^c f(x)\,dx + \int_c^b f(x)\,dx$$

therefore

$$\int_a^b f(x)\,dx = A_1 - A_2$$

Note that, if f is continuous, then the area between the graph of f on $[a,b]$ and the x-axis is given by

$$\int_a^b |f(x)|\,dx$$

This implies that, over any interval within $[a,b]$ for which $f(x) < 0$ (for which its graph dips below the x-axis), $|f(x)| = -f(x)$. The area between the graph of f and the x-axis in Figure 6.3 equals

$$\int_a^b |f(x)|\,dx = \int_a^c f(x)\,dx - \int_c^b f(x)\,dx$$

This topic is discussed further in Chapter 7.

Figure 6.3

E. Approximations of the Definite Integral; Riemann Sums

It is always possible to approximate the value of a definite integral, even when an integrand cannot be expressed in terms of elementary functions. If f is nonnegative on $[a,b]$, we interpret $\int_a^b f(x)\,dx$ as the area bounded above by $y = f(x)$, below by the x-axis, and vertically by the lines $x = a$ and $x = b$. The value of the definite integral is then approximated by dividing the area into n strips, approximating the area of each strip by a rectangle or other geometric figure, then summing these approximations. We often divide the interval from a to b into n strips of equal width, but any strips will work.

E1. Using Rectangles

We may approximate $\int_a^b f(x)\,dx$ by any of the following sums, where Δx represents the subinterval widths:

Left sum

1. Left sum: $f(x_0)\,\Delta x_1 + f(x_1)\,\Delta x_2 + \cdots + f(x_{n-1})\,\Delta x_n$, using the value of f at the left endpoint of each subinterval.

Right sum

2. Right sum: $f(x_1)\,\Delta x_1 + f(x_2)\,\Delta x_2 + \cdots + f(x_n)\,\Delta x_n$, using the value of f at the right end of each subinterval.

Midpoint sum

3. Midpoint sum: $f\left(\dfrac{x_0 + x_1}{2}\right)\Delta x_1 + f\left(\dfrac{x_1 + x_2}{2}\right)\Delta x_2 + \ldots + f\left(\dfrac{x_{n-1} + x_n}{2}\right)\Delta x_n$, using the value of f at the midpoint of each subinterval.

These approximations are illustrated in Figures 6.4 and 6.5, which accompany Example 24.

❯ Example 24

Approximate $\displaystyle\int_0^2 x^3\,dx$ by using four subintervals of equal width and calculating:

(a)   the left sum

(b)   the right sum

(c)   the midpoint sum

(d)   the integral

✓ **Solutions**

Here $\Delta x = \dfrac{2 - 0}{4} = \dfrac{1}{2}$.

(a)   For a left sum we use the left-hand altitudes at $x = 0, \frac{1}{2}, 1$, and $\frac{3}{2}$. The approximating sum is

$$(0)^3 \cdot \frac{1}{2} + \left(\frac{1}{2}\right)^3 \cdot \frac{1}{2} + (1)^3 \cdot \frac{1}{2} + \left(\frac{3}{2}\right)^3 \cdot \frac{1}{2} = \frac{9}{4}$$

The dashed lines in Figure 6.4 show the inscribed rectangles used.

(b)   For the right sum we use right-hand altitudes at $x = \frac{1}{2}, 1, \frac{3}{2}$, and 2. The approximating sum is

$$\left(\frac{1}{2}\right)^3 \cdot \frac{1}{2} + (1)^3 \cdot \frac{1}{2} + \left(\frac{3}{2}\right)^3 \cdot \frac{1}{2} + (2)^3 \cdot \frac{1}{2} = \frac{25}{4}$$

This sum uses the circumscribed rectangles shown in Figure 6.4.

Figure 6.4   **Figure 6.5**

(c)   The midpoint sum uses the heights at the midpoints of the subintervals, as shown in Figure 6.5. The approximating sum is

$$\left(\frac{1}{4}\right)^3 \cdot \frac{1}{2} + \left(\frac{3}{4}\right)^3 \cdot \frac{1}{2} + \left(\frac{5}{4}\right)^3 \cdot \frac{1}{2} + \left(\frac{7}{4}\right)^3 \cdot \frac{1}{2} = \frac{31}{8} \text{ or } \frac{15.5}{4}$$

(d)   Since the exact value of $\displaystyle\int_0^2 x^3\, dx$ is $\dfrac{x^4}{4} \Big|_0^2$ or 4, the midpoint sum is the best of the three approximations. This is usually the case.

We will denote the three Riemann Sums, with n subintervals, by $L(n)$, $R(n)$, and $M(n)$. (These sums are also sometimes called "rules.")

E2. Using Trapezoids

We now find the areas of the strips in Figure 6.6 by using trapezoids. We denote the bases of the trapezoids by $y_0, y_1, y_2, \ldots, y_n$ and the heights by $\Delta x = h_1, h_2, \ldots, h_n$.

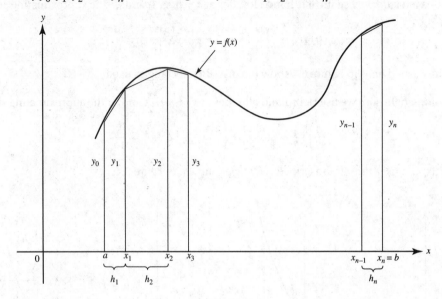

Figure 6.6

The following sum, denoted $T(n)$, approximates the area between f and the x-axis from a to b:

Trapezoidal sum

$$T(n) = \frac{y_0 + y_1}{2} \cdot h_1 + \frac{y_1 + y_2}{2} \cdot h_2 + \frac{y_2 + y_3}{2} \cdot h_3 + \ldots + \frac{y_{n-1} + y_n}{2} \cdot h_n$$

> **Example 25**

Use $T(4)$ to approximate $\displaystyle\int_0^2 x^3 \, dx$.

✓ **Solution**

From Example 24, $h = \dfrac{1}{2}$. Then,

$$T(4) = \frac{0^3 + (1/2)^3}{2} \cdot \frac{1}{2} + \frac{(1/2)^3 + 1^3}{2} \cdot \frac{1}{2} + \frac{1^3 + (3/2)^3}{2} \cdot \frac{1}{2} + \frac{(3/2)^3 + 2^3}{2} \cdot \frac{1}{2} = \frac{17}{4}$$

This is better than either $L(4)$ or $R(4)$, but $M(4)$ is the best approximation here.

> **Example 26**

A function f passes through the five points shown. Estimate the area $A = \displaystyle\int_2^{12} f(x)\, dx$ using

(a)   a left rectangular approximation

(b)   a trapezoidal approximation

✓ **Solutions**

Note that the subinterval widths are not equal.

(a)   In each subinterval, we sketch the rectangle with height determined by the point on f at the left endpoint. Our estimate is the sum of the areas of these rectangles:

$$A \approx 1(7) + 3(11) + 2(13) + 4(12) \approx 114$$

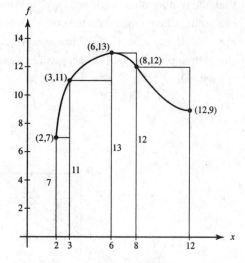

(b)   In each subinterval, we sketch trapezoids by drawing segments connecting the points on f. Our estimate is the sum of the areas of these trapezoids:

$$A \approx \left(\frac{7+11}{2}\right) \cdot 1 + \left(\frac{11+13}{2}\right) \cdot 3 +$$

$$\left(\frac{13+12}{2}\right) \cdot 2 + \left(\frac{12+9}{2}\right) \cdot 4 \approx 112$$

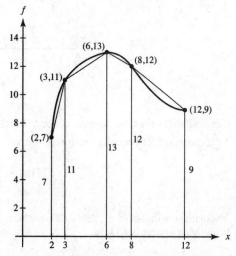

E3. Comparing Approximating Sums

If f is an increasing function on $[a,b]$, then $L(n) \leq \int_a^b f(x)\,dx \leq R(n)$, while if f is decreasing, then $R(n) \leq \int_a^b f(x)\,dx \leq L(n)$.

From Figure 6.7 we infer that the area of a trapezoid is less than the true area if the graph of f is concave down but is more than the true area if the graph of f is concave up.

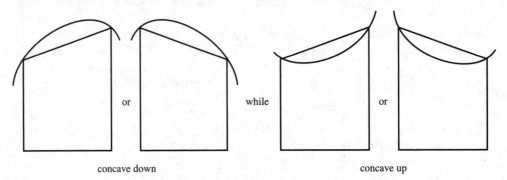

concave down   concave up

Figure 6.7

Figure 6.8 is helpful in showing how the area of a midpoint rectangle compares with that of a trapezoid and with the true area. Our graph here is concave down. If M is the midpoint of AB, then the midpoint rectangle is AM_1M_2B. We've drawn T_1T_2 tangent to the curve at T (where the midpoint ordinate intersects the curve). Since the shaded triangles have equal areas, we see that area AM_1M_2B = area AT_1T_2B.[†] But area AT_1T_2B clearly exceeds the true area, as does the area of the midpoint rectangle. This fact justifies the right half of the inequality below; Figure 6.7 verifies the left half.

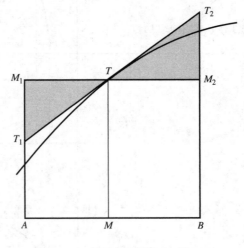

Figure 6.8

Generalizing to n subintervals, we conclude:

If the graph of f is concave down, then

$$T(n) \leq \int_a^b f(x)\,dx \leq M(n)$$

[†] Note that the trapezoid AT_1T_2B is *different* from the trapezoids in Figure 6.7, which are like the ones we use in applying a trapezoidal sum.

If the graph of f is concave up, then

$$M(n) \leq \int_a^b f(x)\,dx \leq T(n)$$

> **Example 27**

Write an inequality including $L(n)$, $R(n)$, $M(n)$, $T(n)$, and $\int_a^b f(t)\,dt$ for the graph of f shown in Figure 6.9.

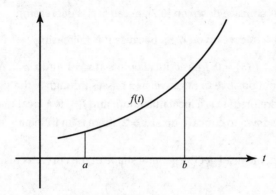

Figure 6.9

✓ **Solution**

Since f increases on $[a,b]$ and is concave up, the inequality is

$$L(n) \leq M(n) \leq \int_a^b f(x)\,dx \leq T(n) \leq R(n)$$

F. Graphing a Function from Its Derivative; Another Look

> **Example 28**

Figure 6.10 is the graph of function $f'(x)$; it consists of two line segments and a semicircle. If $f(0) = 1$, sketch the graph of $f(x)$. Identify any critical or inflection points of f and give their coordinates.

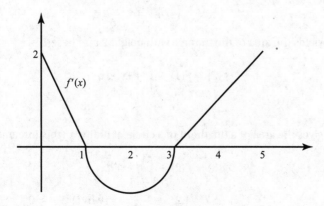

Figure 6.10

✓ **Solution**

We know that if $f' > 0$ on an interval then f increases on the interval, while if $f' < 0$ then f decreases; also, if f' is increasing on an interval then the graph of f is concave up on the interval, while if f' is decreasing then the graph of f is concave down. These statements lead to the following conclusions:

f increases on [0,1] and [3,5], because $f' > 0$ there

but f decreases on [1,3], because $f' < 0$ there

also   the graph of f is concave down on [0,2], because f' is decreasing

but   the graph of f is concave up on [2,5], because f' is increasing

Additionally, since $f'(1) = f'(3) = 0$, f has critical points at $x = 1$ and $x = 3$. As x passes through 1, the sign of f' changes from positive to negative; as x passes through 3, the sign of f' changes from negative to positive. Therefore $f(1)$ is a local maximum and $f(3)$ is a local minimum. Since f changes from concave down to concave up at $x = 2$, there is an inflection point on the graph of f there.

These conclusions enable us to get the general shape of the curve, as displayed in Figure 6.11a.

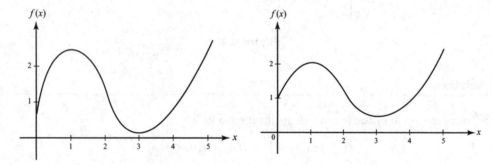

Figure 6.11a   **Figure 6.11b**

All that remains is to evaluate $f(x)$ at $x = 1$, 2, and 3. We use the Fundamental Theorem of Calculus to accomplish this, finding f also at $x = 4$ and 5 for completeness.
We are given that $f(0) = 1$. Then

$$f(1) = f(0) + \int_0^1 f'(x)\,dx$$
$$= \quad 1 \quad + \quad 1 \quad = 2$$

where the integral yields the area of the triangle with height 2 and base 1;

$$f(2) = f(1) + \int_1^2 f'(x)\,dx$$
$$= \quad 2 \quad - \quad \frac{\pi}{4} \quad \approx 1.2$$

where the integral gives the area of a quadrant of a circle of radius 1 (this integral is negative!);

$$f(3) = f(2) + \int_2^3 f'(x)\,dx$$
$$= 1.2 \quad - \quad \frac{\pi}{4} \quad \text{(why?)} \quad \approx 0.4$$

$$f(4) = f(3) + \int_3^4 f'(x)\, dx$$

$$\approx 0.4 + \quad \frac{1}{2} \quad\quad \approx 0.9$$

where the integral is the area of the triangle with height 1 and base 1;

$$f(5) = f(4) + \int_4^5 f'(x)\, dx$$

$$\approx 0.9 + \quad 1.5 \quad \text{(why?)} \ \approx 2.4$$

So the function $f(x)$ has a local maximum at (1,2), a point of inflection at (2,1.2), and a local minimum at (3,0.4) where we have rounded to one decimal place when necessary.

In Figure 6.11b, the graph of f is shown again, but now it incorporates the information just obtained using the FTC.

❯ Example 29

Readings from a car's speedometer at 10-minute intervals during a 1-hour period are given in the table; $t = $ time in minutes, $v = $ speed in miles per hour:

t	0	10	20	30	40	50	60
v	26	40	55	10	60	32	45

(a)   Draw a graph that could represent the car's speed during the hour.
(b)   Approximate the distance traveled, using $L(6)$, $R(6)$, and $T(6)$.
(c)   Draw a graph that could represent the distance traveled during the hour.

✓ Solutions

(a)   Any number of curves will do. The graph only has to pass through the points given in the table of speeds, as does the graph in Figure 6.12a.

Figure 6.12a

(b) $L(6) = 26 \cdot \dfrac{1}{6} + 40 \cdot \dfrac{1}{6} + 55 \cdot \dfrac{1}{6} + 10 \cdot \dfrac{1}{6} + 60 \cdot \dfrac{1}{6} + 32 \cdot \dfrac{1}{6} = 37\dfrac{1}{6}\,\text{mi}$

$R(6) = 40 \cdot \dfrac{1}{6} + 55 \cdot \dfrac{1}{6} + 10 \cdot \dfrac{1}{6} + 60 \cdot \dfrac{1}{6} + 32 \cdot \dfrac{1}{6} + 45 \cdot \dfrac{1}{6} = 40\dfrac{1}{3}\,\text{mi}$

$T(6) = \dfrac{26+40}{2} \cdot \dfrac{1}{6} + \dfrac{40+55}{2} \cdot \dfrac{1}{6} + \dfrac{55+10}{2} \cdot \dfrac{1}{6} +$

$\qquad\quad \dfrac{10+60}{2} + \dfrac{60+32}{2} \cdot \dfrac{1}{6} + \dfrac{32+45}{2} \cdot \dfrac{1}{6} = 38\dfrac{3}{4}\,\text{mi}$

(c)   To calculate the distance traveled during the hour, we use the methods demonstrated in Example 28. We know that, since $v(t) > 0$, $s = \displaystyle\int_{a}^{b} v(t)\,dt$ is the distance covered from time a to time b, where $v(t)$ is the speed or velocity. Thus,

$s(0) = 0$

$s\left(\dfrac{1}{6}\right) = 0 + \displaystyle\int_{0}^{1/6} v(t)\,dt = 0 + (26+40) \cdot \dfrac{1}{12} = \dfrac{66}{12}$

$s\left(\dfrac{2}{6}\right) = \dfrac{66}{12} + \displaystyle\int_{1/6}^{2/6} v(t)\,dt = \dfrac{66}{12} + (40+55) \cdot \dfrac{1}{12} = \dfrac{161}{12}$

$\qquad \vdots \qquad \vdots$

$s\left(\dfrac{6}{6}\right) = \dfrac{388}{12} + \displaystyle\int_{5/6}^{6/6} v(t)\,dt = \dfrac{388}{12} + (32+45) \cdot \dfrac{1}{12} = \dfrac{465}{12}$

It is left to the student to complete the missing steps above and to verify the distances in the following table ($t =$ time in minutes, $s =$ distance in miles):

t	0	10	20	30	40	50	60
s	0	5.5	13.4	18.8	24.7	32.3	38.8

Figure 6.12b is one possible graph for the distance covered during the hour.

Figure 6.12b

> ### Example 30

The graph of $f(t)$ is given in Figure 6.13. If $F(x) = \int_0^x f(t)\, dt$, fill in the values for $F(x)$ in the table:

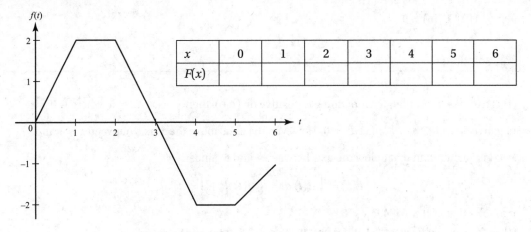

x	0	1	2	3	4	5	6
$F(x)$							

Figure 6.13

✓ Solution

We evaluate $F(x)$ by finding areas of appropriate regions.

$$F(0) = \int_0^1 f(t)dt = 0$$

$$F(1) = \int_0^1 f(t)dt = \frac{1}{2}(1)(2) = 1 \text{ (the area of a triangle)}$$

$$F(2) = \int_0^2 f(t)dt = \frac{1}{2}(1)(2) + (1)(2) = 3 \text{ (a triangle plus a rectangle)}$$

$$F(4) = \int_0^3 f(t)dt = \frac{1}{2}(3+1)(2) - \frac{1}{2}(1)(2) = 3 \text{ (a trapezoid minus a triangle)}$$

Here is the completed table:

x	0	1	2	3	4	5	6
$F(x)$	0	1	3	4	3	1	−0.5

> ### Example 31

The graph of the function $f(t)$ is shown in Figure 6.14.

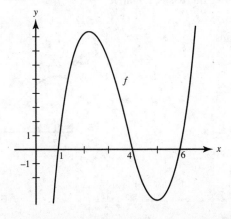

Figure 6.14

Let $F(x) = \int_1^x f(t)\, dt$. Decide whether each statement is true or false; justify your answers:

    (I) If $4 < x < 6$, $F(x) > 0$

    (II) If $4 < x < 5$, $F'(x) > 0$

    (III) $F''(6) > 0$

✓ Solutions

(I) is true. We know that, if a function g is positive on (a,b), then $\int_a^b g(x)\, dx > 0$, whereas if g is negative on (a,b), then $\int_a^b g(x)\, dx < 0$. However, the area above the x-axis between $x = 1$ and $x = 4$ is greater than that below the axis between 4 and 6. Since

$$F(x) = \int_1^x f(t)\, dt = \int_1^4 f(t)\, dt + \int_4^x f(t)\, dt$$

it follows that $F(x) > 0$ if $4 < x < 6$.

(II) is false. Since $F'(x) = f(x)$ and $f(x) < 0$ if $4 < x < 5$, then $F'(x) < 0$.

(III) is true. Since $F'(x) = f(x)$, $F''(x) = f'(x)$. At $x = 6$, $f'(x) > 0$ (because f is increasing). Therefore, $F''(6) > 0$.

❯ Example 32

Graphs of functions $f(x)$, $g(x)$, and $h(x)$ are given in Figures 6.15a, 6.15b, and 6.15c. Consider the following statements:

    (I) $f(x) = g'(x)$ (II) $h(x) = f'(x)$ (III) $g(x) = \int_{-2.5}^x f(t)\, dt$

Which of these statements is (are) true?

(A)   I only

(B)   II only

(C)   III only

(D)   all three

✓ Solution

The correct answer is (D).

I is true since, for example, $f(x) = 0$ for the critical values of g: f is positive where g increases, negative where g decreases, and so on.

    II is true for similar reasons.

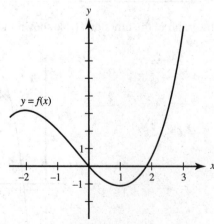

Figure 6.15a

III is also true. Verify that the value of the integral $g(x)$ increases on the interval $-2.5 < x < 0$ (where $f > 0$), decreases between the zeros of f (where $f < 0$), then increases again when f becomes positive.

Figure 6.15b   Figure 6.15c

❯ Example 33

Assume the world use of copper has been increasing at a rate given by $f(t) = 15e^{0.015t}$, where t is measured in years, with $t = 0$ the beginning of 2000, and $f(t)$ is measured in millions of tons per year.

(a)   What definite integral gives the total amount of copper that was used for the 5-year period from $t = 0$ to the beginning of the year 2005?

(b)   Write out the terms in the left sum $L(5)$ for the integral in (a). What do the individual terms of $L(5)$ mean in terms of the world use of copper?

(c)   How good an approximation is $L(5)$ for the definite integral in (a)?

✓ Solutions

(a) $\displaystyle\int_0^5 15e^{0.015t}\, dt$

(b) $L(5) = 15e^{0.015 \cdot 0} + 15e^{0.015 \cdot 1} + 15e^{0.015 \cdot 2} + 15e^{0.015 \cdot 3} + 15e^{0.015 \cdot 4}$. The five terms on the right represent the world's use of copper for the 5 years from 2000 until 2005.

(c)   The answer to (a), using our calculator, is 77.884 million tons. $L(5) = 77.301$ million tons, so $L(5)$ underestimates the projected world use of copper during the 5-year period by approximately 583,000 tons.

Example 33 is an excellent instance of the FTC: if $f = F'$, then $\displaystyle\int_a^b f(x)\, dx$ gives the total change in F as x varies from a to b.

> **Example 34**

Suppose $\int_{-1}^{4} f(x)\,dx = 6$, $\int_{-1}^{4} g(x)\,dx = -3$, and $\int_{-1}^{0} g(x)\,dx = -1$. Evaluate

(a) $\int_{-1}^{4} (f - g)(x)\,dx$ (b) $\int_{0}^{4} g(x)\,dx$ (c) $\int_{2}^{7} f(x - 3)\,dx$

✓ **Solutions**

(a) 9

(b) $\int_{0}^{4} g(x)\,dx = \int_{0}^{-1} g(x)\,dx + \int_{-1}^{4} g(x)\,dx = -\int_{-1}^{0} g(x)\,dx + \int_{-1}^{4} g(x)\,dx$

$$= +1 + (-3) = -2$$

(c)   To evaluate $\int_{2}^{7} f(x - 3)\,dx$, let $u = x - 3$. Then $du = dx$ and, when $x = 2$,

$u = -1$; when $x = 7$, $u = 4$. Therefore $\int_{2}^{7} f(x - 3)\,dx = \int_{-1}^{4} f(u)\,du = 6$.

G. Interpreting ln *x* as an Area

It is quite common to define ln *x*, the natural logarithm of *x*, as a definite integral, as follows:

$$\ln x = \int_{1}^{x} \frac{1}{t}\,dt \qquad (x > 0)$$

This integral can be interpreted as the area bounded above by the curve $y = \frac{1}{t}$ $(t > 0)$, below by the *t*-axis, at the left by $t = 1$, and at the right by $t = x$ $(x > 1)$. See Figure 6.16.

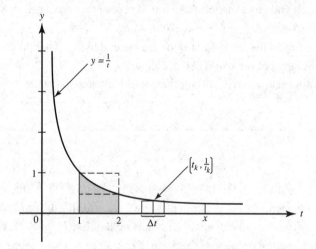

Figure 6.16

Note that if $x = 1$ the above definition yields $\ln 1 = 0$, and if $0 < x < 1$ we can rewrite as follows:

$$\ln x = -\int_{x}^{1} \frac{1}{t}\,dt$$

showing that $\ln x < 0$ if $0 < x < 1$.

With this definition of ln *x* we can approximate ln *x* using rectangles or trapezoids.

> **Example 35**

Show that $\frac{1}{2} < \ln 2 < 1$.

✓ **Solution**

Using the definition of $\ln x$ yields $\ln 2 = \int_1^2 \frac{1}{t}\,dt$, which we interpret as the area under $y = \frac{1}{t}$, above the t-axis, and bounded at the left by $t = 1$ and at the right by $t = 2$ (the shaded region in Figure 6.16). Since $y = \frac{1}{t}$ is strictly decreasing, the area of the inscribed rectangle (height $\frac{1}{2}$, width 1) is less than $\ln 2$, which, in turn, is less than the area of the circumscribed rectangle (height 1, width 1). Thus

$$\frac{1}{2} \cdot 1 < \ln 2 < 1 \cdot 1 \text{ or } \frac{1}{2} < \ln 2 < 1$$

> **Example 36**

Find $L(5)$, $R(5)$, and $T(5)$ for $\int_1^6 \frac{120}{x}\,dx$.

✓ **Solution**

Noting that for $n = 5$ subintervals on the interval $[1,6]$ we have $\Delta x = 1$, we make a table of values for $f(x) = \frac{120}{x}$:

x	1	2	3	4	5	6
$f(x)$	120	60	40	30	24	20

Then:

$$L(5) = 120 \cdot 1 + 60 \cdot 1 + 40 \cdot 1 + 30 \cdot 1 + 24 \cdot 1 = 274$$

$$R(5) = 60 \cdot 1 + 40 \cdot 1 + 30 \cdot 1 + 24 \cdot 1 + 20 \cdot 1 = 174$$

$$T(5) = \frac{120 + 60}{2} \cdot 1 + \frac{60 + 40}{2} \cdot 1 + \frac{40 + 30}{2} \cdot 1 + \frac{30 + 24}{2} \cdot 1 + \frac{24 + 20}{2} \cdot 1 = 224$$

NOTE: The calculator finds that $\int_1^6 \frac{120}{x}\,dx$ is approximately 215.011.

H. Average Value

If the function $y = f(x)$ is integrable on the interval $a \leq x \leq b$, then we define the *average value* of f from a to b to be

$$\frac{1}{b-a}\int_a^b f(x)\,dx \qquad (1)$$

Average value of a function

Note that (1) is equivalent to

$$(\text{average value of } f) \cdot (b - a) = \int_a^b f(x)\,dx \qquad (2)$$

If $f(x) \geq 0$ for all x on $[a,b]$, we can interpret (2) in terms of areas as follows: The right-hand expression represents the area under the curve of $y = f(x)$, above the x-axis, and bounded by the vertical lines $x = a$ and $x = b$. The left-hand expression of (2) represents the area of a rectangle with the same base $(b - a)$ and with the average value of f as its height. See Figure 6.17.

CAUTION: The average value of a function is not the same as the *average rate of change* (see page 113). Before answering any question about either of these, be sure to reread the question carefully to be absolutely certain which is called for.

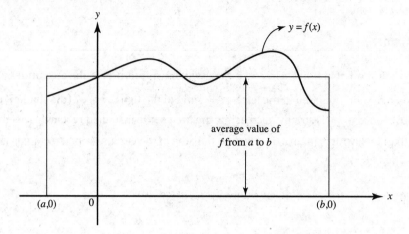

Figure 6.17

> **Example 37**

Find the average value of $f(x) = \ln x$ on the interval $[1,4]$.

✓ **Solution**

$$\frac{1}{4-1}\int_1^4 \ln x \, dx = \frac{1}{3}(x \ln x - x)\Big|_1^4 = \frac{4 \ln 4 - 3}{3}$$

> **Example 38**

Find the average value of y for the semicircle $y = \sqrt{4 - x^2}$ on $[-2,2]$.

✓ **Solution**

$$\frac{1}{2-(-2)}\int_{-2}^2 \sqrt{4-x^2} \, dx = \frac{1}{4}\frac{\pi(2^2)}{2} = \frac{\pi}{2}$$

NOTE: We have used the fact that the definite integral equals exactly the area of a semicircle of radius 2.

> **Example 39**

The graphs (a) through (e) in Figure 6.18 show the velocities of five cars moving along an east-west road (the x-axis) at time t, where $0 \leq t \leq 6$. In each graph, the scales on the two axes are the same.

Which graph shows the car

(1)   with constant acceleration?

(2)   with the greatest initial acceleration?

(3)   back at its starting point when $t = 6$?

(4)   that is farthest from its starting point at $t = 6$?

(5)   with the greatest average velocity?

(6)   with the least average velocity?

(7)   farthest to the left of its starting point when $t = 6$?

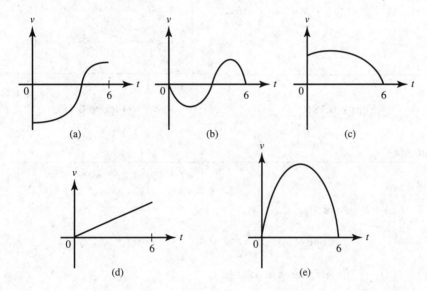

Figure 6.18

✓ **Solutions**

(1)   (d); since acceleration is the derivative of velocity and in (d) v', the slope, is constant.

(2)   (e); when $t = 0$ the slope of this v-curve (which equals acceleration) is greatest.

(3)   (b); since for this car the net distance traveled (given by the net area) equals zero.

(4)   (e); since the area under the v-curve is greatest, this car is farthest east.

(5)   (e); the average velocity equals the total distance divided by 6, which is the net area divided by 6 (see (4)).

(6)   (a); since only for this car is the net area negative.

(7)   (a); again, since net area is negative only for this car.

> **Example 40**

Identify each of the following quantities for the function $f(x)$, whose graph is shown in Figure 6.19a (*NOTE: $F'(x) = f(x)$*):

(a) $f(b) - f(a)$

(b) $\dfrac{f(b) - f(a)}{b - a}$

(c) $F(b) - F(a)$

(d) $\dfrac{F(b) - F(a)''}{b - a}$

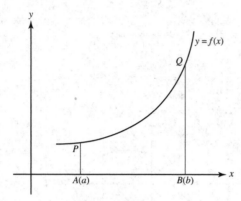

Figure 6.19a   Figure 6.19b

✓ Solutions

See Figure 6.19b.

(a) $f(b) - f(a)$ = length of RQ

(b) $\dfrac{f(b) - f(a)}{b - a} = \dfrac{RQ}{PR}$ = slope of secant PQ

(c) $F(b) - F(a) = \displaystyle\int_a^b f(x)\,dx$ = area of $APDQB$

(d) $\dfrac{F(b) - F(a)''}{b - a}$ = average value of f over $[a,b]$ = length of CD, where $CD \cdot AB$ or $CD \cdot (b - a)$ is equal to the area $F(b) - F(a)$

> **Example 41**

The graph in Figure 6.20 shows the speed $v(t)$ of a car, in miles per hour, at 10-minute intervals during a 1-hour period.

(a)   Give an upper and a lower estimate of the total distance traveled.

(b)   When does the acceleration appear greatest?

(c)   Estimate the acceleration when $t = 20$.

(d)   Estimate the average speed of the car during the interval $30 \leq t \leq 50$.

Figure 6.20

✓ **Solutions**

(a) A lower estimate, using minimum speeds and $\frac{1}{6}$ hour for 10 minutes, is

$$0\left(\frac{1}{6}\right) + 40\left(\frac{1}{6}\right) + 50\left(\frac{1}{6}\right) + 70\left(\frac{1}{6}\right) + 30\left(\frac{1}{6}\right) + 30\left(\frac{1}{6}\right)$$

This yields $36\frac{2}{3}$ miles for the total distance traveled during the hour. An upper estimate uses maximum speeds; it equals

$$40\left(\frac{1}{6}\right) + 50\left(\frac{1}{6}\right) + 70\left(\frac{1}{6}\right) + 70\left(\frac{1}{6}\right) + 70\left(\frac{1}{6}\right) + 30\left(\frac{1}{6}\right)$$

or 55 miles for the total distance.

(b) The acceleration, which is the slope of $v(t)$, appears greatest at $t = 5$ minutes, when the curve is steepest.

(c) To estimate the acceleration $v'(t)$ at $t = 20$, we approximate the slope of the curve at $t = 20$. The slope of the tangent at $t = 20$ appears to be equal to

$$(10 \text{ mph})/(10 \text{ min}) = (10 \text{ mph})/\left(\frac{1}{6}\text{ hr}\right) = 60 \text{ mi/hr}^2.$$

(d) The average speed equals the distance traveled divided by the time. We can approximate the distance from $t = 30$ to $t = 50$ by the area under the curve, or, roughly, by the sum of the areas of a rectangle and a trapezoid:

$$70\left(\frac{1}{6}\right) + \frac{70 + 30}{2}\left(\frac{1}{6}\right) = 20 \text{ mi}$$

Thus the average speed from $t = 30$ to $t = 50$ is

$$\frac{20 \text{ mi}}{20 \text{ min}} = 1 \text{ mi/min} = 60 \text{ mph}$$

> **Example 42**

Given the graph of $G(x)$ in Figure 6.21a, identify the following if $G'(x) = g(x)$:

(a) $g(b)$

(b) $\displaystyle\int_a^b G(x)\,dx$

(c) $\displaystyle\int_a^b g(x)\,dx$

(d) $\dfrac{\displaystyle\int_a^b g(x)\,dx}{b-a}$

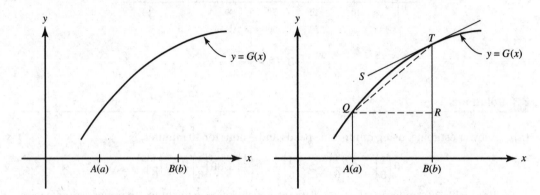

Figure 6.21a   **Figure 6.21b**

✓ **Solutions**

See Figure 6.21b.

(a) $g(b)$ is the slope of $G(x)$ at b, the slope of line ST

(b) $\displaystyle\int_a^b G(x)\,dx$ is equal to the area under $G(x)$ from a to b

(c) $\displaystyle\int_a^b g(x)\,dx = G(b) - G(a) = $ length of $BT - $ length of $BR = $ length of RT

(d) $\dfrac{\displaystyle\int_a^b g(x)\,dx}{b-a} = \dfrac{\text{length of } RT}{\text{length of } QR} = $ slope of QT

> **Example 43**

The function $f(t)$ is graphed in Figure 6.22a. Let

$$F(x) = \int_4^{x/2} f(t)\, dt$$

Figure 6.22a

(a)   What is the domain of F?

(b)   Find x, if $F'(x) = 0$.

(c)   Find x, if $F(x) = 0$.

(d)   Find x, if $F(x) = 1$.

(e)   Find $F'(6)$.

(f)   Find $F(6)$.

(g)   Sketch the complete graph of F.

✓ **Solutions**

(a)   The domain of f is $[-2,1]$ and $[2,6]$. We choose the portion of this domain that contains the lower limit of integration, 4. Thus the domain of F is $2 \le \dfrac{x}{2} \le 6$, or $4 \le x \le 12$.

(b)   Since $F'(x) = f\!\left(\dfrac{x}{2}\right) \cdot 1/2$, $F'(x) = 0$ if $f\!\left(\dfrac{x}{2}\right) = 0$. Then $\dfrac{x}{2} = 2$ and $x = 4$.

(c) $F(x) = 0$ when $\dfrac{x}{2} = 4$ or $x = 8$. $F(8) = \displaystyle\int_4^{8/2} f(t)\, dt = 0$

(d)   For $F(x)$ to equal 1, we need a region under f whose left endpoint is 4 with area equal to 1. The region from 4 to 5 works nicely; so $\dfrac{x}{2} = 5$ and $x = 10$.

(e) $F'(6) = f\!\left(\dfrac{6}{2}\right) \cdot \dfrac{1}{2} = f(3) \cdot \left(\dfrac{1}{2}\right) = 2 \cdot \left(\dfrac{1}{2}\right) = 1$

(f) $F(6) = \displaystyle\int_4^3 f(t)\, dt = -(\text{area of trapezoid}) = -\dfrac{2+1}{2} \cdot 1 = -\dfrac{3}{2}$

(g)   In Figure 6.22b we evaluate the areas in the original graph.

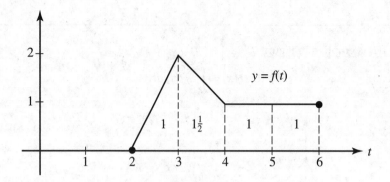

Figure 6.22b

Measured from the lower limit of integration, 4, we have:

$$F(4) = \int_4^2 f(t)\,dt = -2\frac{1}{2} \qquad\qquad F(6) = \int_4^3 f(t)\,dt = -1\frac{1}{2}$$

$$F(8) = \int_4^4 f(t)\,dt = 0 \qquad\qquad F(10) = \int_4^5 f(t)\,dt = 1$$

$$F(12) = \int_4^6 f(t)\,dt = 2$$

We note that, since $F'(=f)$ is linear on $(2,4)$, F is quadratic on $(4,8)$; also, since F' is positive and increasing on $(2,3)$, the graph of F is increasing and concave up on $(4,6)$, while since F' is positive and decreasing on $(3,4)$, the graph of F is increasing but concave down on $(6,8)$. Finally, since F' is constant on $(4,6)$, F is linear on $(8,12)$. (See Figure 6.22c.)

Figure 6.22c

CHAPTER SUMMARY

In this chapter, we reviewed definite integrals, starting with the Fundamental Theorem of Calculus. We looked at techniques for evaluating definite integrals algebraically, numerically, and graphically. We reviewed Riemann Sums, including the left, right, and midpoint approximations as well as the trapezoidal sum. We also looked at the average value of a function.

This chapter also reviewed integrals based on parametrically defined functions, a BC Calculus topic.

Practice Exercises

PART A. NO CALCULATOR—DIRECTIONS: Answer these questions *without* using your calculator.

A1. $\int_{-1}^{1} x^2 - x - 1 \, dx =$

(A) $\frac{2}{3}$

(B) 0

(C) $-\frac{4}{3}$

(D) -2

A2. $\int_{1}^{2} \frac{3x - 1}{3x} \, dx =$

(A) $\frac{3}{4}$

(B) $1 - \frac{1}{3} \ln 2$

(C) $1 - \ln 2$

(D) $-\frac{1}{3} \ln 2$

A3. $\int_{0}^{3} \frac{dt}{\sqrt{4 - t}} =$

(A) -2

(B) 4

(C) -1

(D) 2

A4. $\int_{-1}^{0} \sqrt{3u + 4} \, du =$

(A) 2

(B) $\frac{14}{9}$

(C) $\frac{14}{3}$

(D) $\frac{7}{2}$

A5. $\int_{2}^{3} \frac{dy}{2y - 3} =$

(A) $\ln 3$

(B) $\frac{1}{2} \ln \frac{3}{2}$

(C) $\frac{16}{9}$

(D) $\ln \sqrt{3}$

A6. $\int_{0}^{\sqrt{3}} \frac{x}{\sqrt{4 - x^2}} \, dx =$

(A) 1

(B) $\frac{\pi}{6}$

(C) -1

(D) 2

A7. $\int_{0}^{1} (2t - 1)^3 \, dt =$

(A) $\frac{1}{4}$

(B) $\frac{1}{2}$

(C) 0

(D) 4

A8. $\int_{4}^{9} \frac{2 + x}{2\sqrt{x}} \, dx =$

(A) $\frac{25}{3}$

(B) $\frac{41}{3}$

(C) $\frac{100}{3}$

(D) $\frac{5}{3}$

A9. $\int_{-3}^{3} \frac{dx}{9 + x^2} =$

(A) $\frac{\pi}{2}$

(B) 0

(C) $\frac{\pi}{6}$

(D) $-\frac{\pi}{2}$

A10. $\int_{0}^{1} e^{-x} \, dx =$

(A) $\frac{1}{e} - 1$

(B) $-\frac{1}{e}$

(C) $1 - \frac{1}{e}$

(D) $\frac{1}{e}$

A11. $\int_0^1 xe^{x^2}\,dx =$

(A) $e-1$

(B) $\frac{1}{2}(e-1)$

(C) $2(e-1)$

(D) $\frac{e}{2}-1$

A12. $\int_0^{\pi/4} \sin 2\theta\,d\theta =$

(A) 2

(B) $\frac{1}{2}$

(C) -1

(D) $-\frac{1}{2}$

A13. $\int_1^2 \frac{dz}{3-z} =$

(A) $-\ln 2$

(B) $\frac{3}{4}$

(C) $\frac{1}{2}\ln 2$

(D) $\ln 2$

A14. If we let $x = 2\sin\theta$, then $\int_1^2 \frac{\sqrt{4-x^2}}{x}\,dx$ is equivalent to

(A) $2\int_1^2 \frac{\cos^2\theta}{\sin\theta}\,d\theta$

(B) $\int_{\pi/6}^{\pi/2} \frac{\cos\theta}{\sin\theta}\,d\theta$

(C) $2\int_{\pi/6}^{\pi/2} \frac{\cos^2\theta}{\sin\theta}\,d\theta$

(D) $\int_1^2 \frac{\cos\theta}{\sin\theta}\,d\theta$

A15. $\int_0^\pi \cos^2\theta \sin\theta\,d\theta =$

(A) $-\frac{2}{3}$

(B) 1

(C) $\frac{2}{3}$

(D) 0

A16. $\int_1^e \frac{\ln x}{x}\,dx =$

(A) $\frac{1}{2}$

(B) $\frac{1}{2}\left(e^2-1\right)$

(C) 1

(D) $e-1$

A17. $\int_0^1 xe^x\,dx =$ **BC ONLY**

(A) -1

(B) $e+1$

(C) 1

(D) $e-1$

A18. $\int_0^{\pi/6} \frac{\cos\theta}{1+2\sin\theta}\,d\theta =$

(A) $\ln 2$

(B) $\frac{3}{8}$

(C) $-\frac{1}{2}\ln 2$

(D) $\ln\sqrt{2}$

A19. $\int_{\sqrt{2}}^2 \frac{u}{u^2-1}\,du =$

(A) $\ln\sqrt{3}$

(B) $\frac{8}{9}$

(C) $\ln\frac{3}{2}$

(D) $\ln 3$

A20. $\int_{\sqrt{2}}^2 \frac{u\,du}{(u^2-1)^2} =$

(A) $-\frac{1}{3}$

(B) $-\frac{2}{3}$

(C) $\frac{2}{3}$

(D) $\frac{1}{3}$

A21. $\int_{\pi/12}^{\pi/4} \frac{\cos 2x\,dx}{\sin^2 2x} =$

(A) 1

(B) $\frac{1}{2}$

(C) $-\frac{1}{2}$

(D) -1

A22. $\displaystyle\int_0^1 \frac{e^{-x}+1}{e^{-x}}\,dx =$

(A) e

(B) $2+e$

(C) $\dfrac{1}{e}$

(D) $1+e$

A23. $\displaystyle\int_0^1 \frac{e^x}{e^x+1}\,dx =$

(A) $\ln 2$

(B) $1+e$

(C) $-\ln 2$

(D) $\ln\dfrac{e+1}{2}$

A24. If we let $x = \tan\theta$, then $\displaystyle\int_1^{\sqrt{3}} \sqrt{1+x^2}\,dx$ is equivalent to

(A) $\displaystyle\int_{\pi/4}^{\pi/3} \sec\theta\,d\theta$

(B) $\displaystyle\int_1^{\sqrt{3}} \sec^3\theta\,d\theta$

(C) $\displaystyle\int_{\pi/4}^{\pi/3} \sec^3\theta\,d\theta$

(D) $\displaystyle\int_{\pi/4}^{\pi/3} \sec^2\theta\tan\theta\,d\theta$

A25. If the substitution $u = \sqrt{x+1}$ is used, then $\displaystyle\int_0^3 \frac{dx}{x\sqrt{x+1}}$ is equivalent to

(A) $\displaystyle\int_1^2 \frac{du}{u^2-1}$

(B) $\displaystyle\int_1^2 \frac{2\,du}{u^2-1}$

(C) $\displaystyle 2\int_0^3 \frac{du}{(u-1)(u+1)}$

(D) $\displaystyle 2\int_1^2 \frac{du}{u(u^2-1)}$

x	$f(x)$	$f'(x)$
0	11	3
2	15	2
4	16	−1
6	12	−3
8	7	0

A26. The table above shows some values of continuous function f and its first derivative. Evaluate $\displaystyle\int_8^0 f'(x)\,dx$.

(A) $-1/2$

(B) 3

(C) 4

(D) -1

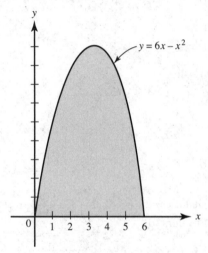

A27. Using a midpoint Riemann Sum with 3 equal width subintervals, find the approximate area of the shaded region above.

(A) 19

(B) 36

(C) 38

(D) 54

A28. The graph of a continuous function f passes through the points (4,2), (6,6), (7,5), and (10,8). Using trapezoids, we estimate that $\displaystyle\int_4^{10} f(x)\,dx \approx$

(A) 30

(B) 32

(C) 33

(D) 41

A29. The area of the shaded region in the figure above is equal to ln 3. If we approximate ln 3 using both a left Riemann Sum and a right Riemann Sum with 2 equal width subintervals as shown, which inequality follows?

(A) $\frac{1}{2} < \int_1^2 \frac{1}{x} dx < 1$

(B) $\frac{1}{3} < \int_1^3 \frac{1}{x} dx < 2$

(C) $\frac{1}{3} < \int_2^3 \frac{1}{x} dx < \frac{1}{2}$

(D) $\frac{5}{6} < \int_1^3 \frac{1}{x} dx < \frac{3}{2}$

A30. Let $A = \int_0^1 \cos x \, dx$. We estimate A using the L, R, and T approximations with $n = 100$ subintervals. Which is true?

(A) $L < A < T < R$

(B) $L < T < A < R$

(C) $R < A < T < L$

(D) $R < T < A < L$

A31. $\int_{-1}^3 |x| \, dx =$

(A) $\frac{7}{2}$

(B) 4

(C) $\frac{9}{2}$

(D) 5

A32. $\int_{-3}^2 |x + 1| \, dx =$

(A) $\frac{5}{2}$

(B) $\frac{7}{2}$

(C) $\frac{11}{2}$

(D) $\frac{13}{2}$

A33. The average value of $y = \sqrt{64 - x^2}$ on its domain is

(A) 2

(B) 4

(C) 2π

(D) 4π

A34. The average value of $\cos x$ over the interval $\frac{\pi}{3} \le x \le \frac{\pi}{2}$ is

(A) $\frac{3}{\pi}$

(B) $\frac{3(2 - \sqrt{3})}{\pi}$

(C) $\frac{3}{2\pi}$

(D) $\frac{2}{3\pi}$

A35. The average value of $\csc^2 x$ over the interval from $x = \frac{\pi}{6}$ to $x = \frac{\pi}{4}$ is

(A) $\frac{3\sqrt{3}}{\pi}$

(B) $\frac{\sqrt{3}}{\pi}$

(C) $\frac{12}{\pi}(\sqrt{3} - 1)$

(D) $3(\sqrt{3} - 1)$

A36. Choose the Riemann Sum whose limit is the integral $\int_1^5 x^3 dx$.

(A) $\lim_{n \to \infty} \sum_{k=1}^n \left(\left(1 + \frac{k}{n}\right)^3 \cdot \left(\frac{4}{n}\right) \right)$

(B) $\lim_{n \to \infty} \sum_{k=1}^n \left(\left(1 + \frac{4k}{n}\right)^3 \cdot \left(\frac{1}{n}\right) \right)$

(C) $\lim_{n \to \infty} \sum_{k=1}^n \left(\left(1 + \frac{4k}{n}\right)^3 \cdot \left(\frac{4}{n}\right) \right)$

(D) $\lim_{n \to \infty} \sum_{k=1}^n \left(\left(1 + \frac{k}{n}\right)^3 \cdot \left(\frac{1}{n}\right) \right)$

A37. Choose the Riemann Sum whose limit is the integral $\int_0^\pi \sin(3x)\,dx$.

(A) $\displaystyle\lim_{n\to\infty}\sum_{k=1}^{n}\left(\sin\left(\frac{3k}{n}\right)\cdot\left(\frac{1}{n}\right)\right)$

(B) $\displaystyle\lim_{n\to\infty}\sum_{k=1}^{n}\left(\sin\left(\frac{3\pi k}{n}\right)\cdot\left(\frac{1}{n}\right)\right)$

(C) $\displaystyle\lim_{n\to\infty}\sum_{k=1}^{n}\left(\sin\left(\frac{3k}{n}\right)\cdot\left(\frac{\pi}{n}\right)\right)$

(D) $\displaystyle\lim_{n\to\infty}\sum_{k=1}^{n}\left(\sin\left(\frac{3\pi k}{n}\right)\cdot\left(\frac{\pi}{n}\right)\right)$

A38. Choose the integral that is the limit of the Riemann Sum $\displaystyle\lim_{n\to\infty}\sum_{k=1}^{n}\left(\sin\left(1+\frac{6k}{n}\right)\cdot\left(\frac{3}{n}\right)\right)$.

(A) $\int_1^4 \sin(6x)\,dx$

(B) $\int_0^3 \sin(2x+1)\,dx$

(C) $\int_1^4 \sin(2x)\,dx$

(D) $\int_0^3 \sin(6x+1)\,dx$

A39. $\int_{\pi/4}^{\pi/2} \sin^3\alpha\cos\alpha\,d\alpha$ is equal to

(A) $\dfrac{3}{16}$

(B) $\dfrac{1}{4}$

(C) $-\dfrac{1}{4}$

(D) $-\dfrac{3}{16}$

PART B. CALCULATOR ACTIVE—DIRECTIONS: Some of the following questions require the use of a graphing calculator.

B1. Find the average value of function f, as shown in the graph below, on the interval [0,5].

(A) 4

(B) 5

(C) 7

(D) 8

B2. The integral $\int_{-4}^{4} \sqrt{16-x^2}\,dx$ gives the area of

(A) a circle of radius 4

(B) a semicircle of radius 4

(C) a quadrant of a circle of radius 4

(D) half of an ellipse

Use the graph of function *f*, shown below, for Questions B3–B6.

B3. Calculate the average value of $f(x)$ on the interval [0,6]. Which of the following intervals contains $x = c$, where $f(c)$ is the average value of f on [0,6]?

   I. [0,2]

   II. [2,4]

   III. [4,6]

(A) I only

(B) II only

(C) III only

(D) I and II only

B4. $\int_0^2 f'(3x)\,dx =$

(A) -2

(B) $-\dfrac{2}{3}$

(C) $\dfrac{2}{3}$

(D) 2

B5. Let $g(x) = \int_0^{2x} f(t)\,dt$; then $g'(1) =$

(A) 3

(B) 4

(C) 6

(D) 8

B6. Let $h(x) = x^2 - f(x)$. Find $\int_0^6 h(x)\,dx$.

(A) 38

(B) 58

(C) 70

(D) 74

B7. If $f(x)$ is continuous on the closed interval $[a,b]$, then there exists at least one number c, $a < c < b$, such that $\int_a^b f(x)\,dx$ is equal to

(A) $\dfrac{f(c)}{b-a}$

(B) $f'(c)(b-a)$

(C) $f(c)(b-a)$

(D) $\dfrac{f'(c)}{b-a}$

B8. If $f(x)$ is continuous on the closed interval $[a,b]$ and k is a constant, then $\int_a^b kf(x)\,dx$ is equal to

(A) $k(b-a)$

(B) $k\big[f(b)-f(a)\big]$

(C) $\dfrac{\big[kf(x)\big]^2}{2}\bigg|_a^b$

(D) $k\int_a^b f(x)\,dx$

B9. $\dfrac{d}{dt}\int_0^t \sqrt{x^3+1}\,dx =$

(A) $\sqrt{t^3+1}$

(B) $\dfrac{\sqrt{t^3+1}}{3t^2}$

(C) $3x^2\sqrt{x^3+1}$

(D) $\sqrt{t^3+1}-1$

B10. If $F(u) = \int_1^u \left(2-x^2\right)^3 dx$, then $F'(u)$ is equal to

(A) $-6u\left(2-u^2\right)^2$

(B) $\left(2-u^2\right)^3 - 1$

(C) $\left(2-u^2\right)^3$

(D) $-2u\left(2-u^2\right)^3$

B11. $\dfrac{d}{dx}\int_{\pi/2}^{x^2} \sqrt{\sin t}\,dt =$

(A) $\sqrt{\sin t^2}$

(B) $2x\sqrt{\sin x^2}-1$

(C) $\sqrt{\sin x^2}-1$

(D) $2x\sqrt{\sin x^2}$

B12. If $x = 4\cos\theta$ and $y = 3\sin\theta$, then $\int_2^4 xy\,dx$ is equivalent to

(A) $48\int_{\pi/3}^{0} \sin\theta\cos^2\theta\,d\theta$

(B) $48\int_{2}^{4} \sin^2\theta\cos\theta\,d\theta$

(C) $48\int_{0}^{\pi/3} \sin\theta\cos^2\theta\,d\theta$

(D) $48\int_{0}^{\pi/3} \sin^2\theta\cos\theta\,d\theta$

B13. A continuous function f takes on the values shown in the table below. Estimate $\int_0^{30} f(x)dx$ using a left rectangular approximation with five subintervals.

x	0	5	12	20	24	30
$f(x)$	9	11	7	3	1	2

(A) 144

(B) 170

(C) 186

(D) 196

B14. Find the value of x at which the function $y = x^2$ reaches its average value on the interval $[0,10]$.

(A) 5.313

(B) 5.774

(C) 6.083

(D) 18.257

B15. The average value of $f(x) = \begin{cases} x^3 & x < 2 \\ 4x & x \geq 2 \end{cases}$ on the interval $0 \leq x \leq 5$ is

(A) 6.25

(B) 6.41

(C) 9.2

(D) 10

Answer Key

A1. **C**	A15. **C**	A29. **D**	B3. **D**
A2. **B**	A16. **A**	A30. **D**	B4. **B**
A3. **D**	A17. **C**	A31. **D**	B5. **D**
A4. **B**	A18. **D**	A32. **D**	B6. **B**
A5. **D**	A19. **A**	A33. **C**	B7. **C**
A6. **A**	A20. **D**	A34. **B**	B8. **D**
A7. **C**	A21. **B**	A35. **C**	B9. **A**
A8. **A**	A22. **A**	A36. **C**	B10. **C**
A9. **C**	A23. **D**	A37. **D**	B11. **D**
A10. **C**	A24. **C**	A38. **B**	B12. **D**
A11. **B**	A25. **B**	A39. **A**	B13. **D**
A12. **B**	A26. **C**	B1. **D**	B14. **B**
A13. **D**	A27. **C**	B2. **B**	B15. **C**
A14. **C**	A28. **C**		

Answer Explanations

A1. **(C)** The integral is equal to

$$\left(\frac{1}{3}x^3 - \frac{1}{2}x^2 - x\right)\Big|_{-1}^{1} = -\frac{7}{6} - \frac{1}{6}$$

A2. **(B)** Rewrite as $\int_{1}^{2}\left(1 - \frac{1}{3}\cdot\frac{1}{x}\right)dx$. This equals

$$\left(x - \frac{1}{3}\ln x\right)\Big|_{1}^{2} = 2 - \frac{1}{3}\ln 2 - 1$$

A3. **(D)** Rewrite as

$$-\int_{0}^{3}(4-t)^{-1/2}(-1\,dt) = -2\sqrt{4-t}\,\Big|_{0}^{3} = -2(1-2)$$

A4. **(B)** This integral equals

$$\frac{1}{3}\int_{-1}^{0}(3u+4)^{1/2}\cdot 3\,du = \frac{1}{3}\cdot\frac{2}{3}(3u+4)^{3/2}\,\Big|_{-1}^{0}$$
$$= \frac{2}{9}\left(4^{3/2} - 1^{3/2}\right)$$

A5. **(D)** $\frac{1}{2}\int_{2}^{3}\frac{2\,dy}{2y-3} = \frac{1}{2}\ln(2y-3)\,\Big|_{2}^{3} = \frac{1}{2}(\ln 3 - \ln 1)$

A6. **(A)** Rewrite as

$$-\frac{1}{2}\int_{0}^{\sqrt{3}}\left(4-x^2\right)^{-1/2}(-2x\,dx) = -\frac{1}{2}\cdot 2\sqrt{4-x^2}\,\Big|_{0}^{\sqrt{3}} = -(1-2)$$

A7. **(C)** $\frac{1}{2}\int_{0}^{1}(2t-1)^3(2\,dt) = \frac{1}{2}\cdot\frac{(2t-1)^4}{4}\,\Big|_{0}^{1} = \frac{1}{2}\left(\frac{(2\cdot 1-1)^4}{4} - \frac{(2\cdot 0-1)^4}{4}\right)$

A8. **(A)** Divide: $\int_{4}^{9}\left(x^{-1/2} + \frac{1}{2}x^{1/2}\right)dx = \left(2x^{1/2} + \frac{1}{2}\cdot\frac{2}{3}x^{3/2}\right)\Big|_{4}^{9}$

$$= \left(2\cdot 3 + \frac{1}{3}\cdot 27\right) - \left(2\cdot 2 + \frac{1}{3}\cdot 8\right)$$

A9. **(C)** $\frac{1}{9}\int_{-3}^{3}\frac{dx}{1+\frac{x^2}{9}} = 3\cdot\frac{1}{9}\int_{-3}^{3}\frac{\frac{1}{3}dx}{1+\left(\frac{x}{3}\right)^2} = \frac{1}{3}\tan^{-1}\frac{x}{3}\,\Big|_{-3}^{3} = \frac{1}{3}\left(\frac{\pi}{4} - \left(-\frac{\pi}{4}\right)\right)$

A10. **(C)** You get $-e^{-x}\,\Big|_{0}^{1} = -\left(e^{-1} - 1\right)$.

A11. **(B)** $\frac{1}{2}\int_{0}^{1}e^{x^2}(2x\,dx) = \frac{1}{2}e^{x^2}\,\Big|_{0}^{1} = \frac{1}{2}(e-1)$

A12. **(B)** Evaluate $-\frac{1}{2}\cos 2\theta\,\Big|_{0}^{\pi/4}$, which equals $-\frac{1}{2}(0-1)$.

A13. **(D)** $-\int_{1}^{2}\frac{-dx}{3-z} = -\ln(3-z)\,\Big|_{1}^{2} = -(\ln 1 - \ln 2)$

A14. **(C)** If $x = 2\sin\theta$, $\sqrt{4-x^2} = 2\cos\theta$, $dx = 2\cos\theta\,d\theta$. When $x=1$, $\theta=\frac{\pi}{6}$; when $x=2$, $\theta=\frac{\pi}{2}$.

The integral is equivalent to $\int_{\pi/6}^{\pi/2}\frac{(2\cos\theta)(2\cos\theta)\,d\theta}{2\sin\theta}$.

A15. **(C)** Evaluate $-\int_{0}^{\pi}\cos^2\theta(-\sin\theta\,d\theta)$. This equals $-\frac{1}{3}\cos^3\theta\,\Big|_{0}^{\pi} = -\frac{1}{3}(-1-1)$.

A16. **(A)** $\int_{1}^{e}(\ln x)\left(\frac{1}{x}dx\right) = \frac{1}{2}\ln^2 x\,\Big|_{1}^{e} = \frac{1}{2}(1-0)$

A17. **(C)** Use the Parts Formula with $u = x$ and $dv = e^x\,dx$. Then $du = dx$ and $v = e^x$.

The result is $\left(xe^x - \displaystyle\int e^x dx\right)\Big|_0^1 = \left(xe^x - e^x\right)\Big|_0^1 = (e - e) - (0 - 1)$.

A18. **(D)** $\dfrac{1}{2}\displaystyle\int_0^{\pi/6} \dfrac{2\cos\theta\,d\theta}{1 + 2\sin\theta} = \dfrac{1}{2}\ln(1 + 2\sin\theta)\Big|_0^{\pi/6}$ and get $\dfrac{1}{2}\Big(\ln(1 + 1) - \ln 1\Big)$

A19. **(A)** $\dfrac{1}{2}\displaystyle\int_{\sqrt{2}}^2 \dfrac{2u}{u^2 - 1}\,du = \dfrac{1}{2}\ln\left(u^2 - 1\right)\Big|_{\sqrt{2}}^2 = \dfrac{1}{2}(\ln 3 - \ln 1)$

A20. **(D)** $\dfrac{1}{2}\displaystyle\int_{\sqrt{2}}^2 \left(u^2 - 1\right)^{-2}\cdot 2u\,du = -\dfrac{1}{2\left(u^2 - 1\right)}\Big|_{\sqrt{2}}^2 = -\dfrac{1}{2}\left(\dfrac{1}{3} - \dfrac{1}{1}\right)$

A21. **(B)** $\dfrac{1}{2}\displaystyle\int_{\pi/12}^{\pi/4} \sin^{-2} 2x \cos 2x\,(2\,dx) = -\dfrac{1}{2}\cdot\dfrac{1}{\sin 2x}\Big|_{\pi/12}^{\pi/4} = -\dfrac{1}{2}\left(\dfrac{1}{1} - \dfrac{1}{1/2}\right)$

A22. **(A)** $\displaystyle\int_0^1 \left(1 + e^x\right) dx = \left(x + e^x\right)\Big|_0^1 = (1 + e) - 1$

A23. **(D)** Evaluate $\ln\left(e^x + 1\right)\Big|_0^1$, getting $\ln(e + 1) - \ln 2$.

A24. **(C)** Note that $dx = \sec^2\theta\,d\theta$ and that $\sqrt{1 + \tan^2\theta} = \sec\theta$. Be sure to express the limits as values of θ: $1 = \tan\theta$ yields $\theta = \dfrac{\pi}{4}$; $\sqrt{3} = \tan\theta$ yields $\theta = \dfrac{\pi}{3}$.

A25. **(B)** If $u = \sqrt{x + 1}$, then $u^2 = x + 1$, and $2u\,du = dx$. When you substitute for the limits, you

get $2\displaystyle\int_1^4 \dfrac{u\,du}{u(u^2 - 1)}$. Since $u \neq 0$ on its interval of integration, you may divide the numerator and denominator by it.

A26. **(C)** $\displaystyle\int_8^0 f'(x)\,dx = f(0) - f(8) = 11 - 7 = 4$

A27. **(C)** On $[0,6]$ with $n = 3$, $\Delta x = 2$. Heights of rectangles at $x = 1$, 3, and 5 are 5, 9, and 5, respectively; $M(3) = (5 + 9 + 5)(2)$.

A28. **(C)** $\displaystyle\int_4^{10} f(x)\,dx \approx \left(\dfrac{2 + 6}{2}\right)\cdot 2 + \left(\dfrac{6 + 5}{2}\right)\cdot 1 + \left(\dfrac{5 + 8}{2}\right)\cdot 3 \approx 33$

A29. **(D)** For $L(2)$ use the circumscribed rectangles: $1 \cdot 1 + \dfrac{1}{2}\cdot 1 = \dfrac{3}{2}$;

for $R(2)$ use the inscribed rectangles: $\dfrac{1}{2}\cdot 1 + \dfrac{1}{3}\cdot 1 = \dfrac{5}{6}$.

A30. **(D)** On $[0,1]$ $f(x) = \cos x$ is decreasing, so $L > A$, $R < A$, and $T > R$. Furthermore, f is concave downward, so $T < A$.

A31. **(D)** On the interval $[-1,3]$ the area under the graph of $y = |x|$ is the sum of the areas of two triangles: $\dfrac{1}{2}(1)(1) + \dfrac{1}{2}(3)(3) = 5$.

A32. **(D)** Note that the graph $y = |x + 1|$ is the graph of $y = |x|$ translated one unit to the left. The area under the graph $y = |x + 1|$ on the interval $[-3,2]$ is the sum of the areas of two triangles: $\dfrac{1}{2}(2)(2) + \dfrac{1}{2}(3)(3) = \dfrac{13}{2}$.

A33. **(C)** Because $y = \sqrt{64 - x^2}$ is a semicircle of radius 8, its area is $\displaystyle\int_{-8}^8 \sqrt{64 - x^2}\,dx = \dfrac{1}{2}\cdot\pi\cdot(8)^2 = 32\pi$. The domain is $[-8,8]$, or 16 units wide. Hence the average height of the function is $\dfrac{1}{8 - (-8)}\displaystyle\int_{-8}^8 \sqrt{64 - x^2}\,dx = \dfrac{1}{16}\cdot 32\pi = 2\pi$.

A34. **(B)** The average value is equal to $\dfrac{1}{\pi/2 - \pi/3}\displaystyle\int_{\pi/3}^{\pi/2}\cos x\,dx$.

A35. **(C)** The average value is equal to $\dfrac{1}{\pi/4 - \pi/6}\displaystyle\int_{\pi/6}^{\pi/4}\csc^2 x\,dx$.

A36. **(C)** From the integral, we get $a = 1$, $b = 5$, so $\Delta x = \dfrac{5-1}{n} = \dfrac{4}{n}$ and $x_k = a + k\cdot\Delta x = 1 + \dfrac{4k}{n} = 1 + \dfrac{4k}{n}$. Replace x with x_k and replace dx with Δx in the integrand to get the general term in the summation.

A37. **(D)** From the integral, we get $a = 0$, $b = \pi$, so $\Delta x = \dfrac{\pi - 0}{n} = \dfrac{\pi}{n}$ and $x_k = a + k\cdot\Delta x = 0 + \dfrac{\pi k}{n} = \dfrac{\pi k}{n}$. Replace x with x_k and replace dx with Δx in the integrand to get the general term in the summation.

A38. **(B)** From the Riemann Sum, we see $\Delta x = \dfrac{3}{n}$, then $k\cdot\Delta x = \dfrac{3k}{n}$. Notice that the term involving k in the Riemann Sum is not equal to $\dfrac{3k}{n}$ but $2\left(\dfrac{3k}{n}\right)$. Thus, we choose $x_k = \dfrac{3k}{n}$, so $a = 0$ and $\Delta x = \dfrac{b-0}{n} = \dfrac{3}{n}$, so $b = 3$. Since x_k replaces x, $f(x) = \sin(2x+1)$ giving the integral $\displaystyle\int_0^3 \sin(2x+1)\,dx$.

A39. **(A)** The integral is of the form $\displaystyle\int u^3\,du$; evaluate $\dfrac{1}{4}\sin^4\alpha\,\Big|_{\pi/4}^{\pi/2}$.

B1. **(D)** The average value is $\dfrac{1}{5-0}\displaystyle\int_0^5 f(x)\,dx$. The integral represents the area of a trapezoid: $\dfrac{1}{2}(5+3)\cdot 10 = 40$. The average value is $\dfrac{1}{5}(40)$.

B2. **(B)** Since $x^2 + y^2 = 16$ is a circle, the given integral equals the area of a semicircle of radius 4.

B3. **(D)** A vertical line at $x = 2$ divides the area under f into a trapezoid and a triangle; hence, $\displaystyle\int_0^6 f(x)\,dx = \dfrac{1}{2}(2+4)(2) + \dfrac{1}{2}(4)(4) = 14$. Thus, the average value of f on $[0,6]$ is $\dfrac{14}{6} = 2\dfrac{1}{3}$. There are points on f with y-values of $2\dfrac{1}{3}$ in the intervals $[0,2]$ and $[2,4]$.

B4. **(B)** $\displaystyle\int_0^2 f'(3x)\,dx = \dfrac{1}{3}\int_0^2 f'(3x)(3\,dx) = \dfrac{1}{3}f(3x)\Big|_0^2 = \dfrac{1}{3}\big(f(6) - f(0)\big) = \dfrac{1}{3}(0-2)$

B5. **(D)** $g'(x) = f(2x)\cdot 2$; thus $g'(1) = f(2)\cdot 2$

B6. **(B)** $\displaystyle\int_0^6 \big(x^2 - f(x)\big)\,dx = \int_0^6 x^2\,dx - \int_0^6 f(x)\,dx = \dfrac{x^3}{3}\Big|_0^6 - 14$. (Why 14? See the Solution for Question B3.)

B7. **(C)** This is the Mean Value Theorem for Integrals (page 242).

B8. **(D)** This is theorem (2) from page 242. Prove by counterexamples that (A), (B), and (C) are false.

B9. **(A)** This is a restatement of the Fundamental Theorem. In theorem (1) from page 242, interchange t and x.

B10. **(C)** Apply theorem (1) from page 242, noting that

$$F'(u) = \dfrac{d}{du}\int_a^u f(x)\,dx = f(u)$$

B11. **(D)** Let $y = \int_{\pi/2}^{x^2} \sqrt{\sin t}\, dt$ and $u = x^2$; then
$$y = \int_{\pi/2}^{u} \sqrt{\sin t}\, dt$$
By the Chain Rule, $\dfrac{dy}{dx} = \dfrac{dy}{du} \cdot \dfrac{du}{dx} = \sqrt{\sin u} \cdot 2x$, where theorem (1) from page 242 is used to find $\dfrac{dy}{du}$. Replace u by x^2.

B12. **(D)** Since $dx = -4 \sin \theta\, d\theta$, you get the new integral $-48 \int_{\pi/3}^{0} \sin^2 \theta \cos \theta\, d\theta$. Use theorem (4) from page 242 to get the correct answer.

B13. **(D)** $L(5) = 9 \cdot 5 + 11 \cdot 7 + 7 \cdot 8 + 3 \cdot 4 + 1 \cdot 6$

B14. **(B)** The average value is $\dfrac{1}{10-0} \int_{0}^{10} x^2 dx = \dfrac{1}{10} \cdot \dfrac{x^3}{3}\Big|_0^{10} = \dfrac{100}{3}$. Solve $x^2 = \dfrac{100}{3}$.

B15. **(C)** $\dfrac{1}{5-0} \int_{0}^{5} f(x) dx = \dfrac{1}{5}\left(\int_{0}^{2} x^3 dx + \int_{2}^{5} 4x\, dx \right) = \dfrac{1}{5}\left(\dfrac{x^4}{4}\Big|_0^2 + 2x^2 \Big|_2^5 \right) = \dfrac{1}{5}(4 + 42)$

7

Applications of Integration to Geometry

Learning Objectives

In this chapter, you will review using definite integrals to find areas and volumes; specifically:

→ Area under a curve
→ Area between two curves
→ Volumes of solids with known cross sections
→ Volumes of solids of revolution (using disks and washers)

You will also review related BC topics, including:

→ Lengths of curve (arc lengths), areas, and volumes involving parametrically defined functions
→ Area and length of curve (arc length) for polar curves

In addition, BC Calculus students will review the topic of improper integrals, including:

→ Recognizing when an integral is improper
→ Techniques for determining whether an improper integral converges or diverges

A. Area

To find an area, we

(1) draw a sketch of the given region and of a typical element
(2) write the expression for the area of a typical rectangle
(3) set up the definite integral that is the limit of the Riemann Sum of n areas as $n \to \infty$.

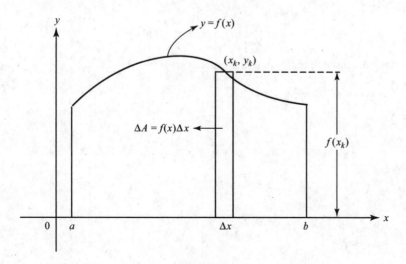

Figure 7.1

If $f(x)$ is nonnegative on $[a,b]$, as in Figure 7.1, then $f(x_k)\,\Delta x$ can be regarded as the area of a typical approximating rectangle, and the area bounded by the x-axis, the curve, and the vertical lines $x = a$ and $x = b$ is given exactly by

$$\lim_{n\to\infty} \sum_{k=1}^{n} f(x_k)\,\Delta x \qquad \text{and hence by} \qquad \int_{a}^{b} f(x)\,dx$$

See Questions A1, A5, and A10 in the Practice Exercises at the end of this chapter.

If $f(x)$ changes sign on the interval (Figure 7.2), we find the values of x for which $f(x) = 0$ and note where the function is positive and where it is negative. The total area bounded by the x-axis, the curve, $x = a$, and $x = b$ is here given exactly by

$$\int_{a}^{c} f(x)\,dx - \int_{c}^{d} f(x)\,dx + \int_{d}^{b} f(x)\,dx$$

where we have taken into account that $f(x_k)\,\Delta x$ is a negative number if $c < x < d$.

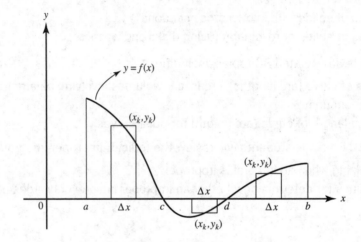

Figure 7.2

See Question A11 in the Practice Exercises.

If x is given as a function of y, say $x = g(y)$, then (Figure 7.3) the subdivisions are made along the y-axis, and the area bounded by the y-axis, the curve, and the horizontal lines $y = a$ and $y = b$ is given exactly by

$$\lim_{n\to\infty} \sum_{k=1}^{n} g(y_k)\,\Delta y = \int_{a}^{b} g(y)\,dy$$

See Questions A3 and A13 in the Practice Exercises.

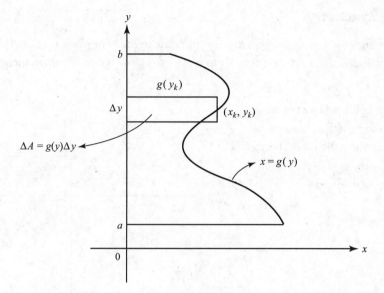

Figure 7.3

A1. Area Between Curves

To find the area between curves (Figure 7.4), we first find where they intersect and then write the area of a typical element for each region between the points of intersection. For the total area bounded by the curves $y = f(x)$ and $y = g(x)$ between $x = a$ and $x = e$, we see that, if they intersect at $[c,d]$, the total area is given exactly by

$$\int_a^c [f(x) - g(x)] \, dx + \int_c^e [g(x) - f(x)] \, dx$$

See Questions A4, A6, A7, and A9 in the Practice Exercises.

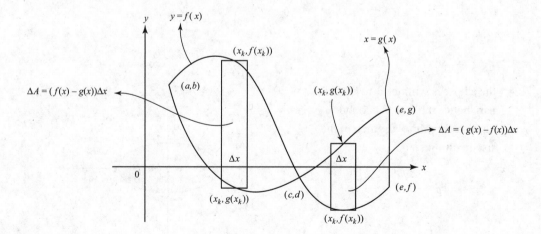

Figure 7.4

A2. Using Symmetry

Frequently we seek the area of a region that is symmetric to the *x*- or *y*-axis (or both) or to the origin. In such cases it is almost always simpler to make use of this symmetry when integrating. For example:

- The area bounded by the *x*-axis and this arch of the cosine curve is symmetric to the *y*-axis; hence it is twice the area of the region to the right of the *y*-axis.

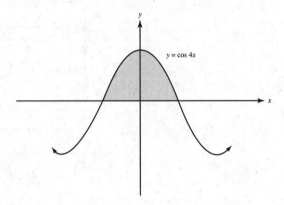

- The area bounded by the parabola and the line is symmetric to the *x*-axis; hence it is twice the area of the region above the *x*-axis.

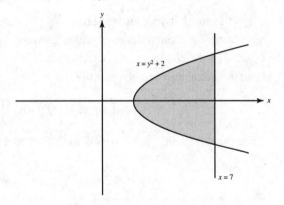

- The ellipse is symmetric to both axes; hence the area inside the ellipse is four times the area in the first quadrant.

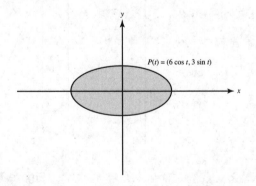

Evaluating $\displaystyle\int_{a}^{b} f(x)\,dx$ Using a Graphing Calculator

The calculator is especially useful in evaluating definite integrals when the *x*-intercepts are not easily determined otherwise or when an explicit antiderivative of *f* is not obvious (or does not exist).

> **Example 1**

Evaluate $\int_0^1 e^{-x^2}\, dx$.

✓ **Solution**

The integrand $f(x) = e^{-x^2}$ has no easy antiderivative. The calculator estimates the value of the integral to be 0.747 to three decimal places.

> **Example 2**

In Figure 7.5, find the area under $f(x) = -x^4 + x^2 + x + 10$ and above the x-axis.

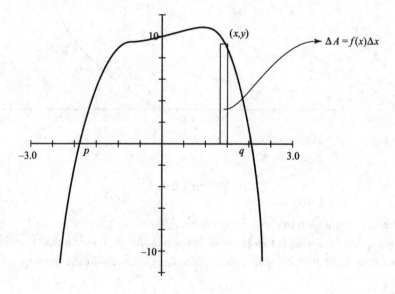

Figure 7.5

✓ **Solution**

To get an accurate answer for the area $\int_p^q f(x)\, dx$, use the calculator to find the two intercepts, storing them as p and q, and then evaluate the integral:

$$\int_p^q \left(-x^4 + x^2 + x + 10\right) dx = 32.832$$

which is accurate to three decimal places.

A3. Region Bounded by Polar Curve

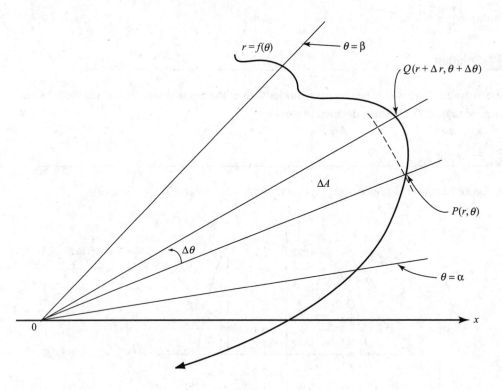

Figure 7.6

To find the area A bounded by the polar curve $r = f(\theta)$ and the rays $\theta = \alpha$ and $\theta = \beta$ (see Figure 7.6), we divide the region into n sectors like the one shown. If we think of that element of area, ΔA, as a circular sector with radius r and central angle $\Delta \theta$, its area is given by

$$\Delta A = \frac{1}{2} r^2 \Delta \theta.$$

Summing the areas of all such sectors yields the area of the entire region:

$$A = \lim_{n \to \infty} \sum_{k=1}^{n} \frac{1}{2} r_k^2 \Delta \theta_k \tag{1}$$

The expression above is a Riemann Sum, equivalent to this definite integral:

$$A = \int_{\alpha}^{\beta} \frac{1}{2} r^2 d\theta \tag{2}$$

We have assumed above that $f(\theta) \geq 0$ on $[\alpha, \beta]$. We must be careful in determining the limits α and β in (2); often it helps to think of the required area as that "swept out" (or generated) as the radius vector (from the pole) rotates from $\theta = \alpha$ to $\theta = \beta$. It is also useful to exploit symmetry of the curve wherever possible.

The relations between rectangular and polar coordinates, some common polar equations, and graphs of polar curves are given in the Appendix, starting on page 650.

> **Example 3**

Find the area inside both the circle $r = 3 \sin \theta$ and the cardioid $r = 1 + \sin \theta$.

✓ **Solution**

Choose an appropriate window, and graph the curves on your calculator. See Figure 7.7, where one-half of the required area is shaded. Since $3 \sin \theta = 1 + \sin \theta$ when $\theta = \dfrac{\pi}{6}$ or $\dfrac{5\pi}{6}$, we see that the desired area is twice the sum of two parts: the area of the circle swept out by θ as it varies from 0 to $\dfrac{\pi}{6}$ plus the area of the cardioid swept out by a radius vector as θ varies from $\dfrac{\pi}{6}$ to $\dfrac{\pi}{2}$. Consequently

$$A = 2 \left[\int_0^{\pi/6} \frac{9}{2} \sin^2 \theta \, d\theta + \int_{\pi/6}^{\pi/2} \frac{1}{2}(1 + \sin \theta)^2 \, d\theta \right] = \frac{5\pi}{4}$$

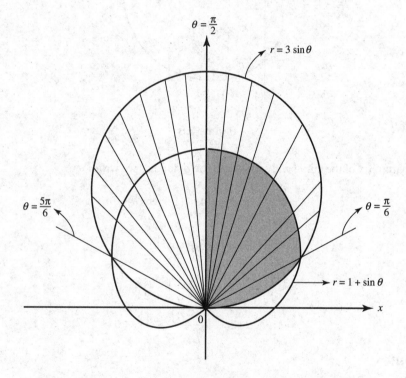

Figure 7.7

See Questions B3 and B4 in the Practice Exercises.

> Example 4

Find the area enclosed by the cardioid $r = 2(1 + \cos \theta)$.

✓ Solution

We graphed the cardioid on our calculator, using polar mode, in the window $[-2,5] \times [-3,3]$ with θ in $[0,2\pi]$.

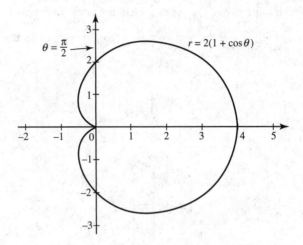

Figure 7.8

Using the symmetry of the curve with respect to the polar axis we write

$$A = 2 \cdot \frac{1}{2} \int_0^\pi r^2 \, d\theta = 4 \int_0^\pi (1 + \cos \theta)^2 \, d\theta$$

$$= 4 \int_0^\pi (1 + 2 \cos \theta + \cos^2 \theta) \, d\theta$$

$$= 4 \int_0^\pi \left(1 + 2 \cos \theta + \frac{1}{2} + \frac{\cos 2\theta}{2}\right) d\theta$$

$$= 4 \left[\theta + 2 \sin \theta + \frac{\theta}{2} + \frac{\sin 2\theta}{4}\right]\Bigg|_0^\pi = 6\pi$$

B. Volume

B1. Solids with Known Cross Sections

If the *area of a cross section* of a solid *is known* and can be expressed in terms of x, then the volume of a typical slice, ΔV, can be determined. The volume of the solid is obtained, as usual, by letting the number of slices increase indefinitely. In Figure 7.9, the slices are taken perpendicular to the x-axis so that $\Delta V = A(x) \, \Delta x$, where $A(x)$ is the area of a cross section and Δx is the thickness of the slice.

Figure 7.9

> **Example 5**

A solid has as its base the circle $x^2 + y^2 = 9$, and all cross sections parallel to the x-axis are squares. Find the volume of the solid.

✓ **Solution**

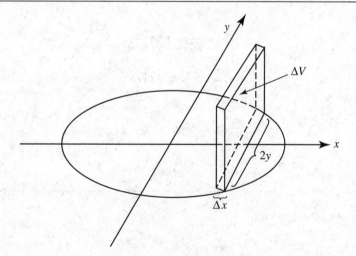

Figure 7.10

In Figure 7.10, the element of volume is a square prism with sides of length $2y$ and thickness Δx, so

$$\Delta V = (2y)^2 \, \Delta x = 4y^2 \, \Delta x = 4(9 - x^2) \, \Delta x$$

Now, using symmetry across the y-axis, we find the volume of the solid:

$$V = 2\int_0^3 4\left(9 - x^2\right) dx = 8\int_0^3 \left(9 - x^2\right) dx = 8\left(9x - \frac{x^3}{3}\right)\bigg|_0^3 = 144$$

Questions A23, A24, and A25 in the Practice Exercises illustrate solids with known cross sections.

When the cross section of a solid is a circle, a typical slice is a disk. When the cross section is the region between two circles, a typical slice is a washer—a disk with a hole in it. Both of these solids, which are special cases of solids with known cross sections, can be generated by revolving a plane area about a fixed line.

B2. Solids of Revolution

A *solid of revolution* is obtained when a plane region is revolved about a fixed line, called the *axis of revolution*. There are two major methods of obtaining the volume of a solid of revolution: "disks" and "washers."

Disks

Disks

The region bounded by a curve and the x-axis is revolved around the x-axis, forming the solid of revolution seen in Figure 7.11. We think of the "rectangular strip" of the region at the left as generating the solid disk, ΔV (an element of the volume), shown at the right.

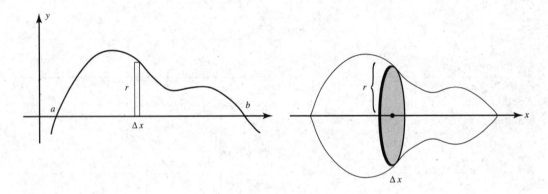

Figure 7.11

This disk is a cylinder whose radius, r, is the height of the rectangular strip and whose height is the thickness of the strip, Δx. Thus

$$\Delta V = \pi r^2 \, \Delta x \quad \text{and} \quad V = \pi \int_a^b r^2 \, dx$$

> **Example 6**

Find the volume of a sphere of radius r.

✓ **Solution**

If the region bounded by a semicircle (with center O and radius r) and its diameter is revolved about the x-axis, the solid of revolution obtained is a sphere of radius r, as seen in Figure 7.12.

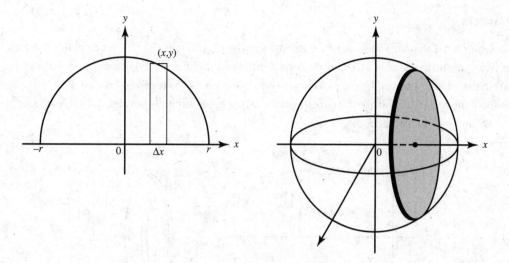

Figure 7.12

The volume ΔV of a typical disk is given by $\Delta V = \pi y^2 \, \Delta x$. The equation of the circle is $x^2 + y^2 = r^2$. To find the volume of the sphere, we form a Riemann Sum whose limit as n becomes infinite is a definite integral. Then,

$$V = \int_{-r}^{r} \pi y^2 \, dx = \pi \int_{-r}^{r} \left(r^2 - x^2 \right) dx = \pi \left[r^2 x - \frac{x^3}{3} \right]\Bigg|_{-r}^{r} = \frac{4}{3} \pi r^3$$

> **Example 7**

Find the volume of the solid generated when the region bounded by $y = x^2$, $x = 2$, and $y = 0$ is rotated about the line $x = 2$ as shown in Figure 7.13.

✓ **Solution**

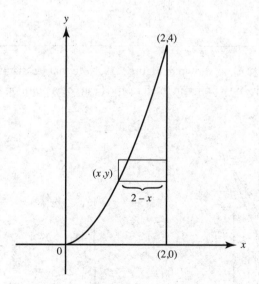

Disk

$$r = 2 - x$$

$$\Delta V = \pi(2 - x)^2 \, \Delta y$$

$$V = \pi \int_{0}^{4} (2 - x)^2 \, dy$$

$$= \pi \int_{0}^{4} \left(2 - \sqrt{y} \right)^2 dy$$

$$= \frac{8\pi}{3}$$

Figure 7.13

See Questions A16, B6, B8, and B9 in the Practice Exercises for examples of finding volumes by disks.

Washers

A washer is a disk with a hole in it. The volume may be regarded as the difference in the volumes of two concentric disks. As an example, consider the volume of the solid of revolution formed when the region bounded by the two curves seen in Figure 7.14 is revolved around the x-axis. We think of the rectangular strip of the region at the left as generating the washer, ΔV (an element of the volume), shown at the right.

Washers

Figure 7.14

This washer's height is the thickness of the rectangular strip, Δx. The washer is a disk whose outer radius, R, is the distance to the top of the rectangular strip, with the disk of inner radius r (the distance to the bottom of the strip) removed. Thus:

$$\Delta V = \pi R^2 \Delta x - \pi r^2 \Delta x = \pi \left(R^2 - r^2\right)\Delta x \quad \text{and} \quad V = \pi \int_a^b \left(R^2 - r^2\right)dx$$

> ### ❯ Example 8

Find the volume obtained when the region bounded by $y = x^2$ and $y = 2x$ is revolved about the x-axis.

✓ Solution

The curves intersect at the origin and at $(2,4)$, as shown in Figure 7.15. Note that we distinguish between the two functions by letting (x, y_1) be a point on the line and (x, y_2) be a point on the parabola.

Washer

$R = y_1 = 2x$

$r = y_2 = x^2$

$\Delta V = \pi \left(y_1^2 - y_2^2\right)\Delta x$

$V = \pi \int_0^2 \left(y_1^2 - y_2^2\right)dx$

$= \pi \int_0^2 \left((2x)^2 - \left(x^2\right)^2\right)dx$

$= \dfrac{64\pi}{15}$

Figure 7.15

> ## Example 9

Find the volume of the solid generated when the region bounded by $y = x^2$, $x = 2$, and $y = 0$ is rotated about the y-axis, as shown in Figure 7.16.

✓ Solution

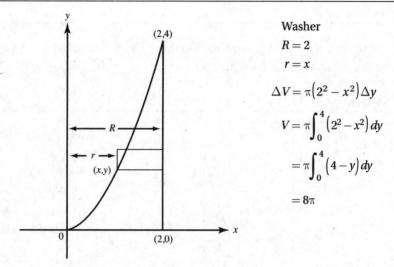

Washer

$R = 2$

$r = x$

$\Delta V = \pi \left(2^2 - x^2 \right) \Delta y$

$V = \pi \displaystyle\int_0^4 \left(2^2 - x^2 \right) dy$

$= \pi \displaystyle\int_0^4 \left(4 - y \right) dy$

$= 8\pi$

Figure 7.16

See Questions A17, A19, B5, B7, and B10 in the Practice Exercises for examples in which washers are regarded as the differences of two disks.

Occasionally when more than one method is satisfactory we try to use the most efficient. In the answers to each question in the Practice Exercises, a sketch is shown and the type and volume of a typical element are given. The required volume is then found by letting the number of elements become infinite and applying the Fundamental Theorem.

Shells‡

A cylindrical shell may be regarded as the outer skin of a cylinder. Its volume is the volume of the rectangular solid formed when this skin is peeled from the cylinder and flattened out. As an example, consider the volume of the solid of revolution formed when the region bounded by the two curves seen in Figure 7.17 is revolved around the y-axis. We think of the rectangular strip of the region at the left as generating the shell, ΔV (an element of the volume), shown at the right.

Figure 7.17

‡Examples 10–12 involve finding volumes by the method of shells. Although shells are not included in the Topic Outline, we include this method here because it is often the most efficient (and elegant) way to find a volume. No question *requiring* shells will appear on the AP exam.

This shell's radius, r, is the distance from the axis to the rectangular strip, and its height is the height of the rectangular strip, h. When the shell is unwound and flattened to form a rectangular solid, the length of the solid is the circumference of the cylinder, $2\pi r$, its height is the height of the cylinder, h, and its thickness is the thickness of the rectangular strip, Δx. Thus:

$$\Delta V = 2\pi\, rh\,\Delta x \quad \text{and} \quad V = 2\pi \int_a^b rh\,dx$$

> **Example 10**

Find the volume of the solid generated when the region bounded by $y = x^2$, $x = 2$, and $y = 0$ is rotated about the line x = 2. See Figure 7.18.

✓ **Solution**

About $x = 2$
Shell
$$r = 2 - x$$
$$h = y$$
$$\Delta V = 2\pi(2 - x)y\,\Delta x$$
$$\qquad = 2\pi(2 - x)x^2\,\Delta x$$
$$2\pi \int_0^2 (2 - x)x^2\,dx = \frac{8\pi}{3}$$

(Note that we obtained the same result using disks in Example 7.)

Figure 7.18

> **Example 11**

The region bounded by $y = 3x - x^2$ and $y = x$ is rotated about the y-axis. Find the volume of the solid obtained. See Figure 7.19.

✓ **Solution**

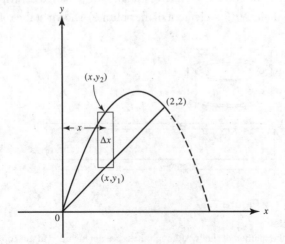

About the y-axis
Shell
$$\Delta V = 2\pi x(y_2 - y_1)\Delta x$$
$$\qquad = 2\pi x\left[\left(3x - x^2\right) - x\right]\Delta x$$
$$\qquad = 2\pi \int_0^2 \left(2x^2 - x^3\right)dx = \frac{8\pi}{3}$$

Figure 7.19

> **Example 12**

Find the volume obtained when the region bounded by $y = x^2$ and $y = 2x$ is revolved about the x-axis.

✓ **Solution**

The curves intersect at the origin and at (2,4), as shown in Figure 7.20. Note that we distinguish between the two functions by letting (x_1, y) be a point on the line and (x_2, y) be a point on the parabola.

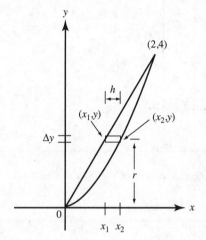

Figure 7.20

Shell
$$r = y$$
$$h = x_2 - x_1$$
$$\Delta V = 2\pi y (x_2 - x_1) \Delta y$$

$$V = 2\pi \int_0^4 y(x_2 - x_1)\, dy$$

$$= 2\pi \int_0^4 y \left(\sqrt{y} - \frac{y}{2} \right) dy$$

$$= \frac{64\pi}{15}$$

(Note that we obtained the same result using washers in Example 8.)

NOTE: On pages 307 and 308 in Examples 32 and 33 we consider finding the volumes of solids using shells that lead to improper integrals.

C. Length of Curve (Arc Length)

BC ONLY

If the derivative of a function $y = f(x)$ is continuous on the interval $a \leq x \leq b$, then the length s of the curve of $y = f(x)$ from the point where $x = a$ to the point where $x = b$ is given by

$$s = \int_a^b \sqrt{1 + \left(\frac{dy}{dx} \right)^2}\, dx \tag{1}$$

Here a small piece of the curve has a length approximately equal to $\sqrt{1 + \left(f'(x) \right)^2}\, \Delta x$.

As $\Delta x \to 0$, the sum of these pieces approaches the definite integral above.

If the derivative of the function $x = g(y)$ is continuous on the interval $c \leq y \leq d$, then the length s of the curve from $y = c$ to $y = d$ is given by

$$s = \int_c^d \sqrt{1 + \left(\frac{dx}{dy} \right)^2}\, dy \tag{2}$$

If a curve is defined parametrically by the equations $x = x(t)$ and $y = y(t)$, if the derivatives of the functions $x(t)$ and $y(t)$ are continuous on $[t_a, t_b]$ (and if the curve does not intersect itself), then the length of the curve from $t = t_a$ to $t = t_b$ is given by

$$s = \int_{t_a}^{t_b} \sqrt{\left(\frac{dx}{dt} \right)^2 + \left(\frac{dy}{dt} \right)^2}\, dt \tag{3}$$

The parenthetical clause on the previous page is equivalent to the requirement that the curve is traced out just once as t varies from t_a to t_b.

As indicated in Equation (4), formulas (1), (2), and (3) can all be derived easily from the very simple relation

$$ds^2 = dx^2 + dy^2 \qquad (4)$$

and can be remembered by visualizing Figure 7.21.

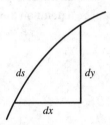

Figure 7.21

> **Example 13**

Find the length, to three decimal places, of the curve of $y = x^{3/2}$ from $x = 1$ to $x = 8$.

✓ **Solution**

Here $\dfrac{dy}{dx} = \dfrac{3}{2}x^{1/2}$, so by (1), $s = \displaystyle\int_1^8 \sqrt{1 + \dfrac{9}{4}x}\ dx \approx 22.803$.

> **Example 14**

Find the length, to three decimal places, of the curve $(x - 2)^2 = 4y^3$ from $y = 0$ to $y = 1$.

✓ **Solution**

Since $x - 2 = 2y^{3/2}$ and $\dfrac{dx}{dy} = 3y^{1/2}$, Equation (2) yields

$$s = \int_0^1 \sqrt{1 + 9y}\ dy \approx 2.268$$

> **Example 15**

The position (x,y) of a particle at time t is given parametrically by $x = t^2$ and $y = \dfrac{t^3}{3} - t$. Find the distance the particle travels between $t = 1$ and $t = 2$.

✓ **Solution**

We can use (4): $ds^2 = dx^2 + dy^2$, where $dx = 2t\ dt$ and $dy = (t^2 - 1)\ dt$. Thus,

$$ds = \sqrt{4t^2 + t^4 - 2t^2 + 1}\ dt, \text{ so}$$

$$s = \int_1^2 \sqrt{(t^2 + 1)^2}\ dt = \int_1^2 (t^2 + 1)\ dt$$
$$= \frac{t^3}{3} + t \Big|_1^2 = \frac{10}{3}$$

> **Example 16**

Find the length of the curve of $y = \ln \sec x$ from $x = 0$ to $x = \dfrac{\pi}{3}$.

✓ **Solution**

$$\frac{dy}{dx} = \frac{\sec x \tan x}{\sec x}, \text{ so}$$

$$s = \int_0^{\pi/3} \sqrt{1 + \tan^2 x}\ dx = \int_0^{\pi/3} \sec x\ dx$$

$$= \ln\left(\sec x + \tan x\right)\Big|_0^{\pi/3} = \ln\left(2 + \sqrt{3}\right)$$

D. Improper Integrals

There are two classes of improper integrals:

(1) those in which at least one of the limits of integration is infinite (the *interval* is not bounded)

(2) those of the type $\displaystyle\int_a^b f(x)\ dx$, where $f(x)$ has a point of discontinuity (becoming infinite) at $x = c$, $a \le c \le b$ (the *function* is not bounded)

Improper integral

Illustrations of improper integrals of class (1) are:

$$\int_0^\infty \frac{dx}{\sqrt[3]{x+1}} \qquad \int_1^\infty \frac{dx}{x} \qquad \int_\infty^\infty \frac{dx}{a^2 + x^2} \qquad \int_\infty^0 e^{-x}\ dx$$

$$\int_{-\infty}^{-1} \frac{dx}{x^n}\ (n \text{ is a real number}) \qquad \int_{-\infty}^0 \frac{dx}{e^x + e^{-x}}$$

$$\int_0^\infty \frac{dx}{(4+x)^2} \qquad \int_{-\infty}^0 e^{-x^2} dx \qquad \int_1^\infty \frac{e^{-x^2}}{x^2}\ dx$$

The following improper integrals are of class (2):

$$\int_0^1 \frac{dx}{x} \qquad \int_1^2 \frac{dx}{(x-1)^n}\ (n \text{ is a real number}) \qquad \int_{-1}^1 \frac{dx}{1 - x^2}$$

$$\int_0^2 \frac{x}{\sqrt{4 - x^2}}\ dx \qquad \int_\pi^{2\pi} \frac{dx}{1 + \sin x} \qquad \int_{-1}^2 \frac{dx}{x(x-1)^2}$$

$$\int_0^{2\pi} \frac{\sin x\ dx}{\cos x + 1} \qquad \int_a^b \frac{dx}{(x-c)^n}\ (n \text{ is real; } a \le c \le b)$$

$$\int_0^1 \frac{dx}{\sqrt{x + x^4}} \qquad \int_{-2}^2 \sqrt{\frac{2 + x}{2 - x}}\ dx$$

Sometimes an improper integral belongs to both classes. Consider, for example,

$$\int_0^\infty \frac{dx}{x} \qquad \int_0^\infty \frac{dx}{\sqrt{x + x^4}} \qquad \int_\infty^1 \frac{dx}{\sqrt{1 - x}}$$

In each case, the interval is not bounded and the integrand fails to exist at some point on the interval of integration.

Note, however, that each integral of the following set is *proper*:

$$\int_{-1}^{3} \frac{dx}{\sqrt{x+2}} \qquad \int_{-2}^{2} \frac{dx}{x^2+4} \qquad \int_{0}^{\pi/6} \frac{dx}{\cos x}$$

$$\int_{0}^{e} \ln(x+1)\,dx \qquad \int_{-3}^{3} \frac{dx}{e^x+1}$$

The integrand, in every example above, is defined at each number on the interval of integration.

Improper integrals of class (1), where the interval is not bounded, are handled as limits:

$$\int_{a}^{\infty} f(x)\,dx = \lim_{b\to\infty} \int_{a}^{b} f(x)\,dx$$

where f is continuous on $[a,b]$. If the limit on the right exists, the improper integral on the left is said to *converge* to this limit; if the limit on the right fails to exist, we say that the improper integral *diverges* (or is *meaningless*).

Converge
Diverge

The evaluation of improper integrals of class (1) is illustrated in Examples 17–23.

> **Example 17**

Find $\displaystyle\int_{1}^{\infty} \frac{dx}{x^2}$.

✓ **Solution**

$\displaystyle\int_{1}^{\infty} \frac{dx}{x^2} = \lim_{b\to\infty} \int_{1}^{b} x^{-2}\,dx = \lim_{b\to\infty} -\frac{1}{x}\Big|_{1}^{b} = \lim_{b\to\infty} -\left(\frac{1}{b}-1\right) = 1$. The given integral thus converges to 1.

In Figure 7.22 we interpret $\displaystyle\int_{1}^{\infty} \frac{dx}{x^2}$ as the area above the x-axis, under the curve of $y = \dfrac{1}{x^2}$, and bounded at the left by the vertical line $x = 1$.

Figure 7.22

> **Example 18**

$$\int_1^\infty \frac{dx}{\sqrt{x}} = \lim_{b\to\infty} \int_1^b x^{-1/2}\,dx = \lim_{b\to\infty} 2\sqrt{x}\,\Big|_1^b = \lim_{b\to\infty} 2\left(\sqrt{b}-1\right) = +\infty.$$

Then $\int_1^\infty \dfrac{dx}{\sqrt{x}}$ diverges.

It can be proved that $\int_1^\infty \dfrac{dx}{x^p}$ converges if $p>1$ but diverges if $p\le 1$. Figure 7.23 gives a geometric interpretation in terms of area of $\int_1^\infty \dfrac{dx}{x^p}$ for $p=\dfrac{1}{2}$, 1, 2. Only the first-quadrant area under $y=\dfrac{1}{x^2}$ bounded at the left by $x=1$ exists. Note that

$$\int_1^\infty \frac{dx}{x} = \lim_{b\to\infty} \ln x\,\Big|_1^b = +\infty$$

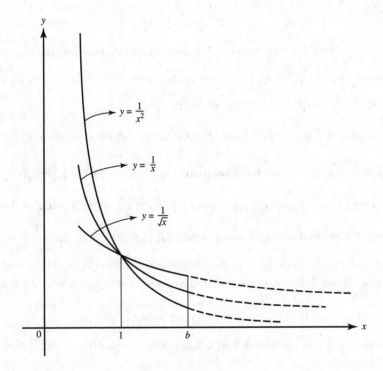

Figure 7.23

> **Example 19**

$$\int_0^\infty \frac{dx}{x^2+9} = \lim_{b\to\infty} \int_0^b \frac{dx}{x^2+9} = \lim_{b\to\infty} \frac{1}{3}\tan^{-1}\frac{x}{3}\,\Big|_0^b = \lim_{b\to\infty} \frac{1}{3}\tan^{-1}\frac{b}{3} = \frac{1}{3}\cdot\frac{\pi}{2} = \frac{\pi}{6}$$

> **Example 20**

$$\int_0^\infty \frac{dy}{e^y} = \lim_{b\to\infty} \int_0^b e^{-y}\,dy = \lim_{b\to\infty} -(e^{-b}-1) = 1$$

> **Example 21**

$$\int_{-\infty}^{0} \frac{dz}{(z-1)^2} = \lim_{a \to -\infty} \int_{a}^{0} (z-1)^{-2}\, dz = \lim_{a \to -\infty} \ -\frac{1}{z-1}\Big|_{a}^{0}$$

$$= \lim_{a \to -\infty} \ -\left(-1 - \frac{1}{a-1}\right) = 1$$

> **Example 22**

$$\int_{-\infty}^{0} e^{-x}\, dx = \lim_{a \to -\infty} \int_{a}^{0} e^{-x}\, dx = \lim_{a \to -\infty} \ -e^{-x}\Big|_{a}^{0} = \lim_{a \to -\infty} \ -(1 - e^{-a}) = +\infty$$

Thus, this improper integral diverges.

> **Example 23**

$$\int_{0}^{\infty} \cos x\, dx = \lim_{b \to \infty} \int_{0}^{b} \cos x\, dx = \lim_{b \to \infty} \ \sin x \Big|_{0}^{b} = \lim_{b \to \infty} \ \sin b.$$ Since this limit does not exist (sin b takes on values between -1 and 1 as $b \to \infty$), it follows that the given integral diverges. Note, however, that it does not become infinite; rather, it diverges by oscillation.

Improper integrals of class (2), where the function has an infinite discontinuity, are handled as follows.

To investigate $\int_{a}^{b} f(x)\, dx$, where f becomes infinite at $x = a$, we define $\int_{a}^{b} f(x)\, dx$ to be

$\lim_{k \to a^+} \int_{k}^{b} f(x)\, dx$. The given integral then converges or diverges according to whether the limit does

or does not exist. If f has its discontinuity at b, we define $\int_{a}^{b} f(x)\, dx$ to be $\lim_{k \to b^-} \int_{a}^{k} f(x)\, dx$; again,

the given integral converges or diverges as the limit does or does not exist. When, finally, the integrand has a discontinuity at an interior point c on the interval of integration ($a < c < b$), we let

$$\int_{a}^{b} f(x)\, dx = \lim_{k \to c^-} \int_{a}^{k} f(x)\, dx + \lim_{m \to c^+} \int_{m}^{b} f(x)\, dx$$

Now the improper integral converges only if both of the limits exist. If *either* limit does not exist, the improper integral diverges.

The evaluation of improper integrals of class (2) is illustrated in Examples 24–31.

> **Example 24**

Find $\displaystyle\int_0^1 \frac{dx}{\sqrt[3]{x}}$.

✓ **Solution**

$$\int_0^1 \frac{dx}{\sqrt[3]{x}} = \lim_{k\to 0^+}\int_k^1 x^{-1/3}\,dx = \lim_{k\to 0^+}\frac{3}{2}x^{2/3}\Big|_k^1 = \lim_{k\to 0^+}\frac{3}{2}\left(1-k^{2/3}\right) = \frac{3}{2}$$

In Figure 7.24 we interpret this integral as the first-quadrant area under $y = \dfrac{1}{\sqrt[3]{x}}$ and to the left of $x = 1$.

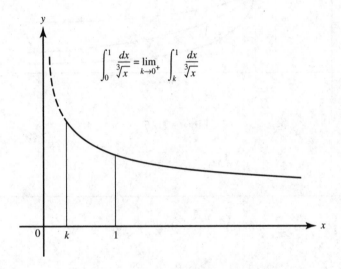

$$\int_0^1 \frac{dx}{\sqrt[3]{x}} = \lim_{k\to 0^+}\int_k^1 \frac{dx}{\sqrt[3]{x}}$$

Figure 7.24

> **Example 25**

Does $\displaystyle\int_0^1 \frac{dx}{x^3}$ converge or diverge?

✓ **Solution**

$$\int_0^1 \frac{dx}{x^3} = \lim_{k\to 0^+}\int_k^1 x^{-3}\,dx = \lim_{k\to 0^+} -\frac{1}{2x^2}\Big|_k^1 = \lim_{k\to 0^+} -\frac{1}{2}\left(1-\frac{1}{k^2}\right) = \infty$$

Therefore, this integral diverges.

It can be shown that $\displaystyle\int_0^a \frac{dx}{x^p}\ (a>0)$ converges if $p < 1$ but diverges if $p \geq 1$. Figure 7.25 shows an interpretation of $\displaystyle\int_0^1 \frac{dx}{x^p}$ in terms of areas where $p = \dfrac{1}{3}$, 1, and 3. Only the first-quadrant area under $y = \dfrac{1}{\sqrt[3]{x}}$ to the left of $x = 1$ exists. Note that

$$\int_0^1 \frac{dx}{x} = \lim_{k\to 0^+}\ln x\Big|_k^1 = \lim_{k\to 0^+}(\ln 1 - \ln k) = +\infty$$

Figure 7.25

> **Example 26**

$$\int_0^2 \frac{dy}{\sqrt{4 - y^2}} = \lim_{k \to 2^-} \int_0^k \frac{dy}{\sqrt{4 - y^2}} = \lim_{k \to 2^-} \sin^{-1} \frac{y}{2} \bigg|_0^k = \sin^{-1} 1 - \sin^{-1} 0 = \frac{\pi}{2}$$

> **Example 27**

$$\int_2^3 \frac{dt}{(3 - t)^2} = \lim_{k \to 3^-} -\int_2^k (3 - t)^{-2}(-dt) = \lim_{k \to 3^-} \frac{1}{3 - t} \bigg|_2^k = +\infty$$

This integral diverges.

> **Example 28**

$$\int_0^2 \frac{dx}{(x - 1)^{2/3}} = \lim_{k \to 1^-} \int_0^k (x - 1)^{-2/3} \, dx + \lim_{m \to 1^+} \int_m^2 (x - 1)^{-2/3} \, dx$$

$$= \lim_{k \to 1^-} 3(x - 1)^{1/3} \bigg|_0^k + \lim_{m \to 1^+} 3(x - 1)^{1/3} \bigg|_m^2 = 3(0 + 1) + 3(1 - 0) = 6$$

> **Example 29**

$$\int_{-2}^2 \frac{dx}{x^2} = \lim_{k \to 0^-} \int_{-2}^k x^{-2} \, dx + \lim_{m \to 0^+} \int_m^2 x^{-2} \, dx = \lim_{k \to 0^-} -\frac{1}{x} \bigg|_{-2}^k + \lim_{m \to 0^+} -\frac{1}{x} \bigg|_m^2$$

Neither limit exists; the integral diverges.

NOTE: This example demonstrates how careful one must be to notice a discontinuity at an interior point. If it were overlooked, one might proceed as follows:

$$\int_{-2}^{2} \frac{dx}{x^2} = -\frac{1}{x}\Big|_{-2}^{2} = -\left(\frac{1}{2} + \frac{1}{2}\right) = -1$$

Since this integrand is positive except at zero, the result obtained is clearly meaningless. Figure 7.26 shows the impossibility of this answer.

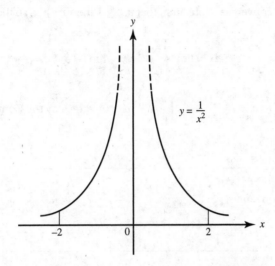

Figure 7.26

The Comparison Test

We can often determine whether an improper integral converges or diverges by comparing it to a known integral on the same interval. This method is especially helpful when it is not easy to actually evaluate the appropriate limit by finding an antiderivative for the integrand. There are two cases.

(1) Convergence. If on the interval of integration $f(x) \le g(x)$ and $\int g(x)\,dx$ is known to

converge, then $\int f(x)\,dx$ also converges. For example, consider $\int_{1}^{\infty} \frac{1}{x^3 + 1}\,dx$. We know that

$\int_{1}^{\infty} \frac{1}{x^3}\,dx$ converges. Since $\frac{1}{x^3 + 1} < \frac{1}{x^3}$, the improper integral $\int_{1}^{\infty} \frac{1}{x^3 + 1}\,dx$ must also converge.

(2) Divergence. If on the interval of integration $f(x) \ge g(x)$ and $\int g(x)\,dx$ is known to diverge,

then $\int f(x)\,dx$ also diverges. For example, consider $\int_{0}^{1} \frac{\sec x}{x^3}\,dx$. We know that $\int_{0}^{1} \frac{1}{x^3}\,dx$

diverges. Since $\sec x \ge 1$, it follows that $\frac{\sec x}{x^3} \ge \frac{1}{x^3}$; hence the improper integral

$\int_{0}^{1} \frac{\sec x}{x^3}\,dx$ must also diverge.

BC ONLY

> **Example 30**

Determine whether or not $\displaystyle\int_{1}^{\infty} e^{-x^2}\, dx$ converges.

✓ **Solution**

Although there is no elementary function whose derivative is e^{-x^2}, we can still show that the given improper integral converges. Note, first, that if $x \geq 1$ then $x^2 \geq x$, so that $-x^2 \leq -x$ and $e^{-x^2} \leq e^{-x}$. Furthermore,

$$\int_{1}^{\infty} e^{-x}\, dx = \lim_{k\to\infty} \int_{1}^{k} e^{-x}\, dx = \lim_{k\to\infty} \left. -e^{-x}\right|_{1}^{k} = \frac{1}{e}$$

Since $\displaystyle\int_{1}^{\infty} e^{-x}\, dx$ converges and $e^{-x^2} \leq e^{-x}$, $\displaystyle\int_{1}^{\infty} e^{-x^2}\, dx$ converges by the Comparison Test.

> **Example 31**

Show that $\displaystyle\int_{0}^{\infty} \frac{dx}{\sqrt{x + x^4}}$ converges.

✓ **Solution**

$$\int_{0}^{\infty} \frac{dx}{\sqrt{x + x^4}} = \int_{0}^{1} \frac{dx}{\sqrt{x + x^4}} + \int_{1}^{\infty} \frac{dx}{\sqrt{x + x^4}}$$

We will use the Comparison Test to show that both of these integrals converge.
Since if $0 < x \leq 1$, then $x + x^4 > x$ and $\sqrt{x + x^4} > \sqrt{x}$, it follows that

$$\frac{1}{\sqrt{x + x^4}} < \frac{1}{\sqrt{x}} \quad (0 < x \leq 1)$$

We know that $\displaystyle\int_{0}^{1} \frac{dx}{\sqrt{x}}$ converges; hence $\displaystyle\int_{0}^{1} \frac{dx}{\sqrt{x + x^4}}$ must converge. Further, if $x \geq 1$ then $x + x^4 \geq x^4$ and $\sqrt{x + x^4} \geq \sqrt{x^4} = x^2$, so

$$\frac{1}{\sqrt{x + x^4}} \leq \frac{1}{x^2} \quad (x \geq 1)$$

We know that $\displaystyle\int_{1}^{\infty} \frac{1}{x^2}\, dx$ converges; hence $\displaystyle\int_{1}^{\infty} \frac{dx}{\sqrt{x + x^4}}$ also converges.

Thus the given integral, $\displaystyle\int_{0}^{\infty} \frac{dx}{\sqrt{x + x^4}}$, converges.

NOTE: Examples 32 and 33 involve finding the volumes of solids. Both lead to improper integrals.

> **Example 32**

Find the volume, if it exists, of the solid generated in Figure 7.27 by rotating the region in the first quadrant bounded above by $y = \frac{1}{x}$, at the left by $x = 1$, and below by $y = 0$, about the x-axis.

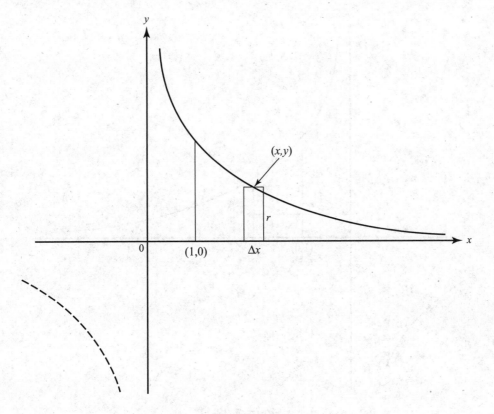

Figure 7.27

✓ **Solution**

About the x-axis
Disk

$$\Delta V = \pi y^2 \, \Delta x$$

$$V = \pi \int_{1}^{\infty} y^2 \, dx = \pi \int_{1}^{\infty} \frac{1}{x^2} \, dx$$

$$= \pi \lim_{k \to \infty} \int_{1}^{k} \frac{1}{x^2} \, dx = \pi$$

⟩ **Example 33‡**

Find the volume, if it exists, of the solid generated in Figure 7.28 by rotating the region in the first quadrant bounded above by $y = \frac{1}{x}$, at the left by $x = 1$, and below by $y = 0$, about the y-axis.

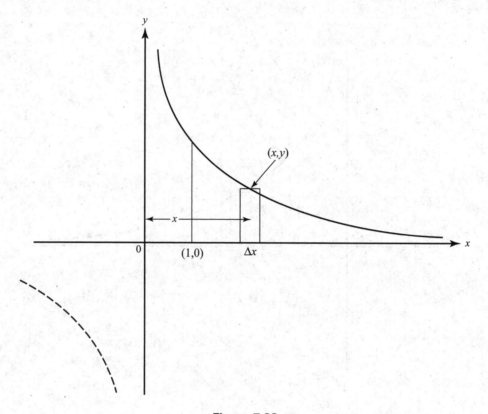

Figure 7.28

✓ **Solution**

About the y-axis
Shell
$\Delta V = 2\pi xy\, \Delta x = 2\pi\, \Delta x$

Note that $2\pi \displaystyle\int_{1}^{\infty} dx$ diverges to infinity.

CHAPTER SUMMARY

In this chapter, we reviewed how to find areas and volumes using definite integrals. We looked at area under a curve and between two curves. We reviewed volumes of solids with known cross sections and the methods of disks and washers for finding volumes of solids of revolution.

For BC Calculus students, we applied these techniques to parametrically defined functions and polar curves and added methods for finding lengths of curve. We also looked at improper integrals and tests for determining convergence and divergence.

‡No question requiring the use of shells will appear on the AP exam.

Practice Exercises

PART A. NO CALCULATOR—DIRECTIONS: Answer these questions *without* using your calculator.

Area

In Questions A1–A11, choose the alternative that gives the area of the region whose boundaries are given.

A1. The curve $y = x^2$, $y = 0$, $x = -1$, and $x = 2$.

(A) $\dfrac{11}{3}$

(B) $\dfrac{7}{3}$

(C) 3

(D) 5

A2. The parabola $y = x^2 - 3$ and the line $y = 1$.

(A) $\dfrac{8}{3}$

(B) 32

(C) $\dfrac{32}{3}$

(D) $\dfrac{16}{3}$

A3. The curve $x = y^2 - 1$ and the y-axis.

(A) $\dfrac{4}{3}$

(B) $\dfrac{2}{3}$

(C) $\dfrac{8}{3}$

(D) $\dfrac{1}{2}$

A4.

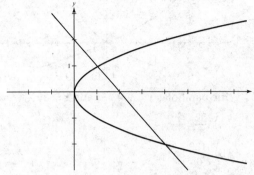

The parabola $y^2 = x$ and the line $x + y = 2$.

(A) $\dfrac{15}{2}$

(B) $\dfrac{3}{2}$

(C) $\dfrac{21}{6}$

(D) $\dfrac{9}{2}$

A5. The curve $y = \dfrac{4}{x^2 + 4}$, the x-axis, and the vertical lines $x = -2$ and $x = 2$.

(A) $\dfrac{\pi}{4}$

(B) $\dfrac{\pi}{2}$

(C) 2π

(D) π.

A6.

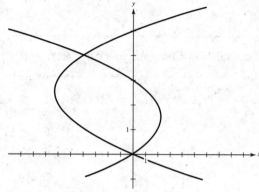

The parabolas $x = y^2 - 5y$ and $x = 3y - y^2$.

(A) $\dfrac{32}{3}$

(B) $\dfrac{139}{6}$

(C) $\dfrac{64}{3}$

(D) $\dfrac{128}{3}$

A7. The curve $y = \dfrac{2}{x}$ and $x + y = 3$.

(A) $\dfrac{1}{2} - 2\ln 2$

(B) $\dfrac{3}{2}$

(C) $\dfrac{1}{2} - \ln 4$

(D) $\dfrac{3}{2} - \ln 4$

A8. In the first quadrant, bounded below by the x-axis and above by the curves of $y = \sin x$ and $y = \cos x$.

(A) $2 - \sqrt{2}$

(B) $2 + \sqrt{2}$

(C) 2

(D) $\sqrt{2}$

A9. Bounded above by the curve $y = \sin x$ and below by $y = \cos x$ from $x = \dfrac{\pi}{4}$ to $x = \dfrac{5\pi}{4}$.

(A) $2\sqrt{2}$

(B) $\dfrac{2}{\sqrt{2}}$

(C) $2\left(\sqrt{2} - 1\right)$

(D) $2\left(\sqrt{2} + 1\right)$

A10. The curve $y = \cot x$, the line $x = \dfrac{\pi}{4}$, and the x-axis.

(A) $\ln 2$

(B) $\dfrac{1}{2}\ln\dfrac{1}{2}$

(C) 1

(D) $\dfrac{1}{2}\ln 2$

A11. The curve $y = x^3 - 2x^2 - 3x$ and the x-axis.

(A) $\dfrac{28}{3}$

(B) $\dfrac{79}{6}$

(C) $\dfrac{45}{4}$

(D) $\dfrac{71}{6}$

A12.

The total area bounded by the cubic $x = y^3 - y$ and the line $x = 3y$ is equal to

(A) 4

(B) $\dfrac{16}{3}$

(C) 8

(D) 16

A13. The area bounded by $y = e^x$, $y = 2$, and the y-axis is equal to

(A) $3 - e$

(B) $e^2 + 1$

(C) $2\ln 2 - 1$

(D) $2\ln 2 - 3$

CHALLENGE A14. The area enclosed by the curve $y^2 = x(1-x)$ is given by

(A) $2\displaystyle\int_0^1 x\sqrt{1-x}\,dx$

(B) $2\displaystyle\int_0^1 \sqrt{x-x^2}\,dx$

(C) $4\displaystyle\int_0^1 \sqrt{x-x^2}\,dx$

(D) π

A15. The figure below shows part of the curve $y = x^3$ and a rectangle with two vertices at $(0,0)$ and $(c,0)$. What is the ratio of the area of the rectangle to the shaded part of it above the cubic?

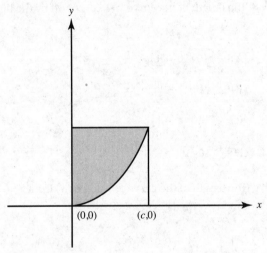

(A) $3:4$

(B) $5:4$

(C) $4:3$

(D) $3:1$

Volume

In Questions A16–A22, the region whose boundaries are given is rotated about the line indicated. Choose the alternative that gives the volume of the solid generated.

A16. $y = x^2$, $x = 2$, and $y = 0$; about the x-axis.

(A) $\dfrac{64\pi}{3}$

(B) 8π

(C) $\dfrac{8\pi}{3}$

(D) $\dfrac{32\pi}{5}$

A17. $y = x^2$, $x = 2$, and $y = 0$; about the y-axis.

(A) $\dfrac{16\pi}{3}$

(B) 4π

(C) $\dfrac{32\pi}{5}$

(D) 8π

A18. The first-quadrant region bounded by $y = x^2$, the y-axis, and $y = 4$; about the y-axis.

(A) 8π

(B) 4π

(C) $\dfrac{32\pi}{3}$

(D) $\dfrac{16\pi}{3}$

A19. $y = x^2$ and $y = 4$; about the x-axis.

(A) $\dfrac{64\pi}{5}$

(B) $\dfrac{512\pi}{15}$

(C) $\dfrac{256\pi}{5}$

(D) $\dfrac{128\pi}{5}$

A20. $y = x^2$ and $y = 4$; about the line $y = 4$.

(A) $\dfrac{256\pi}{15}$

(B) $\dfrac{256\pi}{5}$

(C) $\dfrac{512\pi}{5}$

(D) $\dfrac{512\pi}{15}$

A21. An arch of $y = \sin x$ and the x-axis; about the x-axis.   **CHALLENGE**

(A) $\dfrac{\pi}{2}\left(\pi - \dfrac{1}{2}\right)$

(B) $\dfrac{\pi^2}{2}$

(C) $\dfrac{\pi^2}{4}$

(D) π^2

A22. A trapezoid with vertices at $(2,0)$, $(2,2)$, $(4,0)$, and $(4,4)$; about the x-axis.

(A) $\dfrac{56\pi}{3}$

(B) $\dfrac{128\pi}{3}$

(C) $\dfrac{92\pi}{3}$

(D) $\dfrac{112\pi}{3}$

A23. The base of a solid is a circle of radius a and is centered at the origin. Each cross section perpendicular to the x-axis is a square. The solid has volume

(A) $\dfrac{8}{3}a^3$

(B) $2\pi a^3$

(C) $4\pi a^3$

(D) $\dfrac{16}{3}a^3$

A24. The base of a solid is the region bounded by the parabola $x^2 = 8y$ and the line $y = 4$, and each plane section perpendicular to the y-axis is an equilateral triangle. The volume of the solid is

(A) $\dfrac{64\sqrt{3}}{3}$

(B) $64\sqrt{3}$

(C) $32\sqrt{3}$

(D) 32

A25. The base of a solid is the region bounded by $y = e^{-x}$, the x-axis, the y-axis, and the line $x = 1$. Each cross section perpendicular to the x-axis is a square. The volume of the solid is

(A) $e^2 - 1$

(B) $1 - \dfrac{1}{e^2}$

(C) $\dfrac{e^2 - 1}{2}$

(D) $\dfrac{1}{2}\left(1 - \dfrac{1}{e^2}\right)$

Length of Curve (Arc Length)

A26.

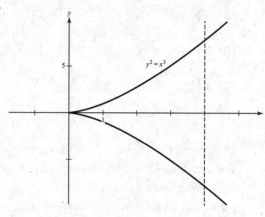

The length of the curve $y^2 = x^3$ cut off by the line $x = 4$ is

(A) $\dfrac{4}{3}\left(10\sqrt{10} - 1\right)$

(B) $\dfrac{8}{27}\left(10^{3/2} - 1\right)$

(C) $\dfrac{16}{27}\left(10^{3/2} - 1\right)$

(D) $\dfrac{16}{27}10\sqrt{10}$

A27. The length of the curve $y = \ln(\cos x)$ from $x = \dfrac{\pi}{4}$ to $x = \dfrac{\pi}{3}$ equals

(A) $\displaystyle\int_{\pi/4}^{\pi/3} \sqrt{1 + \tan^2 x}\ dx$

(B) $\displaystyle\int_{\pi/4}^{\pi/3} \sqrt{1 + \left(\ln(\cos x)\right)^2}\ dx$

(C) $\displaystyle\int_{\pi/4}^{\pi/3} \sqrt{1 + \dfrac{1}{\cos^2 x}}\ dx$

(D) $\displaystyle\int_{\pi/4}^{\pi/3} \sqrt{1 + \cot^2 x}\ dx$

Improper Integrals

A28. $\displaystyle\int_0^{\infty} e^{-x}\ dx =$

(A) 1

(B) $\dfrac{1}{e}$

(C) -1

(D) divergent

A29. $\int_0^e \dfrac{du}{u} =$

(A) 1

(B) $\dfrac{1}{e}$

(C) $-\dfrac{1}{e^2}$

(D) divergent

A30. $\int_1^2 \dfrac{dt}{\sqrt[3]{t-1}} =$

(A) $\dfrac{2}{3}$

(B) $\dfrac{3}{2}$

(C) 3

(D) divergent

A31. $\int_2^4 \dfrac{dx}{(x-3)^{2/3}} =$

(A) 6

(B) 0

(C) $\dfrac{2}{3}$

(D) divergent

A32. $\int_2^4 \dfrac{dx}{(x-3)^2} =$

(A) 2

(B) -2

(C) 0

(D) divergent

A33. $\int_0^{\pi/2} \dfrac{\sin x}{\sqrt{1-\cos x}}\, dx =$

(A) -2

(B) $\dfrac{1}{2}$

(C) 2

(D) divergent

A34. Find the area in the first quadrant under the curve $y = e^{-x}$.

(A) 1

(B) e

(C) $\dfrac{1}{e}$

(D) divergent

A35.

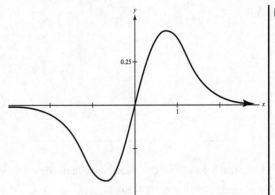

Find the area in the first quadrant under the curve $y = xe^{-x^2}$.

(A) 2

(B) $\dfrac{2}{e}$

(C) $\dfrac{1}{2}$

(D) divergent

A36. Find the area in the first quadrant above $y = 1$, between the y-axis and the curve $y = \dfrac{1}{x}$.

(A) 1

(B) 2

(C) $\dfrac{1}{2}$

(D) divergent

A37.

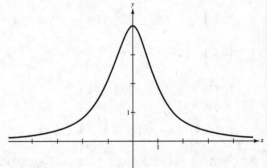

Find the area between the curve $y = \dfrac{4}{1+x^2}$ and the x-axis.

(A) 2π

(B) 4π

(C) 8π

(D) divergent

A38.

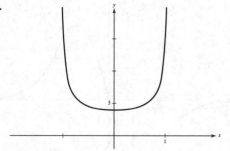

Find the area above the x-axis, between the curve $y = \dfrac{4}{\sqrt{1-x^2}}$ and its asymptotes.

(A) π

(B) 2π

(C) 4π

(D) divergent

A39. Find the volume of the solid generated when the region bounded above by $y = \dfrac{1}{x}$, at the left by $x = 1$, and below by $y = 0$ is rotated about the x-axis.

(A) $\dfrac{\pi}{2}$

(B) π

(C) 2π

(D) divergent

A40. Find the volume of the solid generated when the first-quadrant region under $y = e^{-x}$ is rotated about the x-axis.

(A) $\dfrac{\pi}{2}$

(B) π

(C) 2π

(D) divergent

PART B. CALCULATOR ACTIVE—DIRECTIONS: Some of the following questions require the use of a graphing calculator.

Area

In Questions B1–B4, choose the alternative that gives the area of the region whose boundaries are given.

B1. The area bounded by the parabola $y = 2 - x^2$ and the line $y = x - 4$ is given by

(A) $\displaystyle\int_{-2}^{3} \left(6 - x - x^2\right) dx$

(B) $\displaystyle\int_{-2}^{1} \left(2 + x + x^2\right) dx$

(C) $\displaystyle\int_{-3}^{2} \left(6 - x - x^2\right) dx$

(D) $2\displaystyle\int_{0}^{\sqrt{2}} \left(2 - x^2\right) dx + \int_{-3}^{2} (4 - x) \, dx$

B2. Suppose the following is a table of coordinates for $y = f(x)$, given that f is continuous on $[1,8]$:

x	1	2	5	6	8
y	1.62	4.15	7.5	9.0	12.13

If a trapezoidal sum in used, with $n = 4$, then the approximate area under the curve, from $x = 1$ to $x = 8$, to two decimal places, is

(A) 24.87

(B) 39.57

(C) 49.74

(D) 59.91

B3. The area A enclosed by the four-leaved rose $r = \cos 2\theta$ equals, to three decimal places,

(A) 0.785

(B) 1.571

(C) 3.142

(D) 6.283

B4. The area bounded by the small loop of the limaçon $r = 1 - 2 \sin \theta$ is given by the definite integral

(A) $\displaystyle\int_{\pi/3}^{5\pi/3} \left[\frac{1}{2}(1 - 2 \sin \theta)\right]^2 d\theta$

(B) $\displaystyle\int_{7\pi/6}^{3\pi/2} (1 - 2 \sin \theta)^2 \, d\theta$

(C) $\displaystyle\int_{\pi/6}^{\pi/2} (1 - 2 \sin \theta)^2 \, d\theta$

(D) $\displaystyle\int_{0}^{\pi/6} \left[\frac{1}{2}(1 - 2 \sin \theta)\right]^2 d\theta +$
$\displaystyle\int_{5\pi/6}^{\pi} \left[\frac{1}{2}(1 - 2 \sin \theta)\right]^2 d\theta$

Volume

In Questions B5–B10, the region whose boundaries are given is rotated about the line indicated. Choose the alternative that gives the volume of the solid generated.

B5. $y = x^2$ and $y = 4$; about the line $y = -1$.

(A) $4\pi \int_{-1}^{4} (y+1)\sqrt{y}\, dy$

(B) $2\pi \int_{0}^{2} (4 - x^2)^2\, dx$

(C) $\pi \int_{-2}^{2} (16 - x^4)\, dx$

(D) $2\pi \int_{0}^{2} (24 - 2x^2 - x^4)\, dx$

B6. $y = 3x - x^2$ and $y = 0$; about the x-axis.

(A) $\pi \int_{0}^{3} (9x^2 + x^4)\, dx$

(B) $\pi \int_{0}^{3} (3x - x^2)^2\, dx$

(C) $\pi \int_{0}^{\sqrt{3}} (3x - x^2)\, dx$

(D) $2\pi \int_{0}^{3} y\sqrt{9 - 4y}\, dy$

B7. $y = 3x - x^2$ and $y = x$; about the x-axis.

(A) $\pi \int_{0}^{3/2} \left[(3x - x^2)^2 - x^2 \right] dx$

(B) $\pi \int_{0}^{2} (9x^2 - 6x^3)\, dx$

(C) $\pi \int_{0}^{2} \left[(3x - x^2)^2 - x^2 \right] dx$

(D) $\pi \int_{0}^{3} (2x - x^2)^2\, dx$

B8. $y = \ln x$, $y = 0$, $x = e$; about the line $x = e$.

(A) $\pi \int_{1}^{e} (e - x) \ln x\, dx$

(B) $\pi \int_{0}^{1} (e - e^y)^2\, dy$

(C) $2\pi \int_{1}^{e} (e - \ln x)\, dx$

(D) $\pi \int_{0}^{e} \left(e^2 - 2e^{y+1} + e^{2y} \right) dy$

B9. A sphere of radius r is divided into two parts by a plane at distance h $(0 < h < r)$ from the center. The volume of the smaller part equals   **CHALLENGE**

(A) $\frac{\pi}{3}\left(2r^3 + h^3 - 3r^2 h\right)$

(B) $\frac{\pi h}{3}\left(3r^2 - h^2\right)$

(C) $\frac{4}{3}\pi r^3 + \frac{h^3}{3} - r^2 h$

(D) $\frac{\pi}{3}\left(2r^3 + 3r^2 h - h^3\right)$

B10. If the curves of $f(x)$ and $g(x)$ intersect for $x = a$ and $x = b$ and if $f(x) > g(x) > 0$ for all x on (a,b), then the volume obtained when the region bounded by the curves is rotated about the x-axis is equal to

(A) $\pi \int_{a}^{b} (f(x))^2\, dx - \int_{a}^{b} (g(x))^2\, dx$

(B) $\pi \int_{a}^{b} \left[f(x) - g(x) \right]^2\, dx$

(C) $2\pi \int_{a}^{b} x \left[f(x) - g(x) \right]\, dx$

(D) $\pi \int_{a}^{b} \left[(f(x))^2 - (g(x))^2 \right] dx$

Length of Curve (Arc Length)   **BC ONLY**

B11. The length of one arch of the cycloid $\begin{aligned} x &= t - \sin t \\ y &= 1 - \cos t \end{aligned}$ equals

(A) $\int_{0}^{\pi} \sqrt{1 - \cos t}\, dt$

(B) $\int_{0}^{2\pi} \sqrt{\dfrac{1 - \cos t}{2}}\, dt$

(C) $\int_{0}^{\pi} \sqrt{2 - 2\cos t}\, dt$

(D) $\int_{0}^{2\pi} \sqrt{2 - 2\cos t}\, dt$

B12. The length of the curve of the parabola $4x = y^2$ cut off by the line $x = 2$ is given by the integral   **CHALLENGE**

(A) $\int_{-1}^{1} \sqrt{x^2 + 1}\, dx$

(B) $\frac{1}{2}\int_{0}^{2} \sqrt{4 + y^2}\, dy$

(C) $\int_{-1}^{1} \sqrt{1 + x}\, dx$

(D) $\int_{0}^{2\sqrt{2}} \sqrt{4 + y^2}\, dy$

BC ONLY

B13. The length of $x = e^t \cos t$, $y = e^t \sin t$ from $t = 2$ to $t = 3$ is equal to

(A) 17.956

(B) 25.393

(C) 26.413

(D) 37.354

Improper Integrals

B14. Which one of the following is an improper integral?

(A) $\displaystyle\int_0^2 \frac{dx}{\sqrt{x+1}}$

(B) $\displaystyle\int_{-1}^1 \frac{dx}{1+x^2}$

(C) $\displaystyle\int_0^2 \frac{x\,dx}{1-x^2}$

(D) $\displaystyle\int_0^{\pi/3} \frac{\sin x\,dx}{\cos^2 x}$

BC ONLY

B15. Which one of the following improper integrals diverges?

(A) $\displaystyle\int_1^\infty \frac{dx}{x^2}$

(B) $\displaystyle\int_0^\infty \frac{dx}{e^x}$

(C) $\displaystyle\int_{-1}^1 \frac{dx}{\sqrt[3]{x}}$

(D) $\displaystyle\int_{-1}^1 \frac{dx}{x^2}$

B16. Which one of the following improper integrals diverges?

(A) $\displaystyle\int_0^\infty \frac{dx}{1+x^2}$

(B) $\displaystyle\int_0^1 \frac{dx}{x^{1/3}}$

(C) $\displaystyle\int_0^\infty \frac{dx}{e^x+2}$

(D) $\displaystyle\int_0^\infty \frac{dx}{x^{1/3}}$

Answer Key

A1.	**C**	A13.	**C**	A25.	**D**	A37.	**B**	B9.	**A**
A2.	**C**	A14.	**B**	A26.	**C**	A38.	**C**	B10.	**D**
A3.	**A**	A15.	**C**	A27.	**A**	A39.	**B**	B11.	**D**
A4.	**D**	A16.	**D**	A28.	**A**	A40.	**A**	B12.	**D**
A5.	**D**	A17.	**D**	A29.	**D**	B1.	**C**	B13.	**A**
A6.	**C**	A18.	**A**	A30.	**B**	B2.	**C**	B14.	**C**
A7.	**D**	A19.	**C**	A31.	**A**	B3.	**B**	B15.	**D**
A8.	**A**	A20.	**D**	A32.	**D**	B4.	**C**	B16.	**D**
A9.	**A**	A21.	**B**	A33.	**C**	B5.	**D**		
A10.	**D**	A22.	**A**	A34.	**A**	B6.	**B**		
A11.	**D**	A23.	**D**	A35.	**C**	B7.	**C**		
A12.	**C**	A24.	**B**	A36.	**D**	B8.	**B**		

Answer Explanations

Area

For each of Questions A1–A15, we give a sketch of the region and indicate a typical element of area. The area of the region is given by the definite integral. We exploit symmetry wherever possible.

A1. **(C)**

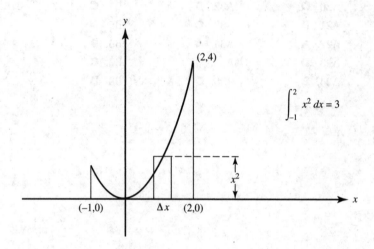

$$\int_{-1}^{2} x^2 \, dx = 3$$

A2. **(C)**

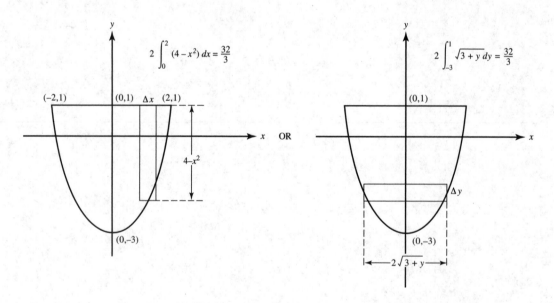

$$2 \int_{0}^{2} (4 - x^2) \, dx = \frac{32}{3}$$

OR

$$2 \int_{-3}^{1} \sqrt{3 + y} \, dy = \frac{32}{3}$$

A3. **(A)**

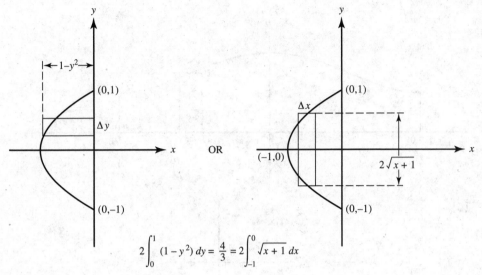

$$2\int_0^1 (1-y^2)\,dy = \frac{4}{3} = 2\int_{-1}^0 \sqrt{x+1}\,dx$$

A4. **(D)**

$$\int_{-2}^1 (2-y-y^2)\,dy = \frac{9}{2}$$

A5. **(D)**

$$2\int_0^2 \frac{4}{x^2+4}\,dx = \pi$$

A6. **(C)**

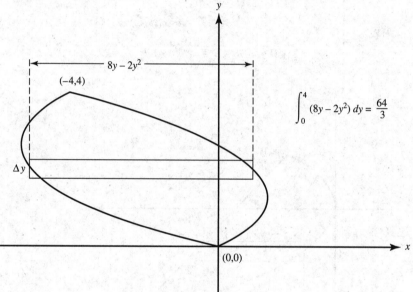

$$\int_0^4 (8y-2y^2)\,dy = \frac{64}{3}$$

A7. **(D)**

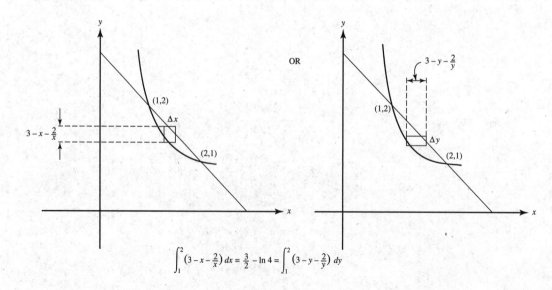

$$\int_1^2 \left(3-x-\frac{2}{x}\right)dx = \frac{3}{2}-\ln 4 = \int_1^2 \left(3-y-\frac{2}{y}\right)dy$$

A8. **(A)**

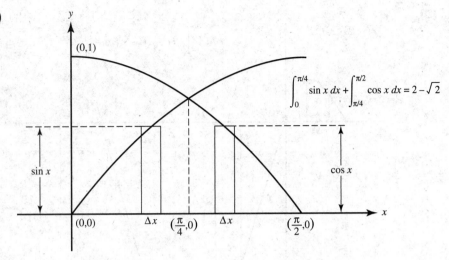

$$\int_0^{\pi/4} \sin x \, dx + \int_{\pi/4}^{\pi/2} \cos x \, dx = 2 - \sqrt{2}$$

A9. **(A)**

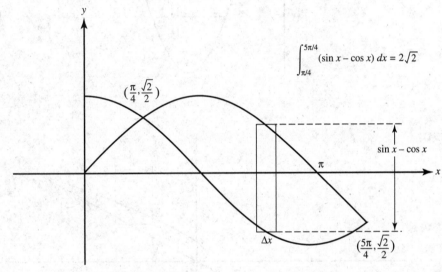

$$\int_{\pi/4}^{5\pi/4} (\sin x - \cos x) \, dx = 2\sqrt{2}$$

A10. **(D)**

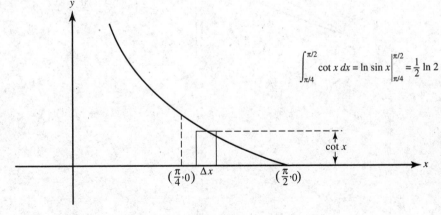

$$\int_{\pi/4}^{\pi/2} \cot x \, dx = \ln \sin x \Big|_{\pi/4}^{\pi/2} = \frac{1}{2} \ln 2$$

A11. **(D)**

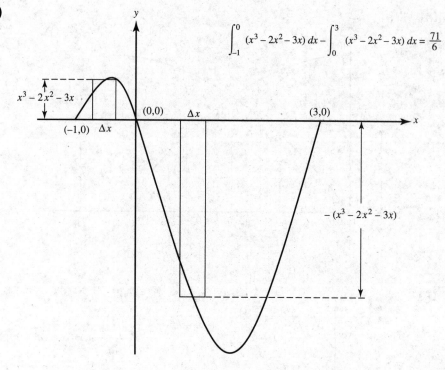

$$\int_{-1}^{0} (x^3 - 2x^2 - 3x)\, dx - \int_{0}^{3} (x^3 - 2x^2 - 3x)\, dx = \frac{71}{6}$$

A12. **(C)**

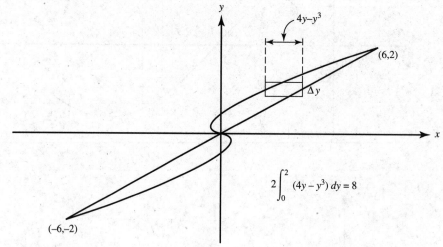

$$2\int_{0}^{2} (4y - y^3)\, dy = 8$$

A13. **(C)**

$$\int_0^{\ln 2} (2 - e^x)dx = (2x - e^x)\Big|_0^{\ln 2}$$
$$= (2 \ln 2 - e^{\ln 2}) - (2 \cdot 0 - e^0)$$
$$= 2 \ln 2 - 1$$

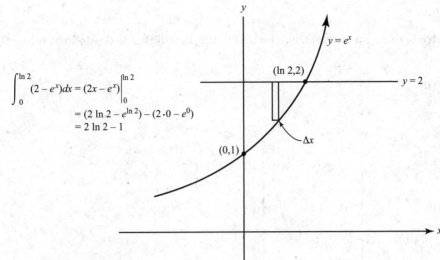

A14. **(B)**

$$\int_0^1 2\sqrt{x - x^2}\,dx$$

A15. **(C)**

$\int_0^c (c^3 - x^3)\,dx = \frac{3}{4}c^4$; thus area of rectangle is to area of shaded region as 4 is to 3.

Volume

A sketch is given, for each of Questions A16–A25, in addition to the definite integral for each volume.

A16. **(D)**

About the x-axis.
Disk.
$\Delta V = \pi y^2 \, \Delta x = \pi x^4 \, \Delta x$

$$\int_0^2 \pi x^4 \, dx = \frac{32\pi}{5}$$

A17. **(D)**

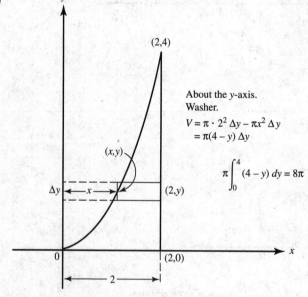

About the y-axis.
Washer.
$V = \pi \cdot 2^2 \, \Delta y - \pi x^2 \, \Delta y$
$\quad = \pi(4 - y) \, \Delta y$

$$\pi \int_0^4 (4 - y) \, dy = 8\pi$$

A18. **(A)**

About the y-axis.
Disk.
$\Delta V = \pi x^2\, \Delta y$
$\quad = \pi y\, \Delta y$

$\pi \displaystyle\int_0^4 y\, dy = 8\pi$

A19. **(C)**

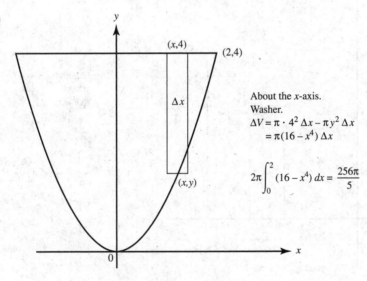

About the x-axis.
Washer.
$\Delta V = \pi \cdot 4^2\, \Delta x - \pi y^2\, \Delta x$
$\quad = \pi(16 - x^4)\, \Delta x$

$2\pi \displaystyle\int_0^2 (16 - x^4)\, dx = \dfrac{256\pi}{5}$

A20. **(D)**

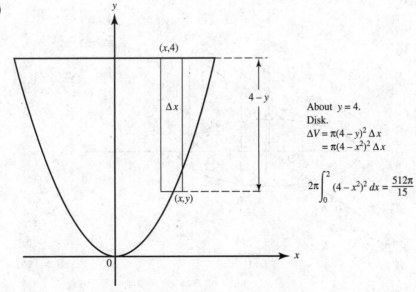

About $y = 4$.
Disk.
$\Delta V = \pi(4 - y)^2\, \Delta x$
$\quad = \pi(4 - x^2)^2\, \Delta x$

$2\pi \displaystyle\int_0^2 (4 - x^2)^2\, dx = \dfrac{512\pi}{15}$

A21. **(B)**

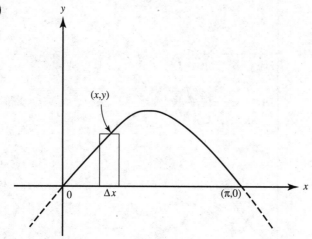

About the x-axis.
Disk.
$$\Delta V = \pi y^2 \, \Delta x$$
$$= \pi \sin^2 x \, \Delta x$$

$$\pi \int_0^\pi \sin^2 x \, dx = \frac{\pi^2}{2}$$

A22. **(A)**

About the x-axis.
Disk.
$$\Delta V = \pi y^2 \, \Delta x$$

$$\pi \int_2^4 x^2 \, dx = \frac{56\pi}{3}$$

A23. **(D)**

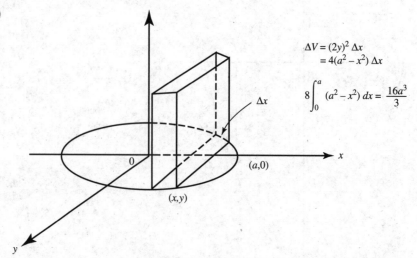

$$\Delta V = (2y)^2 \, \Delta x$$
$$= 4(a^2 - x^2) \, \Delta x$$

$$8 \int_0^a (a^2 - x^2) \, dx = \frac{16a^3}{3}$$

A24. **(B)**

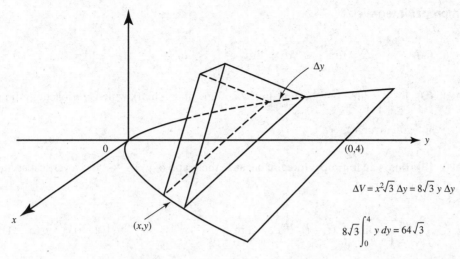

$$\Delta V = x^2\sqrt{3}\,\Delta y = 8\sqrt{3}\,y\,\Delta y$$

$$8\sqrt{3}\int_0^4 y\,dy = 64\sqrt{3}$$

A25. **(D)**

$$\int_0^1 (e^{-x})^2\,dx = \tfrac{1}{2}\left(1 - \tfrac{1}{e^2}\right)$$

Length of Curve (Arc Length)

A26. **(C)** The curve above the x-axis is $y = \sqrt{x^3} = x^{3/2}$. The curve below the x-axis is
$y = -\sqrt{x^3} = -x^{3/2}$. The curves are symmetric with respect to the x-axis, so we can double
the length of the portion above the x-axis. Find the derivative of the curve above the

x-axis: $\dfrac{d}{dx}\left(x^{3/2}\right) = \dfrac{3}{2}\cdot x^{1/2}$. Using the length of curve formula, (1) on page 297, gives

$L = 2\displaystyle\int_0^4 \sqrt{1 + \left(\dfrac{3}{2}x^{1/2}\right)^2}\,dx = 2\displaystyle\int_0^4 \sqrt{1 + \dfrac{9}{4}x}\,dx$. Using u-substitution, with

$u = 1 + \dfrac{9}{4}x$, yields $2\displaystyle\int_0^4 \sqrt{1 + \dfrac{9}{4}x}\,dx = \dfrac{8}{9}\displaystyle\int_1^{10} u^{1/2}\,du = \dfrac{8}{9}\cdot\dfrac{2}{3}x^{3/2}\Big|_1^{10} = \dfrac{16}{27}\left(10^{3/2} - 1\right).$

A27. **(A)** $\dfrac{d}{dx}\big(\ln(\cos x)\big) = \dfrac{1}{\cos x}\cdot(-\sin x) = -\tan x$. Using the length of curve formula, (1) on

page 297, gives $L = \displaystyle\int_{\pi/4}^{\pi/3} \sqrt{1 + (-\tan x)^2}\,dx$.

Improper Integrals

A28. (A) $\int_0^\infty e^{-x}\,dx = \lim_{b\to\infty}\int_0^b e^{-x}\,dx = \lim_{b\to\infty}\left.-\frac{1}{e^x}\right|_0^b = \lim_{b\to\infty}\left(-\frac{1}{e^b}-\left(-\frac{1}{e^0}\right)\right) = (0-(-1)) = 1$

A29. (D) $\int_0^e \frac{du}{u} = \lim_{h\to 0^+}\int_h^e \frac{du}{u} = \lim_{h\to 0^+}\left.\ln|u|\right|_h^e = \lim_{h\to 0^+}(\ln e - \ln h)$, which approaches infinity, so the

integral diverges.

A30. (B) This is an improper integral because the graph of $y = \dfrac{1}{\sqrt[3]{t-1}}$ has a vertical asymptote

at $x = 1$.

$\int_1^2 \dfrac{dt}{\sqrt[3]{t-1}} = \lim_{a\to 1^+}\int_a^2 (t-1)^{-1/3}dt = \lim_{a\to 1^+}\left.\frac{3}{2}(t-1)^{2/3}\right|_a^2 = \frac{3}{2}\lim_{a\to 1^+}((2-1)^{2/3}-(a-1)^{2/3})$

$= \frac{3}{2}(1-0) = \frac{3}{2}$

A31. (A) This is an improper integral because the graph of $y = \dfrac{1}{(x-3)^{2/3}}$ has a vertical asymptote at $x = 3$. Since the asymptote is inside the interval of integration, you must split this into two improper integrals with $x = 3$ as an endpoint of each. Then evaluate each integral individually.

$\int_2^4 \dfrac{dx}{(x-3)^{2/3}} = \lim_{b\to 3^-}\int_2^b (x-3)^{-2/3}\,dx + \lim_{a\to 3^+}\int_a^4 (x-3)^{-2/3}\,dx = 3+3 = 6$

$\lim_{b\to 3^-}\int_2^b (x-3)^{-2/3}\,dx = \lim_{b\to 3^-}\left.3(x-3)^{1/3}\right|_2^b = 3\lim_{b\to 3^-}((b-3)^{1/3}-(2-3)^{1/3})$

$= 3(0-(-1)) = 3$

$\lim_{a\to 3^+}\int_a^4 (x-3)^{-2/3}\,dx = \lim_{a\to 3^+}\left.3(x-3)^{1/3}\right|_a^4 = 3\lim_{a\to 3^+}((4-3)^{1/3}-(a-3)^{1/3}) = 3(1-0) = 3$

A32. (D) This is an improper integral because the graph of $y = \dfrac{1}{(x-3)^2}$ has a vertical asymptote at $x = 3$. Since the asymptote is inside the interval of integration, you must split this into two improper integrals with $x = 3$ as an endpoint of each. Then evaluate each integral individually.

$\int_2^4 \dfrac{dx}{(x-3)^2} = \lim_{b\to 3^-}\int_2^b (x-3)^{-2}\,dx + \lim_{a\to 3^+}\int_a^4 (x-3)^{-2}\,dx \to \infty$

$\lim_{b\to 3^-}\int_2^b (x-3)^{-2}\,dx = \lim_{b\to 3^-}\left.-(x-3)^{-1}\right|_2^b = -\lim_{b\to 3^-}\left(\frac{1}{b-3}-\frac{1}{2-3}\right) \to \infty$

$\lim_{a\to 3^+}\int_a^4 (x-3)^{-2}\,dx = \lim_{a\to 3^+}\left.-(x-3)^{-1}\right|_a^4 = -\lim_{a\to 3^+}\left(\frac{1}{4-3}-\frac{1}{a-3}\right) \to \infty$

A33. (C) This is an improper integral because the graph of $y = \dfrac{\sin x}{\sqrt{1-\cos x}}$ has a vertical asymptote at $x = 0$.

$\int_0^{\pi/2} \dfrac{\sin x}{\sqrt{1-\cos x}}\,dx = \lim_{a\to 0^+}\int_a^{\pi/2} \dfrac{\sin x}{\sqrt{1-\cos x}}\,dx = 2\lim_{a\to 0^+}\left.\sqrt{1-\cos x}\right|_a^{\pi/2}$

$= \lim_{a\to 0^+}2(\sqrt{1-0}-\sqrt{1-\cos a}) = 2(1-0) = 2$

A34. **(A)**

$$\int_0^\infty e^{-x}\,dx = 1$$

A35. **(C)**

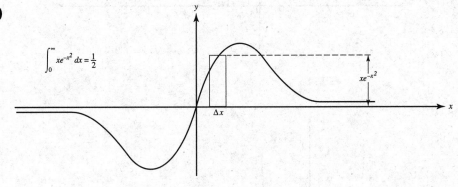

$$\int_0^\infty xe^{-x^2}\,dx = \tfrac{1}{2}$$

A36. **(D)**

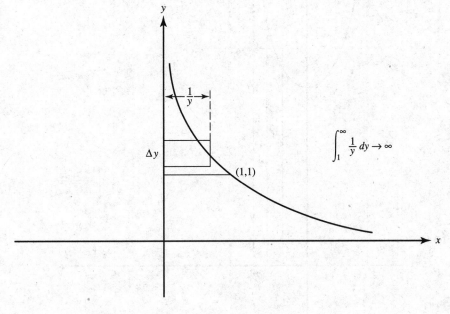

$$\int_1^\infty \frac{1}{y}\,dy \to \infty$$

A37. **(B)**

$$\int_{-\infty}^{\infty} \frac{4}{1+x^2}\, dx = \lim_{b \to \infty}\ 4\tan^{-1}x\ \Big|_{-b}^{b} = 4\pi$$

A38. **(C)**

$$2\int_{0}^{1} \frac{4\, dx}{\sqrt{1-x^2}} = 4\pi$$

A39. **(B)**

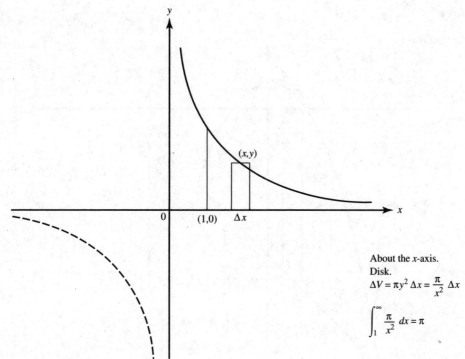

About the x-axis.
Disk.
$\Delta V = \pi y^2 \, \Delta x = \dfrac{\pi}{x^2} \, \Delta x$

$\displaystyle\int_1^\infty \dfrac{\pi}{x^2} \, dx = \pi$

A40. **(A)**

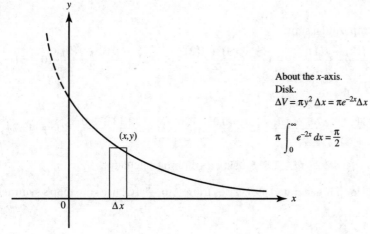

About the x-axis.
Disk.
$\Delta V = \pi y^2 \, \Delta x = \pi e^{-2x} \Delta x$

$\pi \displaystyle\int_0^\infty e^{-2x} \, dx = \dfrac{\pi}{2}$

Area

B1. **(C)**

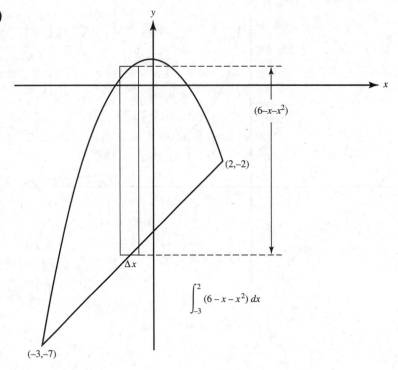

$$\int_{-3}^{2} (6 - x - x^2)\, dx$$

B2. **(C)** Using a trapezoidal sum:

$$\int_{1}^{8} f(x)\, dx \approx \frac{f(2) + f(1)}{2} \cdot (2 - 1) + \frac{f(5) + f(2)}{2} \cdot (5 - 2) + \frac{f(6) + f(5)}{2} \cdot (6 - 5)$$

$$+ \frac{f(8) + f(6)}{2} \cdot (8 - 6)$$

$$= \frac{4.15 + 1.62}{2} \cdot (1) + \frac{7.5 + 4.15}{2} \cdot (3) + \frac{9.0 + 7.5}{2} \cdot (1) + \frac{12.13 + 9.0}{2} \cdot (2) = 49.74$$

B3. **(B)** $A = 8 \int_{0}^{\pi/4} \frac{1}{2} \cos^2 2\theta\, d\theta = 1.571$, using a graphing calculator.

B4. **(C)** The small loop is generated as θ varies from $\frac{\pi}{6}$ to $\frac{5\pi}{6}$. (C) uses the loop's symmetry.

Volume

B5. **(D)**

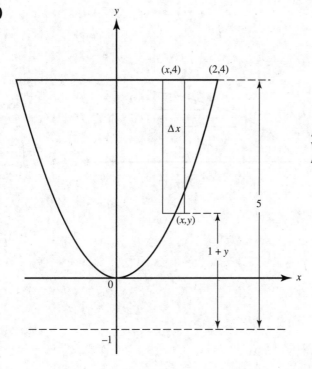

About $y = -1$.
Washer.
$\Delta V = \pi \cdot 5^2 \Delta x - \pi (1 + y)^2 \Delta x$
$\quad = \pi [25 - (1 + y)^2] \Delta x$

$2\pi \displaystyle\int_0^2 (24 - 2x^2 - x^4)\, dx$

B6. **(B)**

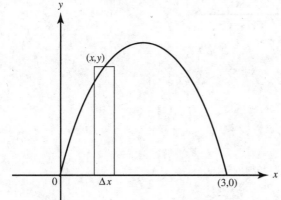

About the x-axis.
Disk.
$\Delta V = \pi y^2 \Delta x$
$\quad = \pi (3x - x^2)^2 \Delta x$

$\pi \displaystyle\int_0^3 (3x - x^2)^2\, dx$

B7. **(C)**

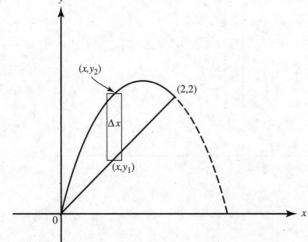

About the x-axis.
Washer.
$\Delta V = \pi y_2^2 \Delta x - \pi y_1^2 \Delta x$
$\quad = \pi [(3x - x^2)^2 - x^2]\, \Delta x$

$\pi \displaystyle\int_0^2 [(3x - x^2)^2 - x^2]\, dx$

B8. **(B)**

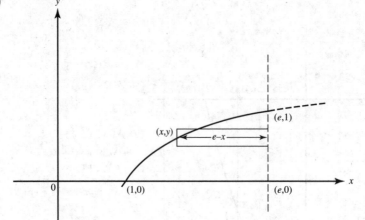

About $x = e$.
Disk.
$$\Delta V = \pi(e - x)^2 \, \Delta y$$
$$= \pi(e - e^y)^2 \, \Delta y$$

$$\pi \int_0^1 (e - e^y)^2 \, dy$$

B9. **(A)**

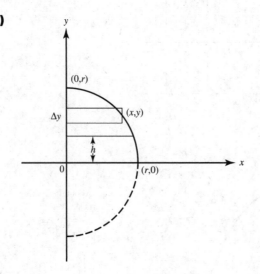

About the y-axis.
Disk.
$$\Delta V = \pi x^2 \, \Delta y$$
$$= \pi(r^2 - y^2) \, \Delta y$$

$$\pi \int_h^r (r^2 - y^2) \, dy = \frac{\pi}{3} \, (2r^3 + h^3 - 3r^2 h)$$

B10. **(D)**

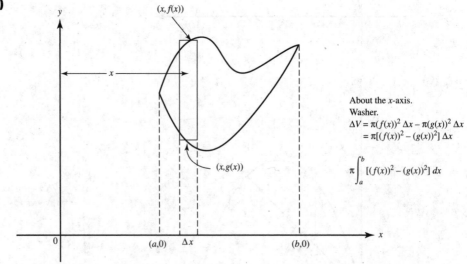

About the x-axis.
Washer.
$$\Delta V = \pi(f(x))^2 \, \Delta x - \pi(g(x))^2 \, \Delta x$$
$$= \pi[(f(x))^2 - (g(x))^2] \, \Delta x$$

$$\pi \int_a^b [(f(x))^2 - (g(x))^2] \, dx$$

Length of Curve (Arc Length)

B11. **(D)** From (3) on page 297, we obtain the length:

$$\int_0^{2\pi} \sqrt{(1 - \cos t)^2 + (\sin t)^2}\, dt = \int_0^{2\pi} \sqrt{2 - 2\cos t}\, dt$$

B12. **(D)** Solve the equation for x: $x = \dfrac{y^2}{4}$. So $\dfrac{dx}{dy} = \dfrac{d}{dy}\left(\dfrac{y^2}{4}\right) = \dfrac{y}{2}$. When $x = 2$, $y = 2\sqrt{2}$ or $y = -2\sqrt{2}$. The curve is symmetric across the x-axis, so double the length from $y = 0$ to $y = 2\sqrt{2}$. Use (2) on page 297:

$$2\int_0^{2\sqrt{2}} \sqrt{1 + \left(\tfrac{y}{2}\right)^2}\, dy = 2\int_0^{2\sqrt{2}} \sqrt{1 + \tfrac{y^2}{4}}\, dy = 2\int_0^{2\sqrt{2}} \sqrt{\tfrac{1}{4}(4 + y^2)}\, dy = 2\int_0^{2\sqrt{2}} \tfrac{1}{2}\sqrt{(4 + y^2)}\, dy$$

B13. **(A)** Use (3) on page 297 to get the integral:

$$\int_2^3 \sqrt{(-e^t \sin t + e^t \cos t)^2 + (e^t \cos t + e^t \sin t)^2}\, dt$$

Improper Integrals

B14. **(C)** The integrand is discontinuous at $x = 1$, which is on the interval of integration.

B15. **(D)** The integral in (D) is the sum of two integrals from -1 to 0 and from 0 to 1. Both diverge (see Example 29, pages 304 and 305). Note that (A), (B), and (C) all converge.

B16. **(D)** Choices (A) and (C) can be shown convergent by the Comparison Test; the convergence of (B) is shown in Example 24, page 303.

8

Further Applications of Integration

Learning Objectives

In this chapter, you will review many ways that definite integrals can be used to solve a variety of problems, notably distance traveled by an object in motion along a line. You'll see that in a variety of settings, accumulated change can be expressed as a Riemann Sum whose limit becomes an integral of the rate of change.

In addition, BC Calculus students will learn more about motion, including objects in motion in a plane along a parametrically defined curve.

A. Motion Along a Straight Line

If the motion of a particle P along a straight line is given by the equation $s = F(t)$, where s is the distance at time t of P from a fixed point on the line, then the velocity and acceleration of P at time t are given respectively by

$$v = \frac{ds}{dt} \quad \text{and} \quad a = \frac{dv}{dt} = \frac{d^2s}{dt^2}$$

This topic was discussed as an application of differentiation on page 180. Here we will apply integration to find velocity from acceleration and distance from velocity.

If we know that particle P has velocity $v(t)$, where v is a continuous function, then the *distance* traveled by the particle during the time interval from $t = a$ to $t = b$ is the definite integral of its speed:

Distance

$$\int_a^b |v(t)|\, dt \tag{1}$$

If $v(t) \geq 0$ for all t on $[a,b]$ (i.e., P moves only in the positive direction), then (1) is equivalent to $\int_a^b v(t)\, dt$; similarly, if $v(t) \leq 0$ on $[a,b]$ (P moves only in the negative direction), then (1) yields $-\int_a^b v(t)\, dt$. If $v(t)$ changes sign on $[a,b]$ (i.e., the direction of motion changes), then (1) gives the total distance traveled. Suppose, for example, that the situation is as follows:

$$a \leq t \leq c \qquad v(t) \geq 0$$
$$c \leq t \leq d \qquad v(t) \leq 0$$
$$d \leq t \leq b \qquad v(t) \geq 0$$

Then the total distance traveled during the time interval from $t = a$ to $t = b$ is exactly

$$\int_a^c v(t)\, dt - \int_c^d v(t)\, dt + \int_d^b v(t)\, dt$$

The *displacement* or *net change* in the particle's position from $t = a$ to $t = b$ is equal, by the Fundamental Theorem of Calculus (FTC), to

Displacement

$$\int_a^b v(t)\, dt$$

> **Example 1**

If a body moves along a straight line with velocity $v = t^3 + 3t^2$, find the distance traveled between $t = 1$ and $t = 4$.

✓ **Solution**

$$\int_1^4 \left(t^3 + 3t^2\right) dt = \left(\frac{t^4}{4} + t^3\right)\Big|_1^4 = \frac{507}{4}$$

Note that $v > 0$ for all t on $[1,4]$.

> **Example 2**

A particle moves along the x-axis so that its velocity at time t is given by $v(t) = 6t^2 - 18t + 12$.
(a)   Find the total distance covered between $t = 0$ and $t = 4$.
(b)   Find the displacement of the particle from $t = 0$ to $t = 4$.

✓ **Solutions**

(a)   Since $v(t) = 6t^2 - 18t + 12 = 6(t - 1)(t - 2)$, we see that:

$$\begin{array}{lll} \text{if} & t < 1, & \text{then } v > 0 \\ \text{if } 1 < t < 2, & & \text{then } v < 0 \\ \text{if } 2 < t, & & \text{then } v > 0 \end{array}$$

Thus, the total distance covered between $t = 0$ and $t = 4$ is

$$\int_0^1 v(t)\, dt - \int_1^2 v(t)\, dt + \int_2^4 v(t)\, dt \qquad (2)$$

When we replace $v(t)$ by $6t^2 - 18t + 12$ in (2) and evaluate, we obtain 34 units for the total distance covered between $t = 0$ and $t = 4$. This can also be verified on your calculator by evaluating

$$\int_0^4 |v(t)|\, dt$$

This example is the same as Example 26 on page 180, in which the required distance is computed by another method.

(b)   To find the *displacement* of the particle from $t = 0$ to $t = 4$, we use the FTC, evaluating

$$\int_0^4 v(t)\, dt = \int_0^4 \left(6t^2 - 18t + 12\right) dt$$

$$= \left(2t^3 - 9t^2 + 12t\right)\Big|_0^4 = 128 - 144 + 48 = 32$$

This is the net change in position from $t = 0$ to $t = 4$, sometimes referred to as "position shift." Here it indicates the particle ended up 32 units to the right of its starting point.

> **Example 3**

The acceleration of an object moving on a line is given at time t by $a = \sin t$; when $t = 0$ the object is at rest. Find the distance s it travels from $t = 0$ to $t = \dfrac{5\pi}{6}$.

✓ **Solution**

Since $a = \dfrac{d^2 s}{dt^2} = \dfrac{dv}{dt} = \sin t$, it follows that

$$v(t) = \frac{ds}{dt} = \int \sin t \, dt \qquad v(t) = -\cos t + C$$

Also, $v(0) = 0$ yields $C = 1$. Thus $v(t) = 1 - \cos t$; and since $\cos t \leq 1$ for all t we see that $v(t) \geq 0$ for all t. Thus, the distance traveled is

$$\int_0^{5\pi/6} (1 - \cos t) \, dt = (t - \sin t) \Big|_0^{5\pi/6} = \frac{5\pi}{6} - \frac{1}{2}$$

B. Motion Along a Plane Curve

In Chapter 4, Section K, it was pointed out that, if the motion of a particle P along a curve is given parametrically by the equations $x = x(t)$ and $y = y(t)$, then at time t the position vector $\mathbf{R}$, the velocity vector $\mathbf{v}$, and the acceleration vector $\mathbf{a}$ are:

$$\mathbf{R} = \langle x, y \rangle$$

$$\mathbf{v} = \frac{d\mathbf{R}}{dt} = \left\langle \frac{dx}{dt}, \frac{dy}{dt} \right\rangle = \langle v_x, v_y \rangle$$

$$\mathbf{a} = \frac{d^2 \mathbf{R}}{dt^2} = \frac{d\mathbf{v}}{dt} = \left\langle \frac{d^2 x}{dt^2}, \frac{d^2 y}{dt^2} \right\rangle = \langle a_x, a_y \rangle$$

The components in the horizontal and vertical directions of $\mathbf{R}$, $\mathbf{v}$, and $\mathbf{a}$ are given, respectively, by the coordinates in the corresponding vector. The slope of $\mathbf{v}$ is $\dfrac{dy}{dx}$; its magnitude,

$$|\mathbf{v}| = \sqrt{\left(\frac{dx}{dt}\right)^2 + \left(\frac{dy}{dt}\right)^2}$$

is the *speed* of the particle, and the velocity vector is tangent to the path. The slope of $\mathbf{a}$ is $\dfrac{d^2 y}{dt^2} \Big/ \dfrac{d^2 x}{dt^2}$. The distance the particle travels from time t_1 to t_2 is given by

$$\int_{t_1}^{t_2} |\mathbf{v}| \, dt = \int_{t_1}^{t_2} \sqrt{\left(\frac{dx}{dt}\right)^2 + \left(\frac{dy}{dt}\right)^2} \, dt$$

How integration may be used to solve problems of curvilinear motion is illustrated in the following examples.

> ## Example 4

Suppose a projectile is launched from the origin at an angle of elevation α and initial velocity v_0. Find the parametric equations for its flight path.

✓ Solution

We have the following initial conditions:

Position: $x(0) = 0$; $y(0) = 0$

Velocity: $\dfrac{dx}{dt}(0) = v_0 \cos \alpha$; $\dfrac{dy}{dt}(0) = v_0 \sin \alpha$

We start with equations representing acceleration due to gravity and integrate each twice, determining the constants as shown:

Acceleration: $\dfrac{d^2x}{dt^2} = 0 \qquad \dfrac{d^2y}{dt^2} = -g$

$$\frac{dx}{dt} = C_1 = v_0 \cos \alpha \qquad\qquad \frac{dy}{dt} = -gt + C_2$$

$$v_0 \sin \alpha = C_2$$

$$x = (v_0 \cos \alpha)t + C_3 \qquad\qquad y = -\frac{1}{2}gt^2 + (v_0 \sin \alpha)t + C_4$$

$$x(0) = 0 \text{ yields } C_3 = 0 \qquad\qquad y(0) = 0 \text{ yields } C_4 = 0$$

Finally, then,

$$x = (v_0 \cos \alpha)t \qquad\qquad y = -\frac{1}{2}gt^2 + (v_0 \sin \alpha)t$$

If desired, t can be eliminated from this pair of equations to yield a parabola in rectangular coordinates.

> ## Example 5

A particle $P(x,y)$ moves along a curve so that

$$\frac{dx}{dt} = 2\sqrt{x} \qquad \text{and} \qquad \frac{dy}{dt} = \frac{1}{x} \text{ at any time } t \geq 0$$

At $t = 0$, $x = 1$ and $y = 0$. Find the parametric equations of motion.

✓ Solution

Since $\dfrac{dx}{\sqrt{x}} = 2\,dt$, we integrate to get $2\sqrt{x} = 2t + C$, and use $x(0) = 1$ to find that $C = 2$. Therefore, $\sqrt{x} = t + 1$ and

$$x = (t+1)^2 \tag{1}$$

Then $\dfrac{dy}{dt} = \dfrac{1}{x} = \dfrac{1}{(t+1)^2}$ by (1), so $dy = \dfrac{dt}{(t+1)^2}$ and $y = -\dfrac{1}{t+1} + C'$. $\tag{2}$

Since $y(0) = 0$, this yields $C' = 1$, and so (2) becomes

$$y = 1 - \frac{1}{t+1} = \frac{t}{t+1}$$

Thus the parametric equations are

$$x = (t+1)^2 \qquad \text{and} \qquad y = \frac{t}{t+1}$$

> **Example 6**

The particle in Example 5 is in motion for 1 second, $0 \le t \le 1$. Find its position, velocity, speed, and acceleration at $t = 1$ and the distance it traveled.

✓ **Solution**

In Example 5 we derived the result $P(t) = \left((t+1)^2, \dfrac{t}{t+1} \right)$, the parametric representation of the

particle's position. Hence its position at $t = 1$ is $P(1) = \left(4, \dfrac{1}{2} \right)$.

From $P(t)$ we write the velocity vector:

$$\mathbf{v} = \left\langle \frac{dx}{dt}, \frac{dy}{dt} \right\rangle = \left\langle 2(t+1), \frac{1}{(t+1)^2} \right\rangle$$

Hence, at $t = 1$ the particle's velocity is $\mathbf{v} = \left\langle 4, \dfrac{1}{4} \right\rangle$.

Speed is the magnitude of the velocity vector, so after 1 second the particle's speed is

$$|\mathbf{v}| = \sqrt{4^2 + \left(\tfrac{1}{4}\right)^2} \approx 4.008 \text{ units/sec}$$

The particle's acceleration vector at $t = 1$ is

$$\mathbf{a} = \left\langle \frac{d^2x}{dt^2}, \frac{d^2y}{dt^2} \right\rangle = \left\langle 2, \frac{-2}{(t+1)^3} \right\rangle = \left\langle 2, -\frac{1}{4} \right\rangle$$

On the interval $0 \le t \le 1$ the distance traveled by the particle is

$$\int_0^1 \sqrt{\left(\frac{dx}{dt}\right)^2 + \left(\frac{dy}{dt}\right)^2} \, dt = \int_0^1 \sqrt{[2(t+1)]^2 + \left[\frac{1}{(t+1)^2}\right]^2} \, dt \approx 3.057 \text{ units}$$

> **Example 7**

A particle $P(x,y)$ moves along a curve so that its acceleration is given by

$$\mathbf{a} = \left\langle -4\cos 2t, -2\sin t \right\rangle \left(-\frac{\pi}{2} \le t \le \frac{\pi}{2} \right)$$

when $t = 0$, the particle is at $(1,0)$ with $\dfrac{dx}{dt} = 0$ and $\dfrac{dy}{dt} = 2$.

(a)   Find the position vector $\mathbf{R}$ at any time t.

(b)   Find a Cartesian equation for the path of the particle, and identify the conic on which
      P moves.

✓ Solutions

(a) $\mathbf{v} = \left\langle -2 \sin 2t + c_1, 2 \cos t + c_2 \right\rangle$, and since $\mathbf{v} = \left\langle 0,2 \right\rangle$ when $t = 0$, it follows that $c_1 = c_2 = 0$.

So $\mathbf{v} = \left\langle -2 \sin 2t, 2 \cos t \right\rangle$. Also $\mathbf{R} = \left\langle \cos 2t + c_3, 2 \sin t + c_4 \right\rangle$; and since $\mathbf{R} = \left\langle 1,0 \right\rangle$

when $t = 0$, we see that $c_3 = c_4 = 0$. Finally, then,

$$\mathbf{R} = \left\langle \cos 2t, 2 \sin t \right\rangle$$

(b)   From (a) the parametric equations of motion are

$$x = \cos 2t \qquad y = 2 \sin t$$

By a trigonometric identity,

$$x = 1 - 2 \sin^2 t = 1 - \frac{y^2}{2}$$

P travels in a counterclockwise direction along *part* of a parabola that has its vertex at $(1,0)$ and opens to the left. The path of the particle is sketched in Figure 8.1; note that $-1 \leq x \leq 1$, $-2 \leq y \leq 2$.

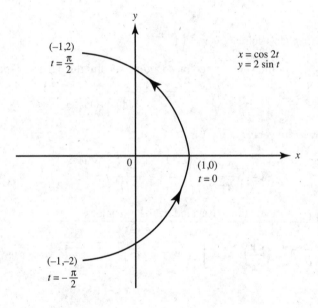

Figure 8.1

C. Other Applications of Riemann Sums

We will continue to set up Riemann Sums to calculate a variety of quantities using definite integrals. In many of these examples, we will partition into n equal subintervals a given interval (or region or ring or solid or the like), approximate the quantity over each small subinterval (and assume it is constant there), then add up all these small quantities. Finally, as $n \to \infty$, we will replace the sum by its equivalent definite integral to calculate the desired quantity.

> **Example 8**

Amount of Leaking Water. Water is draining from a cylindrical pipe of radius 2 inches. At t seconds the water is flowing out with velocity $v(t)$ inches per second. Express the amount of water that has drained from the pipe in the first 3 minutes as a definite integral in terms of $v(t)$.

✓ **Solution**

We first express 3 minutes as 180 seconds. We then partition $[0,180]$ into n subintervals each of length Δt. In Δt seconds, approximately $v(t)\, \Delta t$ inches of water have drained from the pipe. Since a typical cross section has area 4π in.^{2} (Figure 8.2), in Δt seconds the amount that has drained is

$$\left(4\pi \text{ in.}^2\right)(v(t) \text{ in./sec})(\Delta t \text{ sec}) = 4\pi v(t)\, \Delta t \text{ in.}^3$$

The sum of the n amounts of water that drain from the pipe, as $n \to \infty$, is $\displaystyle\int_0^{180} 4\pi v(t)\, dt$; the units are cubic inches $\left(\text{in.}^3\right)$.

Area = 4π in.^{2}

Height = $v(t)\,\Delta t$ in.

Figure 8.2

> **Example 9**

Traffic: Total Number of Cars. The density of cars (the number of cars per mile) on 10 miles of the highway approaching Disney World is equal approximately to $f(x) = 200\big[4 - \ln(2x+3)\big]$, where x is the distance in miles from the Disney World entrance. Find the total number of cars on this 10-mile stretch.

✓ **Solution**

Partition the interval $[0,10]$ into n equal subintervals each of width Δx. In each subinterval the number of cars equals approximately the density of cars $f(x)$ times Δx, where $f(x) = 200\big[4 - \ln(2x+3)\big]$. When we add n of these products we get $\displaystyle\sum f(x)\Delta x$, which is a Riemann Sum. As $n \to \infty$ (or as $\Delta x \to 0$), the Riemann Sum approaches the definite integral

$$\int_0^{10} 200\big[4 - \ln(2x+3)\big]\, dx$$

which, using our calculator, is approximately equal to 3118 cars.

D. FTC: Definite Integral of a Rate Is Net Change

If f is continuous and $f(t) = \dfrac{dF}{dt}$, then we know from the FTC that

$$\int_a^b f(t)\, dt = F(b) - F(a)$$

The definite integral of the rate of change of a quantity over an interval is the *net change* or *net accumulation* of the quantity over that interval. Thus, $F(b) - F(a)$ is the net change in $F(t)$ as t varies from a to b.

We've already illustrated this principle many times. Here are more examples.

❯ Example 10

Let $G(t)$ be the rate of growth of a population at time t. Then the increase in population between times $t = a$ and $t = b$ is given by $\displaystyle\int_a^b G(t)\, dt$. The population may consist of people, deer, fruit flies, bacteria, and so on.

❯ Example 11

Suppose an epidemic is spreading through a city at the rate of $f(t)$ new people per week. Then

$$\int_0^4 f(t)\, dt$$

is the number of people who will become infected during the next 4 weeks (or the total change in the number of infected people).

❯ Example 12

Suppose a rumor is spreading at the rate of $f(t) = 100e^{-0.2t}$ new people per day. Find the number of people who hear the rumor during the 5th and 6th days.

✓ Solution

$$\int_4^6 100e^{-0.2t}\, dt = 74 \text{ people}$$

If we let $F'(t) = f(t)$, then the integral above is the net change in $F(t)$ from $t = 4$ to $t = 6$, or the number of people who hear the rumor from the beginning of the 5th day to the end of the 6th.

❯ Example 13

Economists define the *marginal cost of production* as the additional cost of producing one additional unit at a specified production level. It can be shown that if $C(x)$ is the cost at production level x, then $C'(x)$ is the marginal cost at that production level.

If the marginal cost, in dollars, is $\dfrac{1}{x}$ per unit when x units are being produced, find the change in cost when production increases from 50 to 75 units.

✓ Solution

$$\int_{50}^{75} \frac{1}{x}\, dx \approx \$0.41$$

We replace "cost" by "revenue" or "profit" to find the total change in these quantities.

❯ Example 14

After t minutes, a chemical is decomposing at the rate of $10e^{-t}$ grams per minute. Find the amount that has decomposed during the first 3 minutes.

✓ Solution

$$\int_{0}^{3} 10e^{-t}\, dt \approx 9.5\text{ g}$$

❯ Example 15

An official of the Environmental Protection Agency estimates that t years from now the level of a particular pollutant in the air will be increasing at the rate of $(0.3 + 0.4t)$ parts per million per year (ppm/yr). Based on this estimate, find the change in the pollutant level during the second year.

✓ Solution

$$\int_{1}^{2} (0.3 + 0.4t)\, dt \approx 0.9\text{ ppm}$$

CHAPTER SUMMARY

In this chapter, we reviewed how to find the distance traveled by an object in motion along a line and (for BC Calculus students) along a parametrically defined curve in a plane. We also looked at a broad variety of applications of the definite integral to other situations where definite integrals of rates of change are used to determine accumulated change, using limits of Riemann Sums to create the integrals required.

Practice Exercises

PART A. NO CALCULATOR—DIRECTIONS: Answer these questions *without* using your calculator.

A1. A particle moves along a line in such a way that its position at time t is given by $s = t^3 - 6t^2 + 9t + 3$. Its direction of motion changes when

(A) $t = 1$ only
(B) $t = 2$ only
(C) $t = 3$ only
(D) $t = 1$ and $t = 3$

A2. A body moves along a straight line so that its velocity v at time t is given by $v = 4t^3 + 3t^2 + 5$. The distance the body covers from $t = 0$ to $t = 2$ equals

(A) 34
(B) 49
(C) 24
(D) 44

A3. A particle moves along a line with velocity $v = 3t^2 - 6t$. The total distance traveled from $t = 0$ to $t = 3$ equals

(A) 0
(B) 4
(C) 8
(D) 9

A4. A particle moves along a line with velocity $v(t) = 3t^2 - 6t$. The net change in position of the particle from $t = 0$ to $t = 3$ is

(A) 0
(B) 4
(C) 8
(D) 9

A5. The acceleration of a particle moving on a straight line is given by $a = \cos t$, and when $t = 0$ the particle is at rest. The distance it covers from $t = 0$ to $t = 2$ is

(A) $\sin 2$
(B) $1 - \cos 2$
(C) $2 - \sin 2$
(D) $-\cos 2$

A6. During the worst 4-hour period of a hurricane the wind velocity, in miles per hour, is given by $v(t) = 5t - t^2 + 100$, $0 \leq t \leq 4$. The average wind velocity during this period (in mph) is

(A) 100
(B) 102
(C) $104\frac{2}{3}$
(D) $108\frac{2}{3}$

A7. The acceleration of an object in motion is given by the vector $\vec{a}(t) = \left(2t, e^t\right)$. If the object's initial velocity was $\vec{v}(0) = (2,0)$, which is the velocity vector at any time t?

BC ONLY

(A) $\vec{v}(t) = \left\langle t^2, e^t + 1 \right\rangle$
(B) $\vec{v}(t) = \left\langle t^2 + 2, e^t \right\rangle$
(C) $\vec{v}(t) = \left\langle t^2 + 2, e^t - 1 \right\rangle$
(D) $\vec{v}(t) = \left\langle 2, e^t - 1 \right\rangle$

A8. A stone is thrown upward from the ground with an initial velocity of 96 ft/sec. Its average velocity $\left(\text{given that } a(t) = -32 \text{ ft/sec}^2\right)$ during the first 2 seconds is

(A) 16 ft/sec
(B) 32 ft/sec
(C) 64 ft/sec
(D) 80 ft/sec

A9. Assume that the density of vehicles (number of cars per mile) during morning rush hour, for the 20-mile stretch along the New York State Thruway southbound from the Governor Mario M. Cuomo Bridge, is given by $f(x)$, where x is the distance, in miles, south of the bridge. Which of the following gives the number of vehicles (on this 20-mile stretch) from the bridge to a point x miles south of the bridge?

(A) $\int_0^x f(t)\, dt$

(B) $\int_x^{20} f(t)\, dt$

(C) $\int_0^{20} f(x)\, dx$

(D) $\sum_{k=1}^{n} f(x_k)\, \Delta x$ (where the 20-mile stretch has been partitioned into n equal subintervals)

A10. The center of a city, that we will assume is circular, is on a straight highway (along the diameter of the circle). The radius of the city is 3 miles. (*NOTE*: The equation for a circle centered at the origin is $x^2 + y^2 = r^2$, where r is the radius of the circle.) The density of the population, in thousands of people per square mile, is given by $f(x) = 12 - 2x$ at a distance x miles from the highway. The population of the city (in thousands of people) is given by the integral

(A) $\int_0^3 (12 - 2x)\, dx$

(B) $2\int_0^3 (12 - 2x)\, \sqrt{9 - x^2}\, dx$

(C) $4\int_0^3 (12 - 2x)\, \sqrt{9 - x^2}\, dx$

(D) $\int_0^3 2\pi x(12 - 2x)\, dx$

PART B. CALCULATOR ACTIVE—DIRECTIONS: Some of the following questions require the use of a graphing calculator.

B1. A car accelerates from 0 to 60 mph in 10 seconds, with constant acceleration. (Note that 60 mph = 88 ft/sec.) The acceleration $\left(\text{in ft/sec}^2\right)$ is

(A) 1.76

(B) 5.3

(C) 6

(D) 8.8

For Questions B2–B4, use the following information: The velocity **v** of a particle moving on a curve is given, at time t, by $\mathbf{v} = \langle t, t - 1 \rangle$. When $t = 0$, the particle is at point $(0, 1)$.

BC ONLY

B2. At time t the position vector **R** is

(A) $\left\langle \dfrac{t^2}{2}, -\dfrac{(1 - t^2)}{2} \right\rangle$

(B) $\left\langle \dfrac{t^2}{2}, -\dfrac{(1 - t)^2}{2} \right\rangle$

(C) $\left\langle \dfrac{t^2}{2}, -\dfrac{t^2 - 2t}{2} \right\rangle$

(D) $\left\langle \dfrac{t^2}{2}, \dfrac{t^2 - 2t + 2}{2} \right\rangle$

BC ONLY

B3. The acceleration vector at time $t = 2$ is

(A) $\langle 1,1 \rangle$
(B) $\langle 1,-1 \rangle$
(C) $\langle 1,2 \rangle$
(D) $\langle 2,-1 \rangle$

B4. The speed of the particle is at a minimum when t equals

(A) 0
(B) $\dfrac{1}{2}$
(C) 1
(D) 1.5

CHALLENGE

B5. A particle moves along a curve in such a way that its position vector and velocity vector are perpendicular at all times. If the particle passes through the point (4,3), then the equation of the curve is

(A) $x^2 + y^2 = 5$
(B) $x^2 + y^2 = 25$
(C) $x^2 + 2y^2 = 34$
(D) $x^2 - y^2 = 7$

B6. The velocity of an object is given by $\vec{v}(t) = (3\sqrt{t}, 4)$. If this object is at the origin when $t = 1$, where was it at $t = 0$?

(A) $(-2,-4)$
(B) $(2,4)$
(C) $\left(\dfrac{3}{2}, 0 \right)$
(D) $\left(-\dfrac{3}{2}, 0 \right)$

B7. Suppose the current world population is 8 billion and the population t years from now is estimated to be $P(t) = 8e^{0.0105t}$ billion people. On the basis of this supposition, the average population of the world, in billions, over the next 25 years will be approximately

(A) 8.096
(B) 8.197
(C) 8.827
(D) 9.148

B8. A beach opens at 8 A.M. and people arrive at a rate of $R(t) = 10 + 40t$ people per hour, where t represents the number of hours the beach has been open. Assuming no one leaves before noon, at what time will there be 100 people there?

(A) 9:45
(B) 10:00
(C) 10:15
(D) 10:30

B9. Suppose the amount of a drug in a patient's bloodstream t hours after intravenous administration is $A(t) = \dfrac{30}{(t+1)^2}$ mg. The average amount in the bloodstream during the first 4 hours is

(A) 6.0 mg
(B) 11.0 mg
(C) 16.6 mg
(D) 24.0 mg

B10. A rumor spreads through a town at the rate of $R(t) = t^2 + 10t$ new people per day t days after it was first heard. Approximately how many people hear the rumor during the second week (from the 7th to the 14th days) after it was first heard?

(A) 359
(B) 1535
(C) 1894
(D) 2219

B11. Oil is leaking from a tanker at the rate of $L(t) = 1000e^{-0.3t}$ gal/hr, where t is given in hours. The total number of gallons of oil that will leak out during the first 8 hours is approximately

(A) 1271
(B) 3031
(C) 3161
(D) 4323

CHALLENGE

B12. The population density of Winnipeg, which is located in the middle of the Canadian prairie, drops dramatically as distance from the center of town increases. This is shown in the following table:

x = distance (in mi) from the center	0	2	4	6	8	10
$f(x) = \left(\text{density hundreds of people/mi}^2\right)$	50	45	40	30	15	5

Using a left Riemann Sum, we can calculate the population living within a 10-mile radius of the center to be approximately

(A) 36,000

(B) 226,200

(C) 691,200

(D) 754,000

B13. If a factory continuously dumps pollutants into a river at the rate of $P(t) = \dfrac{\sqrt{t}}{180}$ tons per day, then the amount dumped after 7 weeks is approximately

(A) 0.07 ton

(B) 0.90 ton

(C) 1.55 tons

(D) 1.27 tons

B14. A roast at 160°F is put into a refrigerator whose temperature is 45°F. The temperature of the roast is cooling at time t at the rate of $R(t) = -9e^{-0.08t}$ °F per minute. The temperature, to the nearest degree F, of the roast 20 minutes after it is put into the refrigerator is

(A) 45°

(B) 70°

(C) 81°

(D) 90°

B15. How long will it take to release 9 tons of pollutant if the rate at which the pollutant is being released is $P(t) = te^{-0.3t}$ tons per week?

(A) 10.2 weeks

(B) 11.0 weeks

(C) 12.1 weeks

(D) 12.9 weeks

B16. What is the total area bounded by the curve $f(x) = x^3 - 4x^2 + 3x$ and the x-axis on the interval $0 \leq x \leq 3$?

(A) −2.25

(B) 2.25

(C) 3

(D) 3.083

B17. Water is leaking from a tank at the rate of $W(t) = -0.1t^2 - 0.3t + 2$ gal/hr. The total amount, in gallons, that will leak out in the next 3 hours is approximately

(A) 2.08

(B) 3.13

(C) 3.48

(D) 3.75

B18. A bacterial culture is growing at the rate of $B(t) = 1000e^{0.03t}$ bacteria in t hours. The total increase in bacterial population during the second hour is approximately

(A) 1015

(B) 1046

(C) 1061

(D) 1077

B19. A website went live at noon, and the rate of hits (visitors/hour) increased continuously for the first 8 hours. The graph of $R(t)$, the rate of hits for the website in visitors/hour, is shown below, where t is the number of hours since the website went live.

Hours (P.M.)

Using a midpoint or a trapezoidal sum, approximate the time when the 200th visitor clicked on the site.

(A) between 2 and 3 P.M.

(B) between 3 and 4 P.M.

(C) between 4 and 5 P.M.

(D) after 5 P.M.

B20. An observer recorded the velocity of an object in motion along the x-axis for 10 seconds. Based on the table below, use a trapezoidal approximation, with 5 subintervals indicated in the table, to estimate how far from its starting point the object came to rest at the end of this time.

t (sec)	0	2	3	5	7	10
$v(t)$ (units/sec)	2	3	1	−1	−2	0

(A) 0 units

(B) 1 unit

(C) 3 units

(D) 4 units

B21. An 18-wheeler is traveling at a speed given by $v(t) = \dfrac{80(t+1)}{t+2}$ mph at time t hours. The fuel economy for the diesel fuel in the truck is given by $f(v) = 4 + 0.01v$ miles per gallon. The amount, in gallons, of diesel fuel used during the first 2 hours is approximately

CHALLENGE

(A) 20

(B) 21.5

(C) 23.1

(D) 24

Answer Key

A1.	**D**	A9.	**A**	B7.	**D**	B15.	**A**
A2.	**A**	A10.	**C**	B8.	**B**	B16.	**D**
A3.	**C**	B1.	**D**	B9.	**A**	B17.	**D**
A4.	**A**	B2.	**D**	B10.	**B**	B18.	**B**
A5.	**B**	B3.	**A**	B11.	**B**	B19.	**B**
A6.	**C**	B4.	**B**	B12.	**C**	B20.	**B**
A7.	**C**	B5.	**B**	B13.	**D**	B21.	**C**
A8.	**C**	B6.	**A**	B14.	**B**		

Answer Explanations

A1. **(D)** Velocity $v(t) = \dfrac{ds}{dt} = 3(t-1)(t-3)$, and changes sign both when $t = 1$ and when $t = 3$.

A2. **(A)** Since $v > 0$ for $0 \leq t \leq 2$, the distance is equal to $\displaystyle\int_0^2 \left(4t^3 + 3t^2 + 5\right) dt$.

A3. **(C)** Since the particle reverses direction when $t = 2$, and $v > 0$ for $t > 2$ but $v < 0$ for $t < 2$, therefore, the total distance is

$$-\int_0^2 \left(3t^2 - 6t\, dt\right) + \int_2^3 \left(3t^2 - 6t\right) dt$$

A4. **(A)** $\displaystyle\int_0^3 \left(3t^2 - 6t\right) dt = 0$, so there is no change in position.

A5. **(B)** Since $v = \sin t$ is positive on $0 < t \leq 2$, the distance covered is $\displaystyle\int_0^2 \sin t\, dt = 1 - \cos 2$.

A6. **(C)** Average velocity $= \dfrac{1}{4-0}\displaystyle\int_0^4 \left(5t - t^2 + 100\right) dt = 104\dfrac{2}{3}$ mph

A7. **(C)** $\vec{v}(t) = \langle t^2 + c_1, e^t + c_2 \rangle$; since $\vec{v}(0) = \langle 2,0 \rangle$, $0^2 + c_1 = 2$ and $e^0 + c_2 = 0$; hence $c_1 = 2$ and $c_2 = -1$.

A8. **(C)** Average velocity $= \dfrac{1}{2-0}\displaystyle\int_0^2 (-32t + 96)\, dt = 64$ ft/sec

A9. **(A)** Be careful! The number of cars is to be measured over a distance of x (not 20) miles. The answer to the question is a function, not a number. Note that choice (C) gives the total number of cars on the entire 20-mile stretch.

A10. **(C)** Since the strip of the city shown in the figure is at a distance r miles from the highway, it is $2\sqrt{9 - r^2}$ miles long and its area is $2\sqrt{9 - r^2}\ \Delta r$. The strip's population is approximately $2(12 - 2r)\sqrt{9 - r^2}\ \Delta r$. The total population of the entire city is *twice* the integral $2\displaystyle\int_0^3 (12 - 2r)\sqrt{9 - r^2}\ dr$ as it includes both halves of the city.

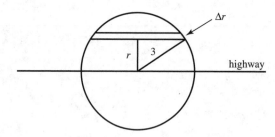

B1. **(D)** The velocity v of the car is linear since its acceleration is constant:

$$a = \frac{dv}{dt} = \frac{(60 - 0)\,\text{mph}}{10\,\text{sec}} = \frac{88\,\text{ft/sec}}{10\,\text{sec}} = 8.8\,\text{ft/sec}^2$$

B2. **(D)** $\mathbf{v} = \dfrac{d\mathbf{R}}{dt}$, so $\mathbf{R}(t) = \left\langle \dfrac{t^2}{2} + c_1, \dfrac{t^2}{2} - t + c_2 \right\rangle$. Since $\mathbf{R}(0) = \langle 0,1 \rangle$, $c_1 = 0$ and $c_2 = 1$.

B3. **(A)** $\mathbf{a} = \mathbf{v}'(t) = \langle 1,1 \rangle$ for all t.

B4. **(B)** $\mathbf{v} = \langle t, t-1 \rangle$. If we define $s(t) = |\mathbf{v}| = \sqrt{t^2 + (t-1)^2} = \sqrt{2t^2 - 2t + 1}$, then

$s'(t) = \dfrac{1}{2}(2t^2 - 2t + 1)^{-1/2} \cdot (4t - 2) = \dfrac{2t - 1}{\sqrt{2t^2 - 2t + 1}}$. Note that $2t^2 - 2t + 1 > 0$ for all t, so

$s'(t)$ is never undefined. Thus, $s'(t) = 0$ when $2t - 1 = 0$ at $t = \dfrac{1}{2}$. A sign analysis based on

the graph of $s'(t)$ indicates that $s'(t) < 0$ for all $t < \dfrac{1}{2}$ and that $s'(t) > 0$ for all $t > \dfrac{1}{2}$. Thus, an

absolute minimum speed occurs at $t = \dfrac{1}{2}$.

B5. **(B)** Since $\mathbf{R} = \langle x, y \rangle$, its slope is $\dfrac{y}{x}$; since $\mathbf{v} = \left\langle \dfrac{dx}{dt}, \dfrac{dy}{dt} \right\rangle$, its slope is $\dfrac{dy}{dx}$.

If $\mathbf{R}$ is perpendicular to $\mathbf{v}$, then $\dfrac{y}{x} \cdot \dfrac{dy}{dx} = -1$, so

$$\frac{y^2}{2} = -\frac{x^2}{2} + C \quad \text{and} \quad x^2 + y^2 = k \, (k > 0)$$

Since $(4,3)$ is on the curve, the equation must be

$$x^2 + y^2 = 25$$

B6. **(A)** The object's position is given by $x(t) = 2t^{\frac{3}{2}} + c_1$, $y(t) = 4t + c_2$. Since the object was at the origin at $t = 1$, $2 \cdot 1^{\frac{3}{2}} + c_1 = 0$ and $4 \cdot 1 + c_2 = 0$, making the position $x(t) = 2t^{\frac{3}{2}} - 2$, $y(t) = 4t - 4$. When $t = 0$, $x(0) = -2$, $y(0) = -4$.

B7. **(D)** $\dfrac{1}{25 - 0} \displaystyle\int_0^{25} P(t)\, dt = \dfrac{1}{25} \int_0^{25} 8e^{0.0105t} \approx 9.148$ (billion people)

B8. **(B)** We want the accumulated number of people to be 100:

$$\int_0^h (10 + 40t)\, dt = 100$$
$$\left. \left(10t + 20t^2\right) \right|_0^h = 100$$
$$20h^2 + 10h - 100 = 0$$
$$10(2h + 5)(h - 2) = 0$$

This occurs at $h = 2$ hours after 8 A.M.

B9. **(A)** Average volume $= \dfrac{1}{4 - 0} \displaystyle\int_0^4 \dfrac{30}{(t + 1)^2}\, dt \approx 6$ mg

B10. **(B)** The number of new people who hear the rumor during the second week is

$$\int_7^{14} \left(t^2 + 10t\right) dt \approx 1535$$

Be careful with the units! The answer is the total change, of course, in $F(t)$ from $t = 7$ to $t = 14$ days, where $F'(t) = t^2 + 10t$.

B11. **(B)** Total gallons $= \displaystyle\lim_{n \to \infty} \sum_{k=1}^{n} 1000e^{-0.3t_k} \, \Delta t = \int_0^8 1000e^{-0.3t} \, dt \approx 3031$

B12. **(C)**

The population equals $\sum$ (area · density). We partition the interval $[0,10]$ along a radius from the center of town into 5 equal subintervals each of width $\Delta r = 2$ miles. We will divide Winnipeg into 5 rings. Each has area equal to (circumference × width), so the area is $2\pi r_k \Delta r$ or $4\pi r_k$. The population in the ring is

$$\left(4\pi r_k\right) \cdot \left(\text{density at } r_k\right) = 4\pi r_k \cdot f\left(r_k\right)$$

A Riemann Sum, using left-hand values, is $4\pi \cdot 0 \cdot 50 + 4\pi \cdot 2 \cdot 45 + 4\pi \cdot 4 \cdot 40 + 4\pi \cdot 6 \cdot 30 + 4\pi \cdot 8 \cdot 15 = 4\pi(90 + 160 + 180 + 120) \approx 6912$ hundred people—or about 691,200 people.

B13. **(D)** The total amount dumped after 7 weeks is

$$\int_0^{49} \frac{\sqrt{t}}{180} \, dt \approx 1.27 \text{ tons}$$

B14. **(B)** The total change in temperature of the roast 20 minutes after it is put into the refrigerator is

$$\int_0^{20} -9e^{-0.08t} \, dt \qquad \text{or} \qquad -89.7°\text{F}$$

Since its temperature was $160°$F when placed into the refrigerator, then 20 minutes later it is $(160 - 89.7)°$F or about $70°$F. Note that the temperature of the refrigerator $(45°$F$)$ is not used in answering the question because it is already "built into" the cooling rate.

B15. **(A)** Let x be the number of weeks required to release 9 tons. Then find the solution to the equation $\int_0^x te^{-0.3t} \, dt = 9$. If your calculator can solve such equations, then use the solve function to solve for x. Otherwise, graph the function $9 - \int_0^x te^{-0.3t} \, dt$ and find the zero (root) of the graph. The answer is about 10.2 weeks.

B16. **(D)** $f(x) = 0$ at $x = 0$, $x = 1$, and $x = 3$. Note that the curve is above the x-axis on $[0,1]$, but below on $[1,3]$, and that the areas for $x < 0$ and $x > 3$ are unbounded.

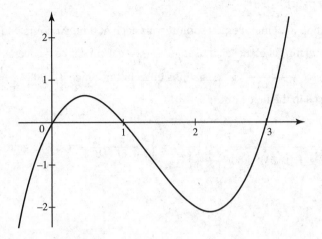

Using the calculator, we get

$$\int_0^3 | x^3 - 4x^2 + 3x | \, dx \approx 3.083$$

B17. **(D)** The FTC yields total change:

$$\int_0^3 \left(-0.1t^2 - 0.3t + 2 \right) dt \approx 3.75 \text{ gal}$$

B18. **(B)** The total change (increase) in population during the second hour is given by

$\int_1^2 1000e^{0.03t} \, dt$. The answer is 1046.

B19. **(B)** The area under the curve represents the number of visitors V. We will estimate the time k when $\int_0^k R(t)dt = 200$ using a Riemann Sum.

The following approximate values can be read from the graph.

t	0	0.5	1	1.5	2	2.5	3	3.5	4
$R(t)$	0	5	10	25	46	70	90	120	150

The table to the right shows one approach, based on midpoint sum.

Hour	Visitors/Hour (midpoint est.)	Total Visitors Since Noon
Noon–1 P.M.	5	5
1–2 P.M.	25	30
2–3 P.M.	70	100
3–4 P.M.	120	220

We estimate that there had been about 100 visitors by 3 P.M. and 220 by 4 P.M., so the 200th visitor arrived between 3 and 4 P.M.

Similarly, using a trapezoidal sum with the above values yields 5, 28, 70, and 115 visitors in the respective time intervals and cumulative totals of 5, 33, 103, and 218, respectively. Thus, this method also gives a time of between 3 and 4 P.M.

B20. **(B)** $\int_0^{10} v(t)dt \approx \left(\dfrac{2+3}{2}\right)2 + \left(\dfrac{3+1}{2}\right)1 + \left(\dfrac{1+(-1)}{2}\right)2 + \left(\dfrac{(-1)+(-2)}{2}\right)2 + \left(\dfrac{(-2)+0}{2}\right)3 =$

1 unit (to the right).

B21. **(C)** We partition [0,2] into n equal subintervals each of time Δt hours. Since the 18-wheeler gets $(4 + 0.01v)$ mi/gal of diesel, it uses $\dfrac{1}{4 + 0.01v}$ gal/mi. Since it covers $v \cdot \Delta t$ miles during Δt hours, it uses $\dfrac{1}{4 + 0.01v} \cdot v \cdot \Delta t$ gallons in Δt hours. Since $v = 80 \cdot \dfrac{t+1}{t+2}$, we see that the diesel fuel used in the first 2 hours is

$$\int_0^2 \frac{80 \cdot \dfrac{t+1}{t+2}}{4 + (0.01) \cdot \left(80 \cdot \dfrac{t+1}{t+2}\right)}\, dt = \int_0^2 \frac{80(t+1)}{4.8t + 8.8}\, dt \approx 23.1 \text{ gallons}$$

9

Differential Equations

Learning Objectives

In this chapter, you will review how to write and solve differential equations, specifically:

→ Writing differential equations to model dynamic situations
→ Understanding a slope field as a graphical representation of a differential equation and its solutions
→ Finding general and particular solutions of separable differential equations
→ Using differential equations to analyze growth and decay

In addition, BC Calculus students will review:

→ Euler's method to estimate numerical solutions
→ Using differential equations to analyze logistic growth and decay

A. Basic Definitions

A *differential equation* (d.e.) is any equation involving a derivative. In Section E of Chapter 5 we solved some simple differential equations. In Example 48, page 225, we were given the velocity at time t of a particle moving along the x-axis:

$$v(t) = \frac{dx}{dt} = 4t^3 - 3t^2 \tag{1}$$

Differential equation

From this we found the antiderivative:

$$x(t) = t^4 - t^3 + C \tag{2}$$

If the initial position (at time $t = 0$) of the particle is $x = 3$, then

$$x(0) = 0 - 0 + C = 3$$

and $C = 3$. So the solution to the initial-value problem is

$$x(t) = t^4 - t^3 + 3 \tag{3}$$

A *solution* of a d.e. is any function that satisfies it. We see from (2) above that the d.e. (1) has an infinite number of solutions—one for each real value of C. We call the family of functions (2) the *general solution* of the d.e. (1). With the given initial condition $x(0) = 3$, we determined C, thus finding the unique solution (3). This is called the *particular solution*.

Note that the particular solution must not only satisfy the differential equation and the initial condition, but the function *must also be differentiable on an interval that contains the initial point*. Features such as vertical tangents or asymptotes restrict the *domain* of the solution. Therefore, even when they are defined by the same algebraic representation, particular solutions with different initial points may have different domains. Determining the proper domain is an important part of finding the particular solution.

Solution

General solution

Particular solution

Domain

In Section A of Chapter 8, we solved more differential equations involving motion along a straight line. In Section B, we found parametric equations for the motion of a particle along a plane curve, given d.e.'s for the components of its acceleration and velocity.

Rate of Change

A differential equation contains derivatives. A derivative gives information about the rate of change of a function. For example:

(1) If P is the size of a population at time t, then we can interpret the d.e.

$$\frac{dP}{dt} = 0.0325P$$

as saying that at any time t the rate at which the population is growing is proportional (3.25%) to its size at that time.

(2) The d.e. $\frac{dQ}{dt} = -(0.000275)Q$ tells us that at any time t the rate at which the quantity Q is decreasing is proportional (0.0275%) to the quantity existing at that time.

(3) In psychology, one typical stimulus-response situation, known as *logarithmic response,* is that in which the response y changes at a rate inversely proportional to the strength of the stimulus x. This is expressed neatly by the differential equation

$$\frac{dy}{dx} = \frac{k}{x} \quad (k \text{ is a constant})$$

If we suppose, further, that there is no response when $x = x_0$, then we have the condition $y = 0$ when $x = x_0$.

(4) We are familiar with the d.e.

$$a = \frac{d^2s}{dt^2} = -32 \text{ ft/sec}^2$$

for the acceleration due to gravity acting on an object at a height s above ground level at time t. The acceleration is the rate of change of the object's velocity.

B. Slope Fields

In this section we investigate differential equations by obtaining a *slope field* or calculator picture that approximates the general solution. We call the graph of a solution of a d.e. a *solution curve.*

The slope field of a d.e. is based on the fact that the d.e. can be interpreted as a statement about the slopes of its solution curves.

> Example 1

The d.e. $\frac{dy}{dx} = y$ tells us that at any point (x,y) on a solution curve the slope of the curve is equal to its y-coordinate. Since the d.e. says that y is a function whose derivative is also y, we know that

$$y = e^x$$

is a solution. In fact, $y = Ce^x$ is a solution of the d.e. for every constant C since $y' = Ce^x = y$. The d.e. $y' = y$ says that, at any point where $y = 1$, say $(0,1)$ or $(1,1)$ or $(5,1)$, the slope of the solution curve is 1; at any point where $y = 3$, say $(0,3)$, $(\ln 3,3)$, or $(\pi,3)$, the slope equals 3; and so on. In Figure 9.1a, we see some small line segments of slope 1 at several points where $y = 1$ and some segments of slope 3 at several points where $y = 3$. In Figure 9.1b, we see the curve of $y = e^x$ with slope segments drawn in as follows:

POINT	SLOPE
$\left(-1, \dfrac{1}{e}\right)$	$\dfrac{1}{e} \approx 0.4$
$\left(-\dfrac{1}{2}, \dfrac{1}{\sqrt{e}}\right)$	$\dfrac{1}{\sqrt{e}} \approx 0.6$
$(0,1)$	1
$\left(\dfrac{1}{2}, \sqrt{e}\right)$	$\sqrt{e} \approx 1.6$
$(1,e)$	$e \approx 2.7$

Figure 9.1a

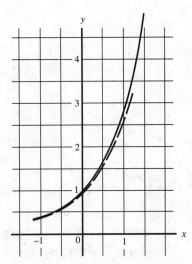

Figure 9.1b

Figure 9.1c is the slope field for the d.e. $\dfrac{dy}{dx} = y$. Slopes at many points are represented by small segments of the tangents at those points. The small segments approximate the solution curves. If we start at any point in the slope field and move so that the slope segments are always tangent to our motion, we will trace a solution curve. The slope field, as mentioned above, closely approximates the family of solutions.

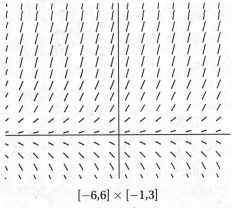

$[-6,6] \times [-1,3]$

Figure 9.1c

> **Example 2**

The slope field for the d.e. $\dfrac{dy}{dx} = \dfrac{1}{x}$ is shown in Figure 9.2.

(a) Carefully draw the solution curve that passes through the point (1,0.5).

(b) Find the general solution for the equation.

$$[0,3] \times [0,3]$$

Figure 9.2

✓ **Solutions**

(a) In Figure 9.2a we started at the point (1,0.5), then moved from segment to segment drawing the curve to which these segments were tangent. The particular solution curve shown is the member of the family of solution curves

$$y = \ln x + C$$

that goes through the point (1,0.5).

Figure 9.2a   **Figure 9.2b**

(b) Since we already know that, if $\dfrac{dy}{dx} = \dfrac{1}{x}$, then $y = \displaystyle\int \dfrac{1}{x}\,dx = \ln x + C$, we are assured of having found the correct general solution in (a). In Figure 9.2b we have drawn several particular solution curves of the given d.e. Note that the vertical distance between any pair of curves is constant.

> **Example 3**

Match each slope field in Figure 9.3 with the proper d.e. from the following set. Find the general solution for each d.e. The particular solution that goes through (0,0) has been sketched in.

(a) $y' = \cos x$

(b) $\dfrac{dy}{dx} = 2x$

(c) $\dfrac{dy}{dx} = 3x^2 - 3$

(d) $y' = -\dfrac{\pi}{2}$

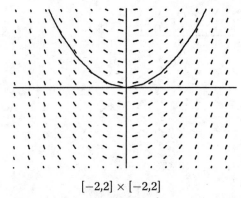

$[-2,2] \times [-2,2]$

Figure 9.3a

$[-2,2] \times [-2,2]$

Figure 9.3b

$[-2\pi,2\pi] \times [-2,2]$

Figure 9.3c

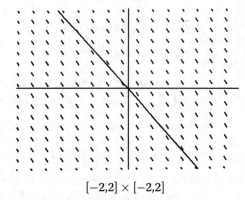

$[-2,2] \times [-2,2]$

Figure 9.3d

✓ **Solutions**

(a)   goes with Figure 9.3c. The solution curves in the family $y = \sin x + C$ are quite obvious.

(b)   goes with Figure 9.3a. The general solution is the family of parabolas $y = x^2 + C$.

(c)   For (c), the slope field is shown in Figure 9.3b. The general solution is the family of cubics
        $y = x^3 - 3x + C.$

(d)   goes with Figure 9.3d; the general solution is the family of lines $y = -\dfrac{\pi}{2}x + C.$

> **Example 4**

(a) Verify that relations of the form $x^2 + y^2 = r^2$ are solutions of the d.e. $\dfrac{dy}{dx} = -\dfrac{x}{y}$.

(b) Using the slope field in Figure 9.4 and your answer to (a), find the particular solution to the d.e. given in (a) that contains point $(4, -3)$.

$$[-7,7] \times [-7,7]$$

Figure 9.4

✓ **Solutions**

(a) By differentiating equation $x^2 + y^2 = r^2$ implicitly, we get $2x + 2y\dfrac{dy}{dx} = 0$, from which $\dfrac{dy}{dx} = -\dfrac{x}{y}$, which is the given d.e.

(b) $x^2 + y^2 = r^2$ describes circles centered at the origin. For initial point $(4, -3)$, $(4)^2 + (-3)^2 = 25$. So $x^2 + y^2 = 25$. However, this is not the particular solution.

A particular solution must be differentiable on an interval containing the initial point. This circle is not differentiable at $(-5,0)$ and $(5,0)$. (The d.e. shows $\dfrac{dy}{dx}$ undefined when $y = 0$, and the slope field shows vertical tangents along the x-axis.) Hence, the particular solution includes only the semicircle in quadrants III and IV.

Solving $x^2 + y^2 = 25$ for y yields $y = \pm\sqrt{25 - x^2}$. The particular solution through point $(4, -3)$ is $y = -\sqrt{25 - x^2}$ with domain $-5 < x < 5$.

Derivatives of Implicitly Defined Functions

In Examples 2 and 3, each d.e. was of the form $\dfrac{dy}{dx} = f(x)$ or $y' = f(x)$. We were able to find the general solution in each case very easily by finding the antiderivative $y = \int f(x)\, dx$.

We now consider d.e.'s of the form $\dfrac{dy}{dx} = f(x,y)$, where $f(x,y)$ is an expression in x and y; that is, $\dfrac{dy}{dx}$ is an implicitly defined function. Example 4 illustrates such a differential equation. Here is another example.

> **Example 5**

Figure 9.5 shows the slope field for

$$\frac{dy}{dx} = x + y \qquad (1)$$

At each point (x,y) the slope is the sum of its coordinates. Three particular solutions have been added, through the points

(a)   (0,0)

(b)   (0,−1)

(c)   (0,−2)

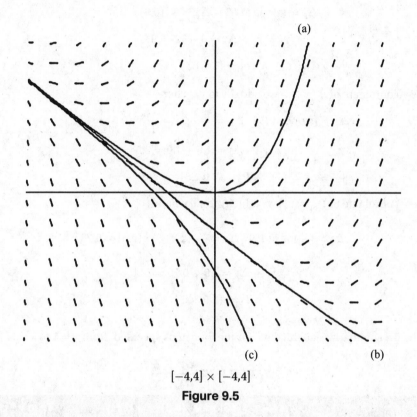

$[-4,4] \times [-4,4]$

Figure 9.5

C. Euler's Method

In Section B, we found solution curves to first-order differential equations *graphically,* using slope fields. Here we will find solutions *numerically,* using *Euler's method* to find points on solution curves.

When we use a slope field we start at an initial point, then move step by step so the slope segments are always tangent to the solution curve. With Euler's method we again select a starting point; but now we *calculate* the slope at that point (from the given d.e.), use the initial point and that slope to locate a new point, use the new point and calculate the slope at it (again from the d.e.) to locate still another point, and so on. The method is illustrated in Example 6.

> ### Example 6

Let $\dfrac{dy}{dx} = \dfrac{3}{x}$. Use Euler's method to approximate the y-values with four steps, starting at point $P_0(1,0)$ and letting $\Delta x = 0.5$.

✓ Solution

The slope at $P_0 = (x_0, y_0) = (1,0)$ is $\dfrac{dy}{dx} = \dfrac{3}{x_0} = \dfrac{3}{1} = 3$. To find the y-coordinate

of $P_1(x_1, y_1)$, we add Δy to y_0. Since $\dfrac{dy}{dx} = \dfrac{\Delta y}{\Delta x}$, we estimate $\Delta y \approx \dfrac{dy}{dx} \cdot \Delta x$:

$$\Delta y = (\text{slope at } P_0) \cdot \Delta x = 3 \cdot (0.5) = 1.5$$

Then
$$y_1 = y_0 + \Delta y = 0 + 1.5 = 1.5$$

and
$$P_1 = (1.5, 1.5)$$

To find the y-coordinate of $P_2(x_2, y_2)$ we add Δy to y_1, where

$$\Delta y = (\text{slope at } P_1) \cdot \Delta x = \dfrac{3}{x_1} \cdot \Delta x = \dfrac{3}{1.5} \cdot (0.5) = 1.0$$

Then
$$y_2 = y_1 + \Delta y = 1.5 + 1.0 = 2.5$$

and
$$P_2 = (2.0, 2.5)$$

To find the y-coordinate of $P_3(x_3, y_3)$ we add Δy to y_2, where

$$\Delta y = (\text{slope at } P_2) \cdot \Delta x = \dfrac{3}{x_2} \cdot \Delta x = \dfrac{3}{2} \cdot (0.5) = 0.75$$

Then
$$y_3 = y_2 + \Delta y = 2.5 + 0.75 = 3.25$$

$$P_3 = (2.5, 3.25)$$

and so on.

The table below summarizes all the data, for the four steps specified, from $x = 1$ to $x = 3$:

TABLE FOR $\dfrac{dy}{dx} = \dfrac{3}{x}$

	x	y	(Slope) · (0.5) = Δy	True y*
P_0	1	0	$(3/1) \cdot (0.5) = 1.5$	0
P_1	1.5	1.5	$(3/1.5) \cdot (0.5) = 1.0$	1.216
P_2	2.0	2.5	$(3/2) \cdot (0.5) = 0.75$	2.079
P_3	2.5	3.25	$(3/2.5) \cdot (0.5) = 0.60$	2.749
P_4	3.0	3.85	$(3/3.0) \cdot (0.5) = 0.50$	3.296

*To three decimal places

The table gives us the numerical solution of $\dfrac{dy}{dx} = \dfrac{3}{x}$ using Euler's method. Figure 9.6a shows the graphical solution, which agrees with the data from the table, for x increasing from 1 to 3 by four steps with Δx equal to 0.5. Figure 9.6b shows this Euler graph and the particular solution of $\dfrac{dy}{dx} = \dfrac{3}{x}$ passing through the point $(1,0)$, which is $y = 3 \ln x$.

Figure 9.6a

Figure 9.6b

We observe that, since y'' for $3 \ln x$ equals $-\dfrac{3}{x^2}$, the true curve is concave down and below the Euler graph.

The last column in the table on page 364 shows the true values (to three decimal places) of y. The Euler approximation for $3 \ln 3$ is 3.85; the true value is 3.296. The Euler approximation with four steps is not very good! However, see what happens as we increase the number n of steps:

n	Euler Approximation	Error
4	3.85	0.554
10	3.505	0.209
20	3.398	0.102
40	3.346	0.050
80	3.321	0.025

Doubling the number of steps cuts the error approximately in half.

❯ Example 7

Given the d.e. $\dfrac{dy}{dx} = x + y$ with initial condition $y(0) = 0$, use Euler's method with $\Delta x = 0.1$ to estimate y when $x = 0.5$.

✓ Solution

Here are the relevant computations:

	x	y	(Slope) $\cdot \Delta x = (x+y) \cdot (0.1) = \Delta y$
P_0	0	0	$0(0.1) = 0$
P_1	0.1	0	$(0.1)(0.1) = 0.01$
P_2	0.2	0.01	$(0.21)(0.1) = 0.021$
P_3	0.3	0.031	$(0.331)(0.1) = 0.033$
P_4	0.4	0.064	$(0.464)(0.1) = 0.046$
P_5	0.5	0.110	

A Caution: Euler's method approximates the solution by substituting short line segments in place of the actual curve. It can be quite accurate when the step sizes are small but only if the curve does not have discontinuities, cusps, or asymptotes.

For example, the reader may verify that the curve $y = \dfrac{1}{2x - 5}$ for the domain $x < \dfrac{5}{2}$ solves the differential equation $\dfrac{dy}{dx} = -2y^2$ with initial condition $y = -1$ when $x = 2$. The domain restriction is important. Recall that a particular solution must be differentiable on an interval containing the initial point. If we attempt to approximate this solution using Euler's method with step size $\Delta x = 1$, the first step carries us to point $(3, -3)$, beyond the discontinuity at $x = \dfrac{5}{2}$ and thus outside the domain of the solution. The accompanying graph (Figure 9.7, page 367) shows that this is nowhere near the solution curve with initial point $y = 1$ when $x = 3$ $\left(\text{and whose domain is } x > \dfrac{5}{2}\right)$.

Here, Euler's method fails because it leaps blindly across the vertical asymptote at $x = \dfrac{5}{2}$.

Always pay attention to the domain of any particular solution.

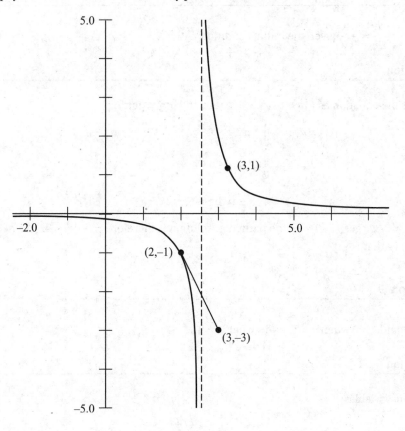

(3,1)

(2,−1)

(3,−3)

−2.0

5.0

5.0

−5.0

Figure 9.7

D. Solving First-Order Differential Equations Analytically

In the preceding sections we solved differential equations graphically, using slope fields, and numerically, using Euler's method. Both methods yield approximations. In this section we review how to solve some differential equations *exactly*.

Separating Variables

A first-order d.e. in *x* and *y* is *separable* if it can be written so that all the terms involving *y* are on one side and all the terms involving *x* are on the other.

Separable

A differential equation has variables separable if it is of the form

$$\frac{dy}{dx} = \frac{f(x)}{g(y)} \quad \text{or} \quad g(y)\,dy - f(x)\,dx = 0$$

The general solution is

$$\int g(y)\,dy - \int f(x)\,dx = C \quad (C \text{ is arbitrary})$$

> **Example 8**

Solve the d.e. $\dfrac{dy}{dx} = -\dfrac{x}{y}$, given the initial condition $y(0) = 2$.

✓ **Solution**

We rewrite the equation as $y\,dy = -x\,dx$. We then integrate, getting

$$\int y\,dy = -\int x\,dx$$

$$\frac{1}{2}y^2 = -\frac{1}{2}x^2 + k$$

$$y^2 + x^2 = C \qquad (\text{where } C = 2k)$$

Since $y(0) = 2$, we get $4 + 0 = C$; the particular solution is therefore $x^2 + y^2 = 4$. (We need to specify above that $y > 0$.)

> **Example 9**

If $\dfrac{ds}{dt} = \sqrt{st}$ and $t = 0$ when $s = 1$, find s when $t = 9$.

✓ **Solution**

We separate variables:

$$\frac{ds}{\sqrt{s}} = \sqrt{t}\,dt$$

then integration yields

$$2s^{1/2} = \frac{2}{3}t^{3/2} + C$$

Using $s = 1$ and $t = 0$, we get $2 = \frac{2}{3} \cdot 0 + C$, so $C = +2$. Then

$$2s^{1/2} = \frac{2}{3}t^{3/2} + 2 \qquad \text{or} \qquad s = \left(\frac{1}{3}t^{3/2} + 1\right)^2$$

When $t = 9$, $s = \left(\frac{1}{3}\left(9^{3/2}\right) + 1\right)^2 = 100$.

> **Example 10**

If $(\ln y)\dfrac{dy}{dx} = \dfrac{y}{x}$, and $y = e$ when $x = 1$, find the value of y greater than 1 that corresponds to $x = e^4$.

✓ **Solution**

Separating, we get $\dfrac{\ln y}{y}\,dy = \dfrac{dx}{x}$. We integrate:

$$\frac{1}{2}\ln^2 y = \ln|x| + C$$

$$\ln y = \pm\sqrt{2\ln|x| + C}$$

$$y = e^{\pm\sqrt{2\ln|x| + C}}$$

Using $y = e$ when $x = 1$, we have $e = e^{\pm\sqrt{2\ln|1| + C}}$, or $e = e^{\pm\sqrt{C}}$. This is true only if $C = 1$, and we choose the positive square root. Hence, $y = e^{\sqrt{2\ln|x| + 1}}$. When $x = e^4$, $y = e^{\sqrt{2\ln|e^4| + 1}} = e^{\sqrt{2 \cdot 4 + 1}} = e^3$.

> **Example 11**

Find the general solution of the differential equation $\dfrac{du}{dv} = e^{v-u}$.

✓ **Solution**

We rewrite $\dfrac{du}{dv} = \dfrac{e^v}{e^u}$ as $e^u\, du = e^v\, dv$.

Taking antiderivatives yields $e^u = e^v + C$, or $u = \ln\left(e^v + C\right)$.

E. Exponential Growth and Decay

We now apply the method of separation of variables to three classes of functions associated with different rates of change. In each of the three cases, we describe the rate of change of a quantity, write the differential equation that follows from the description, then solve—or, in some cases, just give the solution of—the d.e. We list several applications of each case and present relevant problems involving some of the applications.

Case I: Exponential Growth

An interesting special differential equation with wide applications is defined by the following statement: "A positive quantity y increases (or decreases) at a rate that at any time t is proportional to the amount present." It follows that the quantity y satisfies the d.e.

$$\frac{dy}{dt} = ky \qquad (1)$$

Exponential growth

where $k > 0$ if y is increasing and $k < 0$ if y is decreasing.

From (1) it follows that

$$\frac{dy}{y} = k\, dt$$

$$\int \frac{1}{y}\, dy = \int k\, dt$$

$$\ln y = kt + C \qquad (C \text{ is a constant})$$

Then

$$y = e^{kt\, +\, C} = e^{kt} \cdot e^{C}$$

$$= ce^{kt} \qquad \left(\text{where } c = e^{C}\right)$$

If we are given an initial amount y, say y_0 at time $t = 0$, then

$$y_0 = c \cdot e^{k \cdot 0} = c \cdot 1 = c$$

and our law of exponential change

$$y = ce^{kt} \qquad (2)$$

tells us that c is the initial amount of y (at time $t = 0$). If the quantity grows with time, then $k > 0$; if it decays (or diminishes, or decomposes), then $k < 0$. Equation (2) is often referred to as the *law of exponential growth or decay*.

The length of time required for a quantity that is decaying exponentially to be reduced by half is called its *half-life*.

Half-life

> **Example 12**

The population of a country is growing at a rate proportional to its population. If the growth rate per year is 4% of the current population, how long will it take for the population to double?

✓ **Solution**

If the population at time t is P, then we are given that $\dfrac{dP}{dt} = 0.04P$. Substituting in equation (2), we see that the solution is

$$P = P_0 e^{0.04t}$$

where P_0 is the initial population. We seek t when $P = 2P_0$:

$$2P_0 = P_0 e^{0.04t}$$

$$2 = e^{0.04t}$$

$$\ln 2 = 0.04t$$

$$t = \frac{\ln 2}{0.04} \approx 17.33 \text{ years}$$

> **Example 13**

The bacteria in a certain culture increase continuously at a rate proportional to the number present.
(a)   If the number triples in 6 hours, how many will there be in 12 hours?
(b)   In how many hours will the original number quadruple?

✓ **Solutions**

We let N be the number at time t and N_0 the number initially. Then

$$\frac{dN}{dt} = kN \qquad \frac{dN}{N} = k\,dt \qquad \ln N = kt + C \qquad \text{and} \qquad \ln N_0 = 0 + C$$

hence, $C = \ln N_0$. The general solution is then $N = N_0 e^{kt}$, with k still to be determined.

Since $N = 3N_0$ when $t = 6$, we see that $3N_0 = N_0 e^{6k}$ and that $k = \dfrac{1}{6}\ln 3$. Thus

$$N = N_0 e^{(t \ln 3)/6}$$

(a)   When $t = 12$, $N = N_0 e^{2\ln 3} = N_0 e^{\ln 3^2} = N_0 e^{\ln 9} = 9N_0$.
(b)   We let $N = 4N_0$ in the centered equation above, and get

$$4 = e^{(t\ln 3)/6} \qquad \ln 4 = \frac{t}{6}\ln 3 \qquad \text{and} \qquad t = \frac{6\ln 4}{\ln 3} \approx 7.6 \text{ hours}$$

> **Example 14**

Radium-226 decays at a rate proportional to the quantity present. Its half-life is 1612 years. How long will it take for one-quarter of a given quantity of radium-226 to decay?

✓ **Solution**

If $Q(t)$ is the amount present at time t, then it satisfies the equation

$$Q(t) = Q_0 e^{kt} \tag{1}$$

where Q_0 is the initial amount and k is the (negative) factor of proportionality. Since it is given that $Q = \frac{1}{2} Q_0$ when $t = 1612$, equation (1) yields

$$\frac{1}{2} Q_0 = Q_0 e^{k(1612)}$$

$$\frac{1}{2} = e^{1612k}$$

$$k = \frac{\ln \frac{1}{2}}{1612} = -0.00043$$

We now have

$$Q = Q_0 e^{-0.00043t} \tag{2}$$

When one-quarter of Q_0 has decayed, three-quarters of the initial amount remains. We use this fact in equation (2) to find t:

$$\frac{3}{4} Q_0 = Q_0 e^{-0.00043t}$$

$$\frac{3}{4} = e^{-0.00043t}$$

$$t = \frac{\ln \frac{3}{4}}{-0.00043} \approx 669 \text{ years}$$

Applications of Exponential Growth

1. A colony of bacteria may grow at a rate proportional to its size.
2. Other populations, such as those of humans, rodents, or fruit flies, whose supply of food is unlimited may also grow at a rate proportional to the size of the population.
3. Money invested at interest that is compounded continuously accumulates at a rate proportional to the amount present. The constant of proportionality is the interest rate.
4. The demand for certain precious commodities (gas, oil, electricity, valuable metals) has been growing in recent decades at a rate proportional to the existing demand.

    Each of the above quantities (population, amount, demand) is a function of the form ce^{kt} (with $k > 0$). (See Figure 9.8a.)

5. Radioactive isotopes, such as uranium-235, strontium-90, iodine-131, and carbon-14, decay at a rate proportional to the amount still present.
6. If P is the *present value* of a fixed sum of money A due t years from now, where the interest is compounded continuously, then P decreases at a rate proportional to the value of the investment.
7. It is common for the concentration of a drug in the bloodstream to drop at a rate proportional to the existing concentration.
8. As a beam of light passes through murky water or air, its intensity at any depth (or distance) decreases at a rate proportional to the intensity at that depth.

    Each of the above four quantities (5 through 8) is a function of the form ce^{-kt} ($k > 0$). (See Figure 9.8b.)

This is *exponential growth.*

This is *exponential decay.*

As $t \to \infty$, $y \to \infty$.

Figure 9.8a

As $t \to \infty$, $y \to 0$.

Figure 9.8b

> **Example 15**

At a yearly rate of 5% compounded continuously, how long does it take (to the nearest year) for an investment to triple?

✓ **Solution**

If P dollars are invested for t years at 5%, the amount will grow to $A = Pe^{0.05t}$ in t years. We seek t when $A = 3P$:

$$3 = e^{0.05t}$$

$$\frac{\ln 3}{0.05} = t \approx 22 \text{ years}$$

> **Example 16**

One important method of dating fossil remains is to determine what portion of the carbon content of a fossil is the radioactive isotope carbon-14. During life, any organism exchanges carbon with its environment. Upon death this circulation ceases, and the ^{14}C in the organism then decays at a rate proportional to the amount present. The proportionality factor is 0.012% per year.

   When did an animal die, if an archaeologist determines that only 25% of the original amount of ^{14}C is still present in its fossil remains?

✓ **Solution**

The quantity Q of ^{14}C present at time t satisfies the equation

$$\frac{dQ}{dt} = -0.00012Q$$

with solution

$$Q(t) = Q_0 \, e^{-0.00012t}$$

(where Q_0 is the original amount). We are asked to find t when $Q(t) = 0.25Q_0$.

$$0.25Q_0 = Q_0 e^{-0.00012t}$$

$$0.25 = e^{-0.00012t}$$

$$\ln 0.25 = -0.00012t$$

$$-1.386 = -0.00012t$$

$$t \approx 11{,}550$$

Rounding to the nearest 500 years, we see that the animal died approximately 11,500 years ago.

> **Example 17**

In 1970 the world population was approximately 3.5 billion. Since then it has been growing at a rate proportional to the population, and the factor of proportionality has been 1.9% per year. At that rate, in how many years would there be one person per square foot of land? (The land area of Earth is approximately 200,000,000 mi^2, or about 5.5×10^{15} ft^2.)

✓ **Solution**

If $P(t)$ is the population at time t, the problem tells us that P satisfies the equation $\dfrac{dP}{dt} = 0.019P$. Its solution is the exponential growth equation

$$P(t) = P_0 e^{0.019t}$$

where P_0 is the initial population. Letting $t = 0$ for 1970, we have

$$3.5 \times 10^9 = P(0) = P_0 e^0 = P_0$$

Then

$$P(t) = (3.5 \times 10^9)e^{0.019t}$$

The question is: for what t does $P(t)$ equal 5.5×10^{15}? We solve

$$(3.5)(10^9)e^{0.019t} = (5.5)10^{15}$$

$$e^{0.019t} \approx (1.6)10^6$$

Taking the logarithm of each side yields

$$0.019t \approx \ln 1.6 + 6 \ln 10 \approx 14.3$$

$$t \approx 750 \text{ years}$$

where it seems reasonable to round off as we have. Thus, if the human population continued to grow at the present rate, there would be one person for every square foot of land in the year 2720.

Case II: Restricted Growth

The rate of change of a quantity $y = f(t)$ may be proportional, not to the amount present, but to a difference between that amount and a fixed constant. Two situations are to be distinguished: The rate of change is proportional to

Restricted growth

(a) a fixed constant A minus the amount of the quantity present:

$$f'(t) = k\,[A - f(t)]$$

(b) the amount of the quantity present minus a fixed constant A:

$$f'(t) = -k[f(t) - A]$$

where (in both) $f(t)$ is the amount at time t and k and A are both positive. We may conclude that

(a) $f(t)$ is increasing (Figure 9.9a):

$$f(t) = A - ce^{-kt}$$

(b) $f(t)$ is decreasing (Figure 9.9b):

$$f(t) = A + ce^{-kt}$$

for some positive constant c.

$A - c$ is the initial amount; as $t \rightarrow \infty$, $f(t) \rightarrow A$; so A is an upper limit on the size of f.

Figure 9.9a

$A + c$ is the initial amount; as $t \rightarrow \infty$, $f(t) \rightarrow A$; so A is a lower limit on the size of f.

Figure 9.9b

Here is how we solve the d.e. for Case II(a), where $A - y > 0$. If the quantity at time t is denoted by y and k is the positive constant of proportionality, then

$$y' = \frac{dy}{dt} = k(A - y)$$

$$\frac{dy}{A - y} = k \, dt$$

$$-\ln(A - y) = kt + C$$

$$\ln(A - y) = -kt - C$$

$$A - y = e^{-kt} \cdot e^{-C}$$

$$= ce^{-kt}, \text{ where } c = e^{-C}$$

and

$$y = A - ce^{-kt}$$

Case II(b) can be solved similarly.

❯ Example 18

According to Newton's law of cooling, a hot object cools at a rate proportional to the difference between its own temperature and that of its environment. If a roast at room temperature 68°F is put into a 20°F freezer, and if, after 2 hours, the temperature of the roast is 40°F:

(a)   What is its temperature after 5 hours?

(b)   How long will it take for the temperature of the roast to fall to 21°F?

✓ Solutions

This is an example of Case II(b) (the temperature is decreasing toward the limiting temperature 20°F).

(a)   If $R(t)$ is the temperature of the roast at time t, then

$$\frac{dR(t)}{dt} = -k[R(t) - 20] \quad \text{and} \quad R(t) = 20 + ce^{-kt}$$

Since $R(0) = 68°F$, we have

$$68 = 20 + c$$

$$c = 48$$

$$R(t) = 20 + 48e^{-kt}$$

Also, $R(2) = 40°\text{F}$, so $\qquad\qquad 40 = 20 + 48e^{-k \cdot 2}$

and $\qquad\qquad\qquad\qquad\qquad e^{-k} \approx 0.65$

yielding $\qquad\qquad\qquad\qquad R(t) = 20 + 48(0.65)^t \qquad (*)$

and, finally, $\qquad\qquad\qquad R(5) = 20 + 48(0.65)^5 \approx 26°\text{F}$

(b) Equation (*) in part (a) gives the roast's temperature at time t. We must find t when $R = 21$:

$$21 = 20 + 48\,(0.65)^t$$
$$\frac{1}{48} = (0.65)^t$$
$$-\ln 48 = t\ln(0.65)$$
$$t \approx 9 \text{ hours}$$

▶ Example 19

Advertisers generally assume that the rate at which people hear about a product is proportional to the number of people who have not yet heard about it. Suppose that the size of a community is 15,000, that to begin with no one has heard about a product, but that after 6 days 1500 people know about it. How long will it take for 2700 people to have heard of it?

✓ Solution

Let $N(t)$ be the number of people aware of the product at time t. Then we are given that

$$N'(t) = k\big[15,\!000 - N(t)\big]$$

which is Case II(a). The solution of this d.e. is

$$N(t) = 15,\!000 - ce^{-kt}$$

Since $N(0) = 0$, $c = 15,\!000$ and

$$N(t) = 15,\!000(1 - e^{-kt})$$

Since 1500 people know of the product after 6 days, we have

$$1500 = 15,\!000(1 - e^{-6k})$$
$$e^{-6k} = 0.9$$
$$k = \frac{\ln 0.9}{-6} = 0.018$$

We now seek t when $N = 2700$:

$$2700 = 15,\!000(1 - e^{-0.018t})$$
$$0.18 = 1 - e^{-0.018t}$$
$$e^{-0.018t} = 0.82$$
$$t \approx 11 \text{ days}$$

Applications of Restricted Growth

1. Newton's law of heating says that a cold object warms up at a rate proportional to the difference between its temperature and that of its environment. If you put a roast at $68°\text{F}$ into an oven of $400°\text{F}$, then the temperature at time t is $R(t) = 400 - 332e^{-kt}$.

2. Because of air friction, the velocity of a falling object approaches a limiting value L (rather than increasing without bound). The acceleration (rate of change of velocity) is proportional to the difference between the limiting velocity and the object's velocity. If initial velocity is zero, then at time t the object's velocity $V(t) = L\big(1 - e^{-kt}\big)$.

3. If a tire has a small leak, then the air pressure inside drops at a rate proportional to the difference between the inside pressure and the fixed outside pressure O. At time t the inside pressure $P(t) = O + ce^{-kt}$.

BC ONLY

Case III: Logistic Growth

The rate of change of a quantity (for example, a population) may be proportional both to the amount (size) of the quantity and to the difference between a fixed constant A and its amount (size). If $y = f(t)$ is the amount, then

$$\frac{dy}{dt} = ky(A - y) \qquad (1)$$

where k and A are both positive. Equation (1) is called the *logistic differential equation*; it is used to model logistic growth.

The solution of the d.e. (1) is

$$y = \frac{A}{1 + ce^{-Akt}} \qquad (2)$$

for some positive constant c.

In most applications, $c > 1$. In these cases, the initial amount $A/(1 + c)$ is *less* than $A/2$. In all applications, since the exponent of e in the expression for $f(t)$ is negative for all positive t, therefore, as $t \to \infty$,

  (1) $ce^{-Akt} \to 0$
  (2) the denominator of $f(t) \to 1$
  (3) $f(t) \to A$

Thus, A is an upper limit of f in this growth model. When applied to populations, A is called the *carrying capacity* or the *maximum sustainable population*.

Shortly we will solve specific examples of the logistic d.e. (1), instead of obtaining the general solution (2), since the latter is algebraically rather messy. (It is somewhat less complicated to verify that y' in (1) can be obtained by taking the derivative of (2).)

Unrestricted Versus Restricted Growth

In Figures 9.10a and 9.10b, we see the graphs of the growth functions of Cases I and III. The growth function of Case I is known as the *unrestricted* (or *uninhibited* or *unchecked*) model. It is not a very realistic one for most populations. It is clear, for example, that human populations cannot continue endlessly to grow exponentially. Not only is Earth's land area fixed, but also there are limited supplies of food, energy, and other natural resources. The growth function in Case III allows for such factors, which serve to check growth. It is therefore referred to as the *restricted* (or *inhibited*) model.

The two graphs are quite similar at 0. This similarity implies that logistic growth is exponential at the start—a reasonable conclusion since populations are small at the outset.

The S-shaped curve in Case III is often called a *logistic curve*. It shows that the rate of growth y':

  (1) increases slowly for a while; i.e., $y'' > 0$
  (2) attains a maximum when $y = A/2$, at half the upper limit to growth
  (3) then decreases ($y'' < 0$), approaching 0 as y approaches its upper limit

It is not difficult to verify these statements.

Figure 9.10a   **Figure 9.10b**

Applications of Logistic Growth

1. Some diseases spread through a (finite) population P at a rate proportional to the number of people, $N(t)$, infected by time t and the number, $P - N(t)$, not yet infected. Thus $N'(t) = kN(P - N)$ and, for some positive c and k,

$$N(t) = \frac{P}{1 + ce^{-Pkt}}$$

2. A rumor often spreads through a population P according to the formula in (1), where $N(t)$ is the number of people who have heard the rumor and $P - N(t)$ is the number who have not.

3. Bacteria in a culture on a Petri dish grow at a rate proportional to the product of the existing population and the difference between the maximum sustainable population and the existing population. (Replace bacteria on a Petri dish by fish in a small lake, ants confined to a small receptacle, fruit flies supplied with only a limited amount of food, yeast cells, and so on.)

4. Advertisers sometimes assume that sales of a particular product depend on the number of TV commercials for the product and that the rate of increase in sales is proportional both to the existing sales and to the additional sales conjectured as possible.

5. In an autocatalytic reaction a substance changes into a new one at a rate proportional to the product of the amount of the new substance present and the amount of the original substance still unchanged.

> **Example 20**

Because of limited food and space, a squirrel population cannot exceed 1000. It grows at a rate proportional both to the existing population and to the attainable additional population. If there were 100 squirrels 2 years ago and 1 year ago the population was 400, about how many squirrels are there now?

✓ **Solution**

Let P be the squirrel population at time t. It is given that

$$\frac{dP}{dt} = kP(1000 - P) \tag{3}$$

with $P(0) = 100$ and $P(1) = 400$. We seek $P(2)$.

We will find the general solution for the given d.e. (3) by separating the variables:

$$\frac{dP}{P(1000-P)} = k\,dt$$

It can easily be verified, using partial fractions, that

$$\frac{1}{P(1000-P)} = \frac{1}{1000P} + \frac{1}{1000(1000-P)}$$

Now we integrate:

$$\int \frac{dP}{1000P} + \int \frac{dP}{1000(1000-P)} = \int k\,dt$$

getting

$$\ln P - \ln(1000-P) = 1000kt + C$$

or

$$\ln \frac{(1000-P)}{P} = -(1000kt + C)$$

$$\frac{1000-P}{P} = ce^{-1000kt} \quad \left(\text{where } c = e^{-C}\right)$$

$$\frac{1000}{P} - 1 = ce^{-1000kt}$$

$$\frac{1000}{P} = 1 + ce^{-1000kt}$$

$$\frac{P}{1000} = \frac{1}{1+ce^{-1000kt}}$$

and, finally (!),

$$P(t) = \frac{1000}{1+ce^{-1000kt}} \tag{4}$$

Please note that this is precisely the solution "advertised" on page 376 in equation (2), with A equal to 1000.

Now, using our initial condition $P(0) = 100$ in (4), we get

$$\frac{100}{1000} = \frac{1}{1+c} \quad \text{and} \quad c = 9$$

Using $P(1) = 400$, we get

$$400 = \frac{1000}{1+9e^{-1000k}}$$

$$1 + 9e^{-1000k} = 2.5$$

$$e^{-1000k} = \frac{1.5}{9} = \frac{1}{6} \tag{5}$$

Then the particular solution is

$$P(t) = \frac{1000}{1+9(1/6)^t} \tag{6}$$

and $P(2) \approx 800$ squirrels.

$[0,2.5] \times [0,1000]$

Figure 9.11

Figure 9.11 shows the slope field for equation (3), with $k = 0.00179$, which was obtained by solving equation (5). Note that the slopes are the same along any horizontal line, and that they are close to zero initially, reach a maximum at $P = 500$, then diminish again as P approaches its limiting value, 1000. We have superimposed the solution curve for $P(t)$ that we obtained in (6).

▷ Example 21

Suppose a flu-like virus is spreading through a population of 50,000 at a rate proportional both to the number of people already infected and to the number still uninfected. If 100 people were infected yesterday and 130 are infected today:

(a)   write an expression for the number of people $N(t)$ infected after t days

(b)   determine how many will be infected a week from today

(c)   indicate when the virus will be spreading the fastest

✓ Solutions

(a)   We are told that $N'(t) = k \cdot N \cdot (50{,}000 - N)$, that $N(0) = 100$, and that $N(1) = 130$. The d.e. describing logistic growth leads to

$$N(t) = \frac{50{,}000}{1 + ce^{-50{,}000kt}}$$

From $N(0) = 100$, we get

$$100 = \frac{50,000}{1 + c}$$

which yields $c = 499$. From $N(1) = 130$, we get

$$130 = \frac{50,000}{1 + 499e^{-50,000k}}$$

$$130(1 + 499e^{-50,000k}) = 50,000$$

$$e^{-50,000k} = 0.77$$

Then

$$N(t) = \frac{50,000}{1 + 499(0.77)^t}$$

(b)   We must find $N(8)$. Since $t = 0$ represents yesterday:

$$N(8) = \frac{50,000}{1 + 499(0.77)^8} \approx 798 \text{ people}$$

(c)   The virus spreads fastest when $50,000/2 = 25,000$ people have been infected.

CHAPTER SUMMARY AND CAUTION

In this chapter, we considered some simple differential equations and ways to solve them. Our methods have been graphical, numerical, and analytical. Equations that we solved analytically—by antidifferentiation—have been separable.

It is important to realize that, given a first-order differential equation of the type $\frac{dy}{dx} = F(x,y)$, it is the exception, rather than the rule, to be able to find the general solution by analytical methods. Indeed, a great many practical applications lead to differential equations for which no explicit algebraic solution exists.

Practice Exercises

PART A. NO CALCULATOR—DIRECTIONS: Answer these questions *without* using your calculator.

In Questions A1–A10, $a(t)$ denotes the acceleration function, $v(t)$ the velocity function, and $s(t)$ the position or height function at time t. (The acceleration due to gravity is -32 ft/sec^2.)

A1. If $a(t) = 4t - 1$ and $v(1) = 3$, then $v(t)$ equals

   (A) $2t^2 - t$
   (B) $2t^2 - t + 1$
   (C) $2t^2 - t + 2$
   (D) $2t^2 + 1$

A2. If $a(t) = 20t^3 - 6t$, $s(-1) = 2$, and $s(1) = 4$, then $v(t)$ equals

   (A) $5t^4 - 3t^2 + 1$
   (B) $5t^4 - 3t^2 + 3$
   (C) $t^5 - t^3 + t + 3$
   (D) $t^5 - t^3 + 1$

A3. If $a(t) = 20t^3 - 6t$, $s(-1) = 2$, and $s(1) = 4$, then $s(0)$ equals

   (A) 1
   (B) 2
   (C) 3
   (D) 4

A4. A stone is thrown straight up from the top of a building with initial velocity 40 ft/sec and hits the ground 4 seconds later. The height of the building, in feet, is

   (A) 88
   (B) 96
   (C) 112
   (D) 128

A5. A stone is thrown straight up from the top of a building with initial velocity 40 ft/sec and hits the ground 4 seconds later. The maximum height is reached by the stone after

   (A) 4/5 second
   (B) 4 seconds
   (C) 5/4 seconds
   (D) 2 seconds

A6. If a car accelerates from 0 to 60 mph in 10 seconds, what distance does it travel in those 10 seconds? (Assume the acceleration is constant and note that 60 mph = 88 ft/sec.)

   (A) 44 feet
   (B) 88 feet
   (C) 400 feet
   (D) 440 feet

A7. A stone is thrown at a target so that its velocity is $v(t) = 100 - 20t$ ft/sec, where t is measured in seconds. If the stone hits the target in 1 second, then the distance from the sling to the target is

   (A) 80 feet
   (B) 90 feet
   (C) 100 feet
   (D) 110 feet

A8. A stone is thrown straight up from the ground. What should the initial velocity be if you want the stone to reach a height of 100 feet?   `CHALLENGE`

   (A) 50 ft/sec
   (B) 80 ft/sec
   (C) 92 ft/sec
   (D) 96 ft/sec

A9. If the velocity of a car traveling in a straight line at time t is $v(t)$, then the difference in its odometer readings between times $t = a$ and $t = b$ is

(A) $\displaystyle\int_a^b |v(t)|\, dt$

(B) $\displaystyle\int_a^b v(t)\, dt$

(C) the net displacement of the car's position from $t = a$ to $t = b$

(D) the change in the car's position from $t = a$ to $t = b$

A10. If an object is moving up and down along the y-axis with velocity $v(t)$ and $s'(t) = v(t)$, then it is false that $\displaystyle\int_a^b v(t)\, dt$ gives

(A) $s(b) - s(a)$

(B) the total change in $s(t)$ between $t = a$ and $t = b$

(C) the shift in the object's position from $t = a$ to $t = b$

(D) the total distance covered by the object from $t = a$ to $t = b$

A11. Solutions of the differential equation $\dfrac{dy}{dx} = \dfrac{x}{y}$ are of the form

(A) $x^2 - y^2 = C$

(B) $x^2 + y^2 = C$

(C) $x^2 - Cy^2 = 0$

(D) $x^2 = C - y^2$

A12. Find the domain of the particular solution to the differential equation $\dfrac{dy}{dx} = \dfrac{x}{y}$ that passes through point $(-2,1)$.

(A) $x < 0$

(B) $-2 \le x < 0$

(C) $x < -\sqrt{3}$

(D) $|x| > \sqrt{3}$

A13. If $\dfrac{dy}{dx} = \dfrac{y}{2\sqrt{x}}$ and $y = 1$ when $x = 4$, then

(A) $\ln y = 4\sqrt{x} - 8$

(B) $\ln y = \sqrt{x} - 2$

(C) $y = e^{\sqrt{x}}$

(D) $y = e^{\sqrt{x}\, - 2}$

A14. If $\dfrac{dy}{dx} = e^y$ and $y = 0$ when $x = 1$, then

(A) $y = \ln(x)$

(B) $y = \ln(2 - x)$

(C) $y = -\ln(2 - x)$

(D) $y = -\ln(x)$

A15. If $\dfrac{dy}{dx} = \dfrac{x}{\sqrt{9 + x^2}}$ and $y = 5$ when $x = 4$, then y equals

(A) $\sqrt{9 + x^2} - 5$

(B) $\sqrt{9 + x^2}$

(C) $\dfrac{\sqrt{9 + x^2} + 5}{2}$

(D) $\dfrac{\sqrt{9 + x^2}}{2}$

A16. The general solution of the differential equation $x\, dy = y\, dx$ is a family of

(A) circles

(B) hyperbolas

(C) parabolas

(D) none of these

A17. The general solution of the differential equation $\dfrac{dy}{dx} = y$ is a family of

(A) parabolas

(B) lines

(C) ellipses

(D) exponential curves

A18. A function $f(x)$ that satisfies the equations $f(x)f'(x) = x$ and $f(0) = 1$ is

(A) $f(x) = \sqrt{x^2 + 1}$

(B) $f(x) = \sqrt{1 - x^2}$

(C) $f(x) = x$

(D) $f(x) = e^x$

A19. The curve that passes through the point $(1,1)$ and whose slope at any point (x,y) is given by $\dfrac{dy}{dx} = \dfrac{3y}{x}$ has the equation

(A) $3x - 2 = y$

(B) $y^3 = x$

(C) $y = x^3$

(D) $3y^2 = x^2 + 2$

A20. Find the domain of the particular solution to the differential equation $\frac{dy}{dx} = \frac{3y}{x}$ that passes through the point $(1,1)$.

    (A) all real numbers

    (B) $|x| \leq 1$

    (C) $x \neq 0$

    (D) $x > 0$

A21. If $\frac{dy}{dx} = \frac{k}{x}$, where k is a constant, and if $y = 2$ when $x = 1$ and $y = 4$ when $x = e$, then, when $x = 2$, y equals

    (A) 4

    (B) $\ln 8$

    (C) $\ln 2 + 2$

    (D) $\ln 4 + 2$

A22.

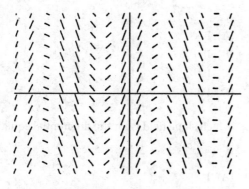

$$[-2\pi, 2\pi] \times [-1.5, 1.5]$$

The slope field shown above is for the differential equation

    (A) $\frac{dy}{dx} = \sin x$

    (B) $\frac{dy}{dx} = -\sin x$

    (C) $\frac{dy}{dx} = \cos x$

    (D) $\frac{dy}{dx} = -\cos x$

A23.

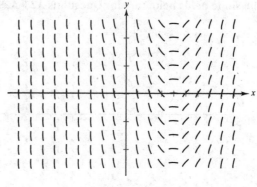

$$[-5,5] \times [-3,3]$$

The slope field shown above is for the differential equation

    (A) $\frac{dy}{dx} = 2x$

    (B) $\frac{dy}{dx} = 2x - 4$

    (C) $\frac{dy}{dx} = 4 - 2x$

    (D) $\frac{dy}{dx} = y$

A24.

$$[-4,4] \times [-4,4]$$

A solution curve has been superimposed on the slope field shown above.
The solution is for the differential equation and initial condition

    (A) $\frac{dy}{dx} = \tan x;\ y(0) = 0$

    (B) $\frac{dy}{dx} = 1 + x^2;\ y(0) = 0$

    (C) $\frac{dy}{dx} = \frac{1}{1+x^2};\ y\left(\frac{\pi}{4}\right) = 1$

    (D) $\frac{dy}{dx} = 1 + y^2;\ y(0) = 0$

The slope fields below are for Questions A25–A29.

I $[-3,3] \times [-3,3]$

II $[-3,3] \times [-3,3]$

III $[-5,5] \times [-5,5]$

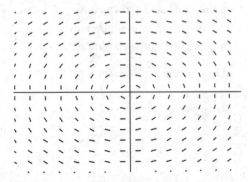

IV $[-3,3] \times [-3,3]$

A25. Which slope field is for the differential equation
$\dfrac{dy}{dx} = y$?

(A) I

(B) II

(C) III

(D) IV

A26. Which slope field is for the differential equation
$\dfrac{dy}{dx} = -\dfrac{x}{y}$?

(A) I

(B) II

(C) III

(D) IV

A27. Which slope field is for the differential equation
$\dfrac{dy}{dx} = \sin x$?

(A) I

(B) II

(C) III

(D) IV

A28. Which slope field is for the differential equation
$\dfrac{dy}{dx} = 2x$?

(A) I

(B) II

(C) III

(D) IV

A29. A particular solution curve of a differential equation whose slope field is shown in II passes through the point $(0,-1)$. The equation is

(A) $y = -e^x$

(B) $y = -e^{-x}$

(C) $y = x^2 - 1$

(D) $y = -\cos x$

BC ONLY

A30. If you use Euler's method with $\triangle x = 0.1$ for the d.e. $\frac{dy}{dx} = x$, with initial value $y(1) = 5$, then, when $x = 1.2$, y is approximately

(A) 5.10

(B) 5.20

(C) 5.21

(D) 6.05

A31. The error in using Euler's method in Question A30 is

(A) 0.005

(B) 0.010

(C) 0.050

(D) 0.500

A32.

Which differential equation has the slope field shown?

(A) $\frac{dy}{dx} = y(y + 2)$

(B) $\frac{dy}{dx} = xy + 2$

(C) $\frac{dy}{dx} = \frac{x}{y + 2}$

(D) $\frac{dy}{dx} = \frac{y}{y + 2}$

A33.

Which function is a possible solution of the slope field shown?

(A) $y = 1 - \ln x$

(B) $y = 1 + \ln x$

(C) $y = 1 + e^x$

(D) $y = 1 + \tan x$

A34.

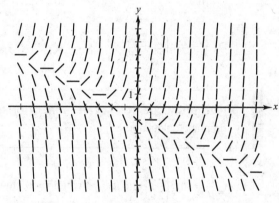

$[-10,10] \times [-7,7]$

Which differential equation has the slope field shown?

(A) $\frac{dy}{dx} = x^2 + 2y^2 + 1$

(B) $\frac{dy}{dx} = 2y + 1$

(C) $\frac{dy}{dx} = x - 2y + 1$

(D) $\frac{dy}{dx} = x + 2y + 1$

A35.

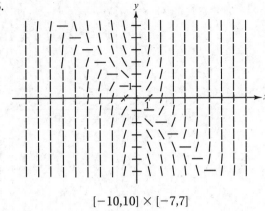

$[-10,10] \times [-7,7]$

A36.

$[-10,10] \times [-4,4]$

Which differential equation has the slope field shown?

(A) $\dfrac{dy}{dx} = x(x+y)$

(B) $\dfrac{dy}{dx} = y(x+y)$

(C) $\dfrac{dy}{dx} = x(x+1)$

(D) $\dfrac{dy}{dx} = x\left(x^2 + y\right)$

Which differential equation has the slope field shown?

(A) $\dfrac{dy}{dx} = \dfrac{2x}{2+y^2}$

(B) $\dfrac{dy}{dx} = \dfrac{2}{2+x^2}$

(C) $\dfrac{dy}{dx} = \dfrac{2y}{2+y^2}$

(D) $\dfrac{dy}{dx} = \dfrac{2}{2+y^2}$

PART B. CALCULATOR ACTIVE—DIRECTIONS: Some of the following questions require the use of a graphing calculator.

B1. If $\dfrac{ds}{dt} = \sin^2\left(\dfrac{\pi}{2}s\right)$ and if $s = 1$ when $t = 0$, then, when $s = \dfrac{3}{2}$, t is equal to

(A) $\dfrac{1}{2}$

(B) $\dfrac{\pi}{2}$

(C) $\dfrac{2}{\pi}$

(D) $-\dfrac{2}{\pi}$

B2. If radium decomposes at a rate proportional to the amount present, then the amount R left after t years, if R_0 is present initially and k is the negative constant of proportionality, is given by

(A) $R = R_0 kt$

(B) $R = R_0 e^{kt}$

(C) $R = e^{R_0 kt}$

(D) $R = e^{R_0 + kt}$

B3. The population of a city increases continuously at a rate proportional, at any time, to the population at that time. The population doubles in 50 years. After 75 years the ratio of the population P to the initial population P_0 is

(A) $\dfrac{9}{4}$

(B) $\dfrac{5}{2}$

(C) $\dfrac{4}{1}$

(D) $\dfrac{2\sqrt{2}}{1}$

B4. If a substance decomposes at a rate proportional to the amount of the substance present,
and if the amount decreases from 40 g to 10 g in
2 hours, then the constant of proportionality is

(A) $-\ln 2$

(B) $-\dfrac{1}{2}$

(C) $-\dfrac{1}{4}$

(D) $\ln \dfrac{1}{4}$

B5. If $\left(g'(x)\right)^2 = g(x)$ for all real x and $g(0) = 0$, $g(4) = 4$,
then $g(1)$ equals

(A) $\dfrac{1}{4}$

(B) $\dfrac{1}{2}$

(C) 2

(D) 4

B6. The solution curve of $\dfrac{dy}{dx} = y$ that passes
through point (2,3) is

(A) $y = e^x + 3$
(B) $y = 0.406e^x$
(C) $y = e^x - \left(e^2 + 3\right)$
(D) $y = e^x/(0.406)$

B7. At any point of intersection of a solution curve
of the d.e. $\dfrac{dy}{dx} = x + y$ and the line $x + y = 0$,
the function y at that point

(A) is equal to 0
(B) is a local maximum
(C) is a local minimum
(D) has a point of inflection

B8.

$$[-2,2] \times [-2,2]$$

The slope field for $F'(x) = e^{-x^2}$ is shown above
with the particular solution $F(0) = 0$ superim-
posed. With a graphing calculator, $\lim\limits_{x \to \infty} F(x)$ to
three decimal places is

(A) 0.886

(B) 0.987

(C) 1.000

(D) ∞

B9. **BC ONLY**

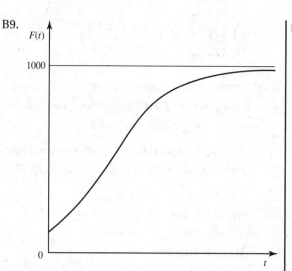

The graph displays logistic growth for a frog
population F. Which differential equation could
be the appropriate model?

(A) $\dfrac{dF}{dt} = 1.5F - 0.003F^2$

(B) $\dfrac{dF}{dt} = 1.5F^2 - 0.003F$

(C) $\dfrac{dF}{dt} = 3F - 0.003F^2$

(D) $\dfrac{dF}{dt} = 0.003F^2 - 3F$

B10. The table shows selected values of the derivative for a differentiable function f.

x	2	3	4	5	6	7
$f'(x)$	2.0	2.5	1.0	−0.5	−1.5	0.5

Given that $f(3) = 100$, use Euler's method with a step size of 2 to estimate $f(7)$.

(A) 102.5
(B) 103
(C) 104
(D) 104.5

B11. A cup of coffee at temperature 180°F is placed on a table in a room at 68°F. The d.e. for its temperature at time t (in minutes) is $\dfrac{dy}{dt} = -0.11(y - 68)$; $y(0) = 180$. After 10 minutes, the temperature (in °F) of the coffee is approximately

(A) 96
(B) 100
(C) 105
(D) 110

B12. A cup of coffee at temperature 180°F is placed on a table in a room at 68°F. The d.e. for its temperature at time t (in minutes) is $\dfrac{dy}{dt} = -0.11(y - 68)$; $y(0) = 180$. Approximately how long does it take the temperature of the coffee to drop to 75°F?

(A) 15 minutes
(B) 18 minutes
(C) 20 minutes
(D) 25 minutes

B13. The concentration of a medication injected into the bloodstream drops at a rate proportional to the existing concentration. If the factor of proportionality is 30% per hour, in approximately how many hours will the concentration be one-tenth of the initial concentration?

(A) 3
(B) $4\dfrac{1}{3}$
(C) $6\dfrac{2}{3}$
(D) $7\dfrac{2}{3}$

B14. Which of the following statements characterize(s) the logistic growth of a population whose limiting value is L and whose initial value is less than $\dfrac{L}{2}$?

I. The rate of growth increases at first.
II. The growth rate attains a maximum when the population equals $\dfrac{L}{2}$.
III. The growth rate approaches 0 as the population approaches L.

(A) I only
(B) I and II only
(C) II and III only
(D) I, II, and III

B15. Which of the following differential equations is not logistic?

(A) $P' = P - P^2$
(B) $\dfrac{dy}{dt} = 0.01y(100 - y)$
(C) $\dfrac{dx}{dt} = 0.8x - 0.004x^2$
(D) $\dfrac{dR}{dt} = 0.16(350 - R)$

B16. Suppose $P(t)$ denotes the size of an animal population at time t and its growth is described by the d.e. $\dfrac{dP}{dt} = 0.002P(1000 - P)$. If the initial population is 200, then the population is growing fastest

(A) initially
(B) when $P = 500$
(C) when $P = 1000$
(D) when $\dfrac{d^2P}{dt^2} > 0$

B17. According to Newton's law of cooling, the temperature of an object decreases at a rate proportional to the difference between its temperature and that of the surrounding air. Suppose a corpse at a temperature of 32°C arrives at a mortuary where the temperature is kept at 10°C. The differential equation is $\frac{dT}{dt} = -k(T - 10)$, where T is the temperature of the corpse (in °C) and t is hours. If the corpse cools to 27°C in 1 hour, then its temperature (in °C) is given by the equation

(A) $T = 22e^{0.205t}$

(B) $T = 10e^{1.163t}$

(C) $T = 10 + 22e^{-0.258t}$

(D) $T = 32e^{-0.169t}$

Answer Key

A1.	**C**	A15.	**B**	A29.	**A**	B7.	**C**
A2.	**A**	A16.	**D**	A30.	**C**	B8.	**A**
A3.	**C**	A17.	**D**	A31.	**B**	B9.	**C**
A4.	**B**	A18.	**A**	A32.	**A**	B10.	**C**
A5.	**C**	A19.	**C**	A33.	**C**	B11.	**C**
A6.	**D**	A20.	**D**	A34.	**D**	B12.	**D**
A7.	**B**	A21.	**D**	A35.	**A**	B13.	**D**
A8.	**B**	A22.	**C**	A36.	**D**	B14.	**D**
A9.	**A**	A23.	**B**	B1.	**C**	B15.	**D**
A10.	**D**	A24.	**D**	B2.	**B**	B16.	**B**
A11.	**A**	A25.	**B**	B3.	**D**	B17.	**C**
A12.	**C**	A26.	**D**	B4.	**A**		
A13.	**D**	A27.	**C**	B5.	**A**		
A14.	**C**	A28.	**A**	B6.	**B**		

Answer Explanations

A1. **(C)** $v(t) = 2t^2 - t + C$; $v(1) = 3$; so $C = 2$.

A2. **(A)** If $a(t) = 20t^3 - 6t$, then

$$v(t) = 5t^4 - 3t^2 + C_1$$

$$s(t) = t^5 - t^3 + C_1 t + C_2$$

Since

$$s(-1) = -1 + 1 - C_1 + C_2 = 2$$

and

$$s(1) = 1 - 1 + C_1 + C_2 = 4$$

therefore

$$2C_2 = 6, C_2 = 3$$
$$C_1 = 1$$

So

$$v(t) = 5t^4 - 3t^2 + 1$$

A3. **(C)** From Answer A2, $s(t) = t^5 - t^3 + t + 3$, so $s(0) = C_2 = 3$.

A4. **(B)** Since $a(t) = -32$, $v(t) = -32t + 40$, and the height of the stone $s(t) = -16t^2 + 40t + C$. When the stone hits the ground, 4 seconds later, $s(t) = 0$, so

$$0 = -16(16) + 40(4) + C$$
$$C = 96 \text{ feet}$$

A5. **(C)** From Answer A4

$$s(t) = -16t^2 + 40t + 96$$

Then

$$s'(t) = -32t + 40$$

which is zero if $t = 5/4$, and that yields maximum height, since $s''(t) = -32$.

A6. **(D)** The velocity $v(t)$ of the car is linear since its acceleration is constant and

$$a(t) = \frac{dv}{dt} = \frac{(60 - 0)\text{ mph}}{10 \text{ sec}} = \frac{88 \text{ ft/sec}}{10 \text{ sec}} = 8.8 \text{ ft/sec}^2$$

$$v(t) = 8.8t + C_1 \quad \text{and} \quad v(0) = 0, \quad \text{so } C_1 = 0$$

$$s(t) = 4.4t^2 + C_2 \quad \text{and} \quad s(0) = 0, \quad \text{so } C_2 = 0$$

$$s(10) = 4.4(10^2) = 440 \text{ feet}$$

A7. **(B)** Since $v = 100 - 20t$, $s = 100t - 10t^2 + C$ with $s(0) = 0$. So $s(1) = 100 - 10 = 90$ feet.

A8. **(B)** Since $v = -32t + v_0$ and $s = -16t^2 + v_0 t$, we solve simultaneously:

$$0 = -32t + v_0$$
$$100 = -16t^2 + v_0 t$$

These yield $t = 5/2$ and $v_0 = 80$ ft/sec.

A9. **(A)** The odometer measures the total trip distance from time $t = a$ to $t = b$ (whether the car moves forward or backward or reverses its direction one or more times from $t = a$ to $t = b$). This total distance is given exactly by $\int_a^b |v(t)|\, dt.$

A10. **(D)** (A), (B), and (C) are all true. (D) is false: see Answer A9.

A11. **(A)** Integrating yields $\frac{y^2}{2} = \frac{x^2}{2} + C$ or $y^2 = x^2 + 2C$ or $y = \pm\sqrt{x^2 + C'}$, where we have replaced the arbitrary constant $2C$ by C'.

A12. **(C)** For initial point $(-2,1)$, $x^2 - y^2 = 3$. The d.e. $\frac{dy}{dx} = \frac{x}{y}$ reveals that the derivative does not exist when $y = 0$, which occurs at $x = \pm\sqrt{3}$. Since the particular solution must be differentiable in an interval containing $x = -2$, the domain is $x < -\sqrt{3}$.

A13. **(D)** We separate variables. $\int\frac{dy}{y} = \frac{1}{2}\int x^{-\frac{1}{2}}\,dx$, so $\ln|y| = \sqrt{x} + C$. The initial point yields $\ln 1 = \sqrt{4} + C$; hence $C = -2$. With $y > 0$, the particular solution is $\ln y = \sqrt{x} - 2$, so $y = e^{\sqrt{x}-2}$.

A14. **(C)** We separate variables. $\int e^{-y}dy = \int dx$, so $-e^{-y} = x + C$. The initial condition yields $-e^0 = 1 + C$, so $C = -2$. Substituting C into the general solution gives $-e^{-y} = x - 2$. Solving for y gives the particular solution $y = -\ln(2 - x)$ with domain $x < 2$.

A15. **(B)** The general solution is $y = \frac{1}{2}\int(9 + x^2)^{-\frac{1}{2}}(2x\,dx) = \sqrt{9 + x^2} + C$; $y = 5$ when $x = 4$ yields $C = 0$.

A16. **(D)** Since $\int\frac{dy}{y} = \int\frac{dx}{x}$, it follows that

$$\ln y = \ln x + C \qquad \text{or} \qquad \ln y = \ln x + \ln k$$

so $y = kx$, with domain $x > 0$ if the initial condition has a positive x-coordinate, or with domain $x < 0$ if the initial condition has a negative x-coordinate. The solution curves are rays that emanate from the origin but do not include the origin since neither x nor y can be zero.

A17. **(D)** $\int\frac{dy}{y} = \int dx$ yields $\ln|y| = x + C$; hence the general solution is $y = ke^x$, $k \neq 0$.

A18. **(A)** We rewrite and separate variables, getting $y\frac{dy}{dx} = x$. The general solution is $y^2 = x^2 + C$ or $f(x) = \pm\sqrt{x^2 + C}$.

A19. **(C)** We are given that $\frac{dy}{dx} = \frac{3y}{x}$. The general solution is $\ln|y| = 3\ln|x| + C$.

Thus, $|y| = c|x^3|$; $y = \pm cx^3$. Since $y = 1$ when $x = 1$, we get $c = 1$.

A20. **(D)** The d.e. $\frac{dy}{dx} = \frac{3y}{x}$ reveals that the derivative does not exist when $x = 0$.

Since the particular solution must be differentiable in an interval containing initial value $x = 1$, the domain is $x > 0$.

A21. **(D)** The general solution is $y = k\ln|x| + C$, and the particular solution is $y = 2\ln|x| + 2$.

$[-2\pi, 2\pi] \times [-1.5, 1.5]$

A22. **(C)** We carefully(!) draw a curve for a solution to the d.e. represented by the slope field. It will be the graph of a member of the family $y = \sin x + C$. Above we have superimposed the graph of the particular solution $y = \sin x - 0.5$.

$[-5, 5] \times [-3, 3]$

A23. **(B)** It's easy to see that the answer must be choice (A), (B), or (C), because the slope field depends only on x: all the slope segments for a given x are parallel. Also, the solution curves in the slope field are all concave up, as they are only for choices (A) and (B). Finally, the solution curves all have a minimum at $x = 2$, which is true only for differential equation (B).

A24. **(D)** The *solution curve* is $y = \tan x$, which we can obtain from the differential equation $\dfrac{dy}{dx} = 1 + y^2$ with the condition $y(0) = 0$ as follows:

$$\frac{dy}{1 + y^2} = dx \qquad \tan^{-1} y = x \qquad y = \tan x + C$$

Since $y(0) = 0$, $C = 0$. Verify that (A) through (C) are incorrect.

NOTE: When matching slope fields and differential equations in Questions A25–A29, keep in mind that if the slope segments along a vertical line are all parallel, signifying equal slopes for a fixed x, then the differential equation can be written as $\dfrac{dy}{dx} = f(x)$. Likewise, if the slope segments along a horizontal line are all parallel, signifying equal slopes for a fixed y, then the differential equation can be written as $\dfrac{dy}{dx} = g(y)$.

A25. **(B)** The slope field for $\dfrac{dy}{dx} = y$ must by II; it is the only one whose slopes are equal along a horizontal line.

A26. **(D)** Of the four slope fields, IV is the only one whose slopes are not equal along either a vertical or a horizontal line (the segments are *not* parallel). Its d.e. therefore cannot be either of type $\dfrac{dy}{dx} = f(x)$ or $\dfrac{dy}{dx} = g(y)$. The d.e. must be implicitly defined—that is, of the form $\dfrac{dy}{dx} = F(x,y)$. So the answer here is IV.

A27. **(C)** Slope fields I and III both have d.e.'s of the type $\dfrac{dy}{dx} = f(x)$. The curves "lurking" in III are trigonometric curves—not so in I.

A28. **(A)** Given $\dfrac{dy}{dx} = 2x$, we immediately obtain the general solution, a family of parabolas, $y = x^2 + C$. (Trace the parabola in I through $(0,0)$, for example.)

A29. **(A)** From Answer A25, we know that the d.e. for slope field II is $\dfrac{dy}{dx} = y$. The general solution is $y = ce^x$. For a solution curve to pass through point $(0,-1)$, we have $-1 = ce^0$ and $c = -1$.

A30. **(C)** Euler's method for $\dfrac{dy}{dx} = x$, starting at $(1,5)$, with $\Delta x = 0.1$, yields

x	y	$(\text{SLOPE})^* \cdot \Delta x = \Delta y$	
1	5	$1 \cdot (0.1) = 0.1$	*The slope is x.
1.1	5.1	$(1.1) \cdot (0.1) = 0.11$	
1.2	5.21		

A31. **(B)** We want to compare the true value of $y(1.2)$ to the estimated value of 5.21 obtained using Euler's method in Solution A30. Solving the d.e. $\dfrac{dy}{dx} = x$ yields $y = \dfrac{x^2}{2} + C$, and initial condition $y(1) = 5$ means that $5 = \dfrac{1^2}{2} + C$, or $C = 4.5$. Hence $y(1.2) = \dfrac{1.2^2}{2} + 4.5 = 5.22$. The error is $5.22 - 5.21 = 0.01$.

A32. **(A)** Slopes depend only on the value of y, and the slope field suggests that $\dfrac{dy}{dx} = 0$ whenever $y = 0$ or $y = -2$.

A33. **(C)** The slope field suggests that the solution function increases (or decreases) without bound as x increases but approaches $y = 1$ as a horizontal asymptote as x decreases.

A34. **(D)** The slope segments are not parallel in either the x or y direction, so the d.e. must include both x and y in the definition; this excludes (B). Next, (A) would result in all positive slopes. This is not the case, so (A) is eliminated. Finally, (C) will have a zero slope at the point $(1,1)$, and (D) will have a zero slope at the point $(1,-1)$. Thus, (D) will create this slope field.

A35. **(A)** The slope segments are not parallel in either the x or y direction, so the d.e. must include both x and y in the definition; this excludes (C). (B) will have zero slopes along the x-axis, so we can eliminate (B). Finally, both (A) and (D) will have zero slopes along the y-axis, and (D) will also have zero slopes at $(1,-1)$ and $(-1,-1)$, eliminating (D). (*NOTE:* (A) will have zero slopes at $(1,-1)$ and $(-1,1)$.)

A36. **(D)** The slope segments are parallel horizontally, meaning that the slopes don't change as x varies; therefore, the d.e. is defined by the y-coordinate only. This excludes (A) and (B). Choice (C) will have zero slopes along the x-axis, whereas (D) will never have zero slopes; thus (D) will create this slope field.

B1. **(C)** We separate variables to get $\csc^2\left(\dfrac{\pi}{2}s\right)ds = dt$. We integrate:

$-\dfrac{2}{\pi}\cot\left(\dfrac{\pi}{2}s\right) = t + C$. With $t = 0$ and $s = 1$, $C = 0$. When $s = \dfrac{3}{2}$, we get $-\dfrac{2}{\pi}\cot\dfrac{3\pi}{4} = t$.

B2. **(B)** Since $\dfrac{dR}{dt} = kR$, $\dfrac{dR}{R} = k\,dt$, and $\ln R = kt + C$. When $t = 0$, $R = R_0$; so $\ln R_0 = C$ or $\ln R = kt + \ln R_0$. Thus

$$\ln R - \ln R_0 = kt; \ln\frac{R}{R_0} = kt \quad \text{or} \quad \frac{R}{R_0} = e^{kt}$$

B3. **(D)** The question gives rise to the differential equation $\dfrac{dP}{dt} = kP$, where $P = 2P_0$ when $t = 50$. We seek $\dfrac{P}{P_0}$ for $t = 75$. We get $\ln\dfrac{P}{P_0} = kt$ with $\ln 2 = 50k$; then

$$\ln\frac{P}{P_0} = \frac{t}{50}\ln 2 \quad \text{or} \quad \frac{P}{P_0} = 2^{t/50}$$

B4. **(A)** We let S equal the amount present at time t; using $S = 40$ when $t = 0$ yields $\ln\dfrac{S}{40} = kt$. Since, when $t = 2$, $S = 10$, we get

$$k = \frac{1}{2}\ln\frac{1}{4} \quad \text{or} \quad \ln\frac{1}{2} \quad \text{or} \quad -\ln 2$$

B5. **(A)** We replace $g(x)$ by y and then solve the equation $\dfrac{dy}{dx} = \pm\sqrt{y}$. We use the constraints given to find the particular solution $2\sqrt{y} = x$ or $2\sqrt{g(x)} = x$.

B6. **(B)** The general solution of $\dfrac{dy}{dx} = y$, or $\dfrac{dy}{y} = dx$ (with $y > 0$), is $\ln y = x + C$ or $y = ce^x$. For a solution to pass through (2,3), we have $3 = ce^2$ and $c = 3/e^2 \approx 0.406$.

B7. **(C)** At a point of intersection, $\dfrac{dy}{dx} = x + y$ and $x + y = 0$. So $\dfrac{dy}{dx} = 0$, which implies that y has a critical point at the intersection. Since $y'' = 1 + \dfrac{dy}{dx} = 1 + (x + y) = 1 + 0 = 1, y'' > 0$ and the function has a local minimum at the point of intersection. [See Figure 9.5, page 363, showing the slope field for $\dfrac{dy}{dx} = x + y$ and the curve $y = e^x - x - 1$ that has a local minimum at (0,0).]

B8. **(A)** Although there is no elementary function (one made up of polynomial, trigonometric, or exponential functions or their inverses) that is an antiderivative of $F'(x) = e^{-x^2}$, we know from the FTC, since $F(0) = 0$, that

$$F(x) = \int_0^x e^{-t^2}\,dt$$

To approximate $\lim\limits_{x\to\infty} F(x)$, use your graphing calculator.

For upper limits of integration $x = 50$ and $x = 60$, answers are identical to 10 decimal places. Rounding to three decimal places yields 0.886.

B9. **(C)** Logistic growth is modeled by equations of the form $\dfrac{dP}{dt} = kP(L - P)$, where L is the upper limit. The graph shows $L = 1000$, so the differential equation must be

$\dfrac{dF}{dt} = kF(1000 - F) = 1000kF - kF^2$. Only equation (C) is of this form ($k = 0.003$).

B10. **(C)** We start with $x = 3$ and $y = 100$. At $x = 3$, $\Delta y \approx \dfrac{dy}{dx} \cdot \Delta x = (2.5)(2) = 5$, moving us to

$x = 3 + 2 = 5$ and $y = 100 + 5 = 105$. From there $\Delta y \approx \dfrac{dy}{dx} \cdot \Delta x = (-0.5)(2) = -1$, so when

$x = 5 + 2 = 7$ we estimate $y = 105 + (-1) = 104$.

B11. **(C)** We separate the variables in the given d.e., then solve:

$$\frac{dy}{y - 68} = -0.11dt$$

$$\ln(y - 68) = -0.11t + c$$

Since $y(0) = 180$, $\ln 112 = c$. Then

$$\ln\frac{y - 68}{112} = -0.11t$$

$$y = 68 + 112e^{-0.11t}$$

When $t = 10$, $y = 68 + 112e^{-1.1} \approx 105°F$.

B12. **(D)** The solution of the d.e. in Question B11, where y is the temperature of the coffee at time t, is

$$y = 68 + 112e^{-0.11t}$$

We find t when $y = 75°F$:

$$75 = 68 + 112e^{-0.11t}$$

$$\frac{7}{112} = e^{-0.11t}$$

$$t = \frac{\ln 7 - \ln 112}{-0.11} \approx 25 \text{ minutes}$$

B13. **(D)** If Q is the concentration at time t, then $\dfrac{dQ}{dt} = -0.30Q$. We separate variables and integrate:

$$\frac{dQ}{Q} = -0.30 \, dt \quad \rightarrow \quad \ln Q = -0.30t + C$$

We let $Q(0) = Q_0$. Then

$$\ln Q = -0.30t + \ln Q_0 \quad \rightarrow \quad \ln\frac{Q}{Q_0} = -0.30t \quad \rightarrow \quad \frac{Q}{Q_0} = e^{-0.30t}$$

We now find t when $Q = 0.1Q_0$:

$$0.1 = e^{-0.30t}$$

$$t = \frac{\ln 0.1}{-0.3} = 7.675 \approx 7\frac{2}{3}\text{hours}$$

B14. **(D)** See pages 376–377 for the characteristics of the logistic model.

B15. **(D)** (A), (B), and (C) are all of the form $y' = ky(A - y)$.

B16. **(B)** The rate of growth, $\dfrac{dP}{dt}$, is greatest when its derivative is 0 and the curve of $y' = \dfrac{dP}{dt}$ is concave down. Since

$$\frac{dP}{dt} = 2P - 0.002P^2$$

therefore

$$\frac{d^2P}{dt^2} = 2 - 0.004P$$

which is equal to 0 if $y'' = \dfrac{2}{0.004}$, or 500, animals. The curve of y' is concave down for all P, since

$$\frac{d}{dt}\left(\frac{d^2P}{dt^2}\right) = -0.004$$

so $P = 500$ is the maximum population.

B17. **(C)** We solve the d.e. given the initial condition $T(0) = 32$:

$$\frac{dT}{T - 10} = -k\,dt$$
$$\ln(T - 10) = -kt + C$$

Using $T(0) = 32$, we get $\ln(22) = C$, so

$$\ln(T - 10) = -kt + \ln 22$$
$$T - 10 = 22e^{-kt}$$

To find k, we use the given information that $T(1) = 27$:

$$27 - 10 = 22e^{-k}$$
$$\frac{17}{22} = e^{-k}$$
$$k = 0.258$$

Therefore $T = 10 + 22e^{-0.258t}$.

10

Sequences and Series

Learning Objectives

In this chapter, you will review infinite series which are for BC Calculus students. Topics include:

→ Tests for determining convergence or divergence
→ Functions defined as power series
→ Maclaurin and Taylor series
→ Estimates of errors

A. Sequences of Real Numbers[‡]

An *infinite sequence* is a function whose domain is the set of positive integers and is often denoted simply by a_n. The sequence defined, for example, by $a_n = \dfrac{1}{n}$ is the set of numbers $1, \dfrac{1}{2}, \dfrac{1}{3}, \dots, \dfrac{1}{n}, \dots$. The elements in this set are called the *terms* of the sequence, and the *nth* or *general* term of this sequence is $\dfrac{1}{n}$.

A sequence a_n *converges* to a finite number L if $\lim\limits_{n\to\infty} a_n = L$. If a_n does not have a (finite) limit, we say the sequence is *divergent*.

⟩ Example 1

Does the sequence $a_n = \dfrac{1}{n}$ converge or diverge?

✓ Solution

$\lim\limits_{n\to\infty} \dfrac{1}{n} = 0$; hence the sequence converges to 0.

⟩ Example 2

Does the sequence $a_n = \dfrac{3n^4 + 5}{4n^4 - 7n^2 + 9}$ converge or diverge?

✓ Solution

$\lim\limits_{n\to\infty} \dfrac{3n^4 + 5}{4n^4 - 7n^2 + 9} = \dfrac{3}{4}$; hence the sequence converges to $\dfrac{3}{4}$.

[‡]Topic will not be tested on the AP examination, but some understanding of the notation and terminology is helpful.

❯ Example 3

Does the sequence $a_n = 1 + \dfrac{(-1)^n}{n}$ converge or diverge?

✓ Solution

$\displaystyle\lim_{n\to\infty} 1 + \dfrac{(-1)^n}{n} = 1$; hence the sequence converges to 1. Note that the terms in the sequence

$0, \dfrac{3}{2}, \dfrac{2}{3}, \dfrac{5}{4}, \dfrac{4}{5}, \dfrac{7}{6}, \ldots$ are alternately smaller and larger than 1. We say this sequence converges to 1 *by oscillation*.

❯ Example 4

Does the sequence $a_n = \dfrac{n^2 - 1}{n}$ converge or diverge?

✓ Solution

Since $\displaystyle\lim_{n\to\infty} \dfrac{n^2 - 1}{n} = \infty$, the sequence diverges (to infinity).

❯ Example 5

Does the sequence $a_n = \sin n$ converge or diverge?

✓ Solution

Because $\displaystyle\lim_{n\to\infty} \sin n$ does not exist, the sequence diverges. However, note that it does not diverge to infinity.

❯ Example 6

Does the sequence $a_n = (-1)^{n+1}$ converge or diverge?

✓ Solution

Because $\displaystyle\lim_{n\to\infty} (-1)^{n+1}$ does not exist, the sequence diverges. Note that the sequence $1, -1, 1, -1, \ldots$ diverges because it oscillates.

B. Infinite Series

B1. Definitions

Infinite series

If a_n is a sequence of real numbers, then an *infinite series* is an expression of the form

$$\sum_{k=1}^{\infty} a_k = a_1 + a_2 + a_3 + \cdots + a_n + \cdots \tag{1}$$

General term

The elements in the sum are called *terms*; a_n is the nth or *general term* of the series.

❯ Example 7

A series of the form $\displaystyle\sum_{k=1}^{\infty}\frac{1}{k^p}$ is called a *p-series*.

p-series

The *p*-series for $p = 2$ is $\displaystyle\sum_{k=1}^{\infty}\frac{1}{k^2}=\frac{1}{1^2}+\frac{1}{2^2}+\frac{1}{3^2}+\cdots+\frac{1}{n^2}+\cdots.$

❯ Example 8

The *p*-series with $p = 1$ is called the *harmonic series*:

Harmonic
series

$$\sum_{k=1}^{\infty}\frac{1}{k}=\frac{1}{1}+\frac{1}{2}+\frac{1}{3}+\cdots+\frac{1}{n}+\cdots$$

❯ Example 9

A *geometric series* has a first term, *a*, and common ratio of terms, *r*:

Geometric
series

$$\sum_{k=1}^{\infty}ar^{k-1}=a+ar+ar^2+\cdots+ar^{n-1}+\cdots$$

If there is a finite number S such that

$$\lim_{n\to\infty}\sum_{k=1}^{\infty}a_k=S$$

then we say that the infinite series is *convergent,* or *converges to S,* or *has the sum S,* and we write, in this case,

Convergent

$$\sum_{k=1}^{\infty}a_k=S$$

When there is no source of confusion, the infinite series (1) may be indicated simply by

$$\sum a_k \quad\text{or}\quad \sum a_n$$

❯ Example 10

Show that the geometric series $1+\frac{1}{2}+\frac{1}{4}+\cdots+\frac{1}{2^n}+\cdots$ converges to 2.

✓ Solution

Let S represent the sum of the series; then:

$$S=\lim_{n\to\infty}\left(1+\frac{1}{2}+\frac{1}{4}+\frac{1}{8}+\cdots+\frac{1}{2^n}\right)$$

$$\frac{1}{2}S=\lim_{n\to\infty}\left(\frac{1}{2}+\frac{1}{4}+\frac{1}{8}+\cdots+\frac{1}{2^{n+1}}\right)$$

Subtraction yields

$$\frac{1}{2}S=\lim_{n\to\infty}\left(1-\frac{1}{2^{n+1}}\right)$$

Hence, $S = 2$.

> **Example 11**

Show that the *harmonic series* $1 + \frac{1}{2} + \frac{1}{3} + \frac{1}{4} + \cdots + \frac{1}{n} + \cdots$ diverges.

✓ Solution

The terms in the series can be grouped as follows:

$$1 + \frac{1}{2} + \left(\frac{1}{3} + \frac{1}{4}\right) + \left(\frac{1}{5} + \frac{1}{6} + \frac{1}{7} + \frac{1}{8}\right) + \left(\frac{1}{9} + \frac{1}{10} + \cdots + \frac{1}{16}\right) + \left(\frac{1}{17} + \cdots + \frac{1}{32}\right) + \cdots$$

This sum clearly exceeds

$$1 + \frac{1}{2} + 2\left(\frac{1}{4}\right) + 4\left(\frac{1}{8}\right) + 8\left(\frac{1}{16}\right) + 16\left(\frac{1}{32}\right) + \cdots$$

which equals

$$1 + \frac{1}{2} + \frac{1}{2} + \frac{1}{2} + \frac{1}{2} + \frac{1}{2} + \cdots$$

Since that sum is not bounded, it follows that $\sum \frac{1}{n}$ diverges to ∞.

B2. Theorems About Convergence or Divergence of Infinite Series

The following theorems are important.

THEOREM 2a.   If $\sum a_k$ converges, then $\lim\limits_{n \to \infty} a_n = 0$.

This provides a convenient and useful test for divergence since it is equivalent to the statement: If a_n does not approach zero, then the series $\sum a_k$ diverges. Note, however, that the converse of Theorem 2a is *not* true. The condition that a_n approach zero is *necessary but not sufficient* for the convergence of the series. The harmonic series $\sum \frac{1}{n}$ is an excellent example of a series whose nth term goes to zero but that diverges (see Example 11 above). The series $\sum \frac{n}{n+1}$ diverges because $\lim\limits_{n \to \infty} a_n = 1$, not zero; the series $\sum \frac{n}{n^2+1}$ does not converge (as will be shown shortly) even though $\lim\limits_{n \to \infty} a_n = 0$.

THEOREM 2b.   A finite number of terms may be added to or deleted from a series without affecting its convergence or divergence; thus

$$\sum_{k=1}^{\infty} a_k \quad \text{and} \quad \sum_{k=m}^{\infty} a_k$$

(where m is any positive integer) both converge or both diverge. (Note that the sums most likely will differ.)

THEOREM 2c.   The terms of a series may be multiplied by a nonzero constant without affecting the convergence or divergence; thus

$$\sum_{k=1}^{\infty} a_k \quad \text{and} \quad \sum_{k=1}^{\infty} c a_k \ (c \neq 0)$$

both converge or both diverge. (Again, the sums will usually differ.)

THEOREM 2d.   If $\sum a_n$ and $\sum b_n$ both converge, so does $\sum \left(a_n + b_n\right)$.

THEOREM 2e.   If the terms of a convergent series are regrouped, the new series converges.

B3. Tests for Convergence of Infinite Series

The *n*th Term Test

If $\lim\limits_{n\to\infty} a_n \neq 0$, then $\sum a_n$ diverges.

   NOTE: When working with series, it's a good idea to start by checking the *n*th Term Test. If the terms don't approach 0, the series cannot converge. This is often the quickest and easiest way to identify a divergent series.

   (Because this is the contrapositive of Theorem 2a on page 402, it's always true. *But beware of the converse! Seeing that the terms do approach 0 does not guarantee that the series must converge. It just means that you need to try other tests.*)

> **Example 12**

Does $\sum \dfrac{n}{2n+1}$ converge or diverge?

✓ **Solution**

Since $\lim\limits_{n\to\infty}\dfrac{n}{2n+1} = \dfrac{1}{2} \neq 0$, the series $\sum \dfrac{n}{2n+1}$ diverges by the *n*th Term Test.

The Geometric Series Test

A geometric series $\sum ar^n$ converges if and only if $|r| < 1$.

   If $|r| < 1$, the sum of the series is $\dfrac{a}{1-r}$.

   The series cannot converge unless it passes the *n*th Term Test; $\lim\limits_{n\to\infty} ar^n = 0$ only if $|r| < 1$. As noted earlier, this is a necessary condition for convergence but may not be sufficient. We now examine the sum using the same technique we employed in Example 10 on page 401:

$$S = \lim_{n\to\infty}\left(a + ar + ar^2 + ar^3 + \cdots + ar^n\right)$$
$$rS = \lim_{n\to\infty}\left(ar + ar^2 + ar^3 + \cdots + ar^n + ar^{n+1}\right)$$
$$(1-r)S = \lim_{n\to\infty}\left(a - ar^{n+1}\right)$$
$$= a - \lim_{n\to\infty} ar^{n+1} \left(\text{and remember: } |r| < 1\right)$$
$$= a$$
$$S = \frac{a}{1-r}$$

> **Example 13**

Does $0.3 + 0.03 + 0.003 + \cdots$ converge or diverge?

✓ Solution

The series $0.3 + 0.03 + 0.003 + \cdots$ is geometric with $a = 0.3$ and $r = 0.1$. Since $|r| < 1$, the series converges, and its sum is

$$S = \frac{a}{1-r} = \frac{0.3}{1-0.1} = \frac{0.3}{0.9} = \frac{1}{3}$$

NOTE: $\frac{1}{3} = 0.333\ldots$, which is the given series.

B4. Tests for Convergence of Nonnegative Series

The series $\sum a_n$ is called a *nonnegative series* if $a_n \geq 0$ for all n.

Integral Test

The Integral Test

Let $\sum a_n$ be a nonnegative series. If $f(x)$ is a continuous, positive, decreasing function and $f(n) = a_n$, then $\sum a_n$ converges if and only if the improper integral $\int_1^\infty f(x)\,dx$ converges.

❯ Example 14

Does $\sum \frac{n}{n^2 + 1}$ converge?

✓ Solution

The associated improper integral is

$$\int_1^\infty \frac{x\,dx}{x^2 + 1}$$

which equals

$$\lim_{b\to\infty} \frac{1}{2}\ln(x^2 + 1)\Big|_1^b = \infty$$

The improper integral and the infinite series both diverge.

❯ Example 15

Test the series $\sum \frac{n}{e^n}$ for convergence.

✓ Solution

$$\int_1^\infty \frac{x}{e^x}dx = \lim_{b\to\infty}\int_1^b xe^{-x}\,dx = \lim_{b\to\infty} -e^{-x}(1+x)\Big|_1^b$$

$$= -\lim_{b\to\infty}\left(\frac{1+b}{e^b} - \frac{2}{e}\right) = \frac{2}{e}$$

by an application of L'Hospital's Rule. Thus $\sum \frac{n}{e^n}$ converges.

The *p*-Series Test

A *p*-series $\sum_{k=1}^{\infty} \dfrac{1}{n^p}$ converges if $p > 1$ but diverges if $p \leq 1$.

p-Series
Test

This follows immediately from the Integral Test and the behavior of improper

integrals of the form $\displaystyle\int_1^{\infty} \dfrac{1}{x^p}\,dx$.

> ### Example 16

Does the series $1 + \dfrac{1}{2^3} + \dfrac{1}{3^3} + \cdots + \dfrac{1}{n^3} + \cdots$ converge or diverge?

✓ Solution

The series $1 + \dfrac{1}{2^3} + \dfrac{1}{3^3} + \cdots + \dfrac{1}{n^3} + \cdots$ is a *p*-series with $p = 3$; hence the series converges by the
p-Series Test.

> ### Example 17

Does the series $\sum \dfrac{1}{\sqrt{n}}$ converge or diverge?

✓ Solution

$\sum \dfrac{1}{\sqrt{n}}$ diverges, because it is a *p*-series with $p = \dfrac{1}{2}$.

The Comparison Test

Comparison
Test

We compare the general term of $\sum a_n$, the nonnegative series we are investigating, with the general term of a series, $\sum u_n$, known to converge or diverge.

(1) If $\sum u_n$ converges and $a_n \leq u_n$, then $\sum a_n$ converges.

(2) If $\sum u_n$ diverges and $a_n \geq u_n$, then $\sum a_n$ diverges.

Any known series can be used for comparison. Particularly useful are *p*-series, which converge if $p > 1$ but diverge if $p \leq 1$, and geometric series, which converge if $|r| < 1$ but diverge if $|r| \geq 1$.

> ### Example 18

Does $\sum \dfrac{1}{1 + n^4}$ converge or diverge?

✓ Solution

Since $\dfrac{1}{1 + n^4} < \dfrac{1}{n^4}$ and the *p*-series $\sum \dfrac{1}{n^4}$ converges, $\sum \dfrac{1}{1 + n^4}$ converges by the Comparison Test.

> **Example 19**

Does the series $\dfrac{1}{\sqrt{2}} + \dfrac{1}{\sqrt{5}} + \dfrac{1}{\sqrt{8}} + \cdots + \dfrac{1}{\sqrt{3n-1}} + \cdots$ converge or diverge?

✓ **Solution**

$\dfrac{1}{\sqrt{2}} + \dfrac{1}{\sqrt{5}} + \dfrac{1}{\sqrt{8}} + \cdots + \dfrac{1}{\sqrt{3n-1}} + \cdots$ diverges, since

$$\frac{1}{\sqrt{3n-1}} > \frac{1}{\sqrt{3n}} = \frac{1}{\sqrt{3} \cdot n^{1/2}}$$

the latter is the general term of the divergent p-series $\sum \dfrac{c}{n^p}$, where $c = \dfrac{1}{\sqrt{3}}$ and $p = \dfrac{1}{2}$.

Remember when using the Comparison Test that you may either discard a finite number of terms or multiply each term by a nonzero constant without affecting the convergence of the series you are testing.

> **Example 20**

Show that $\sum \dfrac{1}{n^n} = 1 + \dfrac{1}{2^2} + \dfrac{1}{3^3} + \cdots + \dfrac{1}{n^n} + \cdots$ converges.

✓ **Solution**

For $n > 2, \dfrac{1}{n^n} < \dfrac{1}{2^n}$ and $\sum \dfrac{1}{2^n}$ is a convergent geometric series with $r = \dfrac{1}{2}$.

The Limit Comparison Test

Limit
Comparison
Test

Let $\sum a_n$ be a nonnegative series that we are investigating. Given $\sum b_n$, a nonnegative series known to be convergent or divergent:

(1) If $\lim\limits_{n\to\infty} \dfrac{a_n}{b_n} = L$, where $0 < L < \infty$, then $\sum a_n$ and $\sum b_n$ both converge or diverge.

(2) If $\lim\limits_{n\to\infty} \dfrac{a_n}{b_n} = 0$ and $\sum b_n$ converges, then $\sum a_n$ converges.

(3) If $\lim\limits_{n\to\infty} \dfrac{a_n}{b_n} = \infty$ and $\sum b_n$ diverges, then $\sum a_n$ diverges.

Any known series can be used for comparison. Particularly useful are p-series, which converge if $p > 1$ but diverge if $p \le 1$, and geometric series, which converge if $|r| < 1$ but diverge if $|r| \ge 1$.

This test is useful when the direct comparisons required by the Comparison Test are difficult to establish or when the behavior of $\sum a_n$ is like that of $\sum b_n$ but the comparison of the individual terms is in the wrong direction necessary for the Comparison Test to be conclusive.

> **Example 21**

Does $\sum \dfrac{1}{2n+1}$ converge or diverge?

✓ **Solution**

This series seems to be related to the divergent harmonic series, but $\frac{1}{2n+1} < \frac{1}{n}$, so the comparison fails. However, the Limit Comparison Test yields:

$$\lim_{n\to\infty} \frac{\frac{1}{2n+1}}{\frac{1}{n}} = \lim_{n\to\infty} \frac{n}{2n+1} = \frac{1}{2}$$

Since $\sum \frac{1}{n}$ diverges, $\sum \frac{1}{2n+1}$ also diverges by the Limit Comparison Test.

The Ratio Test

Ratio Test

Let $\sum a_n$ be a nonnegative series, and let $\lim_{n\to\infty} \frac{a_{n+1}}{a_n} = L$, if it exists. Then $\sum a_n$ converges if $L < 1$ and diverges if $L > 1$.

If $L = 1$, this test is inconclusive; apply one of the other tests.

NOTE: It is good practice, when using the Ratio Test, to first write $\lim_{n\to\infty} \left| \frac{a_{n+1}}{a_n} \right|$; then, if it is known that the ratio is always nonnegative, you may rewrite the limit without the absolute value. However, when using the Ratio Test on a power series (see Examples 33–36), you must retain the absolute value throughout the limit process because it could be possible that $x < 0$.

> **Example 22**

Does $\sum \frac{1}{n!}$ converge or diverge?

✓ **Solution**

$$\lim_{n\to\infty} \frac{a_{n+1}}{a_n} = \lim_{n\to\infty} \frac{\frac{1}{(n+1)!}}{\frac{1}{n!}} = \lim_{n\to\infty} \frac{n!}{(n+1)!} = \lim_{n\to\infty} \frac{1}{n+1} = 0$$

Therefore this series converges by the Ratio Test.

> **Example 23**

Does $\sum \frac{n^n}{n!}$ converge or diverge?

✓ **Solution**

$$\frac{a_{n+1}}{a_n} = \frac{(n+1)^{n+1}}{(n+1)!} \cdot \frac{n!}{n^n} = \frac{(n+1)^n}{n^n}$$

and

$$\lim_{n\to\infty} \left(\frac{n+1}{n} \right)^n = \lim_{n\to\infty} \left(1 + \frac{1}{n} \right)^n = e$$

(See Section E2, page 99.) Since $e > 1$, $\sum \frac{n^n}{n!}$ diverges by the Ratio Test.

> **Example 24**

If the Ratio Test is applied to any *p*-series, $\sum \dfrac{1}{n^p}$, then

$$\frac{a_{n+1}}{a_n} = \frac{\frac{1}{(n+1)^p}}{\frac{1}{n^p}} = \left(\frac{n}{n+1}\right)^p \qquad \text{and} \qquad \lim_{n\to\infty}\left(\frac{n}{n+1}\right)^p = 1 \text{ for all } p$$

But if $p > 1$ then $\sum \dfrac{1}{n^p}$ converges, while if $p \leq 1$ then $\sum \dfrac{1}{n^p}$ diverges. This illustrates the failure of the Ratio Test to resolve the question of convergence when the limit of the ratio is 1.

Root Test

The Root Test

Let $\lim\limits_{n\to\infty} \sqrt[n]{a_n} = L$, if it exists. Then $\sum a_n$ converges if $L < 1$ and diverges if $L > 1$.

If $L = 1$ this test is inconclusive; try one of the other tests.

The decision rule for this test is the same as that for the Ratio Test.

NOTE: The Root Test is not specifically tested on the AP Calculus exam; however, we present it here because it may be helpful in determining convergence.

> **Example 25**

The series $\sum \left(\dfrac{n}{2n+1}\right)^n$ converges by the Root Test since

$$\lim_{n\to\infty} \sqrt[n]{\left(\frac{n}{2n+1}\right)^n} = \lim_{n\to\infty} \frac{n}{2n+1} = \frac{1}{2}$$

Alternating Series

B5. Alternating Series and Absolute Convergence

Any test that can be applied to a nonnegative series can be used for a series all of whose terms are negative. We consider here only one type of series with mixed signs, the so-called *alternating series*. This has the form:

$$\sum_{k=1}^{\infty} (-1)^{k+1} a_k = a_1 - a_2 + a_3 - a_4 + \cdots + (-1)^{k+1} a_k + \cdots$$

where $a_k > 0$. The series

$$1 - \frac{1}{2} + \frac{1}{3} - \frac{1}{4} + \cdots + (-1)^{n+1} \cdot \frac{1}{n} + \cdots$$

is the *alternating harmonic* series.

The Alternating Series Test

Alternating Series Test

An alternating series converges if
(1) $a_{n+1} < a_n$ for all n
(2) $\lim\limits_{n\to\infty} a_n = 0$

> **Example 26**

Does the series $\sum \dfrac{(-1)^{n+1}}{n}$ converge or diverge?

✓ **Solution**

The alternating harmonic series $\sum \frac{(-1)^{n+1}}{n}$ converges since

(1) $\frac{1}{n+1} < \frac{1}{n}$ for all n

(2) $\lim_{n \to \infty} \frac{1}{n} = 0$

▷ **Example 27**

Does the series $\frac{1}{2} - \frac{2}{3} + \frac{3}{4} - \cdots$ converge or diverge?

✓ **Solution**

The series $\frac{1}{2} - \frac{2}{3} + \frac{3}{4} - \cdots$ diverges since we see that $\lim_{n \to \infty} a_n = \lim_{n \to \infty} \frac{n}{n+1}$ is 1, not 0. (By the nth Term Test, page 403, if a_n does not approach 0, then $\sum a_n$ does not converge.)

Absolute Convergence and Conditional Convergence

A series with mixed signs is said to *converge absolutely* (or to be *absolutely convergent*) if the series obtained by taking the absolute values of its terms converges; that is, $\sum a_n$ converges absolutely if $\sum |a_n| = |a_1| + |a_2| + \cdots + |a_n| + \cdots$ converges.

A series that converges but not absolutely is said to *converge conditionally* (or to be *conditionally convergent*). The alternating harmonic series converges conditionally since it converges but does not converge absolutely. (The harmonic series diverges.)

When asked to determine whether an alternating series is absolutely convergent, conditionally convergent, or divergent, it is often advisable to first consider the series of absolute values. Check first for divergence, using the nth Term Test. If that test shows that the series may converge, investigate further, using the tests for nonnegative series. If you find that the series of absolute values converges, then the alternating series is absolutely convergent. If, however, you find that the series of absolute values diverges, then you'll need to use the Alternating Series Test to see whether the series is conditionally convergent.

▷ **Example 28**

Determine whether $\sum \frac{(-1)^n n^2}{n^2 + 9}$ converges absolutely, converges conditionally, or diverges.

✓ **Solution**

We see that $\lim_{n \to \infty} \frac{n^2}{n^2 + 9} = 1$, not 0, so by the nth Term Test the series $\sum \frac{(-1)^n n^2}{n^2 + 9}$ is divergent.

❯ Example 29

Determine whether $\sum \dfrac{\sin \frac{n\pi}{3}}{n^2}$ converges absolutely, converges conditionally, or diverges.

✓ Solution

Note that, since $\left|\sin \dfrac{n\pi}{3}\right| \leq 1$, $\lim\limits_{n\to\infty} \dfrac{\sin \frac{n\pi}{3}}{n^2} = 0$; the series passes the nth Term Test.

Also, $\left|\dfrac{\sin \frac{n\pi}{3}}{n^2}\right| \leq \dfrac{1}{n^2}$ for all n.

But $\dfrac{1}{n^2}$ is the general term of a convergent p-series ($p = 2$), so by the Comparison Test the nonnegative series converges, and therefore the alternating series converges absolutely.

❯ Example 30

Determine whether $\sum \dfrac{(-1)^{n+1}}{\sqrt[3]{n+1}}$ converges absolutely, converges conditionally, or diverges.

✓ Solution

$\sum \dfrac{1}{\sqrt[3]{n+1}}$ is a p-series with $p = \dfrac{1}{3}$, so the nonnegative series diverges.

We see that $\dfrac{1}{\sqrt[3]{(n+1)+1}} < \dfrac{1}{\sqrt[3]{n+1}}$ and $\lim\limits_{n\to\infty} \dfrac{1}{\sqrt[3]{n+1}} = 0$, so the alternating series converges;

hence $\sum \dfrac{(-1)^{n+1}}{\sqrt[3]{n+1}}$ is conditionally convergent.

Approximating the Limit of an Alternating Series

Evaluating the sum of the first n terms of an alternating series, given by $\sum\limits_{k=1}^{\infty} (-1)^{k+1} a_k$, yields an approximation of the limit, L. The error (the difference between the approximation and the true limit) is called the *remainder after n terms* and is denoted by R_n. When an alternating series is first shown to pass the Alternating Series Test, it's easy to place an upper bound on this remainder. Because the terms alternate in sign and become progressively smaller in magnitude, an alternating series converges on its limit by oscillation, as shown in Figure 10.1.

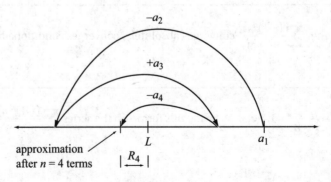

Figure 10.1

Because carrying out the approximation one more term would once more carry us beyond L, we see that the error is always less than that next term. Since $|R_n| < a_{n+1}$, the *Alternating Series Error Bound* for an alternating series is the first term omitted or dropped.

> ### Example 31

The series $\displaystyle\sum_{k=1}^{\infty} \frac{(-1)^{k+1}}{k}$ passes the Alternating Series Test; hence its sum differs from the sum

$$\left(1 - \frac{1}{2} + \frac{1}{3} - \frac{1}{4} + \frac{1}{5} - \frac{1}{6}\right)$$

by less than $\frac{1}{7}$, which is the error bound.

> ### Example 32

Use the Alternating Series Error Bound to determine how many terms must be summed to approximate to three decimal places the value of $1 - \dfrac{1}{4} + \dfrac{1}{9} - \dfrac{1}{16} + \cdots + \dfrac{(-1)^{n+1}}{n^2} + \cdots$.

✓ Solution

Since $\dfrac{1}{(n+1)^2} < \dfrac{1}{n^2}$ and $\displaystyle\lim_{n\to\infty} \dfrac{1}{n^2} = 0$, the series converges by the Alternating Series Test; therefore after summing a number of terms the remainder (Alternating Series Error Bound) will be less than the first omitted term.

We seek n such that $R_n = \dfrac{1}{(n+1)^2} < 0.001$. Thus n must satisfy $(n+1)^2 > 1000$, or $n > 30.623$. Therefore 31 terms are needed for the desired accuracy.

C. Power Series

C1. Definitions; Convergence

An expression of the form

$$\sum_{k=0}^{\infty} a_k x^k = a_0 + a_1 x + a_2 x^2 + \cdots + a_n x^n + \cdots \qquad (1)$$

where the as are constants, is called a *power series in x*; and

$$\sum_{k=0}^{\infty} a_k (x-a)^k = a_0 + a_1(x-a) + a_2(x-a)^2 + \cdots + a_n(x-a)^n + \cdots \qquad (2)$$

Power series

is called a *power series in $(x-a)$*.

If in (1) or (2) x is replaced by a specific real number, then the power series becomes a series of constants that either converges or diverges. Note that series (1) converges if $x = 0$ and series (2) converges if $x = a$.

Radius and Interval of Convergence

Radius of convergence

If power series (1) converges when $|x| < r$ and diverges when $|x| > r$, then r is called the *radius of convergence*. Similarly, r is the radius of convergence of power series (2) if (2) converges when $|x - a| < r$ and diverges when $|x - a| > r$.

BC ONLY

Interval of convergence

The set of *all* values of x for which a power series converges is called its *interval of convergence*. To find the interval of convergence, first determine the radius of convergence by applying the Ratio Test to the series of absolute values. Then check each endpoint to determine whether the series converges or diverges there.

> **Example 33**

Find all x for which the following series converges:

$$1 + x + x^2 + \cdots + x^n + \cdots \qquad (3)$$

✓ **Solution**

By the Ratio Test, the series converges if

$$\lim_{n \to \infty} \left| \frac{u_{n+1}}{u_n} \right| = \lim_{n \to \infty} \left| \frac{x^{n+1}}{x^n} \right| = \lim_{n \to \infty} |x| = |x| < 1$$

Thus, the radius of convergence is 1. The endpoints must be tested separately since the Ratio Test fails when the limit equals 1. When $x = 1$, (3) becomes $1 + 1 + 1 + \cdots$ and diverges; when $x = -1$, (3) becomes $1 - 1 + 1 - 1 + \cdots$ and diverges. Thus the interval of convergence is $-1 < x < 1$.

> **Example 34**

For what x does $\displaystyle\sum_{n=1}^{\infty} \frac{(-1)^{n-1} x^{n-1}}{n+1}$ converge?

✓ **Solution**

$$\lim_{n \to \infty} \left| \frac{u_{n+1}}{u_n} \right| = \lim_{n \to \infty} \left| \frac{x^n}{n+2} \cdot \frac{n+1}{x^{n-1}} \right| = \lim_{n \to \infty} |x| = |x| < 1$$

The radius of convergence is 1. When $x = 1$, we have $\dfrac{1}{2} - \dfrac{1}{3} + \dfrac{1}{4} - \dfrac{1}{5} + \cdots$, an alternating convergent series; when $x = -1$, the series is $\dfrac{1}{2} + \dfrac{1}{3} + \dfrac{1}{4} + \cdots$, which diverges. Thus, the series converges if $-1 < x \le 1$.

> **Example 35**

For what values of x does $\displaystyle\sum_{n=1}^{\infty} \frac{x^n}{n!}$ converge?

✓ **Solution**

$$\lim_{n \to \infty} \left| \frac{u_{n+1}}{u_n} \right| = \lim_{n \to \infty} \left| \frac{x^{n+1}}{(n+1)!} \cdot \frac{n!}{x^n} \right| = \lim_{n \to \infty} \frac{|x|}{n+1} = 0$$

which is always less than 1. Thus the series converges for all x.

> **Example 36**

Find all x for which the following series converges:

$$1 + \frac{x-2}{2^1} + \frac{(x-2)^2}{2^2} + \cdots + \frac{(x-2)^{n-1}}{2^{n-1}} + \cdots \tag{4}$$

✓ **Solution**

$$\lim_{n\to\infty}\left|\frac{u_{n+1}}{u_n}\right| = \lim_{n\to\infty}\left|\frac{(x-2)^n}{2^n} \cdot \frac{2^{n-1}}{(x-2)^{n-1}}\right| = \lim_{n\to\infty}\frac{|x-2|}{2} = \frac{|x-2|}{2}$$

which is less than 1 if $|x-2| < 2$, that is, if $0 < x < 4$. Series (4) converges on this interval and diverges if $|x-2| > 2$, that is, if $x < 0$ or $x > 4$. When $x = 0$, (4) is $1 - 1 + 1 - 1 + \cdots$ and diverges. When $x = 4$, (4) is $1 + 1 + 1 + \cdots$ and diverges. Thus, the interval of convergence is $0 < x < 4$.

> **Example 37**

Find all x for which the series $\displaystyle\sum_{n=1}^{\infty} n!\, x^n$ converges.

✓ **Solution**

$\displaystyle\sum_{n=1}^{\infty} n!\, x^n$ converges only at $x = 0$, since

$$\lim_{n\to\infty}\frac{u_{n+1}}{u_n} = \lim_{n\to\infty}(n+1)x = \infty$$

unless $x = 0$.

C2. Functions Defined by Power Series

Let the function f be defined by

$$f(x) = \sum_{k=0}^{\infty} a_k(x-a)^k$$

$$= a_0 + a_1(x-a) + \cdots + a_n(x-a)^n + \cdots \tag{1}$$

its domain is the interval of convergence of the series.

Functions defined by power series behave very much like polynomials, as indicated by the following properties:

PROPERTY 2a. The function defined by (1) is continuous for each x in the interval of convergence of the series.

PROPERTY 2b. The series formed by differentiating the terms of series (1) converges to $f'(x)$ for each x within the radius of convergence of (1); that is,

$$f'(x) = \sum_{1}^{\infty} ka_k(x-a)^{k-1}$$

$$= a_1 + 2a_2(x-a) + \cdots + na_n(x-a)^{n-1} + \cdots \tag{2}$$

Note that power series (1) and its derived series (2) have the same radius of convergence but not necessarily the same interval of convergence.

> **Example 38**

Let $f(x) = \sum_{k=1}^{\infty} \dfrac{x^k}{k(k+1)} = \dfrac{x}{1 \cdot 2} + \dfrac{x^2}{2 \cdot 3} + \cdots + \dfrac{x^n}{n(n+1)} + \cdots$.

Find the intervals of convergence of the power series for $f(x)$ and $f'(x)$.

✓ **Solution**

$$\lim_{n \to \infty} \left| \dfrac{x^{n+1}}{(n+1)(n+2)} \cdot \dfrac{n(n+1)}{x^n} \right| = |x|$$

also,

$$f(1) = \dfrac{1}{1 \cdot 2} + \dfrac{1}{2 \cdot 3} + \cdots + \dfrac{1}{n(n+1)} + \cdots$$

and

$$f(-1) = -\dfrac{1}{1 \cdot 2} + \dfrac{1}{2 \cdot 3} - \cdots + \dfrac{(-1)^n}{n(n+1)} + \cdots$$

Hence, the power series for f converges if $-1 \le x \le 1$.

For the derivative $f'(x) = \sum_{k=1}^{\infty} \dfrac{x^{k-1}}{k+1} = \dfrac{1}{2} + \dfrac{x}{3} + \dfrac{x^2}{4} + \cdots + \dfrac{x^{n-1}}{n+1} + \cdots$

$$\lim_{n \to \infty} \left| \dfrac{x^n}{n+2} \cdot \dfrac{n+1}{x^{n-1}} \right| = |x|$$

also,

$$f'(1) = \dfrac{1}{2} + \dfrac{1}{3} + \dfrac{1}{4} + \cdots$$

and

$$f'(-1) = \dfrac{1}{2} - \dfrac{1}{3} + \dfrac{1}{4} - \cdots$$

Hence, the power series for f' converges if $-1 \le x < 1$.

Thus, the series given for $f(x)$ and $f'(x)$ have the same radius of convergence, but their intervals of convergence differ.

PROPERTY 2c. The series obtained by integrating the terms of the given series (1) converges to $\int_a^x f(t)\, dt$ for each x within the interval of convergence of (1); that is,

$$\int_a^x f(t)dt = a_0\,(x-a) + \dfrac{a_1\,(x-a)^2}{2} + \dfrac{a_2\,(x-a)^3}{3} + \cdots + \dfrac{a_n(x-a)^{n+1}}{n+1} + \cdots$$

$$= \sum_{k=0}^{\infty} \dfrac{a_k(x-a)^{k+1}}{k+1}$$

> **Example 39**

Let $f(x) = \dfrac{1}{(1-x)^2}$. Show that the power series for $\int f(x)dx$ converges for all values of x in the interval of convergence of the power series for $f(x)$.

✓ **Solution**

Obtain a series for $\dfrac{1}{(1-x)^2}$ by long division.

$$
1 - 2x + x^2 \overline{)1} \quad
\begin{array}{l}
1 + 2x + 3x^2 + 4x^3 + \cdots \\
\hline
\end{array}
$$

$$
\begin{array}{r}
1 - 2x + x^2 \\ \hline
2x - x^2 \\
2x - 4x^2 + 2x^3 \\ \hline
+3x^2 - 2x^3 \\
3x^2 - 6x^3 + 3x^4 \\ \hline
4x^3 - 3x^4
\end{array}
$$

Then,

$$\frac{1}{(1-x)^2} = 1 + 2x + 3x^2 + \cdots + (n+1)x^n + \cdots$$

It can be shown that the interval of convergence is $-1 < x < 1$.

Then by Property 2c

$$\int \frac{1}{(1-x)^2}\,dx = \int \left[1 + 2x + 3x^2 + \cdots + (n+1)x^n + \cdots \right] dx$$

$$\frac{1}{1-x} = c + x + x^2 + x^3 + \cdots + x^{n+1} + \cdots$$

Since when $x = 0$ we see that $c = 1$, we have

$$\frac{1}{1-x} = 1 + x + x^2 + x^3 + \cdots + x^n + \cdots$$

Note that this is a geometric series with ratio $r = x$ and with $a = 1$; if $|x| < 1$,

its sum is $\dfrac{a}{1-r} = \dfrac{1}{1-x}$.

C3. Finding a Power Series for a Function: Taylor and Maclaurin Series

If a function $f(x)$ is representable by a power series of the form

$$c_0 + c_1(x-a) + c_2(x-a)^2 + \cdots + c_n(x-a)^n + \cdots$$

on an interval $|x-a| < r$, then the coefficients are given by

$$c_n = \frac{f^{(n)}(a)}{n!}$$

The series

$$f(x) = f(a) + f'(a)(x-a) + \frac{f''(a)}{2!}(x-a)^2 + \cdots + \frac{f^{(n)}(a)}{n!}(x-a)^n + \cdots$$

Taylor Series

is called the *Taylor series* of the function f about the number a. There is never more than one power series in $(x-a)$ for $f(x)$. It is required that the function and all its derivatives exist at $x = a$ if the function $f(x)$ is to generate a Taylor series expansion.

When $a = 0$ we have the special series

$$f(x) = f(0) + f'(0)x + \frac{f''(0)}{2!}x^2 + \cdots + \frac{f^{(n)}(0)}{n!}x^n + \cdots$$

Maclaurin Series

called the *Maclaurin series* of the function f; this is the expansion of f about $x = 0$.

❯ Example 40

Find the Maclaurin series for $f(x) = e^x$.

✓ Solution

Here $f'(x) = e^x, \ldots, f^{(n)}(x) = e^x, \ldots$, for all n. Then

$$f'(0) = 1, \ldots, f^{(n)}(0) = 1, \ldots$$

for all n, making the coefficients $c_n = \dfrac{1}{n!}$:

$$e^x = 1 + x + \frac{x^2}{2!} + \frac{x^3}{3!} + \cdots + \frac{x^n}{n!} + \cdots$$

❯ Example 41

Find the Maclaurin expansion for $f(x) = \sin x$.

✓ Solution

$$f(x) = \sin x \qquad\qquad f(0) = 0$$

$$f'(x) = \cos x \qquad\qquad f'(0) = 1$$

$$f''(x) = -\sin x \qquad\qquad f''(0) = 0$$

$$f^{(3)}(x) = -\cos x \qquad\qquad f^{(3)}(0) = -1$$

$$f^{(4)}(x) = \sin x \qquad\qquad f^{(4)}(0) = 0$$

Thus,

$$\sin x = x - \frac{x^3}{3!} + \frac{x^5}{5!} - \cdots + (-1)^{n-1}\frac{x^{2n-1}}{(2n-1)!} + \cdots$$

❯ Example 42

Find the Maclaurin series for $f(x) = \dfrac{1}{1-x}$.

✓ Solution

$$f(x) = (1-x)^{-1} \qquad\qquad f(0) = 1$$

$$f'(x) = (1-x)^{-2} \qquad\qquad f'(0) = 1$$

$$f''(x) = 2(1-x)^{-3} \qquad\qquad f''(0) = 2$$

$$f'''(x) = 3!(1-x)^{-4} \qquad\qquad f'''(0) = 3!$$

$$\vdots \qquad\qquad\qquad \vdots$$

$$f^{(n)}(x) = n!(1-x)^{-(n+1)} \qquad\qquad f^{(n)}(0) = n!$$

Then

$$\frac{1}{1-x} = 1 + x + x^2 + x^3 + \cdots + x^n + \cdots$$

Note that this agrees exactly with the power series in x obtained by different methods in Example 39.

> **Example 43**

Find the Taylor series for the function $f(x) = \ln x$ about $x = 1$.

✓ **Solution**

$$f(x) = \ln x \qquad\qquad f(1) = \ln 1 = 0$$

$$f'(x) = \frac{1}{x} \qquad\qquad f'(1) = 1$$

$$f''(x) = -\frac{1}{x^2} \qquad\qquad f''(1) = -1$$

$$f^{(3)}(x) = \frac{2}{x^3} \qquad\qquad f^{(3)}(1) = 2$$

$$f^{(4)}(x) = \frac{-3!}{x^4} \qquad\qquad f^{(4)}(1) = -3!$$

$$\vdots \qquad\qquad \vdots$$

$$f^{(n)}(x) = \frac{(-1)^{n-1}(n-1)!}{x^n} \qquad f^{(n)}(1) = (-1)^{n-1}(n-1)!$$

Then

$$\ln x = (x-1) - \frac{(x-1)^2}{2} + \frac{(x-1)^3}{3} - \frac{(x-1)^4}{4}$$

$$+ \cdots + \frac{(-1)^{n-1}(x-1)^n}{n} + \cdots$$

Common Maclaurin Series

We list here for reference some frequently used Maclaurin series expansions together with their intervals of convergence.

For the AP exam, students are expected to know the Maclaurin series for $f(x) = \sin x$, $f(x) = \cos x$, $f(x) = e^x$, and the geometric series $f(x) = \frac{1}{1-x}$. These Maclaurin series are listed in the table on page 418 as (1), (2), (3), and (4), respectively.

BC ONLY

Function	Maclaurin Series	Interval of Convergence	
$\sin x$	$x - \dfrac{x^3}{3!} + \dfrac{x^5}{5!} - \cdots + \dfrac{(-1)^{n-1} x^{2n-1}}{(2n-1)!} + \cdots$	$-\infty < x < \infty$	(1)
$\cos x$	$1 - \dfrac{x^2}{2!} + \dfrac{x^4}{4!} - \cdots + \dfrac{(-1)^{n-1} x^{2n-2}}{(2n-2)!} + \cdots$	$-\infty < x < \infty$	(2)
e^x	$1 + x + \dfrac{x^2}{2!} + \dfrac{x^3}{3!} + \cdots + \dfrac{x^n}{n!} + \cdots$	$-\infty < x < \infty$	(3)
$\dfrac{1}{1-x}$	$1 + x + x^2 + x^3 + \cdots + x^n + \cdots$	$-1 < x < 1$	(4)
$\ln(1+x)$	$x - \dfrac{x^2}{2} + \dfrac{x^3}{3} - \dfrac{x^4}{4} + \cdots + \dfrac{(-1)^{n-1} x^n}{n} + \cdots$	$-1 < x \le 1$	(5)
$\tan^{-1} x$	$x - \dfrac{x^3}{3} + \dfrac{x^5}{5} - \dfrac{x^7}{7} + \cdots + \dfrac{(-1)^{n-1} x^{2n-1}}{2n-1} + \cdots$	$-1 \le x \le 1$	(6)

Functions That Generate No Series

Note that the following functions are among those that fail to generate a specific series in $(x - a)$ because the function and/or one or more derivatives do not exist at $x = a$:

Function	Series It Fails to Generate
$\ln x$	about 0
$\ln(x-1)$	about 1
$\sqrt{x-2}$	about 2
$\sqrt{x-2}$	about 0
$\tan x$	about $\dfrac{\pi}{2}$
$\sqrt{1+x}$	about -1

C4. Approximating Functions with Taylor and Maclaurin Polynomials

The function $f(x)$ at the point $x = a$ is approximated by a *Taylor polynomial* $P_n(x)$ of order n:

$$f(x) \approx P_n(x) = f(a) + f'(a)(x - a) + \frac{f''(a)}{2!}(x - a)^2 + \cdots + \frac{f^{(n)}(a)}{n!}(x - a)^n$$

The Taylor polynomial $P_n(x)$ and its first n derivatives all agree at a with f and its first n derivatives. The *order* of a Taylor polynomial is the order of the highest derivative, which is also the polynomial's last term.

In the special case where $a = 0$, the *Maclaurin polynomial* of order n that approximates $f(x)$ is

$$P_n(x) = f(0) + f'(0)x + \frac{f''(0)}{2!}x^2 + \cdots + \frac{f^{(n)}(0)}{n!}x^n$$

The Taylor polynomial $P_1(x)$ at $x = 0$ is the tangent-line approximation to $f(x)$ near zero given by

$$f(x) \approx P_1(x) = f(0) + f'(0)x$$

It is the "best" linear approximation to f at 0, discussed at length in Chapter 4, Section L.

A NOTE ON ORDER AND DEGREE

A Taylor polynomial has *degree* n if it has powers of $(x - a)$ up through the nth. If $f^{(n)}(a) = 0$, then the degree of $P_n(x)$ is less than n. Note, for instance, in Example 45, that the second-order polynomial $P_2(x)$ for the function $\sin x$ (which is identical with $P_1(x)$) is $x + 0 \cdot \dfrac{x^2}{2!}$, or just x, which has degree 1, not 2.

> ## Example 44

Find the Taylor polynomial of order 4 at 0 for $f(x) = e^{-x}$. Use this to approximate $f(0.25)$.

✓ Solution

The first four derivatives are $-e^{-x}$, e^{-x}, $-e^{-x}$, and e^{-x}; at $a = 0$, these equal -1, 1, -1, and 1, respectively. The approximating Taylor polynomial of order 4 is therefore

$$e^{-x} \approx 1 - x + \frac{1}{2!}x^2 - \frac{1}{3!}x^3 + \frac{1}{4!}x^4$$

With $x = 0.25$ we have

$$e^{-0.25} \approx 1 - 0.25 + \frac{1}{2!}(0.25)^2 - \frac{1}{3!}(0.25)^3 + \frac{1}{4!}(0.25)^4$$

$$\approx 0.7788$$

This approximation of $e^{-0.25}$ is correct to four places.

In Figure 10.2 we see the graphs of $f(x)$ and of the Taylor polynomials:

$$P_0(x) = 1$$
$$P_1(x) = 1 - x$$
$$P_2(x) = 1 - x + \frac{x^2}{2!}$$
$$P_3(x) = 1 - x + \frac{x^2}{2!} - \frac{x^3}{3!}$$
$$P_4(x) = 1 - x + \frac{x^2}{2!} - \frac{x^3}{3!} + \frac{x^4}{4!}$$

$[0,1.3] \times [0,1]$

Figure 10.2

Notice how closely $P_4(x)$ hugs $f(x)$ even as x approaches 1. Since the series can be shown to converge for $x > 0$ by the Alternating Series Test, the error in $P_4(x)$ is less than the magnitude of the first omitted term, $\dfrac{x^5}{5!}$, or $\dfrac{1}{120}$ at $x = 1$. In fact, $P_4(1) = 0.375$ to three decimal places, close to $e^{-1} \approx 0.368$.

> **Example 45**

(a)   Find the Taylor polynomials P_1, P_3, P_5, and P_7 at $x = 0$ for $f(x) = \sin x$.

(b)   Graph f and all four polynomials in $[-2\pi, 2\pi] \times [-2,2]$.

(c)   Approximate $\sin \dfrac{\pi}{3}$ using each of the four polynomials.

✓ **Solutions**

(a)   The derivatives of the sine function at 0 are given by the following table:

order of deriv	0	1	2	3	4	5	6	7
deriv of $\sin x$	$\sin x$	$\cos x$	$-\sin x$	$-\cos x$	$\sin x$	$\cos x$	$-\sin x$	$-\cos x$
deriv of $\sin x$ at 0	0	1	0	-1	0	1	0	-1

From the table we know that

$$P_1(x) = x$$

$$P_3(x) = x - \frac{x^3}{3!}$$

$$P_5(x) = x - \frac{x^3}{3!} + \frac{x^5}{5!}$$

$$P_7(x) = x - \frac{x^3}{3!} + \frac{x^5}{5!} - \frac{x^7}{7!}$$

BC ONLY

(b) Figure 10.3a shows the graphs of $\sin x$ and the four polynomials. In Figure 10.3b we see graphs only of $\sin x$ and $P_7(x)$, to exhibit how closely P_7 "follows" the sine curve.

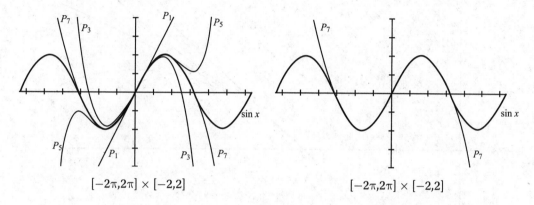

$[-2\pi, 2\pi] \times [-2,2]$

Figure 10.3a

$[-2\pi, 2\pi] \times [-2,2]$

Figure 10.3b

(c) To four decimal places, $\sin \frac{\pi}{3} = 0.8660$. Evaluating the polynomials at $\frac{\pi}{3}$, we get

$$P_1\left(\frac{\pi}{3}\right) = 1.0472 \quad P_3\left(\frac{\pi}{3}\right) = 0.8558 \quad P_5\left(\frac{\pi}{3}\right) = 0.8663 \quad P_7\left(\frac{\pi}{3}\right) = 0.8660$$

We see that P_7 is correct to four decimal places.

> **Example 46**

(a) Find the Taylor polynomials of degrees 0, 1, 2, and 3 generated by $f(x) = \ln x$ at $x = 1$.
(b) Graph f and the four polynomials on the same set of axes.
(c) Using P_2, approximate $\ln 1.3$, and find a bound on the error.

✓ **Solutions**

(a) The derivatives of $\ln x$ at $x = 1$ are given in the table:

order of deriv	0	1	2	3
deriv of $\ln x$	$\ln x$	$\frac{1}{x}$	$-\frac{1}{x^2}$	$\frac{2}{x^3}$
deriv at $x = 1$	0	1	-1	2

From the table we have

$$P_0(x) = 0$$

$$P_1(x) = (x - 1)$$

$$P_2(x) = (x - 1) - \frac{(x-1)^2}{2}$$

$$P_3(x) = (x - 1) - \frac{(x-1)^2}{2} + \frac{(x-1)^3}{3}$$

(b)   Figure 10.4 shows the graphs of $\ln x$ and the four Taylor polynomials above, in $[0,2.5] \times [-1,1]$.

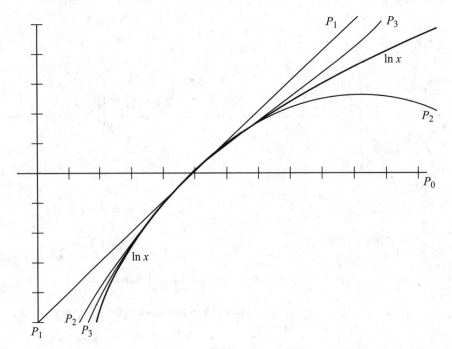

$$[0,2.5] \times [-1,1]$$

Figure 10.4

(c) $\ln 1.3 \approx P_2(1.3) = (1.3 - 1) - \dfrac{(1.3 - 1)^2}{2} = 0.3 - 0.045 = 0.255.$

For $x = 1.3$ the Taylor series converges by the Alternating Series Test, so the error is less than the magnitude of the first omitted term:

$$R_2(1.3) < \left| \frac{(1.3 - 1)^3}{3} \right| = 0.009$$

> **Example 47**

For what positive values of x is the approximate formula

$$\ln(1 + x) = x - \frac{x^2}{2}$$

correct to three decimal places?

✓ Solution

We can use series (5), page 418:

$$\ln(1 + x) = x - \frac{x^2}{2} + \frac{x^3}{3} - \cdots$$

For $x > 0$, this is an alternating series with terms decreasing in magnitude and approaching 0, so the error committed by using the first two terms is less than $\frac{|x|^3}{3}$. If $\frac{|x|^3}{3} < 0.0005$, then the given approximation formula will yield accuracy to three decimal places. We therefore require that $|x|^3 < 0.0015$ or that $|x| < 0.114$.

C5. Taylor's Formula with Remainder; Lagrange Error Bound

When we approximate a function using a Taylor polynomial, it is important to know how large the remainder (error) may be. If at the desired value of x the Taylor series is alternating, this issue is easily resolved: the first omitted term serves as an upper bound on the error. However, when the approximation involves a nonnegative Taylor series, placing an upper bound on the error is more difficult. This issue is resolved by the Lagrange remainder.

TAYLOR'S THEOREM.   If a function f and its first $(n + 1)$ derivatives are continuous on the interval $|x - a| < r$, then for each x in this interval

$$f(x) = f(a) + f'(a)(x - a) + \frac{f''(a)}{2!}(x - a)^2 + \cdots + \frac{f^{(n)}(a)}{n!}(x - a)^n + R_n(x)$$

where

$$R_n(x) = \frac{f^{(n+1)}(c)(x - a)^{n+1}}{(n + 1)!}$$

and c is some number between a and x. $R_n(x)$ is called the *Lagrange remainder*.

Note that the equation above expresses $f(x)$ as the sum of the Taylor polynomial $P_n(x)$ and the error that results when that polynomial is used as an approximation for $f(x)$.

When we truncate a series after the $(n + 1)$st term, we can compute the error bound R_n, according to Lagrange, if we know what to substitute for c. In practice we find, not R_n exactly, but only an upper bound for it by assigning to c the value between a and x that determines the largest possible value of R_n. Hence:

$$R_n(x) < \max \left| \frac{f^{(n+1)}(c)(x - a)^{n+1}}{(n + 1)!} \right|$$

the *Lagrange error bound*.

> ### Example 48

Estimate the error in using the Maclaurin series generated by e^x to approximate the value of e.

✓ Solution

From Example 40 we know that $f(x) = e^x$ generates the Maclaurin series

$$e^x = 1 + x + \frac{x^2}{2!} + \cdots + \frac{x^n}{n!} + \cdots$$

The Lagrange error bound is

$$R_n(x) < \max \left| \frac{e^c (x)^{n+1}}{(n+1)!} \right| \quad (0 < c < x)$$

To estimate e, we use $x = 1$. For $0 < c < 1$, the maximum value of e^c is e. Thus:

$$e \approx 1 + 1 + \frac{1}{2!} + \frac{1}{3!} + \cdots + \frac{1}{n!} \quad \text{with error less than } \frac{e}{(n+1)!}$$

> ## Example 49

Find the Maclaurin series for $\ln(1 + x)$ and the associated Lagrange error bound.

✓ Solution

$$f(x) = \ln(1 + x) \qquad\qquad f(0) = 0$$

$$f'(x) = \frac{1}{1 + x} \qquad\qquad f'(0) = 1$$

$$f''(x) = \frac{1}{(1 + x)^2} \qquad\qquad f''(0) = -1$$

$$f'''(x) = \frac{2}{(1 + x)^3} \qquad\qquad f'''(0) = 2!$$

$$\vdots \qquad\qquad\qquad \vdots$$

$$f^{(n)}(x) = \frac{(-1)^{n-1}(n-1)!}{(1 + x)^n} \qquad f^{(n)}(0) = (-1)^{n-1}(n-1)!$$

$$f^{(n+1)}(x) = \frac{(-1)^n \cdot n!}{(1 + x)^{n+1}}$$

Then

$$\ln(1 + x) = x - \frac{x^2}{2} + \frac{x^3}{3} - \frac{x^4}{4} + \cdots + (-1)^{n-1} \cdot \frac{x^n}{n} + R_n(x)$$

where the Lagrange error bound is

$$R_n(x) < \max \left| \frac{(-1)^n}{(1 + c)^{n+1}} \cdot \frac{x^{n+1}}{n+1} \right|$$

NOTE: For $0 < x < 1$ the Maclaurin series is alternating, and the error bound simplifies to $R_n(x) < \frac{x^{n+1}}{n+1}$, the first omitted term. The more difficult Lagrange error bound applies for $-1 < x < 0$.

> **Example 50**

Find the third-degree Maclaurin polynomial for $f(x) = \cos\left(x + \frac{\pi}{4}\right)$, and determine the upper bound on the error in estimating $f(0.1)$.

✓ **Solution**

We first make a table of the derivatives, evaluated at $x = 0$ and giving us the coefficients.

n	$f^{(n)}(x)$	$f^{(n)}(0)$	$C_n = \dfrac{f^{(n)}(0)}{n!}$
0	$\cos\left(x + \frac{\pi}{4}\right)$	$\dfrac{\sqrt{2}}{2}$	$\dfrac{\sqrt{2}}{2}$
1	$-\sin\left(x + \frac{\pi}{4}\right)$	$-\dfrac{\sqrt{2}}{2}$	$-\dfrac{\sqrt{2}}{2}$
2	$-\cos\left(x + \frac{\pi}{4}\right)$	$-\dfrac{\sqrt{2}}{2}$	$-\dfrac{\sqrt{2}}{2 \cdot 2}$
3	$\sin\left(x + \frac{\pi}{4}\right)$	$\dfrac{\sqrt{2}}{2}$	$\dfrac{\sqrt{2}}{2 \cdot 2}$

Thus $P_3(x) = \dfrac{\sqrt{2}}{2} - \dfrac{\sqrt{2}}{2}x - \dfrac{\sqrt{2}}{2 \cdot 2!}x^2 + \dfrac{\sqrt{2}}{2 \cdot 3!}x^3$.

Since this is not an alternating series for $x = 0.1$, we must use the Lagrange error bound:

$$R_3(x) < \max\left|\frac{f^{(4)}(c)}{4!}x^4\right|, \quad \text{where } x = 0.1 \text{ and } 0 < c < 0.1$$

Note that $f^{(4)}(c) = \cos\left(c + \frac{\pi}{4}\right)$ is decreasing on the interval $0 < c < 0.1$, so its maximum value occurs at $c = 0$. Hence:

$$R_3(x) < \max\left|\frac{\cos\left(c + \frac{\pi}{4}\right)}{4!}x^4\right| = \frac{\cos\left(0 + \frac{\pi}{4}\right)}{4!}0.1^4 = \frac{\sqrt{2}}{2 \cdot 4! \cdot 10^4}$$

C6. Computations with Power Series

The power series expansions of functions may be treated as any other functions for values of x that lie within their intervals of convergence. They may be added, subtracted, multiplied, divided (with division by zero to be avoided), differentiated, or integrated. These properties provide a valuable approach for many otherwise difficult computations. Indeed, power series are often very useful for approximating values of functions, evaluating indeterminate forms of limits, and estimating definite integrals.

> **Example 51**

Compute $\dfrac{1}{\sqrt{e}}$ to four decimal places.

✓ **Solution**

We can use the Maclaurin series,

$$e^x = 1 + x + \frac{x^2}{2!} + \frac{x^3}{3!} + \frac{x^4}{4!} + \cdots$$

and let $x = -\dfrac{1}{2}$ to get

$$e^{-1/2} = 1 - \frac{1}{2} + \frac{1}{4 \cdot 2!} - \frac{1}{8 \cdot 3!} + \frac{1}{16 \cdot 4!} - \frac{1}{32 \cdot 5!} + R_5$$

$$= 1 - 0.50000 + 0.12500 - 0.02083 + 0.00260 - 0.00026 + R_5$$

$$= 0.60651 + R_5$$

Note that, since this series converges by the Alternating Series Test, R_5 is less than the first term dropped:

$$R_5 < \frac{1}{64 \cdot 6!} = 0.0000217$$

The error bound is a positive value, so $0.60651 \le e^{-1/2} \le 0.60651 + 0.0000217$. This gives us $0.60651 \le e^{-1/2} \le 0.606532$, so $\dfrac{1}{\sqrt{e}} = 0.6065$, correct to four decimal places.

> **Example 52**

Estimate the error if the approximate formula

$$\sqrt{1 + x} = 1 + \frac{x}{2}$$

is used and $|x| < 0.02$.

✓ **Solution**

We obtain the first few terms of the Maclaurin series generated by $f(x) = \sqrt{1 + x}$:

$$f(x) = \sqrt{1 + x} \qquad\qquad f(0) = 1$$

$$f'(x) = \frac{1}{2}(1 + x)^{-1/2} \qquad\qquad f'(0) = \frac{1}{2}$$

$$f''(x) = -\frac{1}{4}(1 + x)^{-3/2} \qquad\qquad f''(0) = -\frac{1}{4}$$

$$f'''(x) = \frac{3}{8}(1 + x)^{-5/2} \qquad\qquad f'''(0) = \frac{3}{8}$$

Then

$$\sqrt{1 + x} = 1 + \frac{x}{2} - \frac{1}{4} \cdot \frac{x^2}{2} + \frac{3}{8} \cdot \frac{x^3}{6} - \cdots$$

Note that for $x < 0$, the series is not alternating, so we must use the Lagrange error bound. Here R_1 is $\left|\dfrac{f''(c)x^2}{2!}\right|$, where $-0.02 < c < 0.02$. With $|x| < 0.02$, we see that the upper bound uses $c = -0.02$:

$$|R_1| < \frac{(0.02)^2}{8(1 - 0.02)^{3/2}} < 0.00005$$

> **Example 53**

Use a series to evaluate $\displaystyle\lim_{x \to 0}\frac{\sin x}{x}$.

✓ **Solution**

From series (1), page 418,

$$\sin x = x - \frac{x^3}{3!} + \frac{x^5}{5!} - \cdots$$

Then

$$\lim_{x \to 0}\frac{\sin x}{x} = \lim_{x \to 0}\left(1 - \frac{x^2}{3!} + \frac{x^4}{5!} - \cdots\right) = 1$$

a well-established result obtained previously.

> **Example 54**

Use a series to evaluate $\displaystyle\lim_{x \to 0}\frac{\ln(x + 1)}{3x}$.

✓ **Solution**

We can use series (5), page 418, and write

$$\lim_{x \to 0}\frac{x - \dfrac{x^2}{2} + \dfrac{x^3}{3} - \dfrac{x^4}{4} + \cdots}{3x} = \lim_{x \to 0}\frac{1}{3} - \frac{x}{6} + \frac{x^2}{9} - \cdots$$
$$= \frac{1}{3}$$

> **Example 55**

Use a series to evaluate $\displaystyle\lim_{x \to 0}\frac{e^{-x^2} - 1}{x^2}$.

✓ **Solution**

$$\lim_{x \to 0}\frac{e^{-x^2} - 1}{x^2} = \lim_{x \to 0}\frac{\left(1 - x^2 + \dfrac{x^4}{2!} - \dfrac{x^6}{3!} + \cdots\right) - 1}{x^2}$$

$$= \lim_{x \to 0}\frac{-x^2 + \dfrac{x^4}{2!} - \cdots}{x^2} = \lim_{x \to 0}\left(-1 + \frac{x^2}{2!} - \frac{x^4}{3!} + \cdots\right)$$

$$= -1$$

BC ONLY

> **Example 56**

Show how a series may be used to evaluate π.

✓ **Solution**

Since $\frac{\pi}{4} = \tan^{-1} 1$, a series for $\tan^{-1} x$ may prove helpful. Note that

$$\tan^{-1} x = \int_0^x \frac{dt}{1 + t^2}$$

and that a series for $\frac{1}{1 + t^2}$ is obtainable easily by long division to yield

$$\frac{1}{1 + t^2} = 1 - t^2 + t^4 - t^6 + \cdots$$

If we integrate this series term by term and then evaluate the definite integral, we get

$$\tan^{-1} x = x - \frac{x^3}{3} + \frac{x^5}{5} - \frac{x^7}{7} + \cdots + \frac{(-1)^{n-1} x^{2n-1}}{2n - 1} + \cdots$$

(Compare with series (6) on page 418 and note especially that this series converges on $-1 \le x \le 1$.)

For $x = 1$ we have:

$$\tan^{-1} 1 = 1 - \frac{1}{3} + \frac{1}{5} - \frac{1}{7} + \cdots$$

Then

$$\frac{\pi}{4} = 1 - \frac{1}{3} + \frac{1}{5} - \frac{1}{7} + \cdots$$

and

$$\pi = 4 \left(1 - \frac{1}{3} + \frac{1}{5} - \frac{1}{7} + \cdots \right)$$

Here are some approximations for π using this series:

number of terms	1	2	5	10	25	50	59	60
approximation	4	2.67	3.34	3.04	3.18	3.12	3.16	3.12

Since the series is alternating, the odd sums are greater, the even ones less, than the value of π. It is clear that several hundred terms may be required to get even two-place accuracy. There are series expressions for π that converge much more rapidly. (See Miscellaneous Free-Response Practice Exercises, Problem A14, page 477.)

> **Example 57**

Use a series to evaluate $\int_0^{0.1} e^{-x^2}\,dx$ to four decimal places.

✓ **Solution**

Although $\int e^{-x^2}\,dx$ cannot be expressed in terms of elementary functions, we can write a series for e^u, replace u by $(-x^2)$, and integrate term by term. Thus,

$$e^{-x^2} = 1 - x^2 + \frac{x^4}{2!} - \frac{x^6}{3!} + \cdots$$

so

$$\int_0^{0.1} e^{-x^2}\,dx = x - \frac{x^3}{3} + \frac{x^5}{5 \cdot 2!} - \frac{x^7}{7 \cdot 3!} + \cdots \Big|_0^{0.1}$$

$$= 0.1 - \frac{0.001}{3} + \frac{0.00001}{10} - \frac{0.0000001}{42} + \cdots$$

$$= 0.1 - 0.00033 + 0.000001 + R_6$$

$$= 0.09967 + R_6$$

Since this is a convergent alternating series (with terms decreasing in magnitude and approaching 0), $|R_6| < \dfrac{10^{-7}}{42}$, which will not affect the fourth decimal place. Then, correct to four decimal places,

$$\int_0^{0.1} e^{-x^2}\,dx = 0.0997$$

†C7. Power Series over Complex Numbers

A *complex number* is one of the form $a + bi$, where a and b are real and $i^2 = -1$. If we allow complex numbers as replacements for x in power series, we obtain some interesting results.

Consider, for instance, the series

$$e^x = 1 + x + \frac{x^2}{2!} + \frac{x^3}{3!} + \frac{x^4}{4!} + \cdots + \frac{x^n}{n!} \cdots \tag{1}$$

When $x = yi$, then (1) becomes

$$e^{yi} = 1 + yi + \frac{(yi)^2}{2!} + \frac{(yi)^3}{3!} + \frac{(yi)^4}{4!} + \cdots$$

$$= 1 + yi - \frac{y^2}{2!} - \frac{y^3 i}{3!} + \frac{y^4}{4!} + \cdots$$

$$= \left(1 - \frac{y^2}{2!} + \frac{y^4}{4!} + \cdots\right) + i\left(y - \frac{y^3}{3!} + \frac{y^5}{5!} - \cdots\right) \tag{2}$$

Then

$$e^{yi} = \cos y + i \sin y \tag{3}$$

† This is an optional topic not in the BC Course Description. We include it here because of the dramatic result.

BC ONLY

since the series within the parentheses of equation (2) converge respectively to cos y and sin y. Equation (3) is called *Euler's formula*. It follows from (3) that

$$e^{i\pi} = -1$$

and thus that

$$e^{i\pi} + 1 = 0$$

sometimes referred to as *Euler's magic formula*.

CHAPTER SUMMARY

In this chapter, we reviewed an important BC Calculus topic: infinite series. We looked at a variety of tests to determine whether a series converges or diverges. We worked with functions defined as power series, reviewed how to derive Taylor series, and looked at the Maclaurin series expansions for many commonly used functions. Finally, we reviewed how to find bounds on the errors that arise when series are used for approximations.

Practice Exercises

PART A. NO CALCULATOR—DIRECTIONS: Answer these questions *without* using your calculator.

NOTE: *No* questions on sequences will appear on the BC examination. We have nevertheless chosen to include the topic in Questions A1–A5 because a series and its convergence are defined in terms of sequences. Reviewing sequences will enhance your understanding of series.

A1. Which sequence converges?

(A) $a_n = n + \dfrac{3}{n}$

(B) $a_n = -1 + \dfrac{(-1)^n}{n}$

(C) $a_n = \sin\dfrac{n\pi}{2}$

(D) $a_n = \dfrac{n!}{3^n}$

A2. If $s_n = 1 + \dfrac{(-1)^n}{n}$, then

(A) s_n diverges by oscillation

(B) s_n converges to zero

(C) $\lim\limits_{n\to\infty} s_n = 1$

(D) s_n diverges to infinity

A3. The sequence $a_n = \sin\dfrac{n\pi}{6}$

(A) is unbounded

(B) converges to a number less than 1

(C) is bounded

(D) diverges to infinity

A4. Which of the following sequences diverges?

(A) $a_n = \dfrac{(-1)^{n+1}}{n}$

(B) $a_n = \dfrac{2^n}{e^n}$

(C) $a_n = \dfrac{n^2}{e^n}$

(D) $a_n = \dfrac{n}{\ln n}$

A5. The sequence $\{r^n\}$ converges if and only if

(A) $|r| < 1$

(B) $|r| \le 1$

(C) $-1 < r \le 1$

(D) $0 < r < 1$

A6. $\sum u_n$ is a series of constants for which $\lim\limits_{n\to\infty} u_n = 0$. Which of the following statements is always true?

(A) $\sum u_n$ converges to a finite sum.

(B) $\sum u_n$ does not diverge to infinity.

(C) $\sum u_n$ is a positive series.

(D) None of statements (A), (B), or (C) are always true.

A7. Note that $\dfrac{1}{n(n+1)} = \dfrac{1}{n} - \dfrac{1}{n+1}\,(n \ge 1)$.

$\displaystyle\sum_{n=1}^{\infty} \dfrac{1}{n(n+1)}$ equals

(A) 1

(B) $\dfrac{3}{2}$

(C) $\dfrac{3}{4}$

(D) ∞

A8. The sum of the geometric series

$$2 - 1 + \dfrac{1}{2} - \dfrac{1}{4} + \dfrac{1}{8} - \cdots \text{ is}$$

(A) $\dfrac{4}{3}$

(B) $\dfrac{5}{4}$

(C) 1

(D) $\dfrac{3}{2}$

A9. Which of the following statements about series is true?

(A) If $\lim\limits_{n\to\infty} u_n = 0$, then $\sum u_n$ converges.

(B) If $\lim\limits_{n\to\infty} u_n \ne 0$, then $\sum u_n$ diverges.

(C) If $\sum u_n$ diverges, then $\lim\limits_{n\to\infty} u_n \ne 0$.

(D) $\sum u_n$ converges if and only if $\lim\limits_{n\to\infty} u_n = 0$.

A10. Which of the following series diverges?

(A) $\sum \dfrac{1}{n^2}$

(B) $\sum \dfrac{1}{n^2 + n}$

(C) $\sum \dfrac{n}{n^3 + 1}$

(D) $\sum \dfrac{n}{\sqrt{4n^2 - 1}}$

A11. Which of the following series diverges?

(A) $3 - 1 + \dfrac{1}{9} - \dfrac{1}{27} + \cdots$

(B) $\dfrac{1}{\sqrt{2}} - \dfrac{1}{\sqrt{3}} + \dfrac{1}{\sqrt{4}} - \cdots$

(C) $\dfrac{1}{2^2} - \dfrac{1}{3^2} + \dfrac{1}{4^2} - \cdots$

(D) $1 - 1.1 + 1.21 - 1.331 + \cdots$

A12. Let $S = \sum\limits_{n=1}^{\infty} \left(\dfrac{2}{3}\right)^n$; then S equals

(A) $\dfrac{3}{2}$

(B) $\dfrac{4}{3}$

(C) 2

(D) 3

A13. Which of the following expansions is impossible?

(A) $\sqrt{x - 1}$ in powers of x

(B) $\sqrt{x + 1}$ in powers of x

(C) $\ln x$ in powers of $(x - 1)$

(D) $\tan x$ in powers of $\left(x - \dfrac{\pi}{4}\right)$

A14. The series $\sum\limits_{n=0}^{\infty} n!(x - 3)^n$ converges if and only if

(A) $x = 0$

(B) $2 < x < 4$

(C) $x = 3$

(D) $2 \le x \le 4$

A15. Let $f(x) = \sum\limits_{n=0}^{\infty} x^n$. The radius of convergence of

$\displaystyle\int_0^x f(t)\, dt$ is

(A) 0

(B) 1

(C) 2

(D) e

A16. The coefficient of x^4 in the Maclaurin series for $f(x) = e^{-x/2}$ is

(A) $-\dfrac{1}{24}$

(B) $\dfrac{1}{24}$

(C) $-\dfrac{1}{384}$

(D) $\dfrac{1}{384}$

A17. If an appropriate series is used to evaluate

$\displaystyle\int_0^{0.1} x^2 e^{-x^2}\, dx$, then, correct to four decimal places, the definite integral equals

(A) 0.0002

(B) 0.0003

(C) 0.0032

(D) 0.0033

A18. If the series $\tan^{-1} 1 = 1 - \dfrac{1}{3} + \dfrac{1}{5} - \dfrac{1}{7} + \cdots$ is used to approximate $\dfrac{\pi}{4}$ with an error less than 0.001, then the smallest number of terms needed is

(A) 200

(B) 300

(C) 400

(D) 500

A19. Let $f(x) = \tan^{-1} x$ and $P_7(x)$ be the Taylor polynomial of degree 7 for f about $x = 0$. Given

$\tan^{-1} x = \sum\limits_{n=0}^{\infty} \dfrac{(-1)^n x^{2n+1}}{2n + 1}$, it follows that if $-0.5 < x < 0.5$,

(A) $P_7(x) \le \tan^{-1} x$

(B) $P_7(x) \ge \tan^{-1} x$

(C) $P_7(x) > \tan^{-1} x$ if $x < 0$ but $P_7(x) < \tan^{-1} x$ if $x > 0$

(D) $P_7(x) < \tan^{-1} x$ if $x < 0$ but $P_7(x) > \tan^{-1} x$ if $x > 0$

A20. Let $f(x) = \tan^{-1} x$ and $P_9(x)$ be the Taylor polynomial of degree 9 for f about $x = 0$. Given

$$\tan^{-1} x = \sum_{n=0}^{\infty} \frac{(-1)n\, x^{2n+1}}{2n+1}, \text{ it follows that if}$$

$-0.5 < x < 0.5$,

(A) $P_9(x) \leq \tan^{-1} x$

(B) $P_9(x) \geq \tan^{-1} x$

(C) $P_9(x) > \tan^{-1} x$ if $x < 0$ but
$P_9(x) < \tan^{-1} x$ if $x > 0$

(D) $P_9(x) < \tan^{-1} x$ if $x < 0$ but
$P_9(x) > \tan^{-1} x$ if $x > 0$

A21. Which of the following series converges?

(A) $\sum \dfrac{1}{\sqrt[3]{n}}$

(B) $\sum \dfrac{1}{\sqrt{n}}$

(C) $\sum \dfrac{1}{10n - 1}$

(D) $\sum \dfrac{2}{n^2 - 5}$

A22. Which of the following series diverges?

(A) $\displaystyle\sum_{n=1}^{\infty} \dfrac{n+1}{n!}$

(B) $\displaystyle\sum_{n=2}^{\infty} \dfrac{1}{n \ln n}$

(C) $\displaystyle\sum_{n=1}^{\infty} \dfrac{\ln n}{2^n}$

(D) $\displaystyle\sum_{n=1}^{\infty} \dfrac{n}{2^n}$

A23. For which of the following series does the Ratio Test fail?

(A) $\sum \dfrac{1}{n!}$

(B) $\sum \dfrac{n}{2^n}$

(C) $1 + \dfrac{1}{2^{3/2}} + \dfrac{1}{3^{3/2}} + \dfrac{1}{4^{3/2}} + \cdots$

(D) $\sum \dfrac{n^n}{n!}$

A24. Which of the following alternating series diverges?

(A) $\sum \dfrac{(-1)^{n+1}(n-1)}{n+1}$

(B) $\sum \dfrac{(-1)^{n+1}}{\ln(n+1)}$

(C) $\sum \dfrac{(-1)^{n-1}}{\sqrt{n}}$

(D) $\sum \dfrac{(-1)^{n-1}(n)}{n^2 + 1}$

A25. The power series $x + \dfrac{x^2}{2} + \dfrac{x^3}{3} + \cdots + \dfrac{x^n}{n} + \cdots$ converges if and only if

(A) $-1 < x < 1$

(B) $-1 \leq x \leq 1$

(C) $-1 \leq x < 1$

(D) $-1 < x \leq 1$

A26. The power series

$$(x+1) - \frac{(x+1)^2}{2!} + \frac{(x+1)^3}{3!} - \frac{(x+1)^4}{4!} + \cdots$$

diverges

(A) for no real x

(B) if $-2 < x \leq 0$

(C) if $x < -2$ or $x > 0$

(D) if $-2 \leq x < 0$

A27. The series obtained by differentiating term by term the series

$$(x-2) + \frac{(x-2)^2}{4} + \frac{(x-2)^3}{9} + \frac{(x-2)^4}{16} + \cdots$$

converges for

(A) $1 \leq x \leq 3$

(B) $1 \leq x < 3$

(C) $1 < x \leq 3$

(D) $0 < x < 4$

A28. The Taylor polynomial of degree 3 at $x = 0$ for $f(x) = \sqrt{1+x}$ is

(A) $1 + \dfrac{x}{2} - \dfrac{x^2}{4} + \dfrac{3x^3}{8}$

(B) $1 + \dfrac{x}{2} - \dfrac{x^2}{8} + \dfrac{x^3}{16}$

(C) $1 + \dfrac{x}{2} - \dfrac{x^2}{8} + \dfrac{x^3}{8}$

(D) $1 - \dfrac{x}{2} + \dfrac{x^2}{4} - \dfrac{3x^3}{8}$

A29. The Taylor polynomial of degree 3 at $x = 1$ for e^x is

(A) $e\left[1 + (x-1) + \dfrac{(x-1)^2}{2} + \dfrac{(x-1)^3}{3}\right]$

(B) $e\left[1 + (x+1) + \dfrac{(x+1)^2}{2!} + \dfrac{(x+1)^3}{3!}\right]$

(C) $e\left[1 + (x-1) + \dfrac{(x-1)^2}{2!} + \dfrac{(x-1)^3}{3!}\right]$

(D) $e\left[1 - (x-1) + \dfrac{(x-1)^2}{2!} + \dfrac{(x-1)^3}{3!}\right]$

A30. The coefficient of $\left(x - \dfrac{\pi}{4}\right)^3$ in the Taylor series about $\dfrac{\pi}{4}$ of $f(x) = \cos x$ is

(A) $-\dfrac{1}{12}$

(B) $\dfrac{1}{12}$

(C) $\dfrac{1}{6\sqrt{2}}$

(D) $-\dfrac{1}{3\sqrt{2}}$

A31. The coefficient of x^2 in the Maclaurin series for $e^{\sin x}$ is

(A) 0

(B) 1

(C) $\dfrac{1}{2!}$

(D) $\dfrac{1}{4}$

A32. The coefficient of $(x-1)^5$ in the Taylor series for $x \ln x$ about $x = 1$ is

(A) $-\dfrac{1}{20}$

(B) $-\dfrac{1}{5!}$

(C) $-\dfrac{1}{4!}$

(D) $-\dfrac{1}{5}$

A33. The radius of convergence of the series
$$\sum_{n=1}^{\infty} \frac{x^n}{2^n} \cdot \frac{n^n}{n!} \text{ is}$$

(A) 2

(B) $\dfrac{2}{e}$

(C) $\dfrac{e}{2}$

(D) ∞

A34. The Taylor polynomial of degree 3 at $x = 0$ for $(1+x)^p$, where p is a constant, is

(A) $1 + px + p(p-1)x^2 + p(p-1)(p-2)x^3$

(B) $1 + px + \dfrac{p(p-1)}{2}x^2 + \dfrac{p(p-1)(p-2)}{3}x^3$

(C) $1 + px + \dfrac{p(p-1)}{2!}x^2 + \dfrac{p(p-1)(p-2)}{3!}x^3$

(D) $px + \dfrac{p(p-1)}{2!}x^2 + \dfrac{p(p-1)(p-2)}{3!}x^3$

A35. The Taylor series for $\ln(1+2x)$ about $x = 0$ is

(A) $2x - 2x^2 + \dfrac{8}{3}x^3 - 4x^4 + \cdots$

(B) $2x - 4x^2 + 16x^3 + \cdots$

(C) $2x - x^2 + \dfrac{8}{3}x^3 - 4x^4 + \cdots$

(D) $2x - 2x^2 + \dfrac{4}{3}x^3 - \dfrac{2}{3}x^4 + \cdots$

A36. The set of all values of x for which
$$\sum_{n=1}^{\infty} \frac{n \cdot 2^n}{x^n} \text{ converges is}$$

(A) only $x = 0$

(B) $-2 < x < 2$

(C) $|x| > 2$

(D) $|x| \geq 2$

A37. The third-degree Taylor polynomial $P_3(x)$ for $\sin x$ about $\frac{\pi}{4}$ is

(A) $\frac{1}{\sqrt{2}}\left(\left(x - \frac{\pi}{4}\right) - \frac{1}{3!}\left(x - \frac{\pi}{4}\right)^3\right)$

(B) $\frac{1}{\sqrt{2}}\left(1 + \left(x - \frac{\pi}{4}\right) - \frac{1}{2}\left(x - \frac{\pi}{4}\right)^2 + \frac{1}{3!}\left(x - \frac{\pi}{4}\right)^3\right)$

(C) $\frac{1}{\sqrt{2}}\left(1 + \left(x - \frac{\pi}{4}\right) - \frac{1}{2!}\left(x - \frac{\pi}{4}\right)^2 - \frac{1}{3!}\left(x - \frac{\pi}{4}\right)^3\right)$

(D) $1 + \left(x - \frac{\pi}{4}\right) - \frac{1}{2}\left(x - \frac{\pi}{4}\right)^2 - \frac{1}{6}\left(x - \frac{\pi}{4}\right)^3$

A38. Let h be a function for which all derivatives exist at $x = 1$. If $h(1) = h'(1) = h''(1) = h'''(1) = 6$, which third-degree polynomial best approximates h there?

(A) $6 + 6x + 6x^2 + 6x^3$

(B) $6 + 6(x - 1) + 6(x - 1)^2 + 6(x - 1)^3$

(C) $6 + 6x + 3x^2 + x^3$

(D) $6 + 6(x - 1) + 3(x - 1)^2 + (x - 1)^3$

PART B. CALCULATOR ACTIVE—DIRECTIONS: Some of the following questions require the use of a graphing calculator.

NOTE: Because of the abilities of graphing calculators, Taylor series and convergence are largely tested in a No Calculator environment; as such we offer only a few calculator-active multiple-choice questions here.

B1. Which of the following statements about series is false?

(A) $\sum_{k=1}^{\infty} u_k = \sum_{k=m}^{\infty} u_k$, where m is any positive integer.

(B) If $\sum u_n$ converges, so does $\sum cu_n$ if $c \neq 0$.

(C) If $\sum a_n$ and $\sum b_n$ converge, so does $\sum (ca_n + b_n)$, where $c \neq 0$.

(D) Rearranging the terms of a positive convergent series will not affect its convergence or its sum.

B2. Which of the following statements is always true?

(A) If $\sum u_n$ converges, then so does the series $\sum |u_n|$.

(B) If a series is truncated after the nth term, then the error is less than the first term omitted.

(C) If the terms of an alternating series decrease, then the series converges.

(D) None of statements (A), (B), or (C) are always true.

B3. Which of the following series can be used to compute $\ln 0.8$?

(A) $\ln(x - 1)$ expanded about $x = 0$

(B) $\ln x$ about $x = 0$

(C) $\ln x$ expanded about $x = 1$

(D) $\ln(x - 1)$ expanded about $x = 1$

B4. Let $f(x) = \sum_{n=0}^{\infty} a_n x^n$, $g(x) = \sum_{n=0}^{\infty} b_n x^n$. Suppose both series converge for $|x| < R$. Let x_0 be a number such that $|x_0| < R$. Which of statements (A)–(C) is false?

(A) $\sum_{n=0}^{\infty} (a_n + b_n)(x_0)^n$ converges to $f(x_0) + g(x_0)$.

(B) $f(x) = \sum_{n=0}^{\infty} a_n x^n$ is continuous at $x = x_0$.

(C) $\sum_{n=1}^{\infty} na_n x_0^{n-1}$ converges to $f'(x_0)$.

(D) Statements (A)–(C) are all true.

B5. If the approximate formula $\sin x = x - \dfrac{x^3}{3!}$ is used and $|x| < 1$ (radian), then the error is numerically less than

(A) 0.003

(B) 0.005

(C) 0.008

(D) 0.009

B6. The function $f(x) = \displaystyle\sum_{n=0}^{\infty} a_n x^n$ and $f'(x) = -f(x)$ for all x. If $f(0) = 1$, then how many terms of the series are needed to find $f(0.2)$ correct to three decimal places?

(A) 2

(B) 3

(C) 4

(D) 5

B7. The sum of the series $\displaystyle\sum_{n=1}^{\infty} \left(\dfrac{\pi^3}{3\pi}\right)^n$

(A) $= 1$

(B) $= \dfrac{3^{\pi}}{\pi^3 - 3\pi}$

(C) $= \dfrac{\pi^3}{3\pi - \pi^3}$

(D) diverges

B8. When $\displaystyle\sum_{n=1}^{\infty} \dfrac{(-1)^{n-1}}{3n-1}$ is approximated by the sum of its first 300 terms, the error is closest to

(A) 0.001

(B) 0.002

(C) 0.005

(D) 0.01

B9. You wish to estimate e^x, over the interval $|x| \leq 2$, with an error less than 0.001. The Lagrange error term suggests that you use a Taylor polynomial at 0 with degree at least

(A) 9

(B) 10

(C) 11

(D) 12

Answer Key

A1.	**B**	A13.	**A**	A25.	**C**	A37.	**C**
A2.	**C**	A14.	**C**	A26.	**A**	A38.	**D**
A3.	**C**	A15.	**B**	A27.	**B**	B1.	**A**
A4.	**D**	A16.	**D**	A28.	**B**	B2.	**D**
A5.	**C**	A17.	**B**	A29.	**C**	B3.	**C**
A6.	**D**	A18.	**D**	A30.	**C**	B4.	**D**
A7.	**A**	A19.	**C**	A31.	**C**	B5.	**D**
A8.	**A**	A20.	**D**	A32.	**A**	B6.	**C**
A9.	**B**	A21.	**D**	A33.	**B**	B7.	**C**
A10.	**D**	A22.	**B**	A34.	**C**	B8.	**A**
A11.	**D**	A23.	**C**	A35.	**A**	B9.	**B**
A12.	**C**	A24.	**A**	A36.	**C**		

Answer Explanations

A1. **(B)** $a_n = -1 + \dfrac{(-1)^n}{n}$ converges to -1.

A2. **(C)** Note that $\lim\limits_{n\to\infty} \dfrac{(-1)^n}{n} = 0$.

A3. **(C)** The sine function varies continuously between -1 and 1 inclusive.

A4. **(D)** Note that $a_n = \left(\dfrac{2}{e}\right)^n$ is a sequence of the type $s_n = r^n$ with $|r| < 1$; also that $\lim\limits_{n\to\infty} \dfrac{n^2}{e^n} = 0$ by repeated application of L'Hospital's Rule.

A5. **(C)** $\lim\limits_{n\to\infty} r_n = 0$ for $|r| < 1$; $\lim\limits_{n\to\infty} r_n = 1$ for $r = 1$.

A6. **(D)** The harmonic series $\sum\limits_{1}^{\infty} \dfrac{1}{k}$ is a counterexample for (A), (B), and (C).

A7. **(A)** $\sum\limits_{1}^{\infty} \dfrac{1}{n(n+1)} = \dfrac{1}{1\cdot 2} + \dfrac{1}{2\cdot 3} + \dfrac{1}{3\cdot 4} + \cdots + \dfrac{1}{n(n+1)} + \cdots$; so

$$s_n = 1 - \dfrac{1}{2} + \dfrac{1}{2} - \dfrac{1}{3} + \dfrac{1}{3} - \cdots + \dfrac{1}{n} - \dfrac{1}{n+1} = 1 - \dfrac{1}{n+1},$$

and $\lim\limits_{n\to\infty} s_n = 1$.

A8. **(A)** $S = \dfrac{a}{1-r} = \dfrac{2}{1 - \left(-\frac{1}{2}\right)} = \dfrac{4}{3}$

A9. **(B)** The harmonic series is a counterexample for both (A) and (C). Statement (D) is bidirectional, meaning that it is two if-then statements in one. The first if-then statement contained in (D) is "if a series is convergent, then the nth term goes to zero." This is true. The second if-then statement is "if the nth term goes to zero, then the series converges." This is a restatement of (A), so the harmonic series is a counterexample. Thus, one of the statements in (D) is not always true, making the entire statement false. Statement (B) is true. If the nth term does not go to zero, then the series will not converge.

A10. **(D)** Choice (A) is a p-series with $n = 2$ and is therefore convergent. (B) is convergent by the Direct Comparison Test or by the Limit Comparison Test with $\sum \dfrac{1}{n^2}$. (C) is convergent by the Limit Comparison Test with $\sum \dfrac{1}{n^2}$. (D) is divergent by the nth Term Test:

$$\lim_{n\to\infty} \dfrac{n}{\sqrt{4n^2 - 1}} = \lim_{n\to\infty} \sqrt{\dfrac{n^2}{4n^2 - 1}} = \sqrt{\lim_{n\to\infty} \dfrac{n^2}{4n^2 - 1}} = \sqrt{\dfrac{1}{4}} = \dfrac{1}{2} \neq 0$$

A11. **(D)** (A), (B), and (C) all converge; (D) is the divergent geometric series with $r = -1.1$.

A12. **(C)** $S = \dfrac{a}{1-r} = \dfrac{\frac{2}{3}}{1 - \frac{2}{3}} = \dfrac{\frac{2}{3}}{\frac{1}{3}} = 2$

A13. **(A)** If $f(x) = \sqrt{x-1}$, then $f(0)$ is not defined.

A14. **(C)** $\lim\limits_{n\to\infty} (n+1)(x-3) = \infty$ unless $x = 3$.

A15. **(B)** The integrated series is $\sum\limits_{n=0}^{\infty} \dfrac{x^{n+1}}{n+1}$ or $\sum\limits_{n=1}^{\infty} \dfrac{x^n}{n}$. See Question A25.

A16. **(D)** $e^{-x/2} = 1 + \left(-\dfrac{x}{2}\right) + \left(-\dfrac{x}{2}\right)^2 \cdot \dfrac{1}{2!} + \left(-\dfrac{x}{2}\right)^3 \cdot \dfrac{1}{3!} + \left(-\dfrac{x}{2}\right)^4 \cdot \dfrac{1}{4!} + \cdots$

A17. **(B)** $\displaystyle\int_0^{0.3} x^2 e^{-x^2}\,dx = \int_0^{0.3} x^2\left(1 - x^2 + \frac{x^4}{2!} - \cdots\right)dx$

$$= \int_0^{0.3}\left(x^2 - x^4 + \frac{x^6}{2!} - \cdots\right)dx$$

$$= \frac{x^3}{3} - \frac{x^5}{5} + \cdots \bigg|_0^{0.3}$$

$$= \frac{(0.1)^3}{3} - \frac{(0.1)^5}{5} + \cdots = 0.000\bar{3} - 0.000002 + \cdots$$

The series from the antidifferentiation satisfies the Alternating Series Test. Therefore, the Alternating Series Error Bound applies. If we use the first term as the approximation, then the error is the second term. Thus the value of the integral is between $0.000331\bar{3}$ and $0.000\bar{3}$. Therefore, the value of the integral, to four decimal places, is 0.0003.

A18. **(D)** The series satisfies the Alternating Series Test, so the error is less than the first term dropped, namely, $\dfrac{1}{2(n+1)-1}$, or $\dfrac{1}{2n+1}$ (see (6) on page 418), so $n \geq 500$.

A19. **(C)** Note that the Taylor series for $f(x) = \tan^{-1} x$ satisfies the Alternating Series Test for $-1 \leq x \leq 1$ and $P_7(x) = x - \frac{x^3}{3} + \frac{x^5}{5} - \frac{x^7}{7}$. The first omitted term, $\frac{x^9}{9}$, is the Alternating Series Error Bound. The value of the error term is negative when $x < 0$ and thus $f(x) - P_7(x) < 0$. So $P_7(x) > f(x)$. The value of the error term is positive when $x > 0$ and thus $f(x) - P_7(x) > 0$. So $P_7(x) < f(x)$.

A20. **(D)** Note that the Taylor series for $f(x) = \tan^{-1} x$ satisfies the Alternating Series Test for $-1 \leq x \leq 1$ and $P_9(x) = x - \frac{x^3}{3} + \frac{x^5}{5} - \frac{x^7}{7} + \frac{x^9}{9}$. The first omitted term, $-\frac{x^{11}}{11}$, is the Alternating Series Error Bound. The value of the error term is positive when $x < 0$ and thus $f(x) - P_9(x) > 0$. So $P_9(x) < f(x)$. The value of the error term is negative when $x > 0$ and thus $f(x) - P_9(x) < 0$. So $P_9(x) > f(x)$.

A21. **(D)** Each series given is essentially a p-series. Only in (D) is $p > 1$.

A22. **(B)** Series (A), (C), and (D) converge by the Ratio Test. (B) diverges by the Integral Test (page 404) with $f(x) = \dfrac{1}{x \ln x}$. Since $f(x)$ is continuous, positive, and decreasing for $x > 2$, then series (B) converges if and only if the improper integral $\displaystyle\int_2^\infty \frac{1}{x \ln x}\,dx$ converges:

$$\int_2^\infty \frac{1}{x \ln x}\,dx = \lim_{b\to\infty}\int_2^b \frac{1}{x \ln x}\,dx = \lim_{b\to\infty}\left(\ln(\ln x)\Big|_2^b\right) = \lim_{b\to\infty}\left(\ln(\ln b) - \ln(\ln 2)\right)$$

This limit does not exist. Therefore the integral diverges and so does series (B).

A23. **(C)** The limit of the ratio for the series $\displaystyle\sum \frac{1}{n^{3/2}}$ is 1, so this test fails; note for (D) that

$$\lim_{n\to\infty}\frac{u_{n+1}}{u_n} = \lim_{n\to\infty}\left(\frac{n+1}{n}\right)^n = \lim_{n\to\infty}\left(1 + \frac{1}{n}\right)^n = e.$$

A24. **(A)** $\displaystyle\lim_{n\to\infty}\frac{(-1)^{n+1}(n-1)}{n+1}$ does not equal 0, so the series does not pass the nth Term Test.

A25. **(C)** Since $\lim\limits_{n\to\infty}\left|\dfrac{u_{n+1}}{u_n}\right| = |x|$, the series converges if $|x| < 1$. We must test the endpoints: when $x = 1$, we get the divergent harmonic series; $x = -1$ yields the convergent alternating harmonic series.

A26. **(A)** $\lim\limits_{n\to\infty}\left|\dfrac{x+1}{n+1}\right| = 0$ for all $x \neq -1$; since the given series converges to 0 if $x = -1$, it therefore converges for *all x*.

A27. **(B)** The differentiated series is $\sum\limits_{n=1}^{\infty}\dfrac{(x-2)^{n-1}}{n}$. Perform the Ratio Test:

$$\lim\limits_{n\to\infty}\left|\dfrac{(x-2)^n}{n+1}\cdot\dfrac{n}{(x-2)^{n-1}}\right| = \lim\limits_{n\to\infty}\left|\dfrac{n}{n+1}\cdot(x-2)\right| = |x-2|. \text{ So for convergence,}$$

$|x-2| < 1$, and this gives the open interval $1 < x < 3$. Now check endpoints:

$x = 1$ gives $\sum\limits_{n=1}^{\infty}\dfrac{(-1)^{n-1}}{n}$, which is the alternating harmonic series, convergent;

$x = 3$ gives $\sum\limits_{n=1}^{\infty}\dfrac{1}{n}$, which is the harmonic series, divergent. So the interval of convergence is $1 \le x < 3$.

A28. **(B)** See Example 52 on page 426.

A29. **(C)** Note that every derivative of e^x is e at $x = 1$. The Taylor series is in powers of $(x-1)$ with coefficients of the form $c_n = \dfrac{e}{n!}$.

A30. **(C)** For $f(x) = \cos x$ around $x = \dfrac{\pi}{4}$, $c_3 = \dfrac{f^{(3)}\left(\frac{\pi}{4}\right)}{3!} = \dfrac{\sin\left(\frac{\pi}{4}\right)}{3!} = \dfrac{1/\sqrt{2}}{3!}$.

A31. **(C)** The coefficient of the x^2-term in a Maclaurin series is $\dfrac{f''(0)}{2!}$:

$$f'(x) = e^{\sin x}\cdot\cos x, f''(x) = e^{\sin x}\cdot\cos^2 x - e^{\sin x}\cdot\sin x, f''(0) = e^0\cdot 1^2 - e^0\cdot 0 = 1$$

A32. **(A)** $f(x) = x\ln x$ $\quad f'(x) = 1 + \ln x$ $\quad f''(x) = \dfrac{1}{x}$ $\quad f'''(x) = -\dfrac{1}{x^2}$ $\quad f^{(4)}(x) = \dfrac{2}{x^3}$

$f^{(5)}(x) = -\dfrac{3\cdot 2}{x^4}$ $\quad f^{(5)}(1) = -3\cdot 2$

So the coefficient of $(x-1)^5$ is $-\dfrac{3\cdot 2}{5!} = -\dfrac{1}{20}$.

A33. **(B)** $\lim\limits_{n\to\infty}\left|\dfrac{u_{n+1}}{u_n}\right| = \lim\limits_{n\to\infty}\left|\dfrac{x^{n+1}(n+1)^{n+1}}{2^{n+1}(n+1)!}\cdot\dfrac{2^n\cdot n!}{x^n\cdot n^n}\right|$

$$= \lim\limits_{n\to\infty}\left|\dfrac{x(n+1)^{n+1}}{2(n+1)n^n}\right| = \lim\limits_{n\to\infty}\left|\dfrac{x}{2}\left(\dfrac{n+1}{n}\right)^n\right|$$

$$= \lim\limits_{n\to\infty}\left|\dfrac{x}{2}\left(1+\dfrac{1}{n}\right)^n\right| = \left|\dfrac{x}{2}\cdot e\right|$$

Since the series converges when $\left|\dfrac{x}{2}\cdot e\right| < 1$, that is, when $|x| < \dfrac{2}{e}$, the radius of convergence is $\dfrac{x}{e}$.

A34. **(C)** This polynomial is associated with the binomial series $(1+x)^p$. Verify that $f(0) = 1$, $f'(0) = p, f''(0) = p(p-1), f'''(0) = p(p-1)(p-2)$.

A35. **(A)** For the given function, $f(x) = \ln(1 + 2x)$, the derivative of f is $f'(x) = \dfrac{2}{1 + 2x}$. $f'(x)$ is

in the form of the sum of an infinite geometric series $\left(\dfrac{a}{1 - r} = a + ar + ar^2 + \cdots\right)$, where

$a = 2$ and $r = -2x$. So $\dfrac{2}{1 + 2x} = \dfrac{2}{1 - (-2x)} = 2 - 4x + 8x^2 - 16x^3 + \cdots + (-1)^n 2^{n+1} x^n + \cdots$.

We can now use the Fundamental Theorem of Calculus with the fact that $f(0) = 0$ to find $f(x)$:

$$f(x) = f(0) + \int_0^x f'(t)dt = 0 + \int_0^1 \left(2 - 4t + 8t^2 - 16t^3 + \cdots\right)dt$$

$$= \left(2t - \frac{4t^2}{2} + \frac{8t^3}{3} - \frac{16t^4}{4} + \cdots\right)\Big|_0^1 = 2x - 2x^2 + \frac{8x^3}{3} - 4x^4 + \cdots$$

You could also find the first four derivatives and build the series using the technique similar to the one used in Example 43 on page 417.

A36. **(C)** $\lim\limits_{n \to \infty} \left|\dfrac{u_{n+1}}{u_n}\right| = \left|\dfrac{2}{x}\right|$. The series therefore converges if $\left|\dfrac{2}{x}\right| < 1$. If $x > 0$, $\left|\dfrac{2}{x}\right| = \dfrac{2}{x}$, which is less

than 1 if $2 < x$. If $x < 0$, $\left|\dfrac{2}{x}\right| = -\dfrac{2}{x}$, which is less than 1 if $-2 > x$. Now for the endpoints:

$$x = 2 \text{ yields } 1 + 1 + 1 + 1 + \ldots, \text{ which diverges}$$

$$x = -2 \text{ yields } -1 + 1 - 1 + 1 - \ldots, \text{ which diverges}$$

The answer is $|x| > 2$.

A37. **(C)** The function and its first three derivatives at $\dfrac{\pi}{4}$ are $\sin\dfrac{\pi}{4} = \dfrac{1}{\sqrt{2}}$; $\cos\dfrac{\pi}{4} = \dfrac{1}{\sqrt{2}}$;

$-\sin\dfrac{\pi}{4} = -\dfrac{1}{\sqrt{2}}$; and $-\cos\dfrac{\pi}{4} = -\dfrac{1}{\sqrt{2}}$. $P_3(x)$ is choice (C).

A38. **(D)** $T_3(1) = \dfrac{h(1)}{0!} + \dfrac{h'(1)}{1!}(x - 1) + \dfrac{h''(1)}{2!}(x - 1)^2 + \dfrac{h'''(1)}{3!}(x - 1)^3$

$$= \frac{6}{1} + \frac{6}{1}(x - 1) + \frac{6}{2}(x - 1)^2 + \frac{6}{6}(x - 1)^3$$

B1. **(A)** If $\sum\limits_{k=1}^{\infty} u_k$ converges, so does $\sum\limits_{k=m}^{\infty} u_k$, where m is any positive integer; but their *sums* are probably different.

B2. **(D)** Note the following counterexamples:

(A) $\sum \dfrac{(-1)^{n-1}}{n}$

(B) $\sum \dfrac{1}{n}$

(C) $\sum \dfrac{(-1)^{n-1} \cdot n}{2n - 1}$

(D) $\sum \left(-\dfrac{3}{2}\right)^{n-1}$

B3. **(C)** Note that $\ln q$ is defined only if $q > 0$ and that the derivatives must exist at $x = a$ in the formula for the Taylor series on page 415.

B4. **(D)** (A), (B), and (C) are all true statements.

B5. **(D)** The Maclaurin series $\sin x = x - \dfrac{x^3}{3!} + \dfrac{x^5}{5!} - \dfrac{x^7}{7!} + \cdots$ converges by the Alternating Series

Test, so the error $|R_4|$ is less than the first omitted term. For $x = 1$, we have $\dfrac{1}{5!} < 0.009$.

B6. **(C)** $f(x) = a_0 + a_1 x + a_2 x^2 + a_3 x^3 + \cdots$; if $f(0) = 1$, then $a_0 = 1$.

$f'(x) = a_1 + 2a_2 x + 3a_3 x^2 + 4a_4 x^3 + \cdots$; $f'(0) = -f(0) = -1$,

so $a_1 = -1$. Since $f'(x) = -f(x)$, $f(x) = -f'(x)$:

$$1 - x + a_2 x^2 + a_3 x^3 + \cdots = -\left(-1 + 2a_2 x + 3a_3 x^2 + 4a_4 x^3 + \ldots\right)$$

identically. Thus,

$$-2a_2 = -1 \Rightarrow a_2 = \frac{1}{2}; \; -3a_3 = a_2 \Rightarrow a_3 = -\frac{1}{3!}; \; -4a_4 = a_3 \Rightarrow a_4 = -\frac{1}{4!}; \cdots$$

It is clear, then, that

$$f(x) = 1 - x + \frac{x^2}{2!} - \frac{x^3}{3!} + \frac{x^4}{4!} - \cdots$$

$$f(0.2) = 1 - 0.2 + \frac{(0.2)^2}{2!} - \frac{(0.2)^3}{3!} + R_3 = 0.818\bar{6} + R_3$$

R_3 is the error $f(0.2) - P_3(0.2)$. This series satisfies the Alternating Series Test for $x > 0$. Thus we can use the Alternating Series Error Bound. Therefore, we use the first omitted term to bound the error, in this case $R_3 \leq \dfrac{(0.2)^4}{4!} = 0.0000\bar{6}$. The error bound is a positive value, so $P_3(0.2) \leq f(0.2) \leq P_3(0.2) + 0.0000\bar{6}$. This gives us $0.818\bar{6} \leq f(0.2) \leq 0.8187\bar{3}$. Thus four terms of the series reveal that to three decimal places, $f(0.2) = 0.818$.

NOTE: Since you were told that $f'(x) = -f(x)$, you may determine that $f(x) = e^{-x}$. You may recognize the above series as the Maclaurin series for $f(x) = e^{-x}$.

B7. **(C)** Use a calculator to verify that the ratio $\dfrac{\pi^3}{3\pi}$ (of the given geometric series) equals approximately 0.98. Since the ratio $r < 1$, the sum of the series equals

$\dfrac{a}{1-r}$ or $\dfrac{\pi^3}{3\pi} \cdot \dfrac{1}{1 - \frac{\pi^3}{3\pi}}$. Simplify to get (C).

B8. **(A)** Since the given series converges by the Alternating Series Test, the error is less in absolute value than the first term dropped, that is, less than $\left|\dfrac{(-1)^{300}}{903 - 1}\right| \approx 0.0011$. Choice (A) is closest to this approximation.

B9. **(B)** The Maclaurin expansion is

$$e^x = 1 + x + \frac{x^2}{2!} + \frac{x^3}{3!} + \cdots + \frac{x^n}{n!} + \cdots$$

The Lagrange remainder R, after n terms, for some c in the interval $|x| \leq 2$, is

$$R = \frac{f^{(n+1)}(c) \cdot c^{n+1}}{(n+1)!} = \frac{e^c c^{n+1}}{(n+1)!}$$

Since R is greatest when $c = 2$, n needs to satisfy the inequality

$$\frac{e^2 2^{n+1}}{(n+1)!} < 0.001$$

Using a calculator to evaluate $y = \dfrac{e^2 2^{x+1}}{(x+1)!}$ successively at various integral values of x gives $y(8) > 0.01$, $y(9) > 0.002$, $y(10) < 3.8 \times 10^{-4} < 0.0004$. Thus we achieve the desired accuracy with a Taylor polynomial at 0 of degree at least 10.

11

Miscellaneous Multiple-Choice Practice Questions

These questions provide further practice for Parts A and B of Section I of the examination.
The answer explanations begin on page 459.

PART A. NO CALCULATOR—DIRECTIONS: Answer these questions *without* using your calculator.

A1. Which of the following functions is continuous at $x = 0$?

(A) $f(x) \begin{cases} = \sin \dfrac{1}{x} & \text{for } x \neq 0 \\ = 0 & \text{for } x = 0 \end{cases}$

(B) $f(x) \begin{cases} = \dfrac{x}{x} & \text{for } x \neq 0 \\ = 0 & \text{for } x = 0 \end{cases}$

(C) $f(x) \begin{cases} = x \sin \dfrac{1}{x} & \text{for } x \neq 0 \\ = 0 & \text{for } x = 0 \end{cases}$

(D) $f(x) = \dfrac{x + 1}{x}$

A2. Which of the following statements about the graph of $y = \dfrac{x^2 + 1}{x^2 - 1}$ is *not* true?

(A) The graph is symmetric to the y-axis.

(B) There is no y-intercept.

(C) The graph has one horizontal asymptote.

(D) There is no x-intercept.

A3. $\lim\limits_{x \to 1^-} \left([x] - |x| \right)$

(A) $= -1$

(B) $= 0$

(C) $= 1$

(D) $= 2$

A4. The x-coordinate of the point on the curve $y = x^2 - 2x + 3$ at which the tangent is perpendicular to the line $x + 3y + 3 = 0$ is

(A) $-\dfrac{5}{2}$

(B) $-\dfrac{1}{2}$

(C) $\dfrac{7}{6}$

(D) $\dfrac{5}{2}$

A5. $\lim\limits_{x \to 1} \dfrac{\dfrac{3}{x} - 3}{x - 1}$ is

(A) -3

(B) -1

(C) 1

(D) 3

A6. For polynomial function p, $p''(2) = -4$, $p''(4) = 0$, and $p''(5) = 2$. Which must be true?

(A) p has an inflection point at $x = 4$.

(B) p has a minimum at $x = 4$.

(C) p has a root at $x = 4$.

(D) None of (A)–(C) must be true.

A7. $\displaystyle\int_0^6 |x - 4| \, dx =$

(A) 6

(B) 8

(C) 10

(D) 12

A8. $\lim\limits_{x\to\infty} \dfrac{3+x-2x^2}{4x^2+9}$ is

(A) $-\dfrac{1}{2}$

(B) $\dfrac{1}{2}$

(C) 1

(D) nonexistent

A9. The maximum value of the function $f(x) = x^4 - 4x^3 + 6$ on $[1,4]$ is

(A) 1

(B) 0

(C) 3

(D) 6

A10. Let $f(x) = \begin{cases} \dfrac{\sqrt{x+4}-3}{x-5} & \text{for } x \neq 5 \\ c & \text{for } x = 5 \end{cases}$, and let f be continuous at $x = 5$. Then $c =$

(A) 0

(B) $\dfrac{1}{6}$

(C) 1

(D) 6

A11. $\displaystyle\int_0^{\pi/2} \cos^2 x \sin x \, dx =$

(A) -1

(B) $-\dfrac{1}{3}$

(C) $\dfrac{1}{3}$

(D) 1

A12. If $\sin x = \ln y$ and $0 < x < \pi$, then, in terms of x, $\dfrac{dy}{dx}$ equals

(A) $e^{\sin x} \cos x$

(B) $e^{-\sin x} \cos x$

(C) $\dfrac{e^{\sin x}}{\cos x}$

(D) $e^{\cos x}$

A13. If $f(x) = x \cos x$, then $f'\left(\dfrac{\pi}{2}\right)$ equals

(A) $\dfrac{\pi}{2}$

(B) 0

(C) -1

(D) $-\dfrac{\pi}{2}$

A14. An equation of the tangent to the curve $y = e^x \ln x$, where $x = 1$, is

(A) $y = ex$

(B) $y = e(x - 1)$

(C) $y = ex + 1$

(D) $y = x - 1$

A15. If the displacement from the origin of a particle moving along the x-axis is given by $s = 3 + (t - 2)^4$, then the number of times the particle reverses direction is

(A) 0

(B) 1

(C) 2

(D) 3

A16. $\displaystyle\int_{-1}^{0} e^{-x} \, dx$ equals

(A) $1 - e$

(B) $e - 1$

(C) $1 - \dfrac{1}{e}$

(D) $e + 1$

A17. If $f(x) = \begin{cases} x^2 & \text{for } x \leq 2 \\ 4x - x^2 & \text{for } x > 2 \end{cases}$, then $\displaystyle\int_{-1}^{4} f(x) \, dx$ equals

(A) 7

(B) $\dfrac{23}{3}$

(C) $\dfrac{25}{3}$

(D) 9

A18. If the position of a particle on a line at time t is given by $s = t^3 + 3t$, then the speed of the particle is decreasing when

(A) $-1 < t < 1$

(B) $-1 < t < 0$

(C) $t < 0$

(D) $t > 0$

CHALLENGE A19. A rectangle with one side on the x-axis is inscribed in the triangle formed by the lines $y = x$, $y = 0$, and $2x + y = 12$. The area of the largest such rectangle is

(A) 6

(B) 3

(C) $\frac{5}{2}$

(D) 5

A20. The x-value of the first-quadrant point that is on the curve of $x^2 - y^2 = 1$ and closest to the point $(3,0)$ is

(A) 1

(B) $\frac{3}{2}$

(C) 2

(D) 3

A21. If $y = \ln(4x + 1)$, then $\frac{d^2y}{dx^2}$ is

(A) $\frac{-1}{(4x+1)^2}$

(B) $\frac{-4}{(4x+1)^2}$

(C) $\frac{-16}{(4x+1)^2}$

(D) $\frac{-1}{16(4x+1)^2}$

A22. The region bounded by the parabolas $y = x^2$ and $y = 6x - x^2$ is rotated about the x-axis so that a vertical line segment cut off by the curves generates a ring. The value of x for which the ring of largest area is obtained is

(A) 3

(B) $\frac{5}{2}$

(C) 2

(D) $\frac{3}{2}$

A23. $\int \frac{dx}{x \ln x}$ equals

(A) $\ln|\ln x| + C$

(B) $-\frac{1}{\ln^2 x} + C$

(C) $\frac{(\ln x)^2}{2} + C$

(D) $\ln x + C$

A24. The volume obtained by rotating the region bounded by $x = y^2$ and $x = 2 - y^2$ about the y-axis is equal to

(A) $\frac{16\pi}{3}$

(B) $\frac{32\pi}{3}$

(C) $\frac{32\pi}{15}$

(D) $\frac{64\pi}{15}$

A25. The general solution of the differential equation $\frac{dy}{dx} = \frac{1 - 2x}{y}$ is a family of

(A) circles

(B) hyperbolas

(C) parabolas

(D) ellipses

A26. Estimate $\int_0^4 \sqrt{25 - x^2}\, dx$ using a left rectangular sum and two subintervals of equal width.

(A) $5 + \sqrt{21}$

(B) $6 + 2\sqrt{21}$

(C) $8 + 2\sqrt{21}$

(D) $10 + 2\sqrt{21}$

A27. $\int_0^7 \sin \pi x\, dx =$

(A) $-\frac{2}{\pi}$

(B) 0

(C) $\frac{1}{\pi}$

(D) $\frac{2}{\pi}$

A28. $\lim\limits_{x \to 0} \frac{\tan 3x}{2x} =$

(A) 0

(B) $\frac{2}{3}$

(C) $\frac{3}{2}$

(D) ∞

A29. $\lim\limits_{h \to 0} \frac{\tan(\pi/4 + h) - 1}{h} =$

(A) 0

(B) $\frac{1}{2}$

(C) 1

(D) 2

CHALLENGE A30. The number of values of k for which $f(x) = e^x$ and $g(x) = k\sin x$ have a common point of tangency is

(A) 0
(B) 1
(C) large but finite
(D) infinite

A31. The curve $2x^2y + y^2 = 2x + 13$ passes through $(3,1)$. Use the line tangent to the curve there to find the approximate value of y at $x = 2.8$.

(A) 0.5
(B) 0.9
(C) 0.95
(D) 1.1

CHALLENGE A32. $\displaystyle\int \cos^3 x\, dx =$

(A) $\dfrac{\cos^4 x}{4} + C$

(B) $\sin x - \dfrac{\sin^3 x}{3} + C$

(C) $\sin x + \dfrac{\sin^3 x}{3} + C$

(D) $\cos x - \dfrac{\cos^3 x}{3} + C$

A33. The region bounded by $y = \tan x$, $y = 0$, and $x = \dfrac{\pi}{4}$ is rotated about the x-axis. The volume generated equals

(A) $\pi - \dfrac{\pi^2}{4}$

(B) $\pi\left(\sqrt{2} - 1\right)$

(C) $\dfrac{3\pi}{4}$

(D) $\pi\left(1 + \dfrac{\pi}{4}\right)$

A34. $\displaystyle\lim_{h\to 0}\dfrac{a^h - 1}{h}$, for the constant $a > 0$, equals

(A) 1
(B) a
(C) $\ln a$
(D) $a\ln a$

A35. Solutions of the differential equation whose slope field is shown here are most likely to be

(A) quadratic
(B) cubic
(C) exponential
(D) logarithmic

A36. $\displaystyle\lim_{h\to 0}\dfrac{1}{h}\int_{\frac{\pi}{4}}^{\frac{\pi}{4}+h}\dfrac{\sin x}{x}\, dx =$

(A) 0
(B) 1
(C) $\dfrac{\sqrt{2}}{2}$
(D) $\dfrac{2\sqrt{2}}{\pi}$

A37. The graph of g, shown below, consists of the arcs of two quarter-circles and two straight-line segments. The value of $\int_0^{12} g(x)\, dx$ is

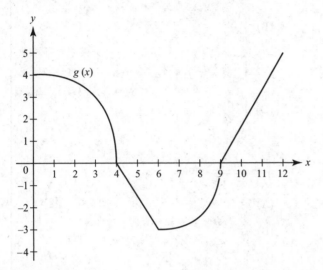

(A) $\pi + 2$

(B) $\dfrac{7\pi}{4} + \dfrac{9}{2}$

(C) $\dfrac{7\pi}{4} + 8$

(D) $\dfrac{25\pi}{4} + \dfrac{21}{2}$

A38. Which of these could be a particular solution of the differential equation whose slope field is shown here?

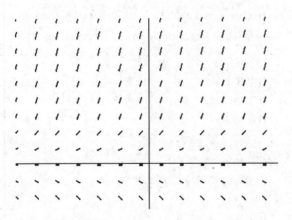

(A) $y = \dfrac{1}{x}$

(B) $y = \ln x$

(C) $y = e^x$

(D) $y = e^{-x}$

A39. What is the domain of the particular solution, $y = f(x)$, for $\dfrac{dy}{dx} = \dfrac{6x}{x^2 - 4}$ containing the point $(-1, \ln 3)$?

(A) $x < 0$

(B) $x > -2$

(C) $-2 < x < 2$

(D) $x \neq \pm 2$

A40. The slope field shown here is for the differential equation

(A) $y' = \dfrac{1}{x}$

(B) $y' = \ln x$

(C) $y' = e^x$

(D) $y' = y^2$

A41. If we substitute $x = \tan \theta$, which of the following is equivalent to $\int_0^1 \sqrt{1 + x^2}\, dx$?

(A) $\int_0^1 \sec \theta\, d\theta$

(B) $\int_0^{\pi/4} \sec \theta\, d\theta$

(C) $\int_0^{\pi/4} \sec^3 \theta\, d\theta$

(D) $\int_0^{\tan 1} \sec^3 \theta\, d\theta$

A42. If $x = 2 \sin u$ and $y = \cos 2u$, then a single equation in x and y is

(A) $x^2 + 4y^2 = 4$

(B) $x^2 + 2y = 2$

(C) $x^2 + y^2 = 4$

(D) $x^2 - 2y = 2$

BC ONLY

CHALLENGE

A43. The area bounded by the lemniscate with polar equation $r^2 = 2\cos 2\theta$ is equal to

(A) 4

(B) 1

(C) $\dfrac{1}{4}$

(D) 2

A44. $\displaystyle\int_{-\infty}^{\infty} \dfrac{dx}{x^2+1}$

(A) $= 0$

(B) $= \dfrac{\pi}{2}$

(C) $= \pi$

(D) $= 2\pi$

A45. The first four nonzero terms of the Maclaurin series (the Taylor series about $x = 0$) for

$f(x) = \dfrac{1}{1-2x}$ are

(A) $1 + 2x + 4x^2 + 8x^3$

(B) $1 - 2x + 4x^2 - 8x^3$

(C) $-1 - 2x - 4x^2 - 8x^3$

(D) $1 - x + x^2 - x^3$

A46. $\displaystyle\int x^2 e^{-x}\, dx =$

(A) $-\dfrac{1}{3}x^3 e^{-x} + C$

(B) $-x^2 e^{-x} + 2xe^{-x} + C$

(C) $-x^2 e^{-x} - 2xe^{-x} - 2e^{-x} + C$

(D) $-x^2 e^{-x} + 2xe^{-x} - 2e^{-x} + C$

A47. $\displaystyle\int_0^{\pi/2} \sin^2 x\, dx$ is equal to

(A) $\dfrac{1}{3}$

(B) $\dfrac{\pi}{4} - \dfrac{1}{4}$

(C) $\dfrac{\pi}{2}$

(D) $\dfrac{\pi}{4}$

A48. If $x = a\cot\theta$ and $y = a\sin^2\theta$, then $\dfrac{dy}{dx}$, when $\theta = \dfrac{\pi}{4}$, is equal to

(A) $\dfrac{1}{2}$

(B) -2

(C) 2

(D) $-\dfrac{1}{2}$

A49. Which of the following improper integrals diverges?

(A) $\displaystyle\int_{-\infty}^{0} e^x\, dx$

(B) $\displaystyle\int_0^1 \dfrac{dx}{x}$

(C) $\displaystyle\int_0^{\infty} e^{-x}\, dx$

(D) $\displaystyle\int_0^1 \dfrac{dx}{\sqrt{x}}$

A50. $\displaystyle\int_2^4 \dfrac{du}{\sqrt{16-u^2}}$ equals

(A) $\dfrac{\pi}{12}$

(B) $\dfrac{\pi}{6}$

(C) $\dfrac{\pi}{3}$

(D) $\dfrac{2\pi}{3}$

A51. $\displaystyle\lim_{x\to 0^+} \left(\dfrac{1}{x}\right)^x$ is

(A) 0

(B) 1

(C) ∞

(D) nonexistent

A52. A particle moves along the parabola $x = 3y - y^2$ so that $\dfrac{dy}{dt} = 3$ at all time t. The speed of the particle when it is at position $(2,1)$ is equal to

(A) 0

(B) 3

(C) $3\sqrt{2}$

(D) $\sqrt{13}$

A53. $\displaystyle\lim_{x\to 0^+} \dfrac{\cot x}{\ln x} =$

(A) $-\infty$

(B) -1

(C) 0

(D) 1

BC ONLY

A54. When rewritten as partial fractions, $\dfrac{3x+2}{x^2-x-12}$ includes which of the following?

    I. $\dfrac{1}{x+3}$

    II. $\dfrac{1}{x-4}$

    III. $\dfrac{2}{x-4}$

(A) none

(B) I only

(C) II only

(D) I and III only

A55. Using three terms of an appropriate Maclaurin series, estimate $\displaystyle\int_0^1 \dfrac{1-\cos x}{x}\,dx$.

(A) $\dfrac{1}{96}$

(B) $\dfrac{23}{96}$

(C) $\dfrac{25}{96}$

(D) undefined; the integral is improper

A56. The slope of the spiral $r=\theta$ at $\theta=\dfrac{\pi}{4}$ is

(A) $-\sqrt{2}$

(B) -1

(C) 1

(D) $\dfrac{4+\pi}{4-\pi}$

A57. $\displaystyle\int_0^{\pi/3} \sec^2 x \tan^2 x\,dx$ equals

(A) $\dfrac{\sqrt{3}}{27}$

(B) $\dfrac{\sqrt{3}}{3}$

(C) $\sqrt{3}$

(D) $3\sqrt{3}$

A58. A particle moves along a line with acceleration $a=6t$. If, when $t=0$, $v=1$, then the total distance traveled between $t=0$ and $t=3$ equals

(A) 30

(B) 28

(C) 27

(D) 26

A59. Air is escaping from a balloon at a rate of $R(t)=\dfrac{60}{1+t^2}$ cubic feet per minute, where t is measured in minutes. How much air, in cubic feet, escapes during the first minute?

(A) 15π

(B) 30

(C) 45

(D) $30\ln 2$

> **PART B. CALCULATOR ACTIVE—DIRECTIONS:** Some of the following questions require the use of a graphing calculator.

B1. The graph of function h is shown here. Which of these statements is (are) true?

$h(x)$

I. The first derivative is never negative.
II. The second derivative is constant.
III. The first and second derivatives equal 0 at the same point.

(A) I only
(B) III only
(C) I and II only
(D) I and III only

B2. Graphs of functions $f(x)$, $g(x)$, and $h(x)$ are shown below.

$y = f(x)$

$y = g(x)$

$y = h(x)$

Consider the following statements:

I. $g(x) = f'(x)$
II. $f(x) = g'(x)$
III. $h(x) = g''(x)$

Which of these statements is (are) true?

(A) I only
(B) II only
(C) II and III only
(D) I, II, and III

B3. If $y = \int_3^x \dfrac{1}{\sqrt{3+2t}}\,dt$, then $\dfrac{d^2y}{dx^2} =$

(A) $-\dfrac{1}{(3+2x)^{\frac{3}{2}}}$

(B) $\dfrac{3}{(3+2x)^{\frac{5}{2}}}$

(C) $\dfrac{\sqrt{3+2x}}{2}$

(D) $\dfrac{(3+2x)^{\frac{3}{2}}}{3}$

B4. If $\int_{-3}^{4} f(x)\,dx = 6$, then $\int_{-4}^{3} f(x+1)\,dx =$

(A) -6

(B) -5

(C) 5

(D) 6

B5. At what point in the interval $[1,1.5]$ is the rate of change of $f(x) = \sin x$ equal to its average rate of change on the interval?

(A) 1.058

(B) 1.239

(C) 1.253

(D) 1.399

B6. Suppose $f'(x) = x^2(x-1)$. Then $f''(x) = x(3x-2)$. Over which interval(s) is the graph of f both increasing and concave up?

I. $x < 0$

II. $0 < x < \dfrac{2}{3}$

III. $\dfrac{2}{3} < x < 1$

IV. $x > 1$

(A) I only

(B) II and IV only

(C) I and III only

(D) IV only

B7. Suppose $f'(x) = x^2(x-1)$. Then $f''(x) = x(3x-2)$. Which of the following statements is true about the graph of $f(x)$?

(A) The graph has one relative extremum and one inflection point.

(B) The graph has one relative extremum and two inflection points.

(C) The graph has two relative extrema and one inflection point.

(D) The graph has two relative extrema and two inflection points.

B8. The nth derivative of $\ln(x+1)$ at $x = 2$ equals

(A) $\dfrac{(-1)^n \cdot n!}{3^{n+1}}$

(B) $\dfrac{(-1)^{n-1}(n-1)!}{3^n}$

(C) $\dfrac{(-1)^{n+1} \cdot n!}{3^{n+1}}$

(D) $\dfrac{(-1)^{n+1}}{3^{n+1}}$

B9. If $f(x)$ is continuous at the point where $x = a$, which of the following statements may be false?

(A) $\lim\limits_{x \to a} f(x) = f(a)$

(B) $f'(a)$ exists

(C) $f(a)$ is defined

(D) $\lim\limits_{x \to a^-} f(x) = \lim\limits_{x \to a^+} f(x)$

B10. Suppose $\int_0^3 f(x+k)\,dx = 4$, where k is a constant. Then $\int_k^{3+k} f(x)\,dx$ equals

(A) 3

(B) $4 - k$

(C) 4

(D) $4 + k$

B11. The volume, in cubic feet, of an "inner tube"   **CHALLENGE**
with inner diameter 4 feet and outer diameter 8 feet is

(A) $6\pi^2$

(B) $12\pi^2$

(C) $24\pi^2$

(D) $48\pi^2$

B12. If $f(u) = \tan^{-1} u^2$ and $g(u) = e^u$, then the derivative of $f(g(u))$ is

(A) $\dfrac{2ue^u}{1 + u^4}$

(B) $\dfrac{2ue^{u^2}}{1 + u^4}$

(C) $\dfrac{2e^u}{1 + 4e^{2u}}$

(D) $\dfrac{2e^{2u}}{1 + e^{4u}}$

B13. If $\sin(xy) = y$, then $\dfrac{dy}{dx}$ equals

(A) $y\cos(xy) - 1$

(B) $\dfrac{1 - y\cos(xy)}{x\cos(xy)}$

(C) $\dfrac{y\cos(xy)}{1 - x\cos(xy)}$

(D) $\cos(xy)$

B14. Let $x > 0$. Suppose $\dfrac{d}{dx} f(x) = g(x)$ and $\dfrac{d}{dx} g(x) = f(\sqrt{x})$; then $\dfrac{d^2}{dx^2} f(x^2) =$

(A) $f(x^2)$

(B) $2xg(x^2)$

(C) $\dfrac{1}{2x} f(x)$

(D) $2g(x^2) + 4x^2 f(x)$

B15. The region bounded by $y = e^x$, $y = 1$, and $x = 2$ is rotated about the x-axis. The volume of the solid generated is given by the integral

(A) $\pi \displaystyle\int_0^2 e^{2x} dx$

(B) $2\pi \displaystyle\int_1^{e^2} (2 - \ln y)(y - 1)\, dy$

(C) $\pi \displaystyle\int_0^2 \left(e^{2x} - 1\right) dx$

(D) $\pi \displaystyle\int_0^2 \left(e^x - 1\right)^2 dx$

B16. Suppose the function f is continuous on $1 \le x \le 2$, that $f'(x)$ exists on $1 < x < 2$, that $f(1) = 3$, and that $f(2) = 0$. Which of the following statements is *not* necessarily true?

(A) $\displaystyle\int_1^2 f(x)\, dx$ exists.

(B) There exists a number c in the open interval $(1,2)$ such that $f'(c) = 0$.

(C) If k is any number between 0 and 3, there is a number c between 1 and 2 such that $f(c) = k$.

(D) If c is any number such that $1 < c < 2$, then $\displaystyle\lim_{x \to c} f(x)$ exists.

B17. The region S in the figure is bounded by $y = \sec x$, the y-axis, and $y = 4$. What is the volume of the solid formed when S is rotated about the y-axis?

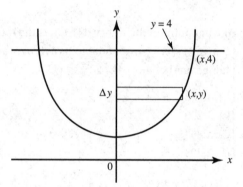

(A) 2.279

(B) 5.692

(C) 11.385

(D) 17.217

B18. If 40 grams of a radioactive substance decomposes to 20 grams in 2 years, then, to the nearest gram, the amount left after 3 years is

(A) 10

(B) 12

(C) 14

(D) 16

B19. An object in motion along a line has acceleration $a(t) = \pi t + \dfrac{2}{1+t^2}$ and is at rest when $t = 1$.
Its average velocity from $t = 0$ to $t = 2$ is

(A) 0.362

(B) 0.274

(C) 3.504

(D) 4.249

B20. Find the area bounded by $y = \tan x$ and
$x + y = 2$, and above the x-axis on the
interval [0,2].

(A) 0.919

(B) 1.013

(C) 1.077

(D) 1.494

CHALLENGE B21. An ellipse has major axis 20 and minor axis 10.
Rounded off to the nearest integer, the maxi-
mum area of an inscribed rectangle is

(A) 50

(B) 79

(C) 82

(D) 100

B22. The average value of $y = x \ln x$ on the interval
$1 \le x \le e$ is

(A) 0.772

(B) 1.221

(C) 1.359

(D) 2.097

B23. Let $f(x) = \displaystyle\int_0^x \left(1 - 2\cos^3 t\right) dt$ for $0 \le x \le 2\pi$. On
which interval is f increasing?

(A) $0 < x < \pi$

(B) $0.654 < x < 5.629$

(C) $0.654 < x < 2\pi$

(D) $\pi < x < 2\pi$

B24. The table shows the speed of an object
(in ft/sec) at certain times during a 6-second
period. Estimate its acceleration $\left(\text{in ft/sec}^2\right)$ at
$t = 2$ seconds.

time, sec	0	1	3	6
speed, ft/sec	30	22	12	6

(A) -10

(B) -6

(C) -5

(D) $\dfrac{-10}{3}$

B25. A maple syrup storage tank 16 feet high hangs
on a wall. The back is in the shape of the parabola $y = x^2$ and all cross sections parallel to the
floor are squares. If syrup is pouring in at the
rate of 12 ft^3/hr, how fast (in ft/hr) is the syrup
level rising when it is 9 feet deep?

(A) $\dfrac{2}{27}$

(B) $\dfrac{1}{3}$

(C) $\dfrac{4}{3}$

(D) 36

B26. In a protected area (no predators, no hunters), the deer population increases at a rate of $\frac{dP}{dt} = k(1000 - P)$, where $P(t)$ represents the population of deer at t years. If 300 deer were originally placed in the area and a census showed the population had grown to 500 in 5 years, how many deer will there be after 10 years?

(A) 608
(B) 643
(C) 700
(D) 833

B27. Shown is the graph of $f(x) = \dfrac{4}{x^2 + 1}$.

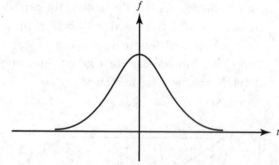

Let $H(x) = \displaystyle\int_0^x f(t)\, dt$. The local linearization of H at $x = 1$ is

(A) $y = 2x$
(B) $y = -2x - 4$
(C) $y = 2x + \pi - 2$
(D) $y = -2x + \pi + 2$

B28. A smokestack 100 feet tall is used to treat industrial emissions. The diameters, measured at selected heights, are shown in the table. Using a left Riemann Sum with the 4 subintervals indicated in the table, estimate the volume of the smokestack to the nearest cubic foot.

ht	dia
100'	5'
70'	7'
50'	9'
10'	15'
0'	17'

(A) 1160
(B) 5671
(C) 8718
(D) 11765

For Questions B29–B33, the table shows the values of differentiable functions f and g.

x	f	f'	g	g'
1	2	$\frac{1}{2}$	-3	5
2	3	1	0	4
3	4	2	2	3
4	6	4	3	$\frac{1}{2}$

B29. If $P(x) = \dfrac{f(x)}{g(x)}$, then $P'(3) =$

(A) -2
(B) $-\dfrac{8}{9}$
(C) $\dfrac{2}{3}$
(D) 2

B30. If $H(x) = f(g(x))$, then $H'(3) =$

  (A) 1
  (B) 2
  (C) 3
  (D) 6

B31. If $M(x) = f(x) \cdot g(x)$, then $M'(3) =$

  (A) 2
  (B) 6
  (C) 8
  (D) 16

B32. If $K(x) = g^{-1}(x)$, then $K'(3) =$

  (A) $-\dfrac{1}{3}$

  (B) $\dfrac{1}{3}$

  (C) $\dfrac{1}{2}$

  (D) 2

B33. If $R(x) = \sqrt{f(x)}$, then $R'(3) =$

  (A) $\dfrac{1}{4}$

  (B) $\dfrac{1}{2\sqrt{2}}$

  (C) $\dfrac{1}{2}$

  (D) $\sqrt{2}$

B34. Water is poured into a spherical tank at a constant rate. If $W(t)$ is the rate of increase of the depth of the water, then W is

  (A) linear and increasing
  (B) linear and decreasing
  (C) concave up
  (D) concave down

B35. The graph of f' is shown below. If $f(7) = 3$ then $f(1) =$

  (A) -10
  (B) -4
  (C) 10
  (D) 16

B36. At an outdoor concert, the crowd stands in front of the stage filling a semicircular disk of radius 100 yards. The approximate density of the crowd x yards from the stage is given by

$$D(x) = \frac{20}{2\sqrt{x} + 1}$$

people per square yard. About how many people are at the concert?

  (A) 200
  (B) 19500
  (C) 21000
  (D) 165000

B37. The Centers for Disease Control announced that, although more AIDS cases were reported this year, the rate of increase is slowing down. If we graph the number of AIDS cases as a function of time, the curve is currently

(A) increasing and concave down
(B) increasing and concave up
(C) decreasing and concave down
(D) decreasing and concave up

The graph below is for Questions B38–B40. It shows the velocity, in feet per second, for $0 < t < 8$, of an object moving along a straight line.

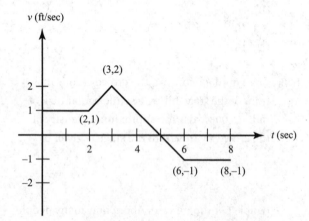

B38. The object's average speed (in ft/sec) for this 8-second interval was

(A) $\dfrac{3}{8}$

(B) 1

(C) $\dfrac{8}{3}$

(D) 8

B39. When did the object return to the position it occupied at $t = 2$?

(A) $t = 6$
(B) $t = 7$
(C) $t = 8$
(D) never

B40. The object's average acceleration (in ft/sec²) for this 8-second interval was

(A) -2

(B) $-\dfrac{1}{4}$

(C) $\dfrac{1}{4}$

(D) 1

B41. If a block of ice melts at the rate of $M(t) = \dfrac{72}{2t+3}$ cm³/min, how much ice melts during the first 3 minutes?

(A) 8 cm³
(B) 16 cm³
(C) 21 cm³
(D) 40 cm³

B42. A particle moves counterclockwise on the circle $x^2 + y^2 = 25$ with a constant speed of 2 ft/sec. Its velocity vector, **v**, when the particle is at $(3,4)$, equals **BC ONLY**

(A) $\left\langle -\dfrac{8}{5}, \dfrac{6}{5} \right\rangle$

(B) $\left\langle \dfrac{8}{5}, -\dfrac{6}{5} \right\rangle$

(C) $\left\langle -2\sqrt{3}, 2 \right\rangle$

(D) $\left\langle 2, -2\sqrt{3} \right\rangle$

B43. Let $\mathbf{R} = \langle a\cos kt, a\sin kt \rangle$ be the (position) vector from the origin to a moving point $P(x,y)$ at time t, where a and k are positive constants. The acceleration vector, **a**, equals

(A) $-k^2\mathbf{R}$
(B) $a^2k^2\mathbf{R}$
(C) $-a\mathbf{R}$
(D) $-ak^2 \langle a\cos kt, a\sin kt \rangle$

B44. The length of the curve $y = 2^x$ between $(0,1)$ and $(2,4)$ is

(A) 3.141
(B) 3.664
(C) 4.823
(D) 7.199

B45. The position of a moving object is given by $P(t) = \left(3t, e^t\right)$. Its acceleration is

(A) constant in both magnitude and direction
(B) constant in magnitude only
(C) constant in direction only
(D) constant in neither magnitude nor direction

B46. Suppose we plot a particular solution of $\dfrac{dy}{dx} = 4y$ from initial point (0,1) using Euler's method. After one step of size $\Delta x = 0.1$, how big is the error?

(A) 0.09
(B) 1.09
(C) 1.49
(D) 1.90

B47. We use the first three terms to estimate $\displaystyle\sum_{n=0}^{\infty} \frac{(-1)^n}{n^2 + 1}$. Which of the following statements is (are) true?

  I. The estimate is 0.7.
  II. The estimate is too low.
  III. The estimate is off by less than 0.1.

(A) I only
(B) III only
(C) I and III only
(D) I, II, and III

B48. Which of these diverges?

(A) $\displaystyle\sum_{n=1}^{\infty} \frac{2}{3^n}$

(B) $\displaystyle\sum_{n=1}^{\infty} \frac{2}{3n}$

(C) $\displaystyle\sum_{n=1}^{\infty} \frac{2}{n^3}$

(D) $\displaystyle\sum_{n=1}^{\infty} \frac{2n}{3^n}$

B49. Find the radius of convergence of $\displaystyle\sum_{n=1}^{\infty} \frac{n!}{n^n} x^n$.

(A) 0
(B) $\dfrac{1}{e}$
(C) 1
(D) e

B50. When we use $e^x \approx 1 + x + \dfrac{x^2}{2}$ to estimate $\sqrt{e}$, the Lagrange remainder is no greater than

(A) 0.021
(B) 0.035
(C) 0.057
(D) 0.063

B51. An object in motion along a curve has position $P(t) = (\tan t, \cos 2t)$ for $0 \le t \le 1$. How far does it travel?

(A) 0.952
(B) 1.726
(C) 1.910
(D) 2.140

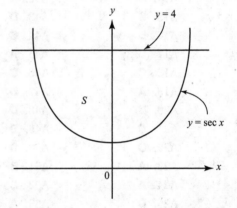

B52. The region S in the figure shown above is bounded by $y = \sec x$ and $y = 4$. What is the volume of the solid formed when S is rotated about the x-axis?

(A) 11.385
(B) 23.781
(C) 53.126
(D) 108.177

Answer Key

A1. **C**	A29. **D**	A57. **C**	B26. **B**
A2. **B**	A30. **D**	A58. **A**	B27. **C**
A3. **A**	A31. **D**	A59. **A**	B28. **D**
A4. **D**	A32. **B**	B1. **D**	B29. **A**
A5. **A**	A33. **A**	B2. **C**	B30. **C**
A6. **D**	A34. **C**	B3. **A**	B31. **D**
A7. **C**	A35. **D**	B4. **D**	B32. **D**
A8. **A**	A36. **D**	B5. **C**	B33. **C**
A9. **D**	A37. **B**	B6. **D**	B34. **C**
A10. **B**	A38. **C**	B7. **B**	B35. **B**
A11. **C**	A39. **C**	B8. **B**	B36. **B**
A12. **A**	A40. **A**	B9. **B**	B37. **A**
A13. **D**	A41. **C**	B10. **C**	B38. **B**
A14. **B**	A42. **B**	B11. **A**	B39. **D**
A15. **B**	A43. **D**	B12. **D**	B40. **B**
A16. **B**	A44. **C**	B13. **C**	B41. **D**
A17. **C**	A45. **A**	B14. **D**	B42. **A**
A18. **C**	A46. **C**	B15. **C**	B43. **A**
A19. **A**	A47. **D**	B16. **B**	B44. **B**
A20. **B**	A48. **D**	B17. **C**	B45. **C**
A21. **C**	A49. **B**	B18. **C**	B46. **A**
A22. **C**	A50. **C**	B19. **A**	B47. **C**
A23. **A**	A51. **B**	B20. **C**	B48. **B**
A24. **A**	A52. **C**	B21. **D**	B49. **D**
A25. **D**	A53. **A**	B22. **B**	B50. **B**
A26. **D**	A54. **D**	B23. **B**	B51. **D**
A27. **D**	A55. **B**	B24. **C**	B52. **D**
A28. **C**	A56. **D**	B25. **B**	

Answer Explanations

A1. (C) If $f(x) = x \sin \frac{1}{x}$ for $x \neq 0$ and $f(0) = 0$ then,

$$\lim_{x \to 0} f(x) = 0 = f(0)$$

Thus, this function is continuous at 0. In (A), $\lim_{x \to 0} \sin \frac{1}{x}$ does not exist; in (B), f has a removable discontinuity; and in (D), f has an infinite discontinuity.

A2. (B) To find the y-intercept, let $x = 0$; $y = -1$.

A3. (A) $\lim_{x \to 1^-} [x] - \lim_{x \to 1^-} |x| = 0 - 1 = -1$

A4. (D) The line $x + 3y + 3 = 0$ has slope $-\frac{1}{3}$; a line perpendicular to it has slope 3.

The slope of the tangent to $y = x^2 - 2x + 3$ at any point is the derivative $2x - 2$. Set $2x - 2$ equal to 3.

A5. (A) $\lim_{x \to 1} \dfrac{\frac{3}{x} - 3}{x - 1}$ is $f'(1)$, where $f(x) = \frac{3}{x}$, $f'(x) = -\frac{3}{x^2}$. Or simplify the given

fraction to $\dfrac{3 - 3x}{x(x - 1)} = \dfrac{3(1 - x)}{x(x - 1)} = \dfrac{-3}{x} \ (x \neq 1)$.

A6. (D) Since $p''(2) = -4 < 0$ and $p''(5) = 2 > 0$, there must be an inflection point in the interval $(2,5)$. Although $p''(4) = 0$, p'' doesn't necessarily change sign at $x = 4$. In other words, there must be an inflection point on $(2,5)$, but it doesn't have to be at $x = 4$. As a counterexample to (A), (B), and (C), a polynomial with the given properties is the polynomial $p(x)$ with second derivative $p''(x) = (x - 3)(x - 4)^2$. For this function, $p''(2) = -4$, $p''(4) = 0$, and $p''(5) = 2$, and there is an inflection point on the interval $(2,5)$, but it occurs at $x = 3$. Also, there is not necessarily a root or minimum at $x = 4$.

A7. (C) $\displaystyle\int_0^6 |x - 4| \, dx = \int_0^4 (4 - x) \, dx + \int_4^6 (x - 4) \, dx = \left(4x - \frac{x^2}{2}\right)\Big|_0^4 + \left(\frac{x^2}{2} - 4x\right)\Big|_4^6$

$$= 8 + [(18 - 24) - (8 - 16)] = 8 + (-6 + 8) = 10$$

Save time by finding the area under $y = |x - 4|$ from a sketch!

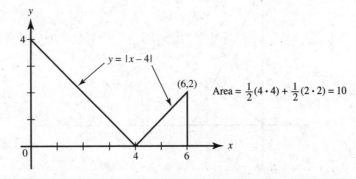

A8. (A) Since the degrees of numerator and denominator are the same, the limit as $x \to \infty$ is the ratio of the coefficients of the terms of highest degree: $\dfrac{-2}{4}$.

A9. (D) On the interval $[1,4]$, $f'(x) = 0$ only for $x = 3$. Since $f(3)$ is a relative minimum, check the endpoints to find that $f(4) = 6$ is the absolute maximum of the function.

A10. **(B)** To find $\lim f$ as $x \to 5$ (if it exists), multiply f by $\dfrac{\sqrt{x+4}+3}{\sqrt{x+4}+3}$.

$$f(x) = \frac{x-5}{(x-5)\left(\sqrt{x+4}+3\right)}$$

and if $x \neq 5$ this equals $\dfrac{1}{\sqrt{x+4}+3}$. So $\lim f(x)$ as $x \to 5$ is $\dfrac{1}{6}$. For f to be continuous at $x = 5$, $f(5)$ or c must also equal $\dfrac{1}{6}$.

A11. **(C)** Evaluate $-\dfrac{1}{3}\cos^3 x \Big|_0^{\pi/2}$.

A12. **(A)** $\cos x = \dfrac{1}{y}\dfrac{dy}{dx}$ and thus $\dfrac{dy}{dx} = y\cos x$. From the equation given, $y = e^{\sin x}$.

A13. **(D)** If $f(x) = x\cos x$, then $f'(x) = -x\sin x + \cos x$, and

$$f'\left(\frac{\pi}{2}\right) = -\frac{\pi}{2} \cdot 1 + 0$$

A14. **(B)** If $y = e^x \ln x$, then $\dfrac{dy}{dx} = \dfrac{e^x}{x} + e^x \ln x$, which equals e when $x = 1$. Since also $y = 0$ when $x = 1$, an equation of the tangent is $y = e(x-1)$.

A15. **(B)** $v = 4(t-2)^3$ and changes sign exactly once, when $t = 2$.

A16. **(B)** Evaluate $-e^{-x}\Big|_{-1}^0$.

A17. **(C)** $\displaystyle\int_{-1}^4 f(x)\,dx = \int_{-1}^2 x^2\,dx + \int_2^4 \left(4x - x^2\right)dx = \dfrac{x^3}{3}\Big|_{-1}^2 + \left(2x^2 - \dfrac{x^3}{3}\right)\Big|_2^4$

A18. **(C)** Since $v = 3t^2 + 3$, it is always positive, while $a = 6t$ and is positive for $t > 0$ but negative for $t < 0$. The speed therefore increases for $t > 0$ but decreases for $t < 0$.

A19. **(A)** Note from the figure that the area, A, of a typical rectangle is

$$A = \left(x_2 - x_1\right) \cdot y = \left(\frac{12-y}{2} - y\right) \cdot y = 6y - \frac{3y^2}{2}$$

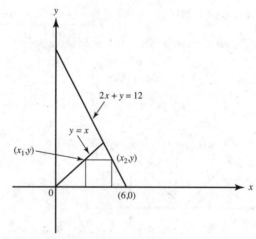

For $y = 2$, $\dfrac{dA}{dy} = 0$. Note that $\dfrac{d^2A}{dy^2}$ is always negative.

A20. **(B)** If S represents the square of the distance from $(3,0)$ to a point (x,y) on the curve, then $S = (3-x)^2 + y^2 = (3-x)^2 + \left(x^2 - 1\right)$. Setting $\dfrac{ds}{dx} = 0$ yields the minimum distance at $x = \dfrac{3}{2}$.

A21. **(C)** $\dfrac{dy}{dx} = \dfrac{4}{4x+1} = 4(4x+1)^{-1}$, so $\dfrac{d^2y}{dx^2} = 4\left(-1(4x+1)^{-2} \cdot 4\right)$

A22. **(C)** See the figure. Since the area, A, of the ring equals $\pi\left(y_2{}^2 - y_1{}^2\right)$,

$$A = \pi\left[\left(6x - x^2\right)^2 - x^4\right] = \pi\left[36x^2 - 12x^3 + x^4 - x^4\right]$$

and $\dfrac{dA}{dx} = \pi\left(72x - 36x^2\right) = 36\pi x\,(2 - x)$.

It can be verified that $x = 2$ produces the maximum area.

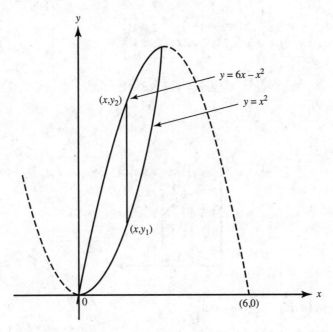

A23. **(A)** This is of type $\displaystyle\int \frac{du}{u}$ with $u = \ln x$: $\displaystyle\int \frac{\frac{1}{x}dx}{\ln x}$

A24. **(A)**

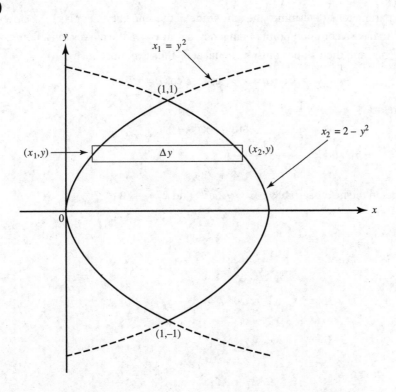

About the y-axis. Washer.

$$\Delta V = \pi\left(x_2{}^2 - x_1{}^2\right)\Delta y, \text{ so } V = 2\pi \int_0^1 \left[\left(2 - y^2\right)^2 - y^4\right] dy = 2\pi \int_0^1 \left(4 - 4y^2\right) dy$$

A25. **(D)** Separating variables, we get $y\, dy = (1 - 2x)\, dx$. Integrating gives

$$\tfrac{1}{2}y^2 = x - x^2 + C$$

or

$$y^2 = 2x - 2x^2 + k$$

or

$$2x^2 + y^2 - 2x = k$$

A26. **(D)** $2(5) + 2\sqrt{21}$

A27. **(D)** $\dfrac{1}{\pi}\displaystyle\int_0^7 \sin \pi x \,(\pi\, dx) = \left.\dfrac{-1}{\pi}\cos\,(\pi x)\right|_0^7$

A28. **(C)** Use L'Hospital's Rule or rewrite the expression as $\displaystyle\lim_{x \to 0}\dfrac{\sin 3x}{3x}\cdot\dfrac{1}{\cos 3x}\cdot\dfrac{3}{2}$.

A29. **(D)** For $f(x) = \tan x$, this is $f'\left(\dfrac{\pi}{4}\right) = \sec^2\left(\dfrac{\pi}{4}\right)$.

A30. **(D)** The parameter k determines the amplitude of the sine curve. For $g(x) = k \sin x$ and $f(x) = e^x$ to have a common point of tangency, say at $x = q$, the curves must both go through (q,y) and their slopes must be equal at q. Thus, we must have

$$k \sin q = e^q \quad \text{and} \quad k \cos q = e^q$$

and therefore

$$\sin q = \cos q$$

Thus, $q = \dfrac{\pi}{4} \pm n\pi$ and $k = \dfrac{e^q}{\sin\left(\dfrac{\pi}{4}\pm n\pi\right)}$.

The figure on the next page shows $k_1 = \sqrt{2}\, e^{\pi/4}$ and $k_2 = -\sqrt{2}\, e^{-3\pi/4}$.

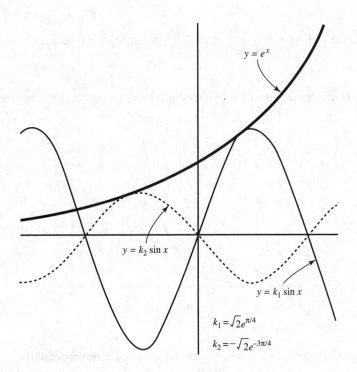

$$k_1 = \sqrt{2}\,e^{\pi/4}$$

$$k_2 = -\sqrt{2}\,e^{-3\pi/4}$$

A31. (D) We differentiate implicitly to find the slope $\dfrac{dy}{dx}$:

$$2\left(x^2 \frac{dy}{dx} + 2xy\right) + 2y \frac{dy}{dx} = 2$$

$$\frac{dy}{dx} = \frac{1 - 2xy}{x^2 + y}$$

At $(3,1)$, $\dfrac{dy}{dx} = -\dfrac{1}{2}$. The linearization is $y \approx -\dfrac{1}{2}(x - 3) + 1$.

A32. (B) $\displaystyle \int \cos^3 x\, dx = \int \cos^2 x \cdot \cos x\, dx = \int \left(1 - \sin^2 x\right) \cos x\, dx$

$$= \int \cos x\, dx - \int \sin^2 x \cdot \cos x\, dx = \sin x - \frac{\sin^3 x}{3} + C$$

A33. (A) About the x-axis. Disk.

$$\Delta V = \pi y^2 \Delta x$$

$$V = \pi \int_0^{\pi/4} \tan^2 x\, dx = \pi \int_0^{\pi/4} \left(\sec^2 x - 1\right) dx = \pi \left[\tan x - x\right]\Big|_0^{\pi/4}$$

$$= \pi\left(1 - \frac{\pi}{4}\right)$$

A34. (C) Let $f(x) = a^x$; then $\displaystyle \lim_{h \to 0} \frac{a^h - 1}{h} = \lim_{h \to 0} \frac{a^{0+h} - a^0}{h} = f'(0) = a^0 \ln a = \ln a$.

A35. (D) $\dfrac{dy}{dx}$ is a function of x alone; curves appear to be asymptotic to the y-axis and to increase more slowly as $|x|$ increases.

A36. (D) The given limit is equivalent to

$$\lim_{h \to 0} \frac{F\left(\dfrac{\pi}{4} + h\right) - F\left(\dfrac{\pi}{4}\right)}{h} = F'\left(\frac{\pi}{4}\right)$$

where $F'(x) = \dfrac{\sin x}{x}$. The answer is $\dfrac{2\sqrt{2}}{\pi}$.

A37. **(B)** $\displaystyle\int_0^{12} g(x)\,dx = \int_0^4 g(x)\,dx + \int_4^6 g(x)\,dx + \int_6^9 g(x)\,dx + \int_9^{12} g(x)\,dx$

$$= 4\pi - 3 - \frac{9\pi}{4} + \frac{15}{2}$$

A38. **(C)** In the figure, the curve for $y = e^x$ has been superimposed on the slope field.

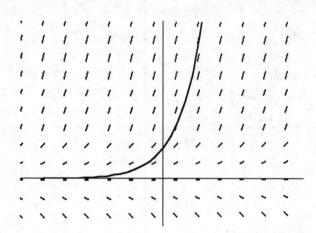

A39. **(C)** The general solution is $y = 3\ln|x^2 - 4| + C$. The differential equation $\dfrac{dy}{dx} = \dfrac{6x}{x^2 - 4}$ reveals that the derivative does not exist for $x = \pm 2$. The particular solution must be differentiable in an interval containing the initial value $x = -1$, so the domain is $-2 < x < 2$.

A40. **(A)** The solution curve shown is $y = \ln x$, so the differential equation is $y' = \dfrac{1}{x}$.

A41. **(C)** $\sqrt{1 + \tan^2\theta} = \sec\theta; \; dx = \sec^2\theta\,d\theta; \; 0 \le x \le 1; \text{ so } 0 \le \theta \le \dfrac{\pi}{4}.$

A42. **(B)** The equations may be rewritten as $\dfrac{x}{2} = \sin u$ and $y = 1 - 2\sin^2 u$, giving $y = 1 - 2 \cdot \dfrac{x^2}{4}.$

A43. **(D)** Use the formula for area in polar coordinates,

$$A = \frac{1}{2}\int_\alpha^\beta r^2\,d\theta$$

then the required area is given by

$$4 \cdot \frac{1}{2}\int_0^{\pi/4} 2\cos 2\theta$$

(See polar graph 63 in the Appendix.)

A44. **(C)** $\displaystyle\int_{-\infty}^{\infty}\frac{dx}{x^2+1}=\lim_{b\to\infty}\ \tan^{-1}x\ \Big|_{-b}^{b}=\frac{\pi}{2}-\left(-\frac{\pi}{2}\right)=\pi$

A45. **(A)** The first three derivatives of $\dfrac{1}{1-2x}$ are $\dfrac{2}{(1-2x)^2}$, $\dfrac{8}{(1-2x)^3}$, and $\dfrac{48}{(1-2x)^4}$.

The first four terms of the Maclaurin series (about $x=0$) are 1, $+2x$, $+\dfrac{8x^2}{2!}$, and $+\dfrac{48x^3}{3!}$.
Note also that $\dfrac{1}{1-2x}$ represents the sum of an infinite geometric series with first term 1 and common ratio $2x$. Hence,

$$\frac{1}{1-2x}=1+2x+(2x)^2+(2x)^3+\cdots$$

A46. **(C)** We use parts, first letting $u=x^2$, $dv=e^{-x}dx$; then $du=2x\,dx$, $v=-e^{-x}$ and

$$\int x^2e^{-x}dx=-x^2e^{-x}+2\int xe^{-x}dx$$

Now we use parts again, letting $u=x$, $dv=e^{-x}dx$; then $du=dx$, $v=-e^{-x}$ and

$$-x^2e^{-x}+2\int xe^{-x}dx=-x^2e^{-x}+2\left(-xe^{-x}+\int e^{-x}dx\right)$$

Alternatively, we could use the tic-tac-toe method (see page 223):

u		dv
x^2	$+$	e^{-x}
$2x$	$-$	$-e^{-x}$
2	$+$	e^{-x}
0		$-e^{-x}$

Then $\displaystyle\int x^2e^{-x}dx=x^2\left(-e^{-x}\right)-(2x)e^{-x}+2\left(-e^{-x}\right)+C.$

A47. **(D)** Use formula (20) in the Appendix to rewrite the integral as

$$\frac{1}{2}\int_0^{\pi/2}(1-\cos 2x)\,dx=\frac{1}{2}\left(x-\frac{\sin 2x}{2}\right)\Bigg|_0^{\pi/2}=\frac{1}{2}\left(\frac{\pi}{2}\right)$$

A48. **(D)** $\dfrac{dy}{dx}=\dfrac{\frac{dy}{d\theta}}{\frac{dx}{d\theta}}=\dfrac{2a\sin\theta\cos\theta}{-a\csc^2\theta}=-2\sin^3\theta\cos\theta$

A49. **(B)** Check to verify that each of the other improper integrals converges.

A50. **(C)** Note that the integral is improper.

$$\lim_{k\to4^-}\int_2^k\frac{du}{\sqrt{16-u^2}}=\lim_{k\to4^-}\frac{1}{4}\int_2^k\frac{du}{\sqrt{1-\frac{u^2}{16}}}=\lim_{k\to4^-}\frac{1}{4}\cdot4\int_2^k\frac{\frac{1}{4}du}{\sqrt{1-\frac{u^2}{16}}}=\lim_{k\to4^-}\sin^{-1}\frac{u}{4}\ \Bigg|\Bigg|_2^k=\frac{\pi}{3}$$

See Example 26, page 304.

A51. **(B)** Let $y=\left(\dfrac{1}{x}\right)^x$. Then $\ln y=-x\ln x$ and

$$\lim_{x\to0^+}\ln y=\lim_{x\to0^+}\frac{-\ln x}{1/x}$$

Now apply L'Hospital's Rule:

$$\lim_{x\to0^+}\ln y=\lim_{x\to0^+}\frac{-1/x}{-1/x^2}=0$$

So, if $\lim\limits_{x\to0^+}\ln y=0$, then $\lim\limits_{x\to0^+}y=1$.

A52. (C) The speed, $|\mathbf{v}|$, equals $\sqrt{\left(\sqrt{\dfrac{dx}{dt}}\right)^2 + \left(\sqrt{\dfrac{dy}{dt}}\right)^2}$, and since $x = 3y - y^2$,

$$\frac{dx}{dt} = (3 - 2y)\frac{dy}{dt} = (3 - 2y)\cdot 3$$

Then $|\mathbf{v}|$ is evaluated, using $y = 1$, and equals $\sqrt{(3)^2 + (3)^2}$.

A53. (A) This is an indeterminate form of type $\dfrac{\infty}{\infty}$; use L'Hospital's Rule:

$$\lim_{x\to 0^+}\frac{\cot x}{\ln x} = \lim_{x\to 0^+}\frac{-\csc^2 x}{1/x} = \lim_{x\to 0^+}\frac{x}{\sin x}\cdot\frac{-1}{\sin x} = -\infty$$

A54. (D) We find A and B such that $\dfrac{3x+2}{(x+3)(x-4)} = \dfrac{A}{x+3} + \dfrac{B}{x-4}$.

After multiplying by the common denominator, we have

$$3x + 2 = A(x - 4) + B(x + 3)$$

Substituting $x = -3$ yields $A = 1$, and $x = 4$ yields $B = 2$; hence,

$$\frac{3x+2}{x^2 - x - 12} = \frac{1}{x+3} + \frac{2}{x-4}$$

A55. (B) Since $\cos x \approx 1 - \dfrac{x^2}{2!} + \dfrac{x^4}{4!}$, $\dfrac{1 - \cos x}{x}\cdot\dfrac{x}{2} - \dfrac{x^3}{4!}$.

Then $\displaystyle\int_0^1 \frac{1 - \cos x}{x}dx \approx \left(\frac{x^2}{4} - \frac{x^4}{96}\right)\Big|_0^1 = \frac{1}{4} - \frac{1}{96}$.

Note that $\displaystyle\lim_{x\to 0^+}\frac{1 - \cos x}{x} = 0$, so the integral is proper.

A56. (D) We represent the spiral as $P(\theta) = (\theta\cos\theta, \theta\sin\theta)$. So

$$\frac{dy}{dx} = \frac{dy/d\theta}{dx/d\theta} = \frac{\theta\cos\theta + \sin\theta}{-\theta\sin\theta + \cos\theta} = \frac{\dfrac{\pi}{4}\cdot\dfrac{\sqrt{2}}{2} + \dfrac{\sqrt{2}}{2}}{-\dfrac{\pi}{4}\cdot\dfrac{\sqrt{2}}{2} + \dfrac{\sqrt{2}}{2}} = \frac{\pi/4 + 1}{-\pi/4 + 1}$$

A57. (C) $\displaystyle\int_0^{\pi/3}(\tan x)^2\left(\sec^2 x\, dx\right) = \frac{\tan^3 x}{3}\Big|_0^{\pi/3} = \frac{1}{3}\left(3\sqrt{3} - 0\right)$

A58. (A) $v(t) = v(0) + \displaystyle\int_0^t a(u)\,du = 1 + (3u^2)\Big|_0^t = 3t^2 + 1$. Total distance is $\displaystyle\int_0^3 |v(t)|\,dt = 30$.

A59. (A) $\displaystyle\int_0^1 \frac{60}{1 + t^2}\,dt = 60\arctan t\Big|_0^1 = 60\arctan 1 = 60\cdot\frac{\pi}{4}$

B1. (D) Since h is increasing, $h' \geq 0$. The graph of h is concave downward for $x < 2$ and upward for $x > 2$, so h'' changes sign at $x = 2$, where it appears that $h' = 0$ also.

B2. (C) I is false since, for example, $f'(-2) = f'(1) = 0$ but neither g(-2) nor g(1) equals zero.

II is true. Note that $f = 0$ where g has relative extrema, and f is positive, negative, then positive on intervals where g increases, decreases, then increases.

III is also true. Check the concavity of g: when the curve is concave down, $h < 0$; when up, $h > 0$.

B3. (A) If $y = \displaystyle\int_3^x \frac{1}{\sqrt{3 + 2t}}\,dt$, then $\dfrac{dy}{dx} = \dfrac{1}{\sqrt{3 + 2x}}$, so $\dfrac{d^2 y}{dx^2} = -\dfrac{1}{2}(3 + 2x)^{-3/2}(2)$.

B4. (D) $\displaystyle\int_{-4}^3 f(x + 1)dx$ represents the area of the same region as $\displaystyle\int_{-3}^4 f(x)dx$, translated one unit to the left.

B5. **(C)** According to the Mean Value Theorem, there exists a number c in the interval $[1,1.5]$ such that $f'(c) = \dfrac{f(1.5) - f(1)}{1.5 - 1}$. Use your calculator to solve the equation $\cos c = \dfrac{\sin 1.5 - \sin 1}{0.5}$ for c (in radians).

B6. **(D)** Here are the relevant sign lines:

We see that f' and f'' are both positive only if $x > 1$.

B7. **(B)** Note from the sign lines in Question B6 that f changes from decreasing to increasing at $x = 1$, so f has a local minimum there. Also, the graph of f changes from concave up to concave down at $x = 0$, then back to concave up at $x = \dfrac{2}{3}$; hence f has two points of inflection.

B8. **(B)** The derivatives of $\ln(x + 1)$ are $\dfrac{1}{x + 1}, \dfrac{-1}{(x + 1)^2}, \dfrac{+2!}{(x + 1)^3}, \dfrac{-(3!)}{(x + 1)^4}, \dots$

The nth derivative at $x = 2$ is $\dfrac{(-1)^{n-1}(n - 1)!}{3^n}$.

B9. **(B)** The absolute-value function $f(x) = |x|$ is continuous at $x = 0$, but $f'(0)$ does not exist.

B10. **(C)** Let $F'(x) = f(x)$; then $F'(x + k) = f(x + k)$;

$$\int_0^3 f(x + k)\,dx = F(3 + k) - F(k)$$
$$\int_k^{3+k} f(x)\,dx = F(3 + k) - F(k)$$

Or let $u = x + k$. Then $dx = du$; when $x = 0$, $u = k$; when $x = 3$, $u = 3 + k$.

B11. **(A)** See the figure. The equation of the generating circle is $(x - 3)^2 + y^2 = 1$, which yields $x = 3 \pm \sqrt{1 - y^2}$.

About the y-axis: $\Delta V = 2\pi \cdot 3 \cdot 2\sqrt{1 - y^2}\,\Delta y$

Thus, $V = 2\displaystyle\int_0^1 12\pi\sqrt{1 - y^2}\,dy$

$= 24\pi$ times the area of a quarter of a unit circle $= 6\pi^2$

B12. **(D)** Note that $f(g(u)) = \tan^{-1}(e^{2u})$; then the derivative is $\dfrac{1}{1 + (e^{2u})^2}(2e^{2u})$.

B13. **(C)** Let $y' = \dfrac{dy}{dx}$. Then $\cos(xy)[xy' + y] = y'$. Solve for y'.

B14. (D) $\dfrac{d^2}{dx^2}f(x^2) = \dfrac{d}{dx}\left[\dfrac{d}{dx}f(x^2)\right] = \dfrac{d}{dx}\left[\dfrac{d}{dx}f(x^2)\cdot\dfrac{dx}{dt}\right] = \dfrac{d}{dx}\left[g(x^2)\cdot 2x\right]$

$= g(x^2)\dfrac{d}{dx}(2x) + 2x\dfrac{d}{dx}g(x^2) = g(x^2)\cdot 2 + 2x\dfrac{d}{dx^2}g(x^2)\dfrac{dx^2}{dx}$

$= 2g(x^2) + 2x\cdot f\left(\sqrt{x^2}\right)\cdot 2x = 2g(x^2) + 4x^2\cdot f(x)$

B15. (C)

About the x-axis. Washer.

$$\Delta V = \pi\left(y^2 - 1^2\right)\Delta x$$

$$V = \pi\int_0^2 (e^{2x} - 1)dx$$

B16. (B) By the Mean Value Theorem, there is a number c in $(1,2)$ such that

$$f'(c) = \frac{f(2) - f(1)}{2 - 1} = -3$$

B17. (C) The enclosed region, S, is bounded by $y = \sec x$, the y-axis, and $y = 4$. It is to be rotated about the y-axis.

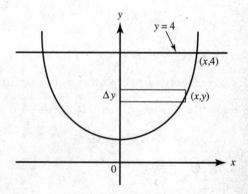

Use disks; then $\Delta V = \pi R^2 H = \pi (\text{arcsec } y)^2 \, \Delta y$. Using the calculator, we find that

$$\pi \int_1^4 \left(\arccos \left(\frac{1}{y} \right) \right)^2 dx \approx 11.385$$

B18. (C) If Q is the amount at time t, then $Q = 40e^{-kt}$. Since $Q = 20$ when $t = 2$, $k = -0.3466$.
Now find Q when $t = 3$, from $Q = 40e^{-(0.3466)(3)}$, getting $Q = 14$ to the nearest gram.

B19. (A) The velocity $v(t)$ is an antiderivative of $a(t)$, where $a(t) = \pi t + \dfrac{2}{1 + t^2}$.

So $v(t) = \dfrac{\pi t^2}{2} + 2 \arctan t + C$. Since $v(1) = 0$, $C = -\pi$.

Required average velocity $= \dfrac{1}{2 - 0} \displaystyle\int_0^2 v(t) dt$

$$= \frac{1}{2} \int_0^2 \left(\frac{\pi t^2}{2} + 2 \arctan t - \pi \right) dt \approx 0.362$$

B20. (C) Graph $y = \tan x$ and $y = 2 - x$ in $[-1,3] \times [-1,3]$ as shown below. Note that

$$\Delta A = \left(x_{\text{line}} - x_{\text{curve}} \right) \Delta y$$

$$= (2 - y - \arctan y) \, \Delta y$$

The limits are $y = 0$ and $y = b$, where b is the ordinate of the intersection of the curve and the line. Using the calculator, solve

$$\arctan y = 2 - y$$

and store the answer in memory as B. Evaluate the desired area:

$$\int_0^B (2 - y - \arctan y) \, dy \approx 1.077$$

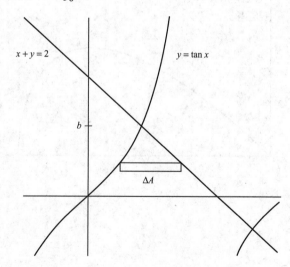

B21. **(D)** Center the ellipse at the origin and let (x,y) be the coordinates of the vertex of the inscribed rectangle in the first quadrant, as shown in the figure.

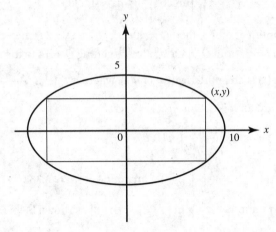

$$\frac{x^2}{100} + \frac{y^2}{25} = 1$$

To maximize the rectangle's area $A = 4xy$, solve the equation of the ellipse, getting

$$x = \sqrt{100 - 4y^2} = 2\sqrt{25 - y^2}$$

So $A = 8y\sqrt{25 - y^2}$. Graph $y = 8x\sqrt{(25 - x^2)}$ in the window $[0,5] \times [0,150]$.

The calculator shows that the maximum area (the y-coordinate) equals 100.

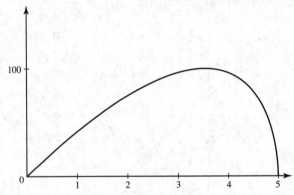

B22. **(B)** $\dfrac{\displaystyle\int_1^e x \ln x \, dx}{e - 1} \approx 1.221$

B23. **(B)** When f' is positive, f increases. By the Fundamental Theorem of Calculus, $f'(x) = 1 - 2 (\cos x)^3$. Graph f' in $[0,2\pi] \times [-2,4]$. It is clear that $f' > 0$ on the interval $a < x < b$. Using the calculator to solve $1 - 2(\cos x)^3 = 0$ yields $a = 0.654$ and $b = 5.629$.

B24. **(C)** $a(2) \approx \dfrac{v(3) - v(1)}{3 - 1} = \dfrac{12 - 22}{2}$

B25. **(B)** The volume is composed of elements of the form $\Delta V = (2x)^2 \, \Delta y$. If h is the depth, in feet, then, after t hours,

$$V(h) = 4\int_0^h y \, dy \text{ and } \frac{dV}{dt} = 4h\frac{dh}{dt}$$

Thus, $12 = 4\,(9)\dfrac{dh}{dt}$ and $\dfrac{dh}{dt} = \dfrac{1}{3}$ ft/hr.

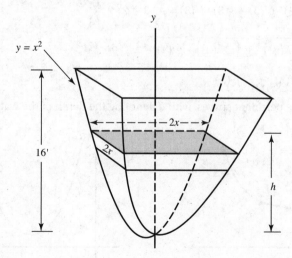

B26. **(B)** Separating variables yields

$$\frac{dP}{1000 - P} = k\, dt$$

$$-\ln(1000 - P) = kt + C$$

$$1000 - P = ce^{-kt}$$

Then

$$P(t) = 1000 - ce^{-kt}$$

$P(0) = 300$ gives $c = 700$. $P(5) = 500$ yields $500 = 1000 - 700e^{-5k}$, so $k \approx +0.0673$. Now $P(10) = 1000 - 700e^{-0.673} \approx 643$.

B27. **(C)** $H(1) = \int_0^1 \frac{4}{x^2 + 1}dx = 4\arctan 1 = \pi.$ $H'(1) = f(1) = 2.$

An equation of the tangent line is $y - \pi = 2(x - 1)$.

B28. **(D)** Horizontal slices of the smokestack are disks with volume $\pi\left(\dfrac{D}{2}\right)^2 \cdot dD$, where D is the diameter of the smokestack. Therefore, the volume of the smokestack is $\displaystyle\int_0^{100} \pi\left(\dfrac{D}{2}\right)^2 dD.$

If we use a left Riemann Sum to approximate the volume,

$$\int_0^{100} \pi\left(\frac{D}{2}\right)^2 dD \approx \pi\left(\frac{17}{2}\right)^2 \cdot (10 - 0) + \pi\left(\frac{15}{2}\right)^2 \cdot (50 - 10) + \pi\left(\frac{9}{2}\right)^2 \cdot (70 - 50) +$$

$$\pi\left(\frac{7}{2}\right)^2 \cdot (100 - 70) = 11{,}765.264$$

B29. **(A)** $\left(\dfrac{f}{g}\right)'(3) = \dfrac{g(3) \cdot f'(3) - f(3) \cdot g'(3)}{\big(g(3)\big)^2} = \dfrac{2(2) - 4(3)}{2^2}$

B30. **(C)** $H'(3) = f'\big(g(3)\big) \cdot g'(3) = f'(2) \cdot g'(3)$

B31. **(D)** $M'(3) = f(3) \cdot g'(3) + g(3) \cdot f'(3) = 4 \cdot 3 + 2 \cdot 2$

B32. **(D)** $K'(3) = \dfrac{1}{g'\big(K(3)\big)} = \dfrac{1}{g'\big(g^{-1}(3)\big)} = \dfrac{1}{g'(4)} = \dfrac{1}{\frac{1}{2}}$

B33. **(C)** $R'(3) = \dfrac{1}{2}\big(f(3)\big)^{-1/2} \cdot f'(3)$

B34. **(C)** Here are the pertinent curves, with d denoting the depth of the water:

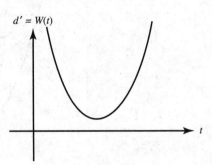

B35. **(B)** Use areas; then $\displaystyle\int_1^7 f' = -3 + 10 = 7.$ Thus, $f(7) - f(1) = 7.$

B36. **(B)** The region x units from the stage can be approximated by the semicircular ring shown; its area is then the product of its circumference and its width.

$$\tfrac{1}{2}(2\pi x)$$

Δx

The number of people standing in the region is the product of the area and the density:

$$\Delta P = (\pi x\,\Delta x)\left(\frac{20}{2\sqrt{x}+1}\right)$$

To find the total number of people, evaluate

$$20\pi\int_0^{100}\frac{x}{2\sqrt{x}+1}\,dx$$

B37. **(A)** $\dfrac{dy}{dt}$ is positive but decreasing; hence $\dfrac{dy^2}{dt^2}<0$.

B38. **(B)** Average speed $=\dfrac{\text{total distance}}{\text{elapsed time}}=\dfrac{\text{total area}}{8}=\dfrac{8}{8}$

B39. **(D)** On $2\le t\le 5$, the object moved $3\dfrac{1}{2}$ feet to the right; then on $5\le t\le 8$, it moved only $2\dfrac{1}{2}$ feet to the left.

B40. **(B)** Average acceleration $=\dfrac{\Delta v}{\Delta t}=\dfrac{v(8)-v(0)}{8-0}=\dfrac{-1-1}{8}$

B41. **(D)** Evaluate $\displaystyle\int_0^3\frac{72}{2t+3}\,dt=36\ln(2t+3)\Big|_0^3=36\ln 3.$

B42. **(A)** $2x\dfrac{dx}{dt}+2y\dfrac{dy}{dt}=0$ and $\dfrac{dy}{dt}=-\dfrac{3}{4}\dfrac{dx}{dt}$ at the point $(3,4)$.

Use, also, the facts that the speed is given by $|\mathbf{v}|=\sqrt{\left(\dfrac{dx}{dt}\right)^2+\left(\dfrac{dy}{dt}\right)^2}$ and that the point moves counterclockwise; then $\left(\dfrac{dx}{dt}\right)^2+\left(\dfrac{dy}{dt}\right)^2=4$, yielding $\dfrac{dx}{dt}=-\dfrac{8}{5}$ and $\dfrac{dy}{dt}=+\dfrac{6}{5}$ at the given point.

The velocity vector, $\mathbf{v}$, at $(3,4)$ must therefore be $\left\langle-\dfrac{8}{5},\dfrac{6}{5}\right\rangle$.

B43. **(A)** $\mathbf{v}=\langle -ak\sin kt, ak\cos kt\rangle$, and

$$\mathbf{a}=\langle -ak^2\cos kt, ak^2\sin kt\rangle=-k^2\mathbf{R}.$$

B44. **(B)** The formula for length of curve is

$$L=\int_a^b\sqrt{1+\left(\frac{dy}{dx}\right)^2}\,dx$$

Since $y=2^x$, we find

$$L=\int_0^2\sqrt{1+(2^x\ln 2)^2}\,dx\approx 3.664$$

B45. **(C)** $\mathbf{a}(t)=\left(0,e^t\right)$; the acceleration is always upward.

B46. **(A)** At $(0,1)$, $\dfrac{dy}{dx} = 4$, so Euler's method yields $(0.1, 1 + 0.1(4)) = (0.1, 1.4)$.

$\dfrac{dy}{dx} = 4y$ has the particular solution $y = e^{4x}$; the error is $e^{4(0.1)} - 1.4$.

B47. **(C)** $1 - \dfrac{1}{2} + \dfrac{1}{5} = 0.7$. Note that the series converges by the Alternating Series Test. Since the first term dropped in the estimate is $-\dfrac{1}{10}$, the estimate is too high but within 0.1 of the true sum.

B48. **(B)** $\displaystyle\sum_{n=1}^{\infty} \dfrac{2}{3n} = \dfrac{2}{3} \sum_{n=1}^{\infty} \dfrac{1}{n}$, which equals a constant times the harmonic series.

B49. **(D)** We seek x such that

$$\lim_{n \to \infty} \left| \frac{(n+1)! \cdot x^{n+1}}{(n+1)^{n+1}} \cdot \frac{n^n}{n! \cdot x^n} \right| < 1$$

or such that $|x| \cdot \displaystyle\lim_{n \to \infty} \left(\dfrac{n}{n+1} \right)^n < 1$

or such that $|x| < \dfrac{1}{\displaystyle\lim_{n \to \infty} \left(\dfrac{n}{n+1} \right)^n}$.

The fraction equals $\displaystyle\lim_{n \to \infty} \left(\dfrac{n+1}{n} \right)^n = \lim_{n \to \infty} \left(1 + \dfrac{1}{n} \right)^n = e$.

Then $|x| < e$ and the radius of convergence is e.

B50. **(B)** The error is less than the maximum value of $\dfrac{e^c}{3!} x^3$ for $0 \le x \le \dfrac{1}{2}$.

This maximum occurs at $c = x = \dfrac{1}{2}$.

B51. **(D)** Distance $= \displaystyle\int_0^1 \sqrt{\left(\dfrac{dx}{dt} \right)^2 + \left(\dfrac{dy}{dt} \right)^2} \, dt$

$= \displaystyle\int_0^1 \sqrt{\left(\sec^2 t \right)^2 + \left(-2\sin 2t \right)^2} \, dt$

Note that the curve is traced exactly once by the parametric equations from $t = 0$ to $t = 1$.

B52. **(D)** In the figure below, S is the region bounded by $y = \sec x$, the y-axis, and $y = 4$.

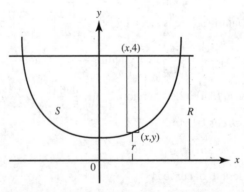

Send region S about the x-axis. Use washers; then $\Delta V = \pi(R^2 = r^2)\, \Delta x$. Symmetry allows you to double the volume generated by the first quadrant of S, so V is

$$2\pi \int_0^{\arccos \frac{1}{4}} \left(16 - \sec^2 x \right) dx$$

A calculator yields 108.177.

12

Miscellaneous Free-Response Practice Exercises

These problems provide further practice for both parts of Section II of the examination. Solutions begin on page 482.

PART A. CALCULATOR ACTIVE—DIRECTIONS: Some of the following questions may require the use of a graphing calculator.

---

x	2.5	3.2	3.5	4.0	4.6	5.0
$f(x)$	7.6	5.7	4.2	3.1	2.2	1.5

A1. A function f is continuous, differentiable, and strictly decreasing on the interval [2.5,5]; some values of f are shown in the table above.

   (a) Estimate $f'(4.0)$ and $f'(4.8)$.

   (b) What does the table suggest may be true of the concavity of f? Explain.

   (c) Estimate $\displaystyle\int_{2.5}^{5} f(x)\,dx$ with a Riemann Sum using left endpoints.

   (d) Set up (but do not evaluate) a Riemann Sum that estimates the volume of the solid formed when f is rotated around the x-axis.

A2. An equation of the tangent line to the curve $x^2 y - x = y^3 - 8$ at the point (0,2) is $12y + x = 24$.

   (a) Given that the point $(0.3, y_0)$ is on the curve, find y_0 approximately, using the tangent line.

   (b) Find the true value of y_0.

   (c) What can you conclude about the curve near $x = 0$ from your answers to parts (a) and (b)?

A3. A differentiable function f defined on $-7 < x < 7$ has $f(0) = 0$ and $f'(x) = 2x \sin x - e^{-x^2} + 1$. (Note: The following questions refer to f, not to f'.)

   (a) Describe the symmetry of f.

   (b) On what intervals is f decreasing?

   (c) For what values of x does f have a relative maximum? Justify your answer.

   (d) How many points of inflection does f have? Justify your answer.

A4. Let C represent the piece of the curve $y = \sqrt[3]{64 - 16x^2}$ that lies in the first quadrant. Let S be the region bounded by C and the coordinate axes.

   (a) Find the slope of the line tangent to C at $y = 1$.

   (b) Find the area of S.

   (c) Find the volume generated when S is rotated about the x-axis.

A5. Let R be the point on the curve of $y = x - x^2$ such that the line OR (where O is the origin) divides the area bounded by the curve and the x-axis into two regions of equal area. Set up (but do not solve) an integral to find the x-coordinate of R.

A6. Suppose $f'' = \sin(2^x)$ for $-1 < x < 3.2$.

  (a) On what intervals is the graph of f concave downward? Justify your answer.
  (b) Find the x-coordinates of all relative minima of f'.
  (c) How many points of inflection does the graph of f' have? Justify your answer.

A7. Let $f(x) = \cos x$ and $g(x) = x^2 - 1$.

  (a) Find the coordinates of any points of intersection of f and g.
  (b) Find the area bounded by f and g.

A8. (a) In order to investigate mail-handling efficiency, at various times one morning a local post office checked the rate (letters/min) at which an employee was sorting mail. Use the results shown in the table and the trapezoid method to estimate the total number of letters he may have sorted that morning.

Time	8:00	8:30	10:00	11:20	12:00
Letters/min	10	12	8	9	11

  (b) Hoping to speed things up a bit, the post office tested a sorting machine that can process mail at the constant rate of 20 letters per minute. The graph shows the rate at which letters arrived at the post office and were dumped into this sorter.

  (i)   When did letters start to pile up?
  (ii) When was the pile the biggest?
  (iii) How big was it then?
  (iv) At about what time did the pile vanish?

A9. Let R represent the region bounded by $y = \sin x$ and $y = x^4$. Find:

  (a) the area of R
  (b) the volume of the solid whose base is R if all cross sections perpendicular to the x-axis are isosceles triangles with height 3
  (c) the volume of the solid formed when R is rotated around the x-axis

A10. The town of East Newton has a water tower whose tank is an ellipsoid, formed by rotating an ellipse about its minor axis. Since the tank is 20 feet tall and 50 feet wide, the equation of the ellipse is $\dfrac{x^2}{625} + \dfrac{y^2}{100} = 1$.

  (a) If there are 7.48 gallons of water per cubic foot, what is the capacity of this tank to the nearest thousand gallons?
  (b) East Newton imposes water rationing whenever the tank is only one-quarter full. Write an equation to find the depth of the water in the tank when rationing becomes necessary. (Do not solve.)

A11.

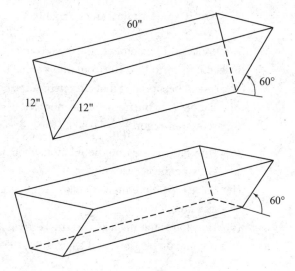

Note: Scales are different on the three figures.

The sides of the watering trough above are made by folding a sheet of metal 24 inches wide and 5 feet (60 inches) long at an angle of 60°, as shown above. Ends are added, and then the trough is filled with water.

(a) If water pours into the trough at the rate of 600 cubic inches per minute, how fast is the water level rising when the water is 4 inches deep?

(b) Suppose, instead, the sheet of metal is folded twice, keeping the sides of equal height and inclined at an angle of 60°, as shown. Where should the folds be in order to maximize the volume of the trough? Justify your answer.

A12. Given the function $f(x) = e^{2x}(x^2 - 2)$:

(a) For what values of x is f decreasing?

(b) Does this decreasing arc reach a local or a global minimum? Justify your answer.

(c) Does f have a global maximum? Justify your answer.

A13. (a) A spherical snowball melts so that its surface area shrinks at the constant rate of 10 square centimeters per minute. What is the rate of change of volume when the snowball is 12 centimeters in diameter?

(b) The snowball is packed most densely nearest the center. Suppose that, when it is 12 centimeters in diameter, its density x centimeters from the center is given by $d(x) = \dfrac{1}{1 + \sqrt{x}}$ grams per cubic centimeter. Set up an integral for the total number of grams (mass) of the snowball then. Do not evaluate.

A14. (a) Using your calculator, verify that

$$\left(4\tan^{-1}(1/5)\right) - \left(\tan^{-1}(1/239)\right) \approx \frac{\pi}{4}$$

(b) Use the Taylor polynomial of degree 7 about 0:

$$\tan^{-1} x \approx x - x^3/3 + x^5/5 - x^7/7$$

to approximate $\tan^{-1} 1/5$ and the polynomial of degree 1 to approximate $\tan^{-1} 1/239$.

(c) Use part (b) to evaluate the expression in (a).

(d) Explain how the approximation for $\pi/4$ given here compares with that obtained using $\pi/4 = \tan^{-1} 1$.

A15. (a) Show that the series $\displaystyle\sum_{n=1}^{\infty} (-1)^{n+1} \frac{1}{\ln(n+1)}$ converges.

(b) How many terms of the series are needed to get a partial sum within 0.1 of the sum of the whole series?

(c) Tell whether the series $\displaystyle\sum_{n=2}^{\infty} (-1)^n \frac{1}{n \ln n}$ is absolutely convergent, conditionally convergent, or divergent. Justify your answer.

A16. Given $\dfrac{dy}{dt} = ky(10 - y)$ with $y = 2$ at $t = 0$ and $y = 5$ at $t = 2$:

(a) Find k.

(b) Express y as a function of t.

(c) For what value of t will $y = 8$?

(d) Describe the long-range behavior of y.

A17. An object P is in motion in the first quadrant along the parabola $y = 18 - 2x^2$ in such a way that at t seconds the x-value of its position is $x = \dfrac{1}{2}t$.

(a) Where is P when $t = 4$?

(b) What is the vertical component of its velocity there?

(c) At what rate is its distance from the origin changing at $t = 4$?

(d) When does it hit the x-axis?

(e) How far did it travel altogether?

A18. A particle moves in the xy-plane in such a way that at any time $t \geq 0$ its position is given by $x(t) = 4 \arctan t$, $y(t) = \dfrac{12t}{t^2 + 1}$.

(a) Sketch the path of the particle, indicating the direction of motion.

(b) At what time t does the particle reach its highest point? Justify.

(c) Find the coordinates of that highest point, and sketch the velocity vector there.

(d) Describe the long-term behavior of the particle.

A19. Let R be the region bounded by the curve $r = 2 + \cos 2\theta$, as shown.

(a) Find the dimensions of the smallest rectangle that contains R and has sides parallel to the x- and y-axes.

(b) Find the area of R.

A20. (a) For what *positive* values of x does
$$f(x) = \sum_{n=1}^{\infty} (-1)^{n+1} \frac{x^n}{\ln(n+1)} \text{ converge?}$$

(b) How many terms are needed to estimate $f(0.5)$ to within 0.01?

(c) Would an estimate for $f(-0.5)$ using the same number of terms be more accurate, less accurate, or the same? Explain.

A21. After pollution-abatement efforts, conservation researchers introduce 100 trout into a small lake. The researchers predict that after m months the population, F, of the trout will be modeled by the differential equation $\dfrac{dF}{dm} = 0.0002F(600 - F)$.

(a) How large is the trout population when it is growing the fastest?

(b) Solve the differential equation, expressing F as a function of m.

(c) How long after the lake was stocked will the population be growing the fastest?

PART B. NO CALCULATOR—DIRECTIONS: Answer these questions *without* using your calculator.

B1. Draw a graph of $y = f(x)$, given that f satisfies all the following conditions:

(a) $f'(-1) = f'(1) = 0$.
(b) If $x < -1$, $f'(x) > 0$ but $f'' < 0$.
(c) If $-1 < x < 0$, $f'(x) > 0$ and $f'' > 0$.
(d) If $0 < x < 1$, $f'(x) > 0$ but $f'' < 0$.
(e) If $x > 1$, $f'(x) < 0$ and $f'' < 0$.

B2. The figure below shows the graph of f', the derivative of f, with domain $-3 \le x \le 9$. The graph of f' has horizontal tangents at $x = 2$ and $x = 4$ and a corner at $x = 6$.

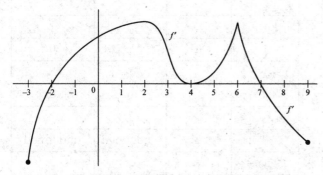

(a) Is f continuous? Explain.
(b) Find all values of x at which f attains a local minimum. Justify.
(c) Find all values of x at which f attains a local maximum. Justify.
(d) At what value of x does f attain its absolute maximum? Justify.
(e) Find all values of x at which the graph of f has a point of inflection. Justify.

B3. Find the area of the largest rectangle (with sides parallel to the coordinate axes) that can be inscribed in the region bounded by the graphs of $f(x) = 8 - 2x^2$ and $g(x) = x^2 - 4$.

B4. Given the graph of $f(x)$, sketch the graph of $f'(x)$.

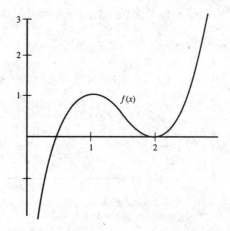

B5. A cube is contracting so that its surface area decreases at the constant rate of 72 in.^{2}/sec. Determine how fast the volume is changing at the instant when the surface area is 54 ft^2.

B6. A square is inscribed in a circle of radius a as shown in the diagram below. Find the volume obtained if the region outside the square but inside the circle is rotated about a diagonal of the square.

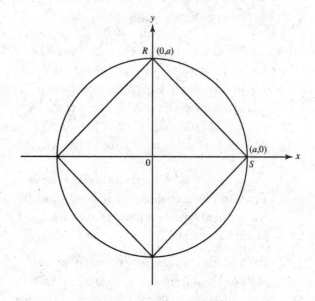

B7. (a) Sketch the region in the first quadrant bounded above by the line $y = x + 4$, below by the line $y = 4 - x$, and to the right by the parabola $y = x^2 + 2$.

(b) Find the area of this region.

B8. The graph shown below is based roughly on data from the U.S. Department of Agriculture.

Food Production in South Asia
(Index was 100 during 1969–71)

(a) During which intervals did food production decrease in South Asia?

(b) During which intervals did the rate of change of food production increase?

(c) During which intervals did the increase in food production accelerate?

B9. A particle moves along a straight line so that its acceleration at any time t is given in terms of its velocity v by $a = -2v$.

(a) Find v in terms of t if $v = 20$ when $t = 0$.

(b) Find the distance the particle travels while v changes from $v = 20$ to $v = 5$.

B10. Let R represent the region bounded above by the parabola $y = 27 - x^2$ and below by the x-axis. Isosceles triangle AOB is inscribed in region R with its vertex at the origin O and its base $\overline{AB}$ parallel to the x-axis. Find the maximum possible area for such a triangle.

B11. Newton's law of cooling states that the rate at which an object cools is proportional to the difference in temperature between the object and its surroundings.

It is 9:00 P.M., time for your milk and cookies. The room temperature is 68° when you pour yourself a glass of 40° milk and start looking for the cookie jar. By 9:03 the milk has warmed to 43°, and the phone rings. It's your friend, with a fascinating calculus problem. Distracted by the conversation, you forget about the glass of milk. If you dislike milk warmer than 60°, how long, to the nearest minute, do you have to solve the calculus problem and still enjoy acceptably cold milk with your cookies?

B12. Let h be a function that is even and continuous on the closed interval $[-4,4]$. The function h and its derivatives have the properties indicated in the table below. Use this information to sketch a possible graph of h on $[-4,4]$.

x	$h(x)$	$h'(x)$	$h''(x)$
0	−	0	+
$0 < x < 1$	−	+	+
1	0	+	0
$1 < x < 2$	+	+	−
2	+	0	0
$2 < x < 3$	+	+	+
3	+	undefined	undefined
$3 < x < 4$	+	−	−

B13. (a) Find the Maclaurin series for $f(x) = \ln(1 + x)$.

(b) What is the radius of convergence of the series in (a)?

(c) Use the first five terms in (a) to approximate $\ln(1.2)$.

(d) Estimate the error in (c), justifying your answer.

BC ONLY

B14. A cycloid is given parametrically by $x = \theta - \sin\theta,\ y = 1 - \cos\theta$.

(a) Find the slope of the curve at the point where $\theta = \dfrac{2\pi}{3}$.

(b) Find an equation of the tangent to the cycloid at the point where $\theta = \dfrac{2\pi}{3}$.

B15. Find the area of the region enclosed by both the polar curves $r = 4 \sin \theta$ and $r = 4 \cos \theta$.

B16. (a) Find the 4th-degree Taylor polynomial about 0 for $\cos x$.

(b) Use part (a) to evaluate $\int_0^1 \cos x \, dx$.

(c) Estimate the error in (b), justifying your answer.

B17. A particle moves on the curve of $y^3 = 2x + 1$ so that its distance from the x-axis is increasing at the constant rate of 2 units/sec. When $t = 0$, the particle is at $(0,1)$.

(a) Find a pair of parametric equations $x = x(t)$ and $y = y(t)$ that describe the motion of the particle for nonnegative t.

(b) Find $|\mathbf{a}|$, the magnitude of the particle's acceleration, when $t = 1$.

B18. Find the area of the region that the polar curves $r = 2 - \cos \theta$ and $r = 3 \cos \theta$ enclose in common.

Answer Explanations

Part A

A1. (a) $f'(4.0) \approx \dfrac{f(4.6) - f(4.0)}{4.6 - 4.0} = \dfrac{2.2 - 3.1}{0.6} = -1.5$

$f'(4.8) \approx \dfrac{f(5) - f(4.6)}{5 - 4.6} = \dfrac{1.5 - 2.2}{0.4} = -1.75$

(b) It appears that the rate of change of f, while negative, is increasing. This implies that the graph of f is concave upward.

(c) $L = 7.6(0.7) + 5.7(0.3) + 4.2(0.5) + 3.1(0.6) + 2.2(0.4) = 11.87$

(d) Using disks $\Delta V = \pi r^2 \Delta x$. One possible answer uses the left endpoints of the sub-intervals as values of r:

$$V \approx \pi(7.6)^2(0.7) + \pi(5.7)^2(0.3) + \pi(4.2)^2(0.5) + \pi(3.1)^2(0.6) + \pi(2.2)^2(0.4)$$

A2. (a) $12y_0 + 0.3 = 24$ yields $y_0 \approx 1.975$

(b) Replace x by 0.3 in the equation of the curve:

$$(0.3)^2 y_0 - (0.3) = y_0{}^3 - 8 \text{ or}$$

$$y_0{}^3 - 0.09 y_0 - 7.7 = 0$$

The calculator's solution to three decimal places is $y_0 = 1.990$.

(c) Since the true value of y_0 at $x = 0.3$ exceeds the approximation, conclude that the given curve is concave up near $x = 0$. (Therefore, it is above the line tangent at $x = 0$.)

A3. Graph $f'(x) = 2x \sin x - e^{\left(-x^2\right)} + 1$ in $[-7,7] \times [-10,10]$.

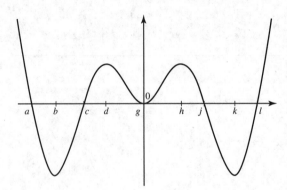

(a) Since f' is even and f contains $(0,0)$, f is odd and its graph is symmetric about the origin.

(b) Since f is decreasing when $f' < 0$, f decreases on the intervals (a,c) and (j,l). Use the calculator to solve $f'(x) = 0$. Conclude that f decreases on $-6.202 < x < -3.294$ and (symmetrically) on $3.294 < x < 6.202$.

(c) f has a relative maximum at $x = q$ if $f'(q) = 0$ and if f changes from increasing ($f' > 0$) to decreasing ($f' < 0$) at q. There are two relative maxima here: at $x = a = -6.202$ and at $x = j = 3.294$.

(d) f has a point of inflection when the graph of f changes its concavity—that is, when f' changes from increasing to decreasing, as it does at points d and h, or when f' changes from decreasing to increasing, as it does at points b, g, and k. So there are five points of inflection altogether.

A4. In the graph below, C is the piece of the curve lying in the first quadrant. S is the region bounded by the curve C and the coordinate axes.

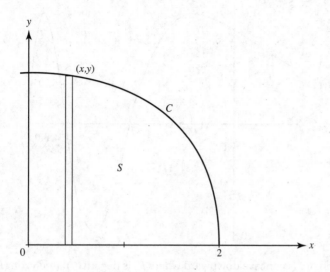

(a) Graph $y = \sqrt[3]{(64 - 16x^2)}$ in $[0,3] \times [0,5]$. Since you want dy/dx, the slope of the tangent, where $y = 1$, use the calculator to solve

$$\sqrt[3]{64 - 16x^2} = 1, \; x = 1.984$$

(storing the answer at B). Then evaluate the slope of the tangent to C at $y = 1$:

$$f'(B) \approx -21.182$$

(b) Since $\Delta A = y \Delta x$, $A = \displaystyle\int_0^2 y \, dx \approx 6.730$.

(c) When S is rotated about the x-axis, its volume can be obtained using disks:

$$\Delta V = \pi R^2 \Delta x = \pi y^2 \, \Delta x$$

$$V = \pi \int_0^2 y^2 \, dx$$

$$= \pi \int_0^2 \left(\sqrt[3]{64 - 16x^2} \right)^2 \, dx \approx 74.310$$

A5. See the figure, where R is the point (a,b), and seek a such that

$$\int_0^a \left(x - x^2 - \frac{b}{a} \cdot x \right) dx = \frac{1}{2} \int_0^1 (x - x^2) \, dx$$

A6.   Graph $y = \sin 2^x$ in $[-1, 3.2] \times [-1, 1]$. Note that $y = f''$.

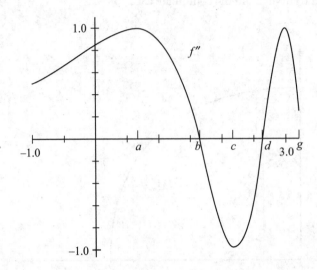

(a) The graph of f is concave downward where f'' is negative, namely, on (b, d). Use the calculator to solve $\sin 2^x = 0$, obtaining $b = 1.651$ and $d = 2.651$. The answer to (a) is therefore $1.651 < x < 2.651$.

(b) f' has a relative minimum at $x = d$, because f'' equals 0 at d, is less than 0 on (b, d), and is greater than 0 on (d, g). Thus f' has a relative minimum (from part (a)) at $x = 2.651$.

(c) The graph of f' has a point of inflection wherever its second derivative f'' changes from positive to negative or vice versa. This is equivalent to f'' changing from increasing to decreasing (as at a and g) or vice versa (as at c). Therefore, the graph of f' has three points of inflection on $[-1, 3.2]$.

A7.   Graph $f(x) = \cos x$ and $g(x) = x^2 - 1$ in $[-2, 2] \times [-2, 2]$. Here, $y_1 = f$ and $y_2 = g$.

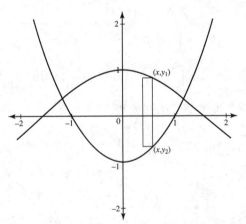

(a) Solve $\cos x = x^2 - 1$ to find the two points of intersection: $(1.177, 0.384)$ and $(-1.177, 0.384)$.

(b) Since $\Delta A = (y_1 - y_2)\,\Delta x = [f(x) - g(x)]\,\Delta x$, the area A bounded by the two curves is

$$A = 2\int_0^{1.177} (y_1 - y_2)\,dx$$

$$= 2\int_0^{1.177} \left(\cos x - (x^2 - 1)\right) dx$$

$$\approx 3.114$$

A8. (a) $\left(\dfrac{10+12}{2}\right) \cdot 30 + \left(\dfrac{12+8}{2}\right) \cdot 90 + \left(\dfrac{8+9}{2}\right) \cdot 80 + \left(\dfrac{9+11}{2}\right) \cdot 40 = 2310$ letters

(b) Draw a horizontal line at $y = 20$ (as shown on the graph below), representing the rate at which letters are processed then.

(i)   Letters began to pile up when they arrived at a rate greater than that at which they were being processed, that is, at $t = 10$ A.M.

(ii) The pile was largest when the letters stopped piling up, at $t = 2$ P.M.

(iii) The number of letters in the pile is represented by the area of the small trapezoid above the horizontal line: $\dfrac{1}{2}(4 \cdot 60 + 1 \cdot 60)(10) = 1500$.

(iv) The pile began to diminish after 2 P.M., when letters were processed at a rate faster than they arrived, and vanished when the area of the shaded triangle represented 1500 letters. At 5 P.M. this area is $\dfrac{1}{2}(3 \cdot 60)(15) = 1350$ letters, so the pile vanished shortly after 5 P.M.

A9. Draw a vertical element of area as shown below.

(a) Let a represent the x-value of the positive point of intersection of $y = x^4$ and $y = \sin x$. Solving $a^4 = \sin a$ with the calculator, we find $a = 0.9496$.

$$\Delta A = \left(y_{\text{top}} - y_{\text{bottom}} \right) \Delta x = \left(\sin x - x^4 \right) \Delta x$$

$$A = \int_0^a \left(\sin x - x^4 \right) dx \approx 0.264$$

(b) Elements of volume are triangular prisms with height $h = 3$ and base $b = \left(\sin x - x^4 \right)$, as shown below.

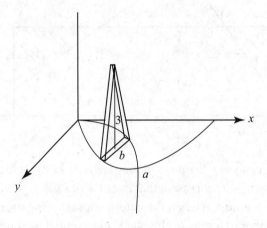

$$\Delta V = \frac{1}{2} \left(\sin x - x^4 \right)(3) \Delta x$$

$$V = \frac{3}{2} \int_0^a \left(\sin x - x^4 \right) dx = 0.395$$

(c) When R is rotated around the x-axis, the element generates washers. If r_1 and r_2 are the radii of the larger and smaller disks, respectively, then

$$\Delta V = \pi \left(r_1^2 - r_2^2 \right) \Delta x = \pi \left((\sin x)^2 - (x^4)^2 \right) \Delta x$$

$$V = \pi \int_0^a \left(\sin^2 x - x^8 \right) dx = 0.529$$

A10.

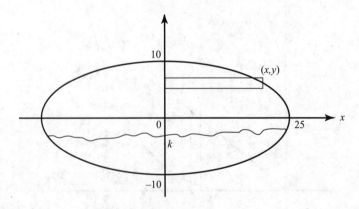

The figure above shows an elliptical cross section of the tank. Its equation is

$$\frac{x^2}{625} + \frac{y^2}{100} = 1$$

(a) The volume of the tank, using disks, is $V = 2\pi \int_0^{10} x^2\, dy$, where the ellipse's symmetry about the x-axis has been exploited. The equation of the ellipse is equivalent to $x^2 = 6.25(100 - y^2)$, so

$$V = 12.5\pi \int_0^{10} \left(100 - y^2\right) dy, \quad V = 26179.939 \text{ gallons}$$

Use the calculator to evaluate this integral, storing the answer as V to have it available for part (b). The capacity of the tank is $7.48V$, or 196,000 gallons of water, rounded to the nearest 1000 gallons.

(b) Let k be the y-coordinate of the water level when the tank is one-fourth full. Then

$$6.25\pi \int_{-10}^{k} \left(100 - y^2\right) dy = \frac{V}{4}, \quad k = -3.473$$

and the depth of the water is $k + 10$.

A11. (a) Let h represent the depth of the water, as shown.

Then h is the altitude of an equilateral triangle, and the base $b = \dfrac{2h}{\sqrt{3}}$.

The volume of water is

$$V = \frac{1}{2}\left(\frac{2h}{\sqrt{3}}\right)h \cdot 60 = \frac{60h^2}{\sqrt{3}} \text{ in.}^3$$

Now $\dfrac{dV}{dt} = \dfrac{120}{\sqrt{3}}h\dfrac{dh}{dt}$, and it is given that $\dfrac{dV}{dt} = 600$. Thus, when $h = 4$,

$$600 = \frac{120}{\sqrt{3}}4\frac{dh}{dt}, \text{ and } \frac{dh}{dt} = \frac{5\sqrt{3}}{4} \text{ in./min}$$

(b) Let x represent the length of one of the sides, as shown. The bases of the trapezoid are $24 - 2x$ and $24 - 2x + 2\dfrac{x}{2}$, and the height is $\dfrac{x}{2}\sqrt{3}$.

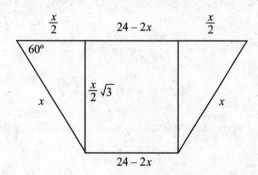

The volume of the trough $\left(\text{in in.}^3\right)$ is given by

$$V = \frac{(24 - 2x) + (24 - x)}{2} \cdot \frac{x}{2} \sqrt{3} \times 60 = 15\sqrt{3}\left(48x - 3x^2\right) \qquad (0 < x < 12)$$
$$V' = 15\sqrt{3}(48 - 6x) = 0 \text{ when } x = 8$$

Since $V'' = 15\sqrt{3}(-6) < 0$, the maximum volume is attained by folding the metal 8 inches from the edges.

A12. (a) $f(x) = e^{2x}(x^2 - 2)$

$$\begin{aligned} f'(x) &= e^{2x}(2x) + 2e^{2x}(x^2 - 2) \\ &= 2e^{2x}(x + 2)(x - 1) \\ &= 0 \text{ at } x = -2, 1 \end{aligned}$$

f is decreasing where $f'(x) < 0$, which occurs for $-2 < x < 1$.

(b) f is decreasing on the interval $-2 < x < 1$, so there is a minimum at $(1, -e^2)$. Note that, as x approaches $\pm\infty$, $f(x) = e^{2x}(x^2 - 2)$ is always positive. Hence $(1, -e^2)$ is the global minimum.

(c) As x approaches $\pm\infty$, $f(x) = e^{2x}(x^2 - 2)$ also approaches $\pm\infty$. There is no global maximum.

A13. (a) $S = 4\pi r^2$, so $\dfrac{dS}{dt} = 8\pi r \dfrac{dr}{dt}$. Substitute given values; then

$$-10 = 8\pi(6)\frac{dr}{dt}, \text{ so } \frac{dr}{dt} = -\frac{5}{24\pi}\text{cm/min}$$

Since $V = \dfrac{4}{3}\pi r^3$, therefore $\dfrac{dV}{dt} = 4\pi r^2 \dfrac{dr}{dt}$. Substituting known values gives
$$\frac{dV}{dt} = 4\pi(6^2) \cdot \frac{-5}{24\pi} = -30 \text{ cm}^3/\text{min}.$$

(b) Regions of consistent density are concentric spherical shells. The volume of each shell is approximated by its surface area $(4\pi x^2)$ times its thickness (Δx). The weight of each shell is its density times its volume $(\text{g/cm}^3 \cdot \text{cm}^3)$. If, when the snowball is 12 centimeters in diameter, ΔG is the weight of a spherical shell x centimeters from the center, then

$$\Delta G = \frac{1}{1 + \sqrt{x}} \cdot 4\pi x^2 \Delta x, \text{ and the integral to find the weight of the snowball is}$$

$$G = \int_0^6 \frac{1}{1 + \sqrt{x}} \cdot 4\pi x^2 dx$$

A14. (a) Both $\pi/4$ and the expression in brackets yield 0.7853981634, which is accurate to ten decimal places.

(b) $\tan^{-1}\dfrac{1}{5} = \dfrac{1}{5} - \dfrac{1}{3}\left(\dfrac{1}{5}\right)^3 + \dfrac{1}{5}\left(\dfrac{1}{5}\right)^5 - \dfrac{1}{7}\left(\dfrac{1}{5}\right)^7 = 0.197396$

$$\tan^{-1}\frac{1}{239} = 0.004184$$

(c) $4\tan^{-1}\dfrac{1}{5} - \tan^{-1}\dfrac{1}{239} = 0.7854$; this agrees with the value of $\dfrac{\pi}{4}$ to four decimal places.

(d) The series

$$\tan^{-1}1 = 1 - \frac{1}{3} + \frac{1}{5} - \frac{1}{7} + \cdots$$

converges *very* slowly. Example 56, page 428, evaluated the sum of 60 terms of the series for π (which equals $4 \tan^{-1}1$). To four decimal places, we get $\pi = 3.1249$, which yields 0.7812 for $\pi/4$—not accurate even to two decimal places.

A15. (a) The given series is alternating. Since $\lim_{n\to\infty} \ln (n+1) = \infty$, $\lim_{n\to\infty} \dfrac{1}{\ln (n+1)} = 0$. Since $\ln x$ is an increasing function,

$$\ln (n+1) > \ln n \quad \text{and} \quad \frac{1}{\ln (n+1)} < \frac{1}{\ln n}$$

The series therefore converges.

(b) Since the series converges by the Alternating Series Test, the error in using the first n terms for the sum of the whole series is less than the absolute value of the $(n+1)$st term. Thus the error is less than $\dfrac{1}{\ln (n+1)}$. Solve for n using $\dfrac{1}{\ln (n+1)} < 0.1$:

$$\ln (n+1) > 10$$
$$(n+1) > e^{10}$$
$$n > e^{10} - 1 > 22{,}025$$

The given series converges very slowly!

(c) The series $\displaystyle\sum_{2}^{\infty} (-1)^n \frac{1}{n \ln n}$ is conditionally convergent. The given alternating series converges since the nth term approaches 0 and $\dfrac{1}{(n+1)\ln (n+1)} < \dfrac{1}{n \ln n}$. However, the *nonnegative* series diverges by the Integral Test since

$$\int_{2}^{\infty} \frac{1}{x \ln x}\, dx = \lim_{b\to\infty} \ln (\ln x)\Big|_{2}^{b} = \infty$$

A16. (a) Solve by separation of variables:

$$\frac{dy}{y(10-y)} = k\, dt$$
$$\frac{1}{10} \int \left(\frac{1}{y} + \frac{1}{10-y} \right) dy = \int k\, dt$$
$$\frac{1}{10} \ln \left(\frac{y}{10-y} \right) = kt + C$$
$$\ln \left(\frac{10-y}{y} \right) = -10\, (kt + C)$$

Let $c = e^{-10C}$; then

$$\frac{10-y}{y} = ce^{-10kt}$$

Now use initial condition $y = 2$ at $t = 0$:

$$\frac{8}{2} = ce^0 \text{ so } c = 4$$

and the other condition, $y = 5$ at $t = 2$, gives

$$\frac{5}{5} = 4e^{-20k} \text{ or } k = \frac{1}{10} \ln 2$$

(b) Since $c = 4$ and $k = \dfrac{1}{10}\ln 2$, then $\dfrac{10 - y}{y} = 4e^{-10\left(\frac{1}{10}\ln 2\right)t}$

Solving for y yields $y = \dfrac{10}{1 + 4 \cdot 2^{-t}}$

(c) $8 = \dfrac{10}{1 + 4 \cdot 2^{-t}}$ means $1 + 4 \cdot 2^{-t} = 1.25$, so $t = 4$.

(d) $\displaystyle\lim_{t \to \infty} \dfrac{10}{1 + 4 \cdot 2^{-t}} = 10$, so the value of y approaches 10.

A17. (a) Since $x = \dfrac{1}{2}t$, $x(4) = \dfrac{1}{2}(4) = 2$. Since $y = 18 - 2 \cdot 2^2 = 10$, P is at $(2,10)$.

(b) Since $y = 18 - 2x^2$, $\dfrac{dy}{dt} = -4x\dfrac{dx}{dt}$. Since $x = \dfrac{1}{2}t$, $\dfrac{dx}{dt} = \dfrac{1}{2}$. Therefore

$$\dfrac{dy}{dt} = -4x\dfrac{dx}{dt} = -4 \cdot 2 \cdot \dfrac{1}{2} = -4 \text{ unit/sec.}$$

(c) Let $D =$ the object's distance from the origin. Then

$$D^2 = x^2 + y^2, \text{ and at } (2,10)\ D = \sqrt{104}$$

$$2D\dfrac{dD}{dt} = 2x\dfrac{dx}{dt} + 2y\dfrac{dy}{dt}$$

$$2\sqrt{104}\dfrac{dD}{dt} = 2 \cdot 2 \cdot \dfrac{1}{2} + 2 \cdot 10(-4)$$

$$\dfrac{dD}{dt} = \dfrac{-78}{2\sqrt{104}} = -3.824 \text{ unit/sec}$$

(d) The object hits the x-axis when $y = 18 - 2x^2 = 0$, or $x = 3$. Since $x = \dfrac{1}{2}t = 3$, $t = 6$.

(e) The length of the curve of $y = 18 - 2x^2$ for $0 \le x \le 3$ is given by

$$L = \int \sqrt{1 + \left(\dfrac{dy}{dx}\right)^2}\ dx = \int_0^3 \sqrt{1 + (-4x)^2}\ dx = 18.460 \text{ units}$$

A18. (a) See graph.

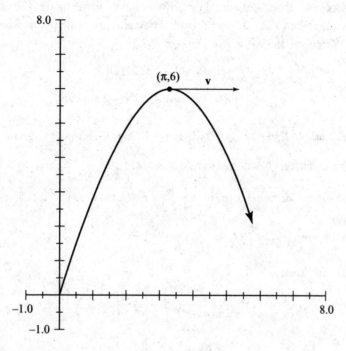

(b) You want to maximize $y(t) = \dfrac{12t}{t^2+1}$.

$$y'(t) = \frac{(t^2+1)(12) - 12t(2t)}{(t^2+1)^2} = \frac{12(1-t)(1+t)}{(t^2+1)^2}$$

See signs analysis.

The maximum y occurs when $t = 1$, because y changes from increasing to decreasing there.

(c) Since $x(1) = 4 \arctan 1 = \pi$ and $y(1) = \dfrac{12}{1+1} = 6$, the coordinates of the highest point are $(\pi, 6)$.

Since $x'(t) = \dfrac{4}{1+t^2}$ and $y'(t) = \dfrac{12(1-t^2)}{(t^2+1)^2}$, therefore $\mathbf{v}(1) = \langle 2, 0 \rangle$. This vector is shown on the graph.

(d) $\displaystyle\lim_{t\to\infty} x(t) = \lim_{t\to\infty} 4 \arctan t = 4\left(\dfrac{\pi}{2}\right) = 2\pi$, and $\displaystyle\lim_{t\to\infty} y(t) = \lim_{t\to\infty} \dfrac{12t}{t^2+1} = 0$. Thus the particle approaches the point $(2\pi, 0)$.

A19. (a) To find the smallest rectangle with sides parallel to the x- and y-axes, you need a rectangle formed by vertical and horizontal tangents as shown in the figure. The vertical tangents are at the x-intercepts, $x = \pm 3$. The horizontal tangents are at the points where y (not r) is a maximum. You need, therefore, to maximize

$$y = r \sin \theta = (2 + \cos 2\theta) \sin \theta$$

$$\frac{dy}{d\theta} = (2 + \cos 2\theta)\cos \theta + \sin \theta(-2 \sin 2\theta)$$

Use the calculator to find that $\dfrac{dy}{d\theta} = 0$ when $\theta = 0.7854$. Therefore, $y = 1.414$,

so the desired rectangle has dimensions 6×2.828.

(b) Since the polar formula for the area is $\dfrac{1}{2}\displaystyle\int_{\theta_1}^{\theta_2} r^2\,d\theta$, the area of R (enclosed by r) is

$4 \cdot \dfrac{1}{2}\displaystyle\int_0^{\pi/2} r^2 d\theta$, which is 14.137.

A20. (a) Use the Ratio Test:

$$\lim_{n\to\infty}\left|\frac{x^{n+1}}{\ln(n+2)}\cdot\frac{\ln(n+1)}{x^n}\right| < 1$$

$$|x|\lim_{n\to\infty}\left|\frac{\ln(n+1)}{\ln(n+2)}\right| < 1$$

$$|x| < 1$$

The radius of convergence is 1. At the endpoint $x = 1$, $f(1) = \displaystyle\sum_{n=1}^{\infty}(-1)^{n+1}\frac{1}{\ln(n+1)}$.

Since

$$\frac{1}{\ln(n+2)} < \frac{1}{\ln(n+1)} \quad \text{and} \quad \lim_{n\to\infty}\frac{1}{\ln(n+1)} = 0$$

this series converges by the Alternating Series Test. Thus $f(x) = \displaystyle\sum_{n=1}^{\infty}(-1)^{n+1}\frac{x^n}{\ln(n+1)}$

converges for positive values $0 < x \le 1$.

(b) Because $f(0.5) = \displaystyle\sum_{n=1}^{\infty}(-1)^{n+1}\frac{(0.5)^n}{\ln(n+1)}$ satisfies the Alternating Series Test, the error in approximation after n terms is less than the magnitude of the next term.

The calculator shows that $\dfrac{(0.5)^{n+1}}{\ln(n+2)} < 0.01$ at $n = 5$ terms.

(c) $f(-0.5) = \displaystyle\sum_{n=1}^{\infty}(-1)^{n+1}\frac{(-0.5)^n}{\ln(n+1)} = \sum_{n=1}^{\infty}\frac{-(0.5)^n}{\ln(n+1)}$ is a negative series. Therefore the error will be larger than the magnitude of the first omitted term and thus less accurate than the estimate for $f(0.5)$.

A21. (a) To find the maximum rate of growth, first find the derivative of $\dfrac{dF}{dm} = 0.0002F(600 - F) = 0.0002(600F - F^2)$. $\dfrac{d^2F}{dm^2} = 0.0002(600 - 2F)$, which equals 0 when $F = 300$. A signs analysis shows that $\dfrac{d^2F}{dm^2}$ changes from positive to negative there, confirming that $\dfrac{dF}{dm}$ is at its maximum when there are 300 trout.

(b) The differential equation $\frac{dF}{dm} = 0.0002F(600 - F)$ is separable.

$$\int \frac{dF}{F(600 - F)} = 0.0002 \int dm$$

To integrate the left side of this equation, use the method of partial fractions.

$$\frac{1}{F(600 - F)} = \frac{A}{F} + \frac{B}{600 - F}$$

$$1 = A(600 - F) + B(F)$$

Let $F = 0$; then $A = \frac{1}{600}$.

Let $F = 600$; then $B = \frac{1}{600}$.

$$\frac{1}{600}\int \left(\frac{1}{F} + \frac{1}{(600 - F)}\right)dF = 0.0002 \int dm$$

$$\int \frac{1}{F}dF + (-1)\int \frac{-1}{(600 - F)}dF = 0.12 \int dm$$

$$\ln F - \ln(600 - F) = 0.12m + C$$

$$\ln\left(\frac{F}{600 - F}\right) = 0.12m + C$$

$$\frac{F}{600 - F} = e^{0.12m + C}$$

$$= ce^{0.12m}, \text{where } c = e^C$$

$$F = \frac{600}{1 + ce^{-0.12m}}$$

Since $F = 100$ when $m = 0$, $100 = \frac{600}{1 + c}$, so c = 5.

The solution is $F = \frac{600}{1 + 5e^{-0.12m}}$.

(c) In (a) the population was found to be growing the fastest when $F = 300$. Then:

$$300 = \frac{600}{1 + 5e^{-0.12m}}$$

$$e^{-0.12m} = \frac{1}{5}$$

$$m = \frac{\ln\frac{1}{5}}{-0.12}\text{months}$$

Part B

B1.   The graph shown below satisfies all five conditions. So do many others!

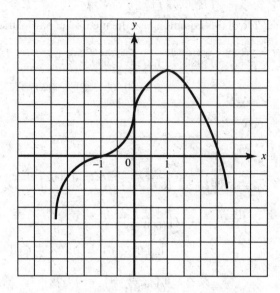

B2.   (a) f' is defined for all x in the interval. Since f is therefore differentiable, it must also be continuous.

(b) Because $f'(-2) = 0$ and f' changes from negative to positive there, f has a local minimum at $x = -2$. To the left of $x = 9$, f' is negative, so f is decreasing as it approaches that endpoint and reaches another local minimum there.

(c) Because f' is negative to the right of $x = -3$, f decreases from its left endpoint, indicating a local max there. Also, $f'(7) = 0$ and f' changes from positive to negative there, so f has a local relative maximum at $x = 7$.

(d) Note that $f(7) - f(-3) = \displaystyle\int_{-3}^{7} f'(x)\, dx$. Since there is more area above the x-axis than below the x-axis on $[-3,7]$, the integral is positive and $f(7) - f(-3) > 0$. This implies that $f(7) > f(-3)$ and that the absolute maximum occurs at $x = 7$.

(e) At $x = 2$ and also at $x = 6$, f' changes from increasing to decreasing, indicating that f changes from concave upward to concave downward at each. At $x = 4$, f' changes from decreasing to increasing, indicating that f changes from concave downward to concave upward there. Hence the graph of f has points of inflection at $x = 2$, 4, and 6.

B3.   Draw a sketch of the region bounded above by $y_1 = 8 - 2x^2$ and below by $y_2 = x^2 - 4$, and inscribe a rectangle in this region as described in the question. If (x, y_1) and (x, y_2) are the vertices of the rectangle in quadrants I and IV, respectively, then the area is:

$$A = 2x(y_1 - y_2) = 2x(12 - 3x^2) \quad \text{or} \quad A(x) = 24x - 6x^3$$

Then $A'(x) = 24 - 18x^2 = 6(4 - 3x^2)$, which equals 0 when $x = \dfrac{2}{\sqrt{3}} = \dfrac{2\sqrt{3}}{3}$.

Check to verify that $A''(x) < 0$ at this point. This assures that this value of x yields maximum area, which is given by $\dfrac{4\sqrt{3}}{3} \times 8$.

B4. The graph of $f'(x)$ is shown here.

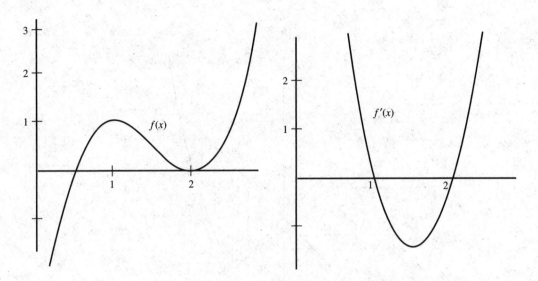

B5. The rate of change in volume when the surface area is 54 ft^2 is $-\dfrac{3}{8}$ ft^3/sec.

B6. See the figure. The equation of the circle is $x^2 + y^2 = a^2$; the equation of RS is $y = a - x$. If y_2 is an ordinate of the circle and y_1 of the line, then,

$$\Delta V = \pi y_2{}^2\, \Delta x - \pi y_1{}^2\, \Delta x$$

$$V = 2\pi \int_0^a \left[(a^2 - x^2) - (a - x)^2\right] dx = \frac{2}{3}\pi a^3$$

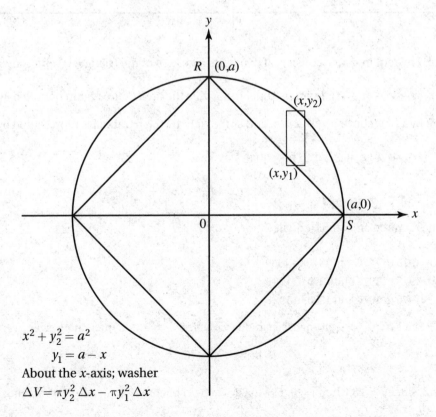

$$x^2 + y_2^2 = a^2$$
$$y_1 = a - x$$

About the x-axis; washer

$$\Delta V = \pi y_2^2\, \Delta x - \pi y_1^2\, \Delta x$$

B7. (a) The region is sketched in the figure. The pertinent points of intersection are labeled.

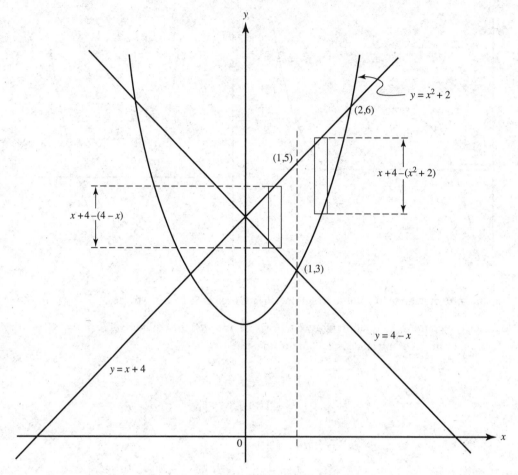

(b) The required area consists of two parts. The area of the triangle is represented by

$\int_0^1 \left[(x+4) - (4-x) \right] dx$ and is equal to 1, while the area of the region bounded at the left by $x=1$, above by $y = x+4$, and at the right by the parabola is represented by

$\int_1^2 \left[(x+4) - (x^2+2) \right] dx$. This equals

$$\int_1^2 \left(x + 2 - x^2\right) dx = \frac{x^2}{2} + 2x - \frac{x^3}{3} \Big|_1^2 = \frac{7}{6}$$

The required area, thus, equals $2\frac{1}{6}$ or $\frac{13}{6}$.

B8. (a) 1975 to 1976 and 1978 to 1980

(b) 1975 to 1977 and 1979 to 1981

(c) 1976 to 1977 and 1980 to 1981

B9. (a) Since $a = \dfrac{dv}{dt} = -2v$, then, separating variables, $\dfrac{dv}{v} = -2dt$. Integrating gives

$$\ln v = -2t + C \tag{1}$$

and, since $v = 20$ when $t = 0$, $C = \ln 20$. Then (1) becomes $\ln \dfrac{v}{20} = -2t$ or, solving for v,

$$v = 20e^{-2t} \tag{2}$$

(b) Note that $v > 0$ for all t. Let s be the required distance traveled (as v decreases from 20 to 5); then

$$s = \int_{v=20}^{v=5} 20e^{-2t}\, dt = \int_{t=0}^{t=\ln 2} 20e^{-2t}\, dt \tag{3}$$

where, when $v = 20$, $t = 0$. Also, when $v = 5$, use (2) to get $\frac{1}{4} = e^{-2t}$ or $-\ln 4 = -2t$.

So $t = \ln 2$. Evaluating s in (3) gives

$$s = -10e^{-2t}\Big|_0^{\ln 2} = -10\left(\frac{1}{4} - 1\right) = \frac{15}{2}$$

B10. Let (x,y) be the point in the first quadrant where the line parallel to the x-axis meets the parabola. The area of the triangle is given by

$$A = xy = x\left(27 - x^2\right) = 27x - x^3 \text{ for } 0 \le x \le 3\sqrt{3}$$

Then $A'(x) = 27 - 3x^2 = 3(3 + x)(3 - x)$, and $A'(x) = 0$ at $x = 3$.

Since A' changes from positive to negative at $x = 3$, the area reaches its maximum there.

The maximum area is $A(3) = 3\left(27 - 3^2\right) = 54$.

B11. Let $M = $ the temperature of the milk at time t. Then

$$\frac{dM}{dt} = k(68 - M)$$

The differential equation is separable:

$$\int \frac{dM}{68 - M} = \int k\, dt$$
$$-\ln|68 - M| = kt + C \quad \text{(note that } 68 - M > 0\text{)}$$
$$\ln(68 - M) = -(kt + C)$$
$$68 - M = e^{-(kt + C)}$$
$$M = 68 - ce^{-kt}$$

where $c = e^{-C}$.

Find c, using the fact that $M = 40°$ when $t = 0$:

$$40 = 68 - ce^0 \quad \text{means} \quad c = 28$$

Find k, using the fact that $M = 43°$ when $t = 3$:

$$43 = 68 - 28e^{-3k}$$
$$e^{-3k} = \frac{25}{28}$$
$$k = -\frac{1}{3}\ln\frac{25}{28}$$

Hence $M = 68 - 28e^{\frac{1}{3}\ln\frac{25}{28}t}$.

Now find t when $M = 60°$:

$$60 = 68 - 28e^{\frac{1}{3}\ln\frac{25}{28}t}$$
$$e^{\frac{1}{3}\ln\frac{25}{28}t} = \frac{8}{28}$$

$$t = \frac{\ln\frac{8}{28}}{\frac{1}{3}\ln\frac{25}{28}} = 33.163$$

Since the phone rang at $t = 3$, you have 30 minutes to solve the problem.

B12. One possible graph of h is shown; it has the following properties:

- continuity on $[-4,4]$
- symmetry about the y-axis
- roots at $x = -1, 1$
- horizontal tangents at $x = -2, 0, 2$
- points of inflection at $x = -3, -2, -1, 1, 2, 3$
- corners at $x = -3, 3$

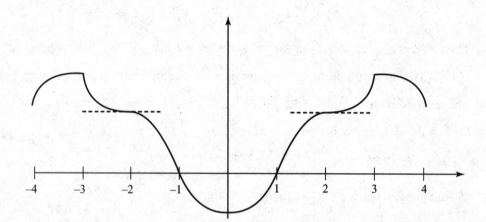

B13. (a) Let $f(x) = \ln(1 + x)$. Then $f'(x) = \dfrac{1}{1+x}, f''(x) = -\dfrac{1}{(1+x)^2}, f'''(x) = \dfrac{2}{(1+x)^3},$

$f^{(4)}(x) = -\dfrac{3!}{(1+x)^4}$, and $f^{(5)}(x) = \dfrac{4!}{(1+x)^5}$. At $x = 0$, $f(0) = 0, f'(0) = 1, f''(0) = -1,$

$f'''(0) = 2, f^{(4)}(0) = -(3!)$, and $f^{(5)}(0) = 4!$. So

$$\ln(1 + x) = x - \frac{x^2}{2} + \frac{x^3}{3} - \frac{x^4}{4} + \frac{x^5}{5} - \cdots$$

(b) Using the Ratio Test, you know that the series converges when $\lim\limits_{x \to} \left| \dfrac{x^{n+1}}{n+1} \cdot \dfrac{n}{x^n} \right| < 1,$

that is, when $|x| < 1$ or $-1 < x < 1$. Thus, the radius of convergence is 1.

(c) $\ln(1.2) = 0.2 - \dfrac{(0.2)^2}{2} + \dfrac{(0.2)^3}{3} - \dfrac{(0.2)^4}{4} + \dfrac{(0.2)^5}{5}$

(d) Since the series converges by the Alternating Series Test, the error in the answer for (c) is less than $\dfrac{(0.2)^6}{6}$.

B14. From the equations for x and y,

$$dx = (1 - \cos\theta)\, d\theta \qquad \text{and} \qquad dy = \sin\theta\, d\theta$$

(a) The slope at any point is given by $\dfrac{dy}{dx}$, which here is $\dfrac{\sin\theta}{1 - \cos\theta}$. When $\theta = \dfrac{2\pi}{3}$,

the slope is $\dfrac{\sqrt{3}}{3}$.

(b) When $\theta = \dfrac{2\pi}{3}$, $x = \dfrac{2\pi}{3} - \dfrac{\sqrt{3}}{2}$ and $y = 1 - \left(-\dfrac{1}{2}\right) = \dfrac{3}{2}$. An equation of the tangent is

$9y - 3\sqrt{3} \cdot x = 18 - 2\pi\sqrt{3}$.

B15. Both curves are circles with centers at, respectively, $(2,0)$ and $\left(2, \dfrac{\pi}{2}\right)$; the circles intersect at $\left(2\sqrt{2}, \dfrac{\pi}{2}\right)$. The common area is given by

$$\int_0^{\pi/4} (4 \sin \theta)^2 \, d\theta \qquad \text{or} \qquad \int_{\pi/4}^{\pi/2} (4 \cos \theta)^2 \, d\theta$$

The answer is $2(\pi - 2)$.

B16. (a) For $f(x) = \cos x, f'(x) = -\sin x, f''(x) = -\cos x, f'''(x) = \sin x, f^{(4)}(x) = \cos x$, $f^{(5)}(x) = -\sin x$, and $f^{(6)}(x) = -\cos x$. The Taylor polynomial of order 4 about 0 is

$$\cos x = 1 - \frac{x^2}{2!} + \frac{x^4}{4!}$$

Note that the next term of the alternating Maclaurin series for $\cos x$ is $-\dfrac{x^6}{6!}$.

(b) $\displaystyle\int_0^1 \cos x \, dx = x - \frac{x^3}{3 \cdot 2!} + \frac{x^5}{5 \cdot 4!} \bigg|_0^1 = 1 - \frac{1}{6} + \frac{1}{120}$

(c) The error in (b), convergent by the Alternating Series Test, is less than the absolute value of the first term dropped:

$$\text{error} < \int_0^1 \frac{x^6}{6!} \, dx = \frac{x^7}{7!} \bigg|_0^1 = \frac{1}{7!}$$

B17. (a) Since $\dfrac{dy}{dx} = 2$, $y = 2t + 1$ and $x = 4t^3 + 6t^2 + 3t$.

(b) Since $\dfrac{d^2y}{dt^2} = 0$ and $\dfrac{d^2x}{dt^2} = 24t + 12$, then, when $t = 1$, $|\mathbf{a}| = 36$.

B18. See the figure. The required area A is twice the sum of the following areas: that of the limaçon from 0 to $\dfrac{\pi}{3}$ and that of the circle from $\dfrac{\pi}{3}$ to $\dfrac{\pi}{2}$. Thus

$$A = 2\left[\frac{1}{2}\int_0^{\pi/3} (2 - \cos \theta)^2 \, d\theta + \frac{1}{2}\int_{\pi/3}^{\pi/2} (3\cos \theta)^2 \, d\theta\right]$$
$$= \frac{9\pi}{4} - 3\sqrt{3}$$

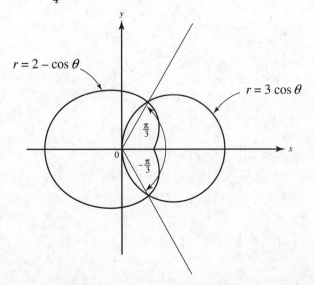

AB Practice Tests

ANSWER SHEET
AB Practice Test 1

SECTION I

Part A

1. Ⓐ Ⓑ Ⓒ Ⓓ
2. Ⓐ Ⓑ Ⓒ Ⓓ
3. Ⓐ Ⓑ Ⓒ Ⓓ
4. Ⓐ Ⓑ Ⓒ Ⓓ
5. Ⓐ Ⓑ Ⓒ Ⓓ
6. Ⓐ Ⓑ Ⓒ Ⓓ
7. Ⓐ Ⓑ Ⓒ Ⓓ
8. Ⓐ Ⓑ Ⓒ Ⓓ

9. Ⓐ Ⓑ Ⓒ Ⓓ
10. Ⓐ Ⓑ Ⓒ Ⓓ
11. Ⓐ Ⓑ Ⓒ Ⓓ
12. Ⓐ Ⓑ Ⓒ Ⓓ
13. Ⓐ Ⓑ Ⓒ Ⓓ
14. Ⓐ Ⓑ Ⓒ Ⓓ
15. Ⓐ Ⓑ Ⓒ Ⓓ
16. Ⓐ Ⓑ Ⓒ Ⓓ

17. Ⓐ Ⓑ Ⓒ Ⓓ
18. Ⓐ Ⓑ Ⓒ Ⓓ
19. Ⓐ Ⓑ Ⓒ Ⓓ
20. Ⓐ Ⓑ Ⓒ Ⓓ
21. Ⓐ Ⓑ Ⓒ Ⓓ
22. Ⓐ Ⓑ Ⓒ Ⓓ
23. Ⓐ Ⓑ Ⓒ Ⓓ
24. Ⓐ Ⓑ Ⓒ Ⓓ

25. Ⓐ Ⓑ Ⓒ Ⓓ
26. Ⓐ Ⓑ Ⓒ Ⓓ
27. Ⓐ Ⓑ Ⓒ Ⓓ
28. Ⓐ Ⓑ Ⓒ Ⓓ
29. Ⓐ Ⓑ Ⓒ Ⓓ
30. Ⓐ Ⓑ Ⓒ Ⓓ

Part B

31. Ⓐ Ⓑ Ⓒ Ⓓ
32. Ⓐ Ⓑ Ⓒ Ⓓ
33. Ⓐ Ⓑ Ⓒ Ⓓ
34. Ⓐ Ⓑ Ⓒ Ⓓ

35. Ⓐ Ⓑ Ⓒ Ⓓ
36. Ⓐ Ⓑ Ⓒ Ⓓ
37. Ⓐ Ⓑ Ⓒ Ⓓ
38. Ⓐ Ⓑ Ⓒ Ⓓ

39. Ⓐ Ⓑ Ⓒ Ⓓ
40. Ⓐ Ⓑ Ⓒ Ⓓ
41. Ⓐ Ⓑ Ⓒ Ⓓ
42. Ⓐ Ⓑ Ⓒ Ⓓ

43. Ⓐ Ⓑ Ⓒ Ⓓ
44. Ⓐ Ⓑ Ⓒ Ⓓ
45. Ⓐ Ⓑ Ⓒ Ⓓ

AB Practice Test 1

Section I

Part A

TIME: 60 MINUTES

> The use of calculators is not permitted for this part of the examination.
> There are 30 questions in Part A, for which 60 minutes are allowed. Because there is no deduction for wrong answers, you should answer every question, even if you need to guess.
>
> **DIRECTIONS:** Choose the best answer for each question.

1. $\lim\limits_{x\to\infty} \dfrac{20x^2 - 13x + 5}{5 - 4x^3}$ is

   (A) -5
   (B) 0
   (C) 1
   (D) ∞

2. $\lim\limits_{h\to 0} \dfrac{\ln(2+h) - \ln 2}{h}$ is

   (A) $\ln 2$
   (B) $\dfrac{1}{2}$
   (C) $\dfrac{1}{\ln 2}$
   (D) ∞

3. If $y = e^{-x^2}$, then $y''(0)$ equals

   (A) 2
   (B) 1
   (C) 0
   (D) -2

Questions 4 and 5. Use the following table, which shows the values of the differentiable functions f and g.

x	f	f'	g	g'
1	2	$\dfrac{1}{2}$	-3	5
2	3	1	0	4
3	4	2	2	3
4	6	4	3	$\dfrac{1}{2}$

4. The average rate of change of function f on $[1,4]$ is

   (A) 7/6
   (B) 4/3
   (C) 15/8
   (D) 15/4

5. If $h(x) = g\big(f(x)\big)$ then $h'(3) =$

   (A) 1/2
   (B) 1
   (C) 4
   (D) 6

GO ON TO THE NEXT PAGE

6. The derivative of a function f is given for all x by

$$f'(x) = x^2(x+1)^3(x-4)^2$$

The set of x-values for which f is a relative minimum is

(A) $\{0, -1, 4\}$

(B) $\{-1\}$

(C) $\{0, 4\}$

(D) $\{0, -1\}$

7. If $y = \dfrac{x-3}{2-5x}$, then $\dfrac{dy}{dx}$ equals

(A) $\dfrac{17 - 10x}{(2-5x)^2}$

(B) $\dfrac{13}{(2-5x)^2}$

(C) $\dfrac{-13}{(2-5x)^2}$

(D) $\dfrac{17}{(2-5x)^2}$

8. The maximum value of the function $f(x) = xe^{-x}$ is

(A) $\dfrac{1}{e}$

(B) 1

(C) -1

(D) $-e$

9. Which equation has the slope field shown below?

(A) $\dfrac{dy}{dx} = \dfrac{5}{y}$

(B) $\dfrac{dy}{dx} = \dfrac{5}{x}$

(C) $\dfrac{dy}{dx} = \dfrac{x}{y}$

(D) $\dfrac{dy}{dx} = 5y$

Questions 10–12. The graph below shows the velocity of an object moving along a line, for $0 \le t \le 9$.

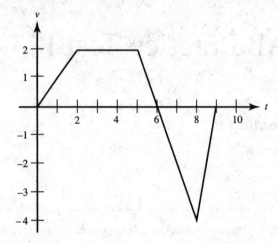

10. At what time does the object attain its maximum acceleration?

(A) $2 < t < 5$

(B) $t = 6$

(C) $t = 8$

(D) $8 < t < 9$

11. The object is farthest from the starting point at $t =$

(A) 5

(B) 6

(C) 8

(D) 9

12. At $t = 8$, the object was at position $x = 10$.
At $t = 5$, the object's position was $x =$

(A) 5

(B) 7

(C) 13

(D) 15

13. $\displaystyle\int_0^2 \dfrac{x^2 - 3x + 7}{x+3}\, dx =$

(A) $\dfrac{4}{3}$

(B) $-10 + 25\ln\dfrac{5}{3}$

(C) $2 + 7\ln\dfrac{5}{3}$

(D) $\dfrac{32}{5}\ln 5$

GO ON TO THE NEXT PAGE

x	$f(x)$	$f'(x)$	$f''(x)$	$g(x)$	$g'(x)$	$g''(x)$
1	1	0	−7	1/3	−2	7

14. You are given two twice-differentiable functions, $f(x)$ and $g(x)$. The table above gives values for $f(x)$ and $g(x)$ and their first and second derivatives at $x = 1$. Find $\lim\limits_{x \to 1} \dfrac{2f(x) - 6g(x)}{4x^2 - 4e^{3(x-1)}}$.

(A) −3

(B) 1

(C) 2

(D) nonexistent

15. A differentiable function has the values shown in this table:

x	2.0	2.2	2.3	2.7	2.8	3.0
$f(x)$	1.39	1.73	2.10	2.48	2.88	3.30

Estimate $f'(2.1)$.

(A) 0.34

(B) 1.56

(C) 1.70

(D) 1.91

16. If $A = \displaystyle\int_0^1 e^{-x}dx$ is approximated using various sums with the same number of subdivisions, and if L, R, and T denote, respectively, left Riemann Sum, right Riemann Sum, and trapezoidal sum, then it follows that

(A) $R \le A \le T \le L$

(B) $R \le T \le A \le L$

(C) $L \le T \le A \le R$

(D) $L \le A \le T \le R$

17. The number of vertical tangents to the graph of $y^2 = x - x^3$ is

(A) 3

(B) 2

(C) 1

(D) 0

18. $\displaystyle\int_0^6 f(x - 1)dx =$

(A) $\displaystyle\int_1^7 f(x)dx$

(B) $\displaystyle\int_{-1}^5 f(x)dx$

(C) $\displaystyle\int_{-1}^5 f(x + 1)dx$

(D) $\displaystyle\int_1^7 f(x + 1)dx$

19. The equation of the curve shown below is $y = \dfrac{4}{1 + x^2}$. What does the area of the shaded region equal?

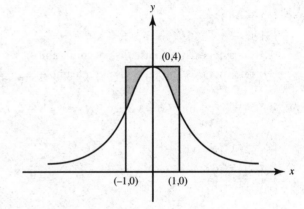

(A) $8 - \pi$

(B) $8 - 2\pi$

(C) $8 - 4\pi$

(D) $8 - 4\ln 2$

20. Over the interval $0 \le x \le 10$, the average value of the function f shown below

(A) is 6.10

(B) is 6.25

(C) does not exist, because f is not continuous

(D) does not exist, because f is not integrable

21. If $f'(x) = 2 f(x)$ and $f(2) = 1$, then $f(x) =$

(A) e^{2x-4}

(B) $e^{2x} + 1 - e^4$

(C) e^{4-2x}

(D) $e^{x^2 - 4}$

22. If $f(t) = \displaystyle\int_0^{t^2} \dfrac{1}{1 + x^2}\, dx$, then $f'(t)$ equals

(A) $\dfrac{1}{1 + t^2}$

(B) $\dfrac{2t}{1 + t^2}$

(C) $\dfrac{1}{1 + t^4}$

(D) $\dfrac{2t}{1 + t^4}$

23. The curve $x^3 + x \tan y = 27$ passes through $(3,0)$. Use the tangent line there to estimate the value of y at $x = 3.1$. The value is

(A) -2.7

(B) -0.9

(C) 0

(D) 0.1

24. $\displaystyle\int \left(\sqrt{x} - 2 \right) x^2 dx =$

(A) $\dfrac{2}{3} x^{3/2} - 2x + C$

(B) $\dfrac{5}{2} x^{3/2} - 4x + C$

(C) $\dfrac{2}{5} x^{5/2} - \dfrac{2}{3} x^3 + C$

(D) $\dfrac{2}{7} x^{7/2} - \dfrac{2}{3} x^3 + C$

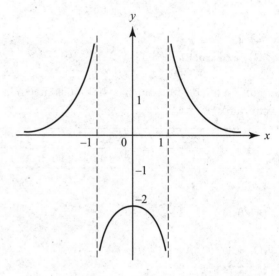

25. The graph of a function $y = f(x)$ is shown above. Which is true?

(A) $\displaystyle\lim_{x \to 1} f(x) = -\infty$

(B) $\displaystyle\lim_{x \to -\infty} f(x) = \pm 1$

(C) $\displaystyle\lim_{x \to -2} f(x) = 0$

(D) $\displaystyle\lim_{x \to \infty} f(x) = 0$

26. A function $f(x)$ equals $\dfrac{x^2 - x}{x - 1}$ for all x except $x = 1$. For the function to be continuous at $x = 1$, the value of $f(1)$ must be

(A) 0

(B) 1

(C) 2

(D) ∞

GO ON TO THE NEXT PAGE

27. $\int_0^1 \dfrac{e^x}{(3-e^x)^2}\,dx$ equals

   (A) $-2\ln(3-e)$

   (B) $\dfrac{1-e}{2(3-e)}$

   (C) $\dfrac{1}{2(3-e)}$

   (D) $\dfrac{e-1}{2(3-e)}$

28. Suppose $f(x)=\displaystyle\int_0^x \dfrac{4+t}{t^2+4}\,dt$. It follows that

   (A) f increases for all x

   (B) f has a critical point at $x=0$

   (C) f has a local min at $x=-4$

   (D) f has a local max at $x=-4$

30. Choose the integral that is the limit of the Riemann Sum: $\displaystyle\lim_{n\to\infty}\sum_{k=1}^{n}\left(\left(\dfrac{3k}{n}+2\right)^2\cdot\left(\dfrac{3}{n}\right)\right)$.

   (A) $\displaystyle\int_2^5 (x+2)^2\,dx$

   (B) $\displaystyle\int_0^3 (3x+2)^2\,dx$

   (C) $\displaystyle\int_2^5 x^2\,dx$

   (D) $\displaystyle\int_0^3 x^2\,dx$

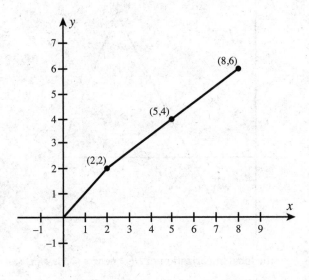

29. The graph of $f(x)$ consists of two line segments as shown above. If $g(x)=f^{-1}(x)$, the inverse function of $f(x)$, find $g'(4)$.

   (A) $\dfrac{1}{5}$

   (B) $\dfrac{2}{3}$

   (C) $\dfrac{3}{2}$

   (D) 5

STOP   End of Part A, Section I

Part B

TIME: 45 MINUTES

Some questions in this part of the examination require the use of a graphing calculator. There are 15 questions in Part B, for which 45 minutes are allowed. Because there is no deduction for wrong answers, you should answer every question, even if you need to guess.

DIRECTIONS: Choose the best answer for each question. If the exact numerical value of the correct answer is not listed as a choice, select the choice that is closest to the exact numerical answer.

31. An object moving along a line has velocity $v(t) = t \cos t - \ln(t+2)$, where $0 \le t \le 10$. The object achieves its maximum speed when t is approximately

    (A) 5.107
    (B) 6.419
    (C) 7.550
    (D) 9.538

32. The graph of f', which consists of a quarter-circle and two line segments, is shown above. At $x = 2$, which of the following statements is true?

    (A) f is not continuous.
    (B) f is continuous but not differentiable.
    (C) f has a local maximum.
    (D) The graph of f has a point of inflection.

33. Let $H(x) = \displaystyle\int_0^x f(t)\,dt$, where f is the function whose graph appears below.

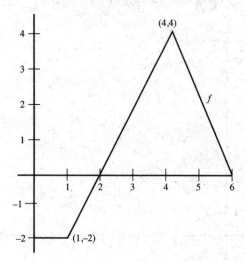

    The local linearization of $H(x)$ near $x = 3$ is $H(x) \approx$

    (A) $-2x + 8$
    (B) $2x - 4$
    (C) $-2x + 4$
    (D) $2x - 8$

34. The table shows the speed of an object, in feet per second, at various times during a 12-second interval.

time (sec)	0	3	6	7	8	10	12
speed (ft/sec)	15	14	11	8	7	3	0

Estimate the distance the object travels, using the midpoint method with 3 subintervals.

(A) 100 feet

(B) 110 feet

(C) 112 feet

(D) 114 feet

35. In a marathon, when the winner crosses the finish line many runners are still on the course, some quite far behind. If the density of runners x miles from the finish line is given by $R(x) = 20\left[1 - \cos\left(1 + 0.03x^2\right)\right]$ runners per mile, how many are within 8 miles of the finish line?

(A) 30

(B) 40

(C) 157

(D) 166

36. Which best describes the behavior of the function $y = \arctan\left(\dfrac{1}{\ln x}\right)$ at $x = 1$?

(A) It has a jump discontinuity.

(B) It has an infinite discontinuity.

(C) It has a removable discontinuity.

(D) It is continuous.

37. Let $G(x) = [f(x)]^2$. In an interval around $x = a$, the graph of f is increasing and concave downward, while G is decreasing. Which describes the graph of G there?

(A) concave downward

(B) concave upward

(C) point of inflection

(D) quadratic

38. The value of c for which $f(x) = x + \dfrac{c}{x}$ has a local minimum at $x = 3$ is

(A) -9

(B) 0

(C) 6

(D) 9

39. The function g is a differentiable function. It is known that $g'(x) \le 4$ for $3 \le x \le 10$ and that $g(7) = 8$. Which of the following could be true?

    I. $g(5) = 0$

    II. $g(8) = -4$

    III. $g(9) = 17$

(A) I only

(B) II only

(C) I and II only

(D) I and III only

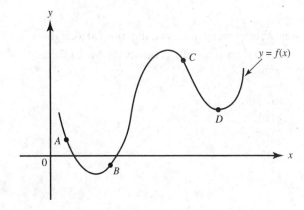

40. At which point on the graph of $y = f(x)$ shown above is $f'(x) < 0$ and $f''(x) > 0$?

(A) A

(B) B

(C) C

(D) D

41. Let $f(x) = x^5 + 1$, and let g be the inverse function of f. What is the value of $g'(0)$?

(A) -1

(B) $\dfrac{1}{5}$

(C) $-\dfrac{1}{5}$

(D) $g'(0)$ does not exist

42. The hypotenuse AB of a right triangle ABC is 5 feet, and one leg, AC, is decreasing at the rate of 2 feet per second. The rate, in square feet per second, at which the area is changing when $AC = 3$ is

(A) $\dfrac{7}{4}$

(B) $-\dfrac{3}{2}$

(C) $-\dfrac{7}{4}$

(D) $-\dfrac{7}{2}$

43. At how many points on the interval $[0,\pi]$ does $f(x) = 2 \sin x + \sin 4x$ satisfy the Mean Value Theorem?

(A) 1

(B) 2

(C) 3

(D) 4

44. If the radius r of a sphere is increasing at a constant rate, then the rate of increase of the volume of the sphere is

(A) constant

(B) increasing

(C) decreasing

(D) decreasing for $r < 1$ and increasing for $r > 1$

45. The rate at which a purification process can remove contaminants from a tank of water is proportional to the amount of contaminant remaining. If 20% of the contaminant can be removed during the first minute of the process and 98% must be removed to make the water safe, approximately how long will the decontamination process take?

(A) 2 minutes

(B) 5 minutes

(C) 7 minutes

(D) 18 minutes

STOP   End of Section I

Section II

Part A

TIME: 30 MINUTES

2 PROBLEMS

> *A graphing calculator is required for some of these problems. See instructions on page 8.*

1. A function f is defined on the interval $[0,4]$ with $f(2) = 3$, $f'(x) = e^{\sin x} - 2\cos(3x)$, and $f''(x) = (\cos x) \cdot e^{\sin x} + 6\sin(3x)$.

   (a) Find all values of x in the interval where f has a critical point. Classify each critical point as a local maximum, local minimum, or neither. Justify your answers.

   (b) On what subinterval(s) of $(0,4)$, if any, is the graph of f concave down? Justify your answer.

   (c) Write an equation for the line tangent to the graph of f at $x = 1.5$.

2. The rate of sales of a new software product is given by the differentiable and increasing function $S(t)$, where S is measured in units per month and t is measured in months from the initial release date. The software company recorded these sales data:

t (months)	1	3	6	8	9
$S(t)$ (units/mo)	154	232	490	763	954

   (a) Use the data in the table to estimate $S'(7)$. Show the work that leads to your answer. Indicate units of measure.

   (b) Use a left Riemann Sum with the four subintervals indicated in the table to estimate the total number of units of software sold during the first 9 months of sales. Is this an underestimate or an overestimate of the total number of software units sold? Give a reason for your answer.

   (c) For $1 \le t \le 9$, must there be a time t when the rate of sales is increasing at $100\ \dfrac{\text{units}}{\text{month}}$ per month? Justify your answer.

   (d) For $9 \le t \le 12$, the rate of sales is modeled by the function $R(t) = 120 \cdot (2)^{x/3}$. Given that the actual sales in the first 9 months is 3636 software units, use this model to find the number of units of the software sold during the first 12 months of sales. Round your answer to the nearest whole unit.

STOP   End of Part A, Section II

Part B

TIME: 60 MINUTES

4 PROBLEMS

> *No calculator is allowed for any of these problems.*
>
> *If you finish Part B before time has expired, you may return to work on Part A, but you may not use a calculator.*

3. The graph of function $y = f(x)$ passes through the point $(1,1)$ and satisfies the differential equation $\dfrac{dy}{dx} = \dfrac{6x^2 - 4}{y}$.

    (a) Sketch the slope field for the differential equation at the 12 indicated points on the axes provided.

    (b) Find an equation of the line tangent to $f(x)$ at the point $(1,1)$ and use the linear equation to estimate $f(1.2)$.

    (c) Solve the differential equation, and find the particular solution for $y = f(x)$ that passes through the point $(1,1)$.

4. Let R represent the first-quadrant region bounded by the y-axis and the curves $y = 2^x$ and $y = 8 \cos \dfrac{\pi x}{6}$, as shown in the graph.

    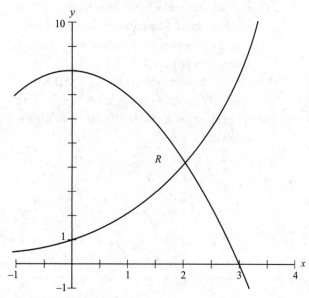

    (a) Find the area of region R.

    (b) Set up, but do not evaluate, an integral expression for the volume of the solid formed when R is rotated around the x-axis.

    (c) Set up, but do not evaluate, an integral expression for the volume of the solid whose base is R and all cross sections in planes perpendicular to the x-axis are squares.

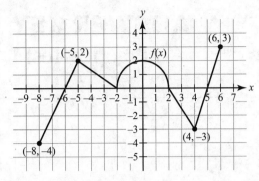

5. The graph of the function f is given above. f is twice differentiable and is defined on the interval $-8 \le x \le 6$. The function g is also twice differentiable and is defined as

$$g(x) = \int_{-2}^{x} f(q)\,dq.$$

(a) Write an equation for the line tangent to the graph of $g(x)$ at $x = -8$. Show the work that leads to your answer.

(b) Using your tangent line from part (a), approximate $g(-7)$. Is your approximation greater than or less than $g(-7)$? Give a reason for your answer.

(c) Evaluate: $\int_{-5}^{4} \left(2f'(x) - 3\right)dx$. Show the work that leads to your answer.

(d) Find the absolute minimum value of g on the interval $-8 \le x \le 6$. Justify your answer.

6. A particle moves along the x-axis from $t = 0$ to $t = 12$. The velocity function of the particle is $v(t) = 1 + 2\sin\left(\frac{\pi}{6}t\right)$. The initial position of the particle is $x = 4$ when $t = 0$.

(a) When is the particle moving to the right from $t = 0$ to $t = 12$?

(b) What is the speed of the particle at $t = 9$?

(c) Find the acceleration function of the particle. Is the speed of the particle increasing or decreasing at time $t = 6$? Explain your reasoning.

(d) Find the position of the particle at time $t = 6$.

STOP   **End of Test**

ANSWER KEY
AB Practice Test 1

SECTION I
Part A

1. **B**	9. **A**	17. **A**	25. **D**
2. **B**	10. **D**	18. **B**	26. **B**
3. **D**	11. **B**	19. **B**	27. **D**
4. **B**	12. **C**	20. **A**	28. **C**
5. **B**	13. **B**	21. **A**	29. **C**
6. **B**	14. **A**	22. **D**	30. **C**
7. **C**	15. **C**	23. **B**	
8. **A**	16. **A**	24. **D**	

Part B

31. **D**	35. **D**	39. **C**	43. **D**
32. **D**	36. **A**	40. **A**	44. **B**
33. **D**	37. **B**	41. **B**	45. **D**
34. **C**	38. **D**	42. **C**	

Answer Explanations

Section I Multiple-Choice

Part A

1. **(B)** Use the Rational Function Theorem on page 98.

2. **(B)** Note that $\lim\limits_{h \to 0} \dfrac{\ln(2+h) - \ln 2}{h} = f'(2)$, where $f(x) = \ln x$.

3. **(D)** Since $y' = -2xe^{-x^2}$, therefore $y'' = -2(x \cdot e^{-x^2} \cdot (-2x) + e^{-x^2})$. Replace x by 0.

4. **(B)** $\dfrac{f(4) - f(1)}{4 - 1} = \dfrac{6 - 2}{4 - 1} = \dfrac{4}{3}$

5. **(B)** $h'(3) = g'\big(f(3)\big) \cdot f'(3) = g'(4) \cdot f'(3) = \dfrac{1}{2} \cdot 2$

6. **(B)** Since $f'(x)$ exists for all x, it must equal 0 for any x_0 for which f is a relative minimum, and it must also change sign from negative to positive as x increases through x_0. For the given derivative, this only occurs at $x = -1$.

7. **(C)** By the Quotient Rule (formula (6) on page 115),

$$\frac{dy}{dx} = \frac{(2 - 5x)(1) - (x - 3)(-5)}{(2 - 5x)^2}$$

8. **(A)** Here, $f'(x)$ is $e^{-x}(1 - x)$; f has maximum value when $x = 1$.

9. **(A)** Note that (1) on a horizontal line the slope segments are all parallel, so the slopes there are all the same and $\dfrac{dy}{dx}$ must depend only on y; (2) along the x-axis (where $y = 0$) the slopes are infinite; and (3) as y increases, the slope decreases.

10. **(D)** Acceleration is the derivative (the slope) of velocity v; v is largest on $8 < t < 9$.

11. **(B)** Velocity v is the derivative of position; because $v > 0$ until $t = 6$ and $v < 0$ thereafter, the position increases until $t = 6$ and then decreases; since the area bounded by the curve above the axis is larger than the area below the axis, the object is farthest from its starting point at $t = 6$.

12. **(C)** From $t = 5$ to $t = 8$, the displacement (the integral of velocity) can be found by evaluating definite integrals based on the areas of two triangles: $\dfrac{1}{2}(1)(2) - \dfrac{1}{2}(2)(4) = -3$. Thus, if K is the object's position at $t = 5$, then $K - 3 = 10$ at $t = 8$.

13. **(B)** When we have a rational function in the integrand and the degree of the numerator is greater than the degree of the denominator, we need to rewrite the integrand using long polynomial division.

$$
\begin{array}{r}
x - 6 \\
x + 3 \overline{\smash{)}\; x^2 - 3x + 7} \\
\underline{-(x^2 + 3x)} \\
-6x + 7 \\
\underline{-(-6x - 18)} \\
25
\end{array}
$$

$$\int_0^2 \left(x - 6 + \frac{25}{x+3}\right) dx = \left(\frac{x^2}{2} - 6x + 25 \ln |x + 3|\right)\Big|_0^2$$

$$= (2 - 12 + 25 \ln 5) - 25 \ln 3 = -10 + 25 \ln \frac{5}{3}$$

14. **(A)** The functions are twice differentiable, meaning the functions and their first derivatives are continuous, so we can use substitution when finding limits:

$$\lim_{x \to 1} (2f(x) - 6g(x)) = 2(1) - 6\left(\frac{1}{3}\right) = 0 \qquad \lim_{x \to 1} \left(4x^2 - 4e^{3(x-1)}\right) = 4(1)^2 - 4e^{3(1-1)} = 0$$

Since the limit of both the numerator and the denominator are zero, we can use L'Hospital's Rule:

$$\lim_{x \to 1} \frac{2f(x) - 6g(x)}{4x^2 - 4e^{3(x-1)}} = \lim_{x \to 1} \frac{2f'(x) - 6g'(x)}{8x - 12e^{3(x-1)}} = \frac{2(0) - 6(-2)}{8(1) - 12e^{3(1-1)}} = \frac{12}{-4} = -3$$

15. **(C)** $f'(2.1) \approx \dfrac{f(2.2) - f(2.0)}{2.2 - 2.0}$

16. **(A)** $f(x) = e^{-x}$ is decreasing and concave upward.

17. **(A)** Implicit differentiation yields $2yy' = 1 - 3x^2$; so $\dfrac{dy}{dx} = \dfrac{1 - 3x^2}{2y}$. At a vertical tangent, $\dfrac{dy}{dx}$ is undefined; y must therefore equal 0 and the numerator must be nonzero. The original equation with $y = 0$ is $0 = x - x^3$, which has three solutions.

18. **(B)** Let $t = x - 1$; then $t = -1$ when $x = 0$, $t = 5$ when $x = 6$, and $dt = dx$.

19. **(B)** The required area, A, is given by the integral

$$2 \int_0^1 \left(4 - \frac{4}{1 + x^2}\right) dx = 2\left(4x - 4\tan^{-1} x\right)\Big|_0^1 = 2\left(4 - 4 \cdot \frac{\pi}{4}\right)$$

20. **(A)** The average value is $\dfrac{1}{10 - 0} \displaystyle\int_0^{10} f(x)dx$. The definite integral represents the sum of the areas of a trapezoid and a rectangle: $\dfrac{1}{2}(8 + 3)(6) = 4(7) = 61$.

21. **(A)** Solve the differential equation $\dfrac{dy}{dx} = 2y$ by separation of variables: $\dfrac{dy}{y} = 2dx$ yields $y = ce^{2x}$. The initial condition yields $1 = ce^{2 \cdot 2}$; so $c = e^{-4}$ and $y = e^{2x-4}$.

22. **(D)** $\dfrac{d}{du} \displaystyle\int_0^u \dfrac{1}{1 + x^2} dx = \dfrac{1}{1 + u^2}$. When $u = t^2$,

$$\frac{d}{dt} \int_0^u \frac{1}{1 + x^2} dx = \frac{1}{1 + u^2} \frac{du}{dt} = \frac{1}{1 + t^4}(2t)$$

23. **(B)** By implicit differentiation, $3x^2 + x\sec^2 y \dfrac{dy}{dx} + \tan y = 0$. At $(3,0)$, $\dfrac{dy}{dx} = -9$; so an equation of the tangent line at $(3,0)$ is $y = -9(x - 3)$.

24. **(D)** $\displaystyle\int \left(\sqrt{x} - 2\right)x^2 \, dx = \int \left(x^{5/2} - 2x^2\right)dx = \frac{2}{7}x^{7/2} - \frac{2}{3}x^3 + C$

25. **(D)** The graph shown has the x-axis as a horizontal asymptote.

26. **(B)** Since $\lim_{x \to 1} f(x) = 1$, to render $f(x)$ continuous at $x = 1$, $f(1)$ must be defined to be 1.

27. **(D)** $-\displaystyle\int_0^1 (3 - e^x)^{-2} \left(-e^x dx\right) = \frac{1}{3 - e^x}\Big|_0^1 = \frac{1}{3 - e} - \frac{1}{2}$

28. **(C)** Note that $f'(x) = \dfrac{4 + x}{x^2 + 4}$, so f has a critical value at $x = -4$. As x passes through -4, the sign of f' changes from $-$ to $+$, so f has a local minimum at $x = -4$.

29. **(C)** Given the point (a,b) on function $f(x)$, $\left(f^{-1}\right)'(b) = \dfrac{1}{f'(a)}$. Note that the slope of the graph of $f(x)$ at $x = 5$ is $\dfrac{2}{3}$, so $g'(4) = \dfrac{1}{f'(5)} = \dfrac{1}{2/3} = \dfrac{3}{2}$.

30. **(C)** From the Riemann Sum, we see $\Delta x = \frac{3}{n}$, then $k \cdot \Delta x = \frac{3k}{n}$. Notice that the term involving k in the Riemann Sum is equal to $\frac{3k}{n}$. Thus, we try $x_k = \frac{3k}{n} + 2$, so $a = 2$ and $\Delta x = \frac{b-2}{n} = \frac{3}{n}$, so $b = 5$. Since x_k replaces x, $f(x) = x^2$ giving the integral $\int_2^5 x^2\,dx$.

Part B

31. **(D)** Use your calculator to graph velocity against time. Speed is the absolute value of velocity. The greatest deviation from $v = 0$ is at $t = c$. With a calculator, $c = 9.538$.

32. **(D)** Because f' changes from increasing to decreasing, f'' changes from positive to negative and thus the graph of f changes concavity.

33. **(D)** $H(3) = \int_0^3 f(t)\,dt$. We evaluate this definite integral by finding the area of a trapezoid (negative) and a triangle: $H(3) = -\frac{1}{2}(2+1)(2) + \frac{1}{2}(1)(2) = -2$, so the tangent line passes through the point $(3, -2)$. The slope of the line is $H'(3) = f(3) = 2$, so an equation of the line is $y - (-2) = 2(x - 3)$.

34. **(C)** The distance is approximately $14(6) + 8(2) + 3(4)$.

35. **(D)** $\int_0^8 R(x)\,dx = 166.396$

36. **(A)** Selecting an answer for this question from your calculator graph is unwise. In some windows the graph may appear continuous; in others there may seem to be cusps, or a vertical asymptote. Put the calculator aside. Find

$$\lim_{x \to 1^+} \left(\arctan\left(\frac{1}{\ln x}\right) \right) = \frac{\pi}{2} \text{ and } \lim_{x \to 1^-} \left(\arctan\left(\frac{1}{\ln x}\right) \right) = -\frac{\pi}{2}$$

These limits indicate the presence of a jump discontinuity in the function at $x = 1$.

37. **(B)** We are given that (1) $f'(a) > 0$; (2) $f''(a) < 0$; and (3) $G'(a) < 0$. Since $G'(x) = 2f(x) \cdot f'(x)$, therefore $G'(a) = 2f(a) \cdot f'(a)$. Conditions (1) and (3) imply that (4) $f(a) < 0$. Since $G''(x) = 2[f(x) \cdot f''(x) + (f'(x))^2]$, therefore $G''(a) = 2[f(a)f''(a) + (f'(a))^2]$. Then the sign of $G''(a)$ is $2[(-) \cdot (-) + (+)]$ or positive, where the minus signs in the parentheses follow from conditions (4) and (2).

38. **(D)** Since $f'(x) = 1 - \dfrac{c}{x^2}$, it equals 0 for $x = \pm\sqrt{c}$. When $x = 3$, $c = 9$; this yields a minimum since $f''(3) > 0$.

39. **(C)** We can define function g as $g(x) = g(7) + \displaystyle\int_7^x g'(t)dt$. We will use this definition for $g(x)$ and the fact that we can find the maximum value of an integral over an interval by using one rectangle with the height equal to the maximum value of the function over the interval and the width equal to the width of the interval. We were given the maximum value of $g'(x)$ on $3 \le x \le 10$.

  I. $g(5) = g(7) + \displaystyle\int_7^5 g'(x)dx \le 8 + (4)(-2) = 0$; therefore $g(5) = 0$ is possible.

  II. $g(8) = g(7) + \displaystyle\int_7^8 g'(x)dx \le 8 + (4)(1) = 12$; therefore $g(8) = -4$ is possible.

  III. $g(9) = g(7) + \displaystyle\int_7^9 g'(x)dx \le 8 + (4)(2) = 16$; therefore $g(9) = 17$ is not possible.

40. **(A)** The curve falls when $f'(x) < 0$ and is concave up when $f''(x) > 0$.

41. **(B)** $g'(y) = \dfrac{1}{f'(x)} = \dfrac{1}{5x^4}$. To find $g'(0)$, find x such that $f(x) = 0$. By inspection, $x = -1$, so $g'(0) = \dfrac{1}{5(-1)^4} = \dfrac{1}{5}$.

42. **(C)**

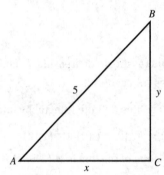

It is given that $\dfrac{dx}{dt} = -2$; you want $\dfrac{dA}{dt}$, where $A = \dfrac{1}{2}xy$.

$$\frac{dA}{dt} = \frac{1}{2}\left(x\frac{dy}{dt} + y\frac{dx}{dy}\right) = \frac{1}{2}\left[3 \cdot \frac{dy}{dt} + y \cdot (-2)\right]$$

Since $y^2 = 25 - x^2$, it follows that $2y\dfrac{dy}{dt} = -2x\dfrac{dx}{dt}$ and, when $x = 3$, $y = 4$ and $\dfrac{dy}{dt} = \dfrac{3}{2}$. Then $\dfrac{dA}{dt} = -\dfrac{7}{4}$.

43. **(D)**

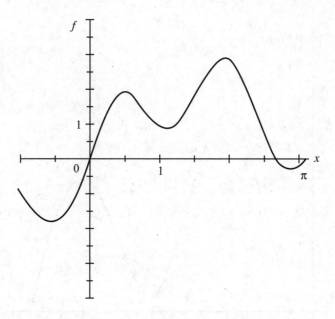

The function $f(x) = 2\sin x + \sin 4x$ is graphed above.

Since $f(0) = f(\pi)$ and f is both continuous and differentiable, Rolle's Theorem predicts at least one c in the interval such that $f'(c) = 0$. There are four points in $[0, \pi]$ of the calculator graph above where the tangent is horizontal.

44. **(B)** Since $\dfrac{dr}{dt} = k$, a positive constant, $\dfrac{dV}{dt} = 4\pi r^2 \dfrac{dr}{dt} = 4\pi r^2 k = cr^2$, where c is a positive constant. Then $\dfrac{d^2V}{dt^2} = 2cr\dfrac{dr}{dt} = 2crk$, which is also positive.

45. **(D)** If $Q(t)$ is the amount of contaminant in the tank at time t and Q_0 is the initial amount, then

$$\frac{dQ}{dt} = kQ \text{ and } Q(t) = Q_0 e^{kt}$$

Since $Q(1) = 0.8Q_0$, $0.8Q_0 = Q_0 e^{k \cdot 1}$, $0.8 = e^k$, and

$$Q(t) = Q_0(0.8)^t$$

We seek t when $Q(t) = 0.02Q_0$. Thus,

$$0.02Q_0 = Q_0(0.8)^t$$

and

$$t \approx 17.53 \text{ minutes}$$

Section II Free-Response

Part A

AB 1. This is the graph of $f'(x)$.

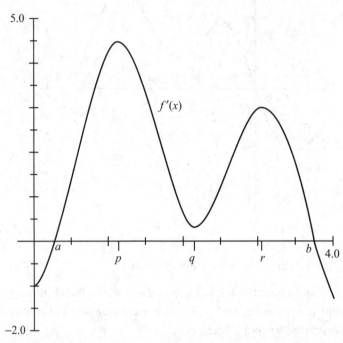

(a) The graph of f has critical points when $f'(x) = 0$. This occurs at points a and b on the graph of $f'(x)$ above, where $a = 0.283$ and $b = 3.760$.

At $x = 0.283$, the graph of f has a local minimum because $f'(x)$ changes from negative to positive at that value.

At $x = 3.760$, the graph of f has a local maximum because $f'(x)$ changes from positive to negative at that value.

(b) $f''(x) = 0$ at points p, q, and r on the graph of $f'(x)$ because that is where the graph of $f'(x)$ has horizontal tangent lines, where $p = 1.108$, $q = 2.166$, and $r = 3.082$. Note, you could also graph $f''(x)$ and find the roots of the graph.

To determine the intervals where f is concave down, look for decreasing intervals on $f'(x)$ or look for intervals on the graph of $f''(x)$ where $f''(x) < 0$.

The intervals where f is concave down are $(p, q) = (1.108, 2.166)$ and $(r, 4) = (3.082, 4)$.

(c) $f'(1.5) = 3.1330726$

$$f(1.5) = f(2) + \int_2^{1.5} f'(x)\,dx = 3 + \int_2^{1.5} f'(x)\,dx = 2.1409738$$

An equation of the tangent line at $x = 1.5$ is $y = 2.141 + 3.133(x - 1.5)$.

AB/BC 2. (a) $S'(7) \approx \dfrac{S(8) - S(6)}{8 - 6} = \dfrac{763 - 490}{2} = 136.5 \dfrac{\text{units}}{\text{month}}$ per month.

(b) The total number of units of software sold is given by $\displaystyle\int_1^9 S(t)dt$.

$$\int_1^9 S(t)dt = S(1) \cdot (3 - 1) + S(3) \cdot (6 - 3) + S(6) \cdot (8 - 6) + S(8) \cdot (9 - 8)$$

$$= 154(2) + 232(3) + 490(2) + 763(1) = 2747$$

This is an underestimate because we used a left Riemann Sum on an increasing function.

(c) Yes, because $\dfrac{S(9) - S(1)}{9 - 1} = \dfrac{954 - 154}{8} = 100$, and $S(t)$ is differentiable and, therefore, continuous, so the Mean Value Theorem guarantees that $S'(t) = 100 \dfrac{\text{units}}{\text{month}}$ per month for some value of t on the interval $(1,9)$.

(d) Total Sales $= 3636 + \displaystyle\int_9^{12} R(t)dt = 7790.961718$

The company sold approximately 7791 software units in the first 12 months of sales.

Part B

AB/BC 3. (a) The slope field for the twelve points indicated is:

The slopes are given in the table below. Be sure that the segments you draw are correct relative to the other slopes in the slope field with respect to the steepness of the segments.

x	-1	-1	-1	-1	0	0	0	0	1	1	1	1
y	-2	-1	1	2	-2	-1	1	2	-2	-1	1	2
dy/dx	-1	-2	2	1	1	2	-2	-1	-1	-2	2	1

(b) At $(1,1)$, $\dfrac{dy}{dx} = 2$, so the tangent line is $y = 1 + 2(x - 1)$. Using this linear equation, $f(1.2) \approx 1 + 2(1.2 - 1) = 1.4$.

(c) The differential equation $\dfrac{dy}{dx} = \dfrac{6x^2 - 4}{y}$ is separable.

$$\int y\, dy = \int (6x^2 - 4)\, dx$$

$$\frac{y^2}{2} = 2x^3 - 4x + C_1$$

$$y = \pm\sqrt{4x^3 - 8x + C_2}, \text{ where } C_2 = 2C_1$$

Since $f(x)$ passes through $(1,1)$, it must be true that $1 = \pm\sqrt{4(1)^3 - 8(1) + C_2}$.

Thus $C_2 = 5$, and the positive square root is used.

The solution is $f(x) = \sqrt{4x^3 - 8x + 5}$.

AB 4. (a) Draw a vertical element of area, as shown.

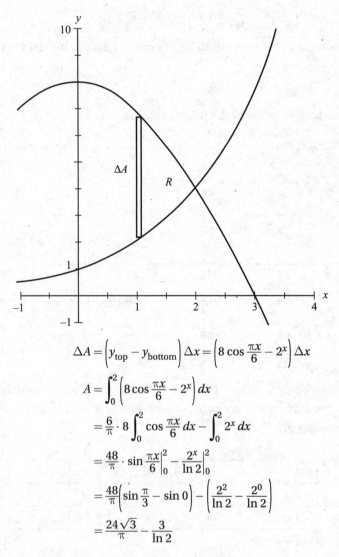

$$\Delta A = \left(y_{\text{top}} - y_{\text{bottom}}\right)\Delta x = \left(8\cos\frac{\pi x}{6} - 2^x\right)\Delta x$$

$$A = \int_0^2 \left(8\cos\frac{\pi x}{6} - 2^x\right) dx$$

$$= \frac{6}{\pi}\cdot 8\int_0^2 \cos\frac{\pi x}{6}\, dx - \int_0^2 2^x\, dx$$

$$= \frac{48}{\pi}\cdot \sin\frac{\pi x}{6}\Big|_0^2 - \frac{2^x}{\ln 2}\Big|_0^2$$

$$= \frac{48}{\pi}\left(\sin\frac{\pi}{3} - \sin 0\right) - \left(\frac{2^2}{\ln 2} - \frac{2^0}{\ln 2}\right)$$

$$= \frac{24\sqrt{3}}{\pi} - \frac{3}{\ln 2}$$

(b) Use washers; then

$$\Delta V = \left(r_2^2 - r_1^2\right)\Delta x = \pi\left(y_{\text{top}}^2 - y_{\text{bottom}}^2\right)\Delta x$$

$$V = \pi\int_0^2 \left[\left(8\cos\frac{\pi x}{6}\right)^2 - (2^x)^2\right] dx$$

(c)

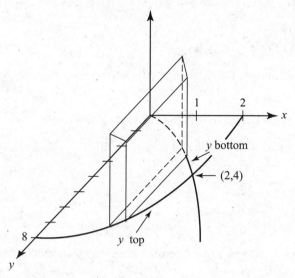

See the figure above.

$$\Delta V = s^2\,\Delta x = \left(y_{\text{top}} - y_{\text{bottom}}\right)^2 \Delta x$$
$$V = \int_0^2 \left(8\cos\frac{\pi x}{6} - 2^x\right)^2 dx$$

AB/BC 5. (a) $g(-8) = \int_{-2}^{-8} f(q)dq = -\int_{-8}^{-2} f(q)dq = -\left(-\frac{1}{2}(2)(4) + \frac{1}{2}(4)(2)\right) = 0$

$g'(x) = f(x) \Rightarrow g'(-8) = f'(-8) = -4$

Tangent line: $y = 0 - 4\big(x - (-8)\big) \Rightarrow y = -4(x+8)$

(b) $g(-7) \approx y(-7) = -4\big((-7)+8\big) = -4$

$g'(x) = f(x)$ is increasing on the interval $[-8,-7]$. Thus, $g''(x) > 0$ on $[-8,-7]$ and $g(x)$ is concave up on this interval. Therefore, the estimate using the tangent line is less than $g(-7)$ because the tangent line lies below the graph on a concave up portion of the graph.

(c) $\int_{-5}^{4}(2f'(x) - 3)dx = \big(2f(x) - 3x\big)\Big|_{-5}^{4} = \big(2f(4) - 3(4)\big) - \big(2f(-5) - 3(-5)\big)$

$$= \big(2(-3) - 12\big) - \big(2(2) + 15\big) = -37$$

(d) We need to perform the Candidates Test to find the absolute minimum on a closed interval.

First, the candidates are the critical points and the endpoints of the interval.

Critical points: $g'(x) = f(x) = 0$ when $x = -6,\ -2, 2, 5$. Endpoints: $x = -8, 6$

We can eliminate $x = -2, 2$ because neither is a local minimum, but we need to check the function value for the remaining candidates.

$g(-8) = 0$ (we calculated this in part (a))

$g(-6) = \int_{-2}^{-6} f(x)dx = -\int_{-6}^{-2} f(x)dx = -\left(\frac{1}{2}(4)(2)\right) = -4$

$g(5) = \int_{-2}^{5} f(x)dx = \frac{1}{2}\cdot\pi(2)^2 - \frac{1}{2}(3)(3) = 2\pi - \frac{9}{2} > 0$

$g(6) = \int_{-2}^{6} f(x)dx = g(5) + \int_{5}^{6} f(x)dx = \left(2\pi - \frac{9}{2}\right) + \frac{1}{2}(1)(3) = 2\pi - 3 > 0$

Therefore, the minimum value of $g(x)$ on the interval $[-8,6]$ is -4, and it occurs at $x = -6$.

AB 6. (a) $v(t) = 1 + 2\sin\left(\frac{\pi}{6}t\right) = 0 \Rightarrow \sin\left(\frac{\pi}{6}t\right) = -\frac{1}{2} \Rightarrow \frac{\pi}{6}t = \frac{7\pi}{6}, \frac{11\pi}{6} \Rightarrow t = 7, 11$

The particle moves to the right when $v(t) > 0$; this will occur when $0 \le t < 7$ and $11 < t \le 12$.

(b) The speed of the particle is the absolute value of the velocity.

$v(9) = 1 + 2\sin\left(\frac{9\pi}{6}\right) = 1 + 2\sin\left(\frac{3\pi}{2}\right) = 1 + 2(-1) = -1$. Therefore, the speed at $t = 9$ is 1.

(c) $a(t) = v'(t) = 2\cos\left(\frac{\pi}{6}t\right) \cdot \frac{\pi}{6} = \frac{\pi}{3}\cos\left(\frac{\pi}{6}t\right)$

$v(6) = 1 + 2\sin\left(\frac{6\pi}{6}\right) = 1 > 0 \qquad a(6) = \frac{\pi}{3}\cos\left(\frac{6\pi}{6}\right) = -\frac{\pi}{3} < 0$

The speed is decreasing at $t = 6$ because the velocity and acceleration have opposite signs.

(d) Using the Fundamental Theorem of Calculus:

$x(6) = x(0) + \int_0^6 x'(t)dt \Rightarrow x(6) = 4 + \int_0^6 v(t)dt$

$= 4 + \int_0^6 \left(1 + 2\sin\left(\frac{\pi}{6}t\right)\right)dt = 4 + \left(t - \frac{2\cos\left(\frac{\pi}{6}t\right)}{\frac{\pi}{6}}\right)\Bigg|_0^6$

$= 4 + \left(t - \frac{12}{\pi}\cos\left(\frac{\pi}{6}t\right)\right)\Bigg|_0^6 = 4 + \left(\left(6 - \frac{12}{\pi}\cos(\pi)\right) - \left(0 - \frac{12}{\pi}\cos(0)\right)\right)$

$= 4 + \left(\left(6 + \frac{12}{\pi}\right) - \left(-\frac{12}{\pi}\right)\right) = 10 + \frac{24}{\pi}$

ANSWER SHEET
AB Practice Test 2

SECTION I

Part A

1. Ⓐ Ⓑ Ⓒ Ⓓ   9. Ⓐ Ⓑ Ⓒ Ⓓ   17. Ⓐ Ⓑ Ⓒ Ⓓ   25. Ⓐ Ⓑ Ⓒ Ⓓ
2. Ⓐ Ⓑ Ⓒ Ⓓ   10. Ⓐ Ⓑ Ⓒ Ⓓ   18. Ⓐ Ⓑ Ⓒ Ⓓ   26. Ⓐ Ⓑ Ⓒ Ⓓ
3. Ⓐ Ⓑ Ⓒ Ⓓ   11. Ⓐ Ⓑ Ⓒ Ⓓ   19. Ⓐ Ⓑ Ⓒ Ⓓ   27. Ⓐ Ⓑ Ⓒ Ⓓ
4. Ⓐ Ⓑ Ⓒ Ⓓ   12. Ⓐ Ⓑ Ⓒ Ⓓ   20. Ⓐ Ⓑ Ⓒ Ⓓ   28. Ⓐ Ⓑ Ⓒ Ⓓ
5. Ⓐ Ⓑ Ⓒ Ⓓ   13. Ⓐ Ⓑ Ⓒ Ⓓ   21. Ⓐ Ⓑ Ⓒ Ⓓ   29. Ⓐ Ⓑ Ⓒ Ⓓ
6. Ⓐ Ⓑ Ⓒ Ⓓ   14. Ⓐ Ⓑ Ⓒ Ⓓ   22. Ⓐ Ⓑ Ⓒ Ⓓ   30. Ⓐ Ⓑ Ⓒ Ⓓ
7. Ⓐ Ⓑ Ⓒ Ⓓ   15. Ⓐ Ⓑ Ⓒ Ⓓ   23. Ⓐ Ⓑ Ⓒ Ⓓ
8. Ⓐ Ⓑ Ⓒ Ⓓ   16. Ⓐ Ⓑ Ⓒ Ⓓ   24. Ⓐ Ⓑ Ⓒ Ⓓ

Part B

31. Ⓐ Ⓑ Ⓒ Ⓓ   35. Ⓐ Ⓑ Ⓒ Ⓓ   39. Ⓐ Ⓑ Ⓒ Ⓓ   43. Ⓐ Ⓑ Ⓒ Ⓓ
32. Ⓐ Ⓑ Ⓒ Ⓓ   36. Ⓐ Ⓑ Ⓒ Ⓓ   40. Ⓐ Ⓑ Ⓒ Ⓓ   44. Ⓐ Ⓑ Ⓒ Ⓓ
33. Ⓐ Ⓑ Ⓒ Ⓓ   37. Ⓐ Ⓑ Ⓒ Ⓓ   41. Ⓐ Ⓑ Ⓒ Ⓓ   45. Ⓐ Ⓑ Ⓒ Ⓓ
34. Ⓐ Ⓑ Ⓒ Ⓓ   38. Ⓐ Ⓑ Ⓒ Ⓓ   42. Ⓐ Ⓑ Ⓒ Ⓓ

AB Practice Test 2

Section I

Part A

TIME: 60 MINUTES

The use of calculators is **not** permitted for this part of the examination.
There are 30 questions in Part A, for which 60 minutes are allowed. Because there is no deduction for wrong answers, you should answer every question, even if you need to guess.

DIRECTIONS: Choose the best answer for each question.

AB PRACTICE TEST 2

1. $\lim\limits_{x \to 2} \dfrac{x^2 - 2}{4 - x^2}$ is

   (A) -1

   (B) $-\dfrac{1}{2}$

   (C) 0

   (D) nonexistent

2. $\lim\limits_{x \to \infty} \dfrac{\sqrt{x} - 4}{4 - 3\sqrt{x}}$ is

   (A) $-\dfrac{1}{3}$

   (B) -1

   (C) 0

   (D) ∞

3. If $y = \dfrac{e^{\ln u}}{u}$, then $\dfrac{dy}{du}$ equals

   (A) $e^{\ln u}$

   (B) $\dfrac{e^{\ln u}}{u^2}$

   (C) $\dfrac{e^{\ln u}(u - 1)}{u^2}$

   (D) 0

4. Using the line tangent to $f(x) = \sqrt{9 + \sin(2x)}$ at $x = 0$, an estimate of $f(0.06)$ is

   (A) 0.02

   (B) 2.98

   (C) 3.01

   (D) 3.02

t (minutes)	0	10	25	30
$S(t)$ (students per minute)	10	18	20	24

5. The rate at which schoolchildren are dropped off to school on a certain day is modeled by the function S, where $S(t)$ is the number of students per minute and t is the number of minutes since the school doors were opened that morning. What is the approximate number of students dropped off at the school in the first 30 minutes the doors were open? Use a trapezoidal sum with three subintervals indicated by the data in the table.

   (A) 470

   (B) 535

   (C) 550

   (D) 1070

6. If $y = \sin^3(1 - 2x)$, then $\dfrac{dy}{dx}$ is

   (A) $3\sin^2(1 - 2x)$

   (B) $-2\cos^3(1 - 2x)$

   (C) $-6\sin^2(1 - 2x)$

   (D) $-6\sin^2(1 - 2x)\cos(1 - 2x)$

7. If $y = x^2 e^{1/x}$ $(x \neq 0)$, then $\dfrac{dy}{dx}$ is

   (A) $xe^{1/x}(x + 2)$

   (B) $e^{1/x}(2x - 1)$

   (C) $-\dfrac{2e^{1/x}}{x}$

   (D) $e^{-x}(2x - x^2)$

8. A point moves along the curve $y = x^2 + 1$ so that the x-coordinate is increasing at the constant rate of $\dfrac{3}{2}$ units per second. The rate, in units per second, at which the distance from the origin is changing when the point has coordinates $(1,2)$ is equal to

   (A) $\dfrac{7\sqrt{5}}{10}$

   (B) $\dfrac{3\sqrt{5}}{2}$

   (C) $\dfrac{3\sqrt{5}}{5}$

   (D) $\dfrac{5}{2}$

9. $\lim\limits_{h \to 0} \dfrac{\sqrt{25 + h} - 5}{h}$

   (A) $= 0$

   (B) $= \dfrac{1}{10}$

   (C) $= 1$

   (D) does not exist

10. The base of a solid is the first-quadrant region bounded by $y = \sqrt[4]{1 - x^2}$. Each cross section perpendicular to the x-axis is a square with one edge in the xy-plane. The volume of the solid is

    (A) $\dfrac{\pi}{4}$

    (B) 1

    (C) $\dfrac{\pi}{2}$

    (D) π

11. $\displaystyle\int \dfrac{x}{\sqrt{9 - x^2}}\, dx$ equals

    (A) $-\dfrac{1}{2}\ln\sqrt{9 - x^2} + C$

    (B) $\sin^{-1}\dfrac{x}{3} + C$

    (C) $-\sqrt{9 - x^2} + C$

    (D) $-\dfrac{1}{4}\sqrt{9 - x^2} + C$

12. $\displaystyle\int \dfrac{(y - 1)^2}{2y}\, dy$ equals

    (A) $\dfrac{y^2}{4} - y + \dfrac{1}{2}\ln|y| + C$

    (B) $y^2 - y + \ln|2y| + C$

    (C) $\dfrac{y^2}{4} - y + 2\ln|2y| + C$

    (D) $\dfrac{(y - 1)^3}{3y^2} + C$

13. $\displaystyle\int_{\pi/6}^{\pi/2} \cot x \, dx$ equals

    (A) 3

    (B) $\ln\dfrac{1}{2}$

    (C) $\ln 2$

    (D) $\ln\dfrac{2}{\sqrt{3}}$

GO ON TO THE NEXT PAGE

14. Given f' as graphed, which could be a graph of f?

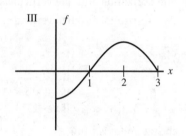

(A) I only

(B) II only

(C) III only

(D) I and III only

15. The first woman officially timed in a marathon was Violet Piercy of Great Britain in 1926. Her record of 3:40:22 stood until 1963, mostly because of a lack of women competitors. Soon after, times began dropping rapidly, but lately they have been declining at a much slower rate. Let $M(t)$ be the curve that best represents winning marathon times in year t. Which of the following is (are) positive for $t > 1963$?

   I. $M(t)$

  II. $M'(t)$

 III. $M''(t)$

(A) I only

(B) I and II only

(C) I and III only

(D) I, II, and III

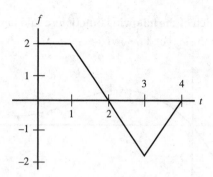

16. The graph of f is shown above. Let $G(x) = \int_0^x f(t)\,dt$ and $H(x) = \int_2^x f(t)\,dt$. Which of the following is true?

(A) $G(x) = H(x)$

(B) $G'(x) = H'(x + 2)$

(C) $G(x) = H(x + 2)$

(D) $G(x) = H(x) + 3$

17. The minimum value of $f(x) = x^2 + \dfrac{2}{x}$ on the interval $\dfrac{1}{2} \le x \le 2$ is

(A) 1

(B) 3

(C) $4\dfrac{1}{2}$

(D) 5

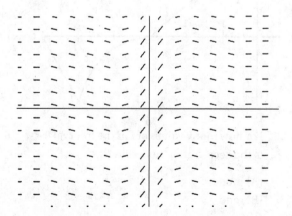

18. Which function could be a particular solution of the differential equation whose slope field is shown above?

(A) $y = \dfrac{2x}{x^2 + 1}$

(B) $y = \dfrac{x^2}{x^2 + 1}$

(C) $y = \sin x$

(D) $y = e^{-x^2}$

GO ON TO THE NEXT PAGE

19. Which of the following functions could have the graph sketched below?

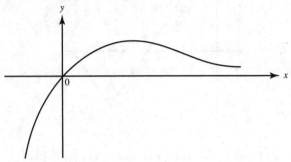

(A) $f(x) = xe^x$

(B) $f(x) = xe^{-x}$

(C) $f(x) = \dfrac{x}{x^2 + 1}$

(D) $f(x) = \dfrac{x^2}{x^3 + 1}$

Questions 20–22. Use the graph below, consisting of two line segments and a quarter-circle. The graph shows the velocity of an object during a 6-second interval.

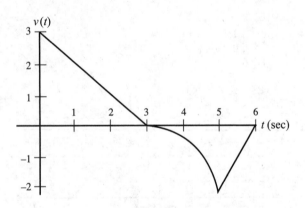

20. For how many values of t in the interval $0 < t < 6$ is the acceleration undefined?

(A) none

(B) one

(C) two

(D) three

21. During what time interval (in seconds) is the speed increasing?

(A) $0 < t < 3$

(B) $3 < t < 5$

(C) $5 < t < 6$

(D) never

22. What is the average acceleration (in units/sec^2) during the first 5 seconds?

(A) $-\dfrac{5}{2}$

(B) -1

(C) $\dfrac{1}{5}$

(D) $\dfrac{1}{2}$

23. The curve of $y = \dfrac{2x^2}{4 - x^2}$ has

(A) two horizontal asymptotes and one vertical asymptote

(B) two vertical asymptotes but no horizontal asymptote

(C) one horizontal and one vertical asymptote

(D) one horizontal and two vertical asymptotes

24. Suppose

$$f(x) = \begin{cases} x^2 & \text{if } x < -2 \\ 4 & \text{if } -2 < x \le 1 \\ 6 - x & \text{if } x > 1 \end{cases}$$

Which statement is true?

(A) f is continuous everywhere.

(B) f is discontinuous only at $x = 1$.

(C) f is discontinuous at $x = -2$ and at $x = 1$.

(D) If $f(-2)$ is defined to be 4, then f will be continuous everywhere.

25. The function $f(x) = x^5 + 3x - 2$ passes through the point $(1,2)$. Let f^{-1} denote the inverse of f. Then $(f^{-1})'(2)$ equals

(A) $\dfrac{1}{83}$

(B) $\dfrac{1}{8}$

(C) 8

(D) 83

26. $\displaystyle\int_1^e \dfrac{(\ln x)^3}{x}\, dx =$

(A) $\dfrac{1}{4}$

(B) 1

(C) $\dfrac{3}{2}$

(D) $\dfrac{e}{4}$

27. Which of the following statements is (are) true about the graph of $y = \ln\left(4 + x^2\right)$?

    I. It is symmetric to the y-axis.

    II. It has a local minimum at $x = 0$.

    III. It has inflection points at $x = \pm 2$.

    (A) I only

    (B) I and II only

    (C) II and III only

    (D) I, II, and III

28. Let $\displaystyle\int_0^x f(t)\, dt = x \sin \pi x$. Then $f(3) =$

    (A) -3π

    (B) -1

    (C) -3

    (D) $-\pi$

29. Choose the integral that is the limit of the Riemann Sum: $\displaystyle\lim_{n \to \infty} \sum_{k=1}^{n} \left(\left(\sqrt{\frac{2k}{n} + 3} \right) \cdot \left(\frac{1}{n} \right) \right)$.

    (A) $\displaystyle\int_3^4 \sqrt{2x}\, dx$

    (B) $\displaystyle\int_0^1 \sqrt{2x + 3}\, dx$

    (C) $\displaystyle\int_0^1 \sqrt{2x}\, dx$

    (D) $\displaystyle\int_3^4 \sqrt{2x + 3}\, dx$

30. The region bounded by $y = e^x$, $y = 1$, and $x = 2$ is rotated about the x-axis. The volume of the solid generated is given by the integral:

    (A) $\pi \displaystyle\int_0^2 e^{2x}\, dx$

    (B) $\pi \displaystyle\int_0^2 (e^x - 1)^2\, dx$

    (C) $\pi \displaystyle\int_0^2 (e^{2x} - 1)\, dx$

    (D) $\pi \displaystyle\int_1^{e^2} \left(4 - \ln^2 y\right)\, dy$

STOP   End of Part A, Section I

Part B

TIME: 45 MINUTES

Some questions in this part of the examination require the use of a graphing calculator. There are 15 questions in Part B, for which 45 minutes are allowed. Because there is no deduction for wrong answers, you should answer every question, even if you need to guess.

DIRECTIONS: Choose the best answer for each question. If the exact numerical value of the correct answer is not listed as a choice, select the choice that is closest to the exact numerical answer.

31. A particle moves on a straight line so that its velocity at time t is given by $v = 12\sqrt{s}$, where s is its distance from the origin. If $s = 1$ when $t = 0$, then, when $t = 1$, s equals

    (A) 6
    (B) 7
    (C) 36
    (D) 49

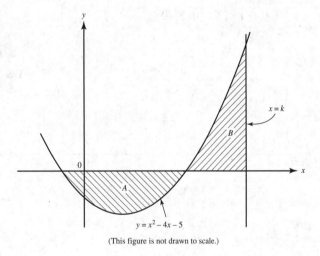

(This figure is not drawn to scale.)

32. The sketch shows the graphs of $f(x) = x^2 - 4x - 5$ and the line $x = k$. The regions labeled A and B have equal areas if $k =$

    (A) 7.899
    (B) 8
    (C) 8.144
    (D) 11

33. Bacteria in a culture increase at a rate proportional to the number present. An initial population of 200 triples in 10 hours. If this pattern of increase continues unabated, then the approximate number of bacteria after 1 full day is

    (A) 1056
    (B) 1440
    (C) 2793
    (D) 3240

34. When the substitution $x = 2t - 1$ is used, the definite integral $\int_3^5 t\sqrt{2t-1}\ dt$ may be expressed in the form $k\int_a^b (x+1)\sqrt{x}\ dx$, where $\{k,a,b\} =$

    (A) $\left\{\dfrac{1}{4}, 2, 3\right\}$
    (B) $\left\{\dfrac{1}{4}, 5, 9\right\}$
    (C) $\left\{\dfrac{1}{2}, 2, 3\right\}$
    (D) $\left\{\dfrac{1}{2}, 5, 9\right\}$

35. The curve defined by $x^3 + xy - y^2 = 10$ has a vertical tangent line when $x =$

    (A) 1.037
    (B) 1.087
    (C) 2.074
    (D) 2.096

GO ON TO THE NEXT PAGE

Questions 36 and 37. Use the graph of f shown on [0,7]. Let $G(x) = \displaystyle\int_{2}^{3x-1} f(t)\, dt.$

36. $G'(1)$ is

    (A) 1

    (B) 2

    (C) 3

    (D) 6

37. G has a local maximum at $x =$

    (A) 1

    (B) $\dfrac{4}{3}$

    (C) 2

    (D) 8

x	$f(x)$	$f'(x)$	$f''(x)$	$g(x)$	$g'(x)$	$g''(x)$
2	6	−1	−2	−2	1/3	−4/3

38. You are given two thrice-differentiable functions, $f(x)$ and $g(x)$. The table above gives values for $f(x)$ and $g(x)$ and their first and second derivatives at $x = 2$. Find $\displaystyle\lim_{x\to 2} \frac{f(x) + 3g(x)}{\frac{1}{2}x^2 - 2e^{x-2}}.$

    (A) 0

    (B) 1

    (C) 6

    (D) nonexistent

39. Using the left rectangular method and four subintervals of equal width, estimate $\displaystyle\int_{0}^{8} |f(t)|\, dt$, where f is the function graphed below.

    (A) 4

    (B) 8

    (C) 12

    (D) 16

40. Suppose $f(3) = 2$, $f'(3) = 5$, and $f''(3) = -2$. Then $\dfrac{d^2}{dx^2}\left(f(x)^2\right)$ at $x = 3$ is equal to

    (A) −20

    (B) −4

    (C) 10

    (D) 42

41. The velocity of a particle in motion along a line (for $t \geq 0$) is $v(t) = \ln\left(2 - t^2\right)$. Find the acceleration when the object is at rest.

    (A) −2

    (B) 0

    (C) $\dfrac{1}{2}$

    (D) 1

42. Suppose $f(x) = \dfrac{1}{3}x^3 + x$, $x > 0$ and x is increasing. The value of x for which the rate of increase of f is 10 times the rate of increase of x is

    (A) 1

    (B) $\sqrt[3]{10}$

    (C) 3

    (D) $\sqrt{10}$

GO ON TO THE NEXT PAGE

43. The rate of change of the surface area, S, of a balloon is inversely proportional to the square of the surface area. Which equation describes this relationship?

(A) $S(t) = \dfrac{k}{t^2}$

(B) $S(t) = \dfrac{k}{S^2}$

(C) $\dfrac{dS}{dt} = \dfrac{k}{S^2}$

(D) $\dfrac{dS}{dt} = \dfrac{k}{t^2}$

44. Two objects in motion from $t = 0$ to $t = 3$ seconds have positions $x_1(t) = \cos\left(t^2 + 1\right)$ and $x_2(t) = \dfrac{e^t}{2t}$, respectively. How many times during the 3 seconds do the objects have the same velocity?

(A) 0

(B) 1

(C) 3

(D) 4

45. After t years, $A(t) = 50e^{-0.015t}$ pounds of a deposit of a radioactive substance remain. The average amount of the radioactive substance that remains during the time from $t = 100$ to $t = 200$ years is

(A) 2.9 pounds

(B) 5.8 pounds

(C) 6.8 pounds

(D) 15.8 pounds

AB PRACTICE TEST 2

Section II

Part A

TIME: 30 MINUTES

2 PROBLEMS

> *A graphing calculator is required for some of these problems. See instructions on page 8.*

1. Let function f be continuous and decreasing, with values as shown in the table:

x	2.5	3.2	3.5	4.0	4.6	5.0
$f(x)$	7.6	5.7	4.2	3.8	2.2	1.6

(a) Use a trapezoidal sum to estimate the area between f and the x-axis on the interval $2.5 \le x \le 5.0$.

(b) Find the average rate of change of f on the interval $2.5 \le x \le 5.0$.

(c) Estimate the instantaneous rate of change of f at $x = 2.5$.

(d) If $g(x) = f^{-1}(x)$, estimate the slope of g at $x = 4$.

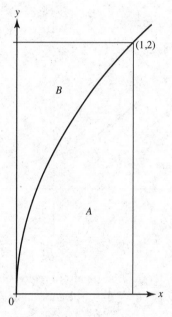

2. The curve $y = \sqrt{8 \sin\left(\dfrac{\pi x}{6}\right)}$ divides a first-quadrant rectangle into regions A and B, as shown in the figure above.

(a) Region A is the base of a solid. Cross sections of this solid perpendicular to the x-axis are rectangles. The height of each rectangle is 5 times the length of its base in region A. Find the volume of this solid.

(b) The other region, B, is rotated around the y-axis to form a different solid. Set up, but do not evaluate, an integral for the volume of this solid.

STOP   End of Part A, Section II

Part B

TIME: 60 MINUTES
4 PROBLEMS

> *No calculator is allowed for any of these problems.*
>
> *If you finish Part B before time has expired, you may return to work on Part A, but you may not use a calculator.*

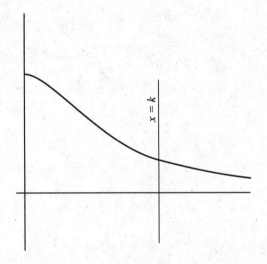

3. Consider the first-quadrant region bounded by the curve $y = \dfrac{18}{9 + x^2}$, the coordinate axes, and the line $x = k$, as shown in the figure above.

   (a) For what value of k will the area of this region equal π?

   (b) What is the average value of the function on the interval $0 \le x \le k$?

   (c) What happens to the area of the region as the value of k increases?

4. Given the following differential equation $\dfrac{dy}{dx} = x - 2y + 1$.

   (a) Sketch the slope field for the differential equation at the nine indicated points on the axes provided.

   (b) Find the second derivative, $\dfrac{d^2y}{dx^2}$, in terms of x and y. The region in the xy-plane where all the solution curves to the differential equation are concave down can be expressed as a linear inequality. Find this region.

   (c) The function $y = f(x)$ is the solution to the differential equation with initial condition $f(-1) = 0$. Determine whether f has a local maximum, local minimum, or neither at $x = -1$. Justify your answer.

   (d) For which values of m and b is the line $y = mx + b$ a solution to the differential equation?

GO ON TO THE NEXT PAGE

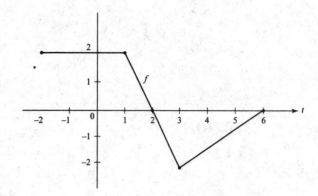

6. The figure above shows the graph of f, whose domain is the closed interval $[-2,6]$. Let

$$F(x) = \int_1^x f(t)\, dt.$$

(a) Find $F(-2)$ and $F(6)$.

(b) For what value(s) of x does $F(x) = 0$?

(c) On what interval(s) is F increasing?

(d) Find the maximum value and the minimum value of F.

(e) At what value(s) of x does the graph of F have points of inflection? Justify your answer.

5. A bungee jumper has reached a point in her exciting plunge where the taut cord is 100 feet long with a 1/2-inch radius and is stretching. She is still 80 feet above the ground and is now falling at 40 feet per second. You are observing her jump from a spot on the ground 60 feet from the potential point of impact, as shown in the diagram above.

(a) Assuming the cord to be a cylinder with volume remaining constant as the cord stretches, at what rate is its radius changing when the radius is 1/2 inch?

(b) From your observation point, at what rate is the angle of elevation to the jumper changing when the radius is 1/2 inch?

🛑 **STOP** _____ **End of Test**

ANSWER KEY
AB Practice Test 2

SECTION I

Part A

1. **D**	9. **B**	17. **B**	25. **B**
2. **A**	10. **A**	18. **A**	26. **A**
3. **D**	11. **C**	19. **B**	27. **D**
4. **D**	12. **A**	20. **C**	28. **A**
5. **B**	13. **C**	21. **B**	29. **B**
6. **D**	14. **D**	22. **B**	30. **C**
7. **B**	15. **C**	23. **D**	
8. **B**	16. **D**	24. **C**	

Part B

31. **D**	35. **C**	39. **D**	43. **C**
32. **B**	36. **D**	40. **D**	44. **D**
33. **C**	37. **B**	41. **A**	45. **B**
34. **B**	38. **C**	42. **C**	

Answer Explanations

Section I Multiple-Choice

Part A

1. **(D)** $\dfrac{x^2 - 2}{4 - x^2} \to +\infty$ as $x \to 2$

2. **(A)** Divide both the numerator and the denominator by $\sqrt{x}$; $\displaystyle\lim_{x\to\infty} \dfrac{1 - \dfrac{4}{\sqrt{x}}}{\dfrac{4}{\sqrt{x}} - 3} = -\dfrac{1}{3}$.

3. **(D)** Since $e^{\ln u} = u$, $y = 1$.

4. **(D)** $f(0) = 3$, and $f'(x) = \dfrac{1}{2}(9 + \sin 2x)^{-1/2} \cdot (2\cos 2x)$, so $f'(0) = \dfrac{1}{3}$; $y \approx \dfrac{1}{3}x + 3$.

5. **(B)** The trapezoidal sum finds three trapezoid areas and adds them together. This procedure gives the following trapezoidal sum.

$$\left(\frac{10 + 18}{2}\right) \cdot 10 + \left(\frac{18 + 20}{2}\right) \cdot 15 + \left(\frac{20 + 24}{2}\right) \cdot 5 = 535$$

6. **(D)** Here $y' = 3\sin^2(1 - 2x)\cos(1 - 2x) \cdot (-2)$.

7. **(B)** $\dfrac{d}{dx}\left(x^2 e^{x^{-1}}\right) = x^2 e^{x^{-1}}\left(-\dfrac{1}{x^2}\right) + 2xe^{x^{-1}}$

8. **(B)** Let s be the distance from the origin; then

$$s = \sqrt{x^2 + y^2} \quad \text{and} \quad \frac{ds}{dt} = \frac{x\dfrac{dx}{dt} + y\dfrac{dy}{dt}}{\sqrt{x^2 + y^2}}$$

Since $\dfrac{dy}{dt} = 2x\dfrac{dx}{dt}$ and $\dfrac{dx}{dt} = \dfrac{3}{2}$, $\dfrac{dy}{dt} = 3x$. Substituting yields $\dfrac{ds}{dt} = \dfrac{3\sqrt{5}}{2}$.

9. **(B)** For $f(x) = \sqrt{x}$, this limit represents $f'(25)$.

10. **(A)** $V = \displaystyle\int_0^1 y^2\, dx = \int_0^1 \sqrt{1 - x^2}\, dx$. This definite integral represents the area of a quadrant of the circle $x^2 + y^2 = 1$; hence $V = \dfrac{\pi}{4}$.

11. **(C)** $-\dfrac{1}{2}\displaystyle\int\left(9 - x^2\right)^{-1/2}(-2x\, dx) = -\dfrac{1}{2}\dfrac{\left(9 - x^2\right)^{\frac{1}{2}}}{1/2} + C$

12. **(A)** The integral is rewritten as

$$\int \frac{(y-1)^2}{y}\, dy = \frac{1}{2}\int \frac{y^2 - 2y + 1}{y}\, dy$$

$$= -\frac{1}{2}\int\left(y - 2 + \frac{1}{y}\right) dy$$

$$= -\frac{1}{2}\left(\frac{y^2}{2} - 2y + \ln|y|\right) + C$$

13. **(C)** $\displaystyle\int_{\pi/6}^{\pi/2} \cot x\, dx = \ln\sin x \,\Big|_{\pi/6}^{\pi/2} = 0 - \ln\frac{1}{2}$

14. **(D)** Note:

15. **(C)** The winning times are positive, decreasing, and concave upward.

16. **(D)** $G(x) = H(x) + \int_0^2 f(t)\,dt$, where $\int_0^2 f(t)dt$ represents the area of a trapezoid.

17. **(B)** $f'(x) = 0$ for $x = 1$ and $f''(1) > 0$

18. **(A)**

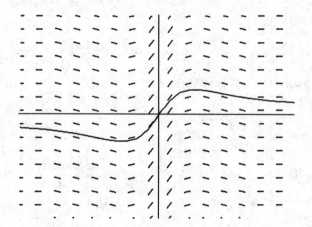

Solution curves appear to represent odd functions with a horizontal asymptote. In the figure above, the curve in (A) of the question has been superimposed on the slope field.

19. **(B)** Note that

$$\lim_{x \to \infty} xe^x = \infty, \quad \lim_{x \to \infty} \frac{e^x}{x} = \infty, \quad \lim_{x \to -\infty} \frac{x}{x^2 + 1} = 0, \text{ and } \frac{x^2}{x^3 + 1} \geq 0 \text{ for } x > -1$$

20. **(C)** v is not differentiable at $t = 3$ or $t = 5$.

21. **(B)**

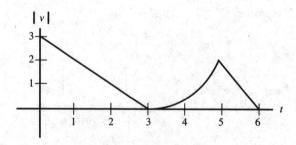

Speed is the magnitude of velocity; its graph is shown above.

22. **(B)** The average rate of change of velocity is $\dfrac{v(5) - v(0)}{5 - 0} = \dfrac{-2 - 3}{5}$.

23. **(D)** The curve has vertical asymptotes at $x = 2$ and $x = -2$ and a horizontal asymptote at $y = -2$.

24. **(C)** The function is not defined at $x = -2$; $\lim\limits_{x \to 1^-} f(x) \neq \lim\limits_{x \to 1^+} f(x)$. Defining $f(-2) = 4$ will make f continuous at $x = -2$, but f will still be discontinuous at $x = 1$.

25. **(B)** Since $\left(f^{-1}\right)'(y) = \dfrac{1}{f'(x)}$,

$$\left(f^{-1}\right)'(y) = \frac{1}{5x^4 + 3} \text{ and } \left(f^{-1}\right)'(2) = \frac{1}{5 \cdot 1 + 3} = \frac{1}{8}$$

26. **(A)** $\displaystyle\int_1^e \frac{\ln^3 x}{x}\,dx = \int_1^e (\ln x)^3 \left(\frac{1}{x}\,dx\right) = \frac{1}{4}\ln^4 x \Big|_1^e = \frac{1}{4}\left(\ln^4 e - 0\right) = \frac{1}{4}$

27. **(D)** $\ln\left(4 + x^2\right) = \ln\left(4 + (-x)^2\right)$; $y' = \dfrac{2x}{4 + x^2}$; $y'' = \dfrac{-2\left(x^2 - 4\right)}{\left(4 + x^2\right)^2}$

28. **(A)** $f(x) = \dfrac{d}{dx}(x \sin \pi x) = \pi x \cos \pi x + \sin \pi x$

29. **(B)** From the Riemann Sum, we see $\Delta x = \dfrac{1}{n}$, then $k \cdot \Delta x = \dfrac{k}{n}$. Notice that the term involving k in the Riemann Sum is not equal to $\dfrac{k}{n}$ but $2\left(\dfrac{k}{n}\right)$. Thus, we will choose $x_k = \dfrac{k}{n}$, so $a = 0$ and $\Delta x = \dfrac{b - 0}{n} = \dfrac{1}{n}$, so $b = 1$. Since x_k replaces x, $f(x) = \sqrt{2x + 3}$ giving the integral $\displaystyle\int_0^1 \sqrt{2x + 3}\,dx$.

30. **(C)** See the figure below. About the x-axis: Washer. $\Delta V = \pi\left(y^2 - 1^2\right)\Delta x$,

$$V = \pi \int_0^2 \left(e^{2x} - 1\right) dx$$

Part B

31. **(D)** We solve the differential equation $\dfrac{ds}{dt} = 12s^{\frac{1}{2}}$ by separation:

$$\int s^{-\frac{1}{2}}\,ds = 12\int dt$$
$$2s^{\frac{1}{2}} = 12t + C$$
$$\sqrt{s} = 6t + C$$

If $s = 1$ when $t = 0$, we have $C = 1$; hence, $\sqrt{s} = 6t + 1$ so $\sqrt{s} = 7$ when $t = 1$.

32. **(B)**

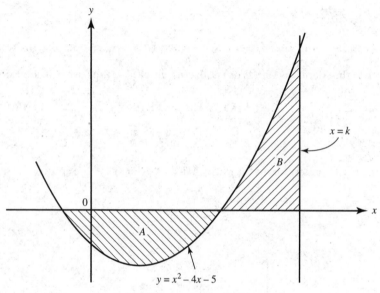

(This figure is not drawn to scale.)

The roots of $f(x) = x^2 - 4x - 5 = (x - 5)(x + 1)$ are $x = -1$ and 5. Since areas

A and B are equal, therefore $\displaystyle\int_{-1}^{k} f(x)\,dx = 0$. Thus,

$$\left(\frac{x^3}{3} - 2x^2 - 5x\right)\Bigg|_{-1}^{k} = \left(\frac{k^3}{3} - 2k^2 - 5k\right) - \left(-\frac{1}{3} - 2 + 5\right)$$
$$= \frac{k^3}{3} - 2k^2 - 5k - \frac{8}{3} = 0$$

Solving on a calculator gives k (or x) equal to 8.

33. **(C)** If N is the number of bacteria at time t, then $N = 200e^{kt}$. It is given that $3 = e^{10k}$.
    When $t = 24$, $N = 200e^{24k}$. Therefore $N = 200\left(e^{10k}\right)^{2.4} = 200(3)^{2.4} \approx 2793$ bacteria.

34. **(B)** Since $t = \dfrac{x+1}{2}$, $dt = \dfrac{1}{2}\,dx$. For $x = 2t - 1$, $t = 3$ yields $x = 5$ and $t = 5$ yields $x = 9$.

35. **(C)** Using implicit differentiation on the equation

$$x^3 + xy - y^2 = 10 \qquad (1)$$

yields

$$3x^2 + x\frac{dy}{dx} + y - 2y\frac{dy}{dx} = 0$$

$$3x^2 + y = (2y - x)\frac{dy}{dx}$$

and

$$\frac{dy}{dx} = \frac{3x^2 + y}{2y - x}$$

The tangent is vertical when $\frac{dy}{dx}$ is undefined—that is, when $2y - x = 0$.

Replacing y by $\frac{x}{2}$ in (1) gives

$$x^3 + \frac{x^2}{2} - \frac{x^2}{4} = 10$$

or

$$4x^3 + x^2 = 40$$

Let $y_1 = 4x^3 + x^2 - 40$. Inspection of the equation $y_1 = f(x) = 0$ reveals that there is a root near $x = 2$. Solving on a calculator yields $x = 2.074$.

36. **(D)** $G'(x) = f(3x - 1) \cdot 3$

37. **(B)** Since f changes from positive to negative at $t = 3$, G' does also where $3x - 1 = 3$.

38. **(C)** The functions are thrice differentiable, meaning the functions and their first and second derivatives are continuous, so we can use substitution when finding limits:

$$\lim_{x \to 2}(f(x) + 3g(x)) = 6 + 3(-2) = 0 \qquad \lim_{x \to 2}\left(\tfrac{1}{2}x^2 - 2e^{x-2}\right) = \tfrac{1}{2}(2)^2 - 2e^{2-2} = 0$$

Since the limit of both the numerator and the denominator are zero, we can use L'Hospital's Rule:

$$\lim_{x \to 2}\frac{f(x) + 3g(x)}{\frac{1}{2}x^2 - 2e^{x-2}} = \lim_{x \to 2}\frac{f'(x) + 3g'(x)}{x - 2e^{x-2}}$$

$$\lim_{x \to 2}(f'(x) + 3g'(x)) = -1 + 3\left(\tfrac{1}{3}\right) = 0 \Rightarrow \lim_{x \to 2}(x - 2e^{x-2}) = 2 - 2e^{2-2} = 0$$

Since the limit of both the numerator and the denominator are zero, we can use L'Hospital's Rule again:

$$\lim_{x \to 2}\frac{f'(x) + 3g'(x)}{x - 2e^{x-2}} = \lim_{x \to 2}\frac{f''(x) + 3g''(x)}{1 - 2e^{x-2}} = \frac{-2 + 3(-4/3)}{1 - 2e^{2-2}} = \frac{-6}{-1} = 6$$

39. **(D)** $2(3) + 2(0) + 2(4) + 2(1)$

40. **(D)** $\dfrac{d}{dx}\left(f^2(x)\right) = 2f(x)f'(x)$

$$\dfrac{d^2}{dx^2}\left(f^2(x)\right) = 2[f(x)f''(x) + f'(x)f'(x)]$$
$$= 2[ff'' + (f')^2]$$

At $x = 3$, the answer is $2\left[2(-2) + 5^2\right] = 42$.

41. **(A)** The object is at rest when $v(t) = \ln\left(2 - t^2\right) = 0$; that occurs when $2 - t^2 = 1$, so $t = 1$.

The acceleration is $a(t) = v'(t) = \dfrac{-2t}{2 - t^2}$; $a(1) = \dfrac{-2(1)}{2 - 1^2}$.

42. **(C)** $\dfrac{df}{dt} = \left(x^2 + 1\right)\dfrac{dx}{dt}$. Find x when $\dfrac{df}{dt} = 10\dfrac{dx}{dt}$.

$$10\dfrac{dx}{dt} = \left(x^2 + 1\right)\dfrac{dx}{dt}$$

implies that $x = 3$.

43. **(C)** $\dfrac{dS}{dt}$ represents the rate of change of the surface area; if y is inversely proportional to x,

then, $y = \dfrac{k}{x}$.

44. **(D)**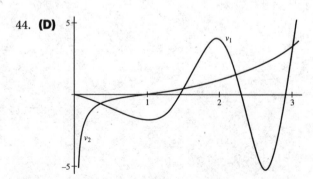

The velocity functions are

$$v_1 = -2t \sin(t^2 + 1)$$

and

$$v_2 = \frac{2t\left(e^t\right) - 2e^t}{(2t)^2} = \frac{e^t\,(t-1)}{2t^2}$$

Graph both functions in $[0,3] \times [-5,5]$. The graphs intersect four times during the first 3 seconds, as shown in the figure on page 546.

45. **(B)** $\dfrac{\displaystyle\int_{100}^{200} 50e^{-0.015t}\,dt}{100} \approx 5.778$ pounds

Section II Free-Response

Part A

AB/BC 1. (a) $T = \left(\dfrac{7.6 + 5.7}{2}\right)(0.7) + \left(\dfrac{5.7 + 4.2}{2}\right)(0.3) + \left(\dfrac{4.2 + 3.8}{2}\right)(0.5) +$

$\left(\dfrac{3.8 + 2.2}{2}\right)(0.6) + \left(\dfrac{2.2 + 1.6}{2}\right)(0.4) = 10.7$

(b) $\dfrac{\Delta y}{\Delta x} = \dfrac{7.6 - 1.6}{2.5 - 5.0} = -2.4$

(c) $f'(2.5) \approx \dfrac{5.7 - 7.6}{3.2 - 2.5} = -2.714$

(d) To work with $g(x) = f^{-1}(x)$, interchange x and y:

x	7.6	5.7	4.2	3.8	2.2	1.6
$g(x)$	2.5	3.2	3.5	4.0	4.6	5.0

Now $g'(4) \approx \dfrac{4.0 - 3.5}{3.8 - 4.2} = -1.25$

AB 2/BC 4. (a) The rectangular slices have base y, height $5y$, and thickness along the x-axis:

$$\Delta V = (y)(5y)\Delta x = 5y^2\Delta x = 5\left(\sqrt{8 \sin\left(\dfrac{\pi x}{6}\right)}\right)^2 \Delta x$$

For the AB exam, this question is calculator active; therefore,

$$V = 40 \int_0^1 \sin\left(\dfrac{\pi x}{6}\right) dx = 10.235 \text{ (or } 10.234)$$

For the BC exam, this question is no calculator; so the integral must be calculated by hand, as follows:

$$V = 40 \int_0^1 \sin\left(\dfrac{\pi x}{6}\right) dx = 40\,\dfrac{6}{\pi}\int_0^1 \sin\left(\dfrac{\pi x}{6}\right)\left(\dfrac{\pi}{6}\,dx\right)$$

$$= -\dfrac{240}{\pi}\cos\left(\dfrac{\pi x}{6}\right)\Big|_0^1 = -\dfrac{240}{\pi}\left(\cos\left(\dfrac{\pi}{6}\right) - \cos(0)\right)$$

$$= \dfrac{240}{\pi}\left(1 - \dfrac{\sqrt{3}}{2}\right)$$

(b) The disks have radius x and thickness along the y-axis:

$$\Delta V = \pi x^2 \Delta y, \text{ so } V = \pi \int_0^2 x^2\, dy$$

Now we solve for x in terms of y:

$$y = \sqrt{8 \sin\left(\dfrac{\pi x}{6}\right)}, \text{ so } y^2 = 8 \sin\left(\dfrac{\pi x}{6}\right) \text{ and } \dfrac{y^2}{8} = \sin\left(\dfrac{\pi x}{6}\right)$$

Then $\arcsin \dfrac{y^2}{8} = \left(\dfrac{\pi x}{6}\right)$, which gives us $x = \dfrac{6}{\pi}\arcsin\dfrac{y^2}{8}$.

Therefore $V = \pi \int_0^2 \left(\dfrac{6}{\pi}\arcsin\dfrac{y^2}{8}\right)^2 dy$.

(*NOTE:* Although the shells method is not a required AP topic, another correct integral for this volume is $V = 2\pi \int_0^1 x\left(2 - \sqrt{8 \sin\dfrac{\pi x}{6}}\right) dx$.)

Part B

AB 3. (a)

$$\int_0^k \frac{18}{9+x^2}\,dx = \pi$$

$$3 \cdot \frac{18}{9} \int_0^k \frac{\frac{1}{3}\,dx}{1+\left(\frac{x}{3}\right)^2} = \pi$$

$$6 \arctan \frac{x}{3}\bigg|_0^k = \pi$$

$$6 \arctan \frac{k}{3} - 6 \arctan \frac{0}{3} = \pi$$

$$\frac{k}{3} = \tan \frac{\pi}{6}$$

$$k = \sqrt{3}$$

See the figure on page 538.

(b) The average value of a function on an interval is the area under the graph of the function divided by the interval width, here $\frac{\pi}{\sqrt{3}}$.

(c) From part (a) you know that the area of the region is given by $\int_0^k \frac{18}{9+x^2}\,dx = 6 \arctan \frac{k}{3}$. Since $\lim\limits_{k \to \infty} 6 \arctan \frac{k}{3} = 6\left(\frac{\pi}{2}\right) = 3\pi$, as k increases the area of the region approaches 3π.

AB 4. (a) The slope field for the nine points indicated is:

The slopes are given in the table below. Be sure that the segments you draw are correct relative to the other slopes in the slope field with respect to the steepness of the segments.

x	-1	-1	-1	0	0	0	1	1	1
y	-1	0	1	-1	0	1	-1	0	1
dy/dx	2	0	-2	3	1	-1	4	2	0

(b)
$$\frac{d^2y}{dx^2} = 1 - 2\frac{dy}{dx} = 1 - 2(x - 2y + 1) = 4y - 2x - 1$$

The solution curves will be concave down when $\frac{d^2y}{dx^2} < 0$.

$$4y - 2x - 1 < 0 \Rightarrow y < \frac{1}{2}x + \frac{1}{4}$$

(c) $(-1,0)$ is a critical point because $\left.\frac{dy}{dx}\right|_{(-1,0)} = -1 - 2(0) + 1 = 0$. And

$\left.\frac{d^2y}{dx^2}\right|_{(-1,0)} = 4(0) - 2(-1) - 1 = 1 > 0$, so by the Second Derivative Test, the solution

curve $f(x)$ has a local minimum.

(d) m is the slope $\left(\frac{dy}{dx}\right)$ so replace into the differential equation:

$$m = x - 2(mx + b) + 1 = (1 - 2m)x + (1 - 2b)$$
$$1 - 2m = 0 \text{ and } 1 - 2b = m, \text{ so } m = \frac{1}{2} \text{ and } b = \frac{1}{4}.$$

AB/BC 5. (a) The volume of the cord is $V = \pi r^2 h$. Differentiate with respect to time, then substitute known values. (Be sure to use consistent units; here, all measurements have been converted to inches.)

$$\frac{dV}{dt} = \pi\left(r^2\frac{dh}{dt} + 2rh\frac{dr}{dt}\right)$$

$$0 = \pi\left(\left(\frac{1}{2}\right)^2 \cdot 480 + 2 \cdot \frac{1}{2} \cdot 1200\frac{dr}{dt}\right)$$

$$\frac{dr}{dt} = -\frac{1}{10} \text{ in./sec}$$

(b)

Let θ represent the angle of elevation and h the height, as shown.

$$\tan \theta = \frac{h}{60}$$

$$\sec^2 \theta \frac{d\theta}{dt} = \frac{1}{60} \frac{dh}{dt}$$

When $h = 80$, your distance to the jumper is 100 feet, as shown.

Then

$$\left(\frac{100}{60}\right)^2 \frac{d\theta}{dt} = \frac{1}{60}(-40)$$

$$\frac{d\theta}{dt} = -\frac{6}{25} \text{ rad/sec}$$

AB/BC 6. (a) $F(-2) = \displaystyle\int_1^{-2} f(t)\, dt = -\int_{-2}^1 f(t)\, dt =$ the negative of the area of the shaded rectangle in the figure. Hence $F(-2) = -(3)(2) = -6$.

$F(6) = \displaystyle\int_1^6 f(t)\, dt$ is represented by the shaded triangles in the figure.

$$\int_1^6 f(t)\, dt = \int_1^2 f(t)\, dt + \int_2^6 f(t)\, dt$$

$$= \frac{1}{2}(1)(2) - \frac{1}{2}(4)(2) = -3$$

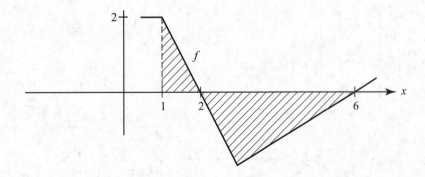

(b) $\displaystyle\int_1^1 f(t)\,dt = 0$, so $F(x) = 0$ at $x = 1$. $\displaystyle\int_1^3 f(t)\,dt = 0$ because the regions above and below the x-axis have the same area. Hence $F(x) = 0$ at $x = 3$.

(c) F is increasing where $F' = f$ is positive: $-2 \le x < 2$.

(d) The maximum value of F occurs at $x = 2$, where $F' = f$ changes from positive to negative.

$$F(2) = \int_1^2 f(t)\,dt = \frac{1}{2}(1)(2) = 1$$

The minimum value of F must occur at one of the endpoints. Since $F(-2) = -6$ and $F(6) = -3$, the minimum is at $x = -2$.

(e) F has points of inflection where F'' changes sign, as occurs where $F' = f$ goes from decreasing to increasing, at $x = 3$.

ANSWER SHEET
AB Practice Test 3

SECTION I

Part A

1. Ⓐ Ⓑ Ⓒ Ⓓ   9. Ⓐ Ⓑ Ⓒ Ⓓ   17. Ⓐ Ⓑ Ⓒ Ⓓ   25. Ⓐ Ⓑ Ⓒ Ⓓ
2. Ⓐ Ⓑ Ⓒ Ⓓ   10. Ⓐ Ⓑ Ⓒ Ⓓ   18. Ⓐ Ⓑ Ⓒ Ⓓ   26. Ⓐ Ⓑ Ⓒ Ⓓ
3. Ⓐ Ⓑ Ⓒ Ⓓ   11. Ⓐ Ⓑ Ⓒ Ⓓ   19. Ⓐ Ⓑ Ⓒ Ⓓ   27. Ⓐ Ⓑ Ⓒ Ⓓ
4. Ⓐ Ⓑ Ⓒ Ⓓ   12. Ⓐ Ⓑ Ⓒ Ⓓ   20. Ⓐ Ⓑ Ⓒ Ⓓ   28. Ⓐ Ⓑ Ⓒ Ⓓ
5. Ⓐ Ⓑ Ⓒ Ⓓ   13. Ⓐ Ⓑ Ⓒ Ⓓ   21. Ⓐ Ⓑ Ⓒ Ⓓ   29. Ⓐ Ⓑ Ⓒ Ⓓ
6. Ⓐ Ⓑ Ⓒ Ⓓ   14. Ⓐ Ⓑ Ⓒ Ⓓ   22. Ⓐ Ⓑ Ⓒ Ⓓ   30. Ⓐ Ⓑ Ⓒ Ⓓ
7. Ⓐ Ⓑ Ⓒ Ⓓ   15. Ⓐ Ⓑ Ⓒ Ⓓ   23. Ⓐ Ⓑ Ⓒ Ⓓ
8. Ⓐ Ⓑ Ⓒ Ⓓ   16. Ⓐ Ⓑ Ⓒ Ⓓ   24. Ⓐ Ⓑ Ⓒ Ⓓ

Part B

31. Ⓐ Ⓑ Ⓒ Ⓓ   35. Ⓐ Ⓑ Ⓒ Ⓓ   39. Ⓐ Ⓑ Ⓒ Ⓓ   43. Ⓐ Ⓑ Ⓒ Ⓓ
32. Ⓐ Ⓑ Ⓒ Ⓓ   36. Ⓐ Ⓑ Ⓒ Ⓓ   40. Ⓐ Ⓑ Ⓒ Ⓓ   44. Ⓐ Ⓑ Ⓒ Ⓓ
33. Ⓐ Ⓑ Ⓒ Ⓓ   37. Ⓐ Ⓑ Ⓒ Ⓓ   41. Ⓐ Ⓑ Ⓒ Ⓓ   45. Ⓐ Ⓑ Ⓒ Ⓓ
34. Ⓐ Ⓑ Ⓒ Ⓓ   38. Ⓐ Ⓑ Ⓒ Ⓓ   42. Ⓐ Ⓑ Ⓒ Ⓓ

AB Practice Test 3

Section I

Part A

TIME: 60 MINUTES

> The use of calculators is not permitted for this part of the examination.
> There are 30 questions in Part A, for which 60 minutes are allowed. Because there is no deduction for wrong answers, you should answer every question, even if you need to guess.
>
> **DIRECTIONS:** Choose the best answer for each question.

1. A cylindrical tank, shown in the figure above, is partially full of water at time $t = 0$, when more water begins flowing in at a constant rate. The tank becomes half full when $t = 4$ and is completely full when $t = 12$. Let h represent the height of the water at time t. During which interval is $\frac{dh}{dt}$ increasing?

   (A) $0 < t < 4$
   (B) $0 < t < 8$
   (C) $0 < t < 12$
   (D) $4 < t < 12$

2. $\lim\limits_{h \to 0} \dfrac{\sin\left(\frac{\pi}{2} + h\right) - 1}{h}$ is

   (A) 1
   (B) −1
   (C) 0
   (D) nonexistent

3. If $f(x) = x \ln x$, then $f'''(e)$ equals

   (A) $\dfrac{1}{e}$
   (B) 2
   (C) $-\dfrac{1}{e^2}$
   (D) $\dfrac{2}{e^3}$

4. An equation of the tangent to the curve $2x^2 - y^4 = 1$ at the point $(-1,1)$ is

   (A) $x + y = 0$
   (B) $x + y = 2$
   (C) $x - y = 0$
   (D) $x - y = -2$

5. On which interval(s) does the function $f(x) = x^4 - 4x^3 + 4x^2 + 6$ increase?

   (A) $x < 0$ and $1 < x < 2$
   (B) $x > 2$ only
   (C) $0 < x < 1$ and $x > 2$
   (D) $0 < x < 1$ only

6. $\int \dfrac{\cos x}{4 + 2 \sin x}\, dx$ equals

    (A) $\ln |4 + 2 \sin x| + C$

    (B) $-2 \ln |4 + 2 \sin x| + C$

    (C) $\ln \sqrt{4 + 2 \sin x} + C$

    (D) $2 \ln |4 + 2 \sin x| + C$

7. Suppose the function f is defined as

$$f(x) = \begin{cases} \sqrt{4x^2 + 9} & \text{if } x \le 2 \\ -\frac{1}{5}(4x^2 - 24x + 7) & \text{if } x > 2 \end{cases}$$

    Which of the following is true?

    (A) The function f is both continuous and differentiable at $x = 2$.

    (B) The function f is continuous but not differentiable at $x = 2$.

    (C) The function f is differentiable but not continuous at $x = 2$.

    (D) The function f is neither continuous nor differentiable at $x = 2$.

8. If a particle moves on a line according to the law $s = t^5 + 2t^3$, then the number of times it reverses direction is

    (A) 3

    (B) 2

    (C) 1

    (D) 0

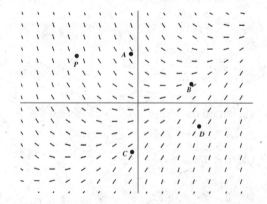

9. A particular solution of the differential equation whose slope field is shown above contains point P. This solution may also contain which other point?

    (A) A

    (B) B

    (C) C

    (D) D

10. Let $F(x) = \displaystyle\int_5^x \dfrac{dt}{1 - t^2}$. Which of the following statements is (are) true?

    I. The domain of F is $x \ne \pm 1$.

    II. $F(2) > 0$.

    III. The graph of F is concave upward.

    (A) I only

    (B) II only

    (C) III only

    (D) II and III only

11. As the tides change, the water level in a bay varies sinusoidally. At high tide today at 8 A.M., the water level was 15 feet; at low tide, 6 hours later at 2 P.M., it was 3 feet. How fast, in feet per hour, was the water level dropping at noon today?

    (A) $\dfrac{\pi}{2}$

    (B) $\dfrac{\pi \sqrt{3}}{2}$

    (C) $3\sqrt{3}$

    (D) $6\sqrt{3}$

12. A smooth curve with equation $y = f(x)$ is such that its slope at each x equals x^2. If the curve goes through the point $(-1, 2)$, then its equation is

    (A) $y = x^3 + 3$

    (B) $y = \dfrac{x^3}{3} + 7$

    (C) $y = \dfrac{x^3 + 7}{3}$

    (D) $y = 3x^3 + 5$

13. $\int \dfrac{e^u}{1 + e^{2u}}\, du$ is equal to

    (A) $\ln (1 + e^{2u}) + C$

    (B) $\frac{1}{2} \ln |1 + e^{2u}| + C$

    (C) $\frac{1}{2} \tan^{-1} e^{2u} + C$

    (D) $\tan^{-1} e^u + C$

14. $\displaystyle\int_{-3}^{-1} \frac{1}{x^2 + 6x + 13}\, dx =$

(A) $\dfrac{\pi}{16}$

(B) $\dfrac{\pi}{8}$

(C) $\dfrac{\pi}{4}$

(D) $\ln 2$

15. If $G(2) = 5$ and $G'(x) = \dfrac{10x}{9 - x^2}$, then an estimate of $G(2.2)$ using a tangent-line approximation is

(A) 4.4

(B) 5.8

(C) 6

(D) 11.6

16. The area bounded by the parabola $y = x^2$ and the lines $y = 1$ and $y = 9$ equals

(A) $\dfrac{20}{3}$

(B) $\dfrac{52}{3}$

(C) 36

(D) $\dfrac{104}{3}$

17. Suppose $f(x) = \dfrac{x^2 + x}{x}$ if $x \neq 0$ and $f(0) = 1$. Which of the following statements is (are) true of f?

   I. f is defined at $x = 0$.

   II. $\lim\limits_{x \to 0} f(x)$ exists.

   III. f is continuous at $x = 0$.

(A) I only

(B) II only

(C) I and II only

(D) I, II, and III

18. Which function could have the graph shown below?

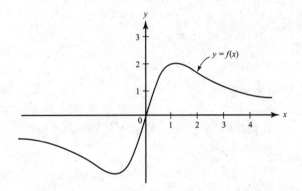

(A) $y = \dfrac{x}{x^2 + 1}$

(B) $y = \dfrac{4x}{x^2 + 1}$

(C) $y = \dfrac{x^2 + 3}{x^2 + 1}$

(D) $y = \dfrac{4x}{x + 1}$

19. Suppose the graph of f is both increasing and concave up on $a \le x \le b$. If $\displaystyle\int_{a}^{b} f(x)\,dx$ is approximated using various sums with the same number of subintervals, and if L, R, M, and T denote, respectively, left Riemann Sum, right Riemann Sum, midpoint Riemann Sum, and trapezoidal sum, then it follows that

(A) $R \le T \le M \le L$

(B) $L \le T \le M \le R$

(C) $R \le M \le T \le L$

(D) $L \le M \le T \le R$

20. $\lim\limits_{x \to 3} \dfrac{x + 3}{x^2 - 9}$ is

(A) $+\infty$

(B) 0

(C) $\dfrac{1}{6}$

(D) nonexistent

21. The only function that does not satisfy the Mean Value Theorem on the interval specified is

(A) $f(x) = x^2 - 2x$ on $[-3,1]$

(B) $f(x) = \frac{1}{x}$ on $[1,3]$

(C) $f(x) = \frac{x^3}{3} - \frac{x^2}{2} + x$ on $[-1,2]$

(D) $f(x) = x + \frac{1}{x}$ on $[-1,1]$

22. Suppose $f'(x) = x(x-2)^2(x+3)$. Which of the following is (are) true?

   I. f has a local maximum at $x = -3$.
   II. f has a local minimum at $x = 0$.
   III. f has neither a local maximum nor a local minimum at $x = 2$.

(A) I only

(B) II only

(C) I and II only

(D) I, II, and III

23. If $y = \ln \dfrac{x}{\sqrt{x^2 + 1}}$, then $\dfrac{dy}{dx}$ is

(A) $\dfrac{1}{x^2 + 1}$

(B) $\dfrac{1}{x(x^2 + 1)}$

(C) $\dfrac{2x^2 + 1}{x(x^2 + 1)}$

(D) $\dfrac{1}{x\sqrt{x^2 + 1}}$

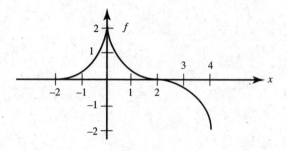

24. The graph of function f shown above consists of three quarter-circles. Which of the following is (are) equivalent to $\displaystyle\int_0^2 f(x)\,dx$?

   I. $\dfrac{1}{2} \displaystyle\int_{-2}^2 f(x)\,dx$

   II. $\displaystyle\int_4^2 f(x)\,dx$

   III. $\dfrac{1}{2} \displaystyle\int_0^4 f(x)\,dx$

(A) I only

(B) II only

(C) I and II only

(D) I, II, and III

25. The base of a solid is the first-quadrant region bounded by $y = \sqrt[4]{4 - 2x}$, and each cross section perpendicular to the x-axis is a semicircle with a diameter in the xy-plane. The volume of the solid is

(A) $\dfrac{\pi}{2} \displaystyle\int_0^2 \sqrt{4 - 2x}\ dx$

(B) $\dfrac{\pi}{4} \displaystyle\int_0^2 \sqrt{4 - 2x}\ dx$

(C) $\dfrac{\pi}{8} \displaystyle\int_0^2 \sqrt{4 - 2x}\ dx$

(D) $\dfrac{\pi}{8} \displaystyle\int_{-2}^2 \sqrt{4 - 2x}\ dx$

26. The average value of $f(x) = 3 + |x|$ on the interval $[-2, 4]$ is

    (A) $\dfrac{1}{3}$

    (B) 2

    (C) $\dfrac{14}{3}$

    (D) 14

27. $\displaystyle\lim_{x \to \infty} \dfrac{3 + x - 2x^2}{4x^2 + 9}$ is

    (A) $-\dfrac{1}{2}$

    (B) $\dfrac{1}{2}$

    (C) 1

    (D) $\dfrac{3}{4}$

28. The area of the region in the xy-plane bounded by the curves $y = e^x$, $y = e^{-x}$, and $x = 1$ is equal to

    (A) $e + \dfrac{1}{e} - 2$

    (B) $e - \dfrac{1}{e}$

    (C) $e + \dfrac{1}{e}$

    (D) $2e - 2$

29. If $f(x) = \displaystyle\int_0^{x^2+2} \sqrt{1 + \cos t}\, dt$, then $f'(x) =$

    (A) $2x \sqrt{1 + \cos(x^2 + 2)}$

    (B) $2x \sqrt{1 - \sin x}$

    (C) $\sqrt{1 + \cos(x^2 + 2)}$

    (D) $\sqrt{[1 + \cos(x^2 + 2)] \cdot 2x}$

30. Choose the Riemann Sum whose limit is the integral $\displaystyle\int_{-1}^{3} \ln(x^2 + 2)\, dx$.

    (A) $\displaystyle\lim_{n \to \infty} \sum_{k=1}^{n} \left(\ln\left(\left(\dfrac{4k}{n} - 1\right)^2 + 2 \right) \cdot \left(\dfrac{4}{n}\right) \right)$

    (B) $\displaystyle\lim_{n \to \infty} \sum_{k=1}^{n} \left(\ln\left(\left(\dfrac{4k}{n} - 1\right)^2 + 2 \right) \cdot \left(\dfrac{1}{n}\right) \right)$

    (C) $\displaystyle\lim_{n \to \infty} \sum_{k=1}^{n} \left(\ln\left(\left(\dfrac{k}{n} - 1\right)^2 + 2 \right) \cdot \left(\dfrac{1}{n}\right) \right)$

    (D) $\displaystyle\lim_{n \to \infty} \sum_{k=1}^{n} \left(\ln\left(\left(\dfrac{k}{n} - 1\right)^2 + 2 \right) \cdot \left(\dfrac{4}{n}\right) \right)$

STOP   End of Part A, Section I

Part B

TIME: 45 MINUTES

Some questions in this part of the examination require the use of a graphing calculator. There are 15 questions in Part B, for which 45 minutes are allowed. Because there is no deduction for wrong answers, you should answer every question, even if you need to guess.

DIRECTIONS: Choose the best answer for each question. If the exact numerical value of the correct answer is not listed as a choice, select the choice that is closest to the exact numerical answer.

31. A particle moves on a line according to the law $s = f(t)$ so that its velocity is $v = ks$, where k is a nonzero constant. The acceleration of the particle is

    (A) k^2v
    (B) k^2s
    (C) k
    (D) 0

32. A cup of coffee placed on a table cools at a rate of $\dfrac{dH}{dt} = -0.05(H - 70)°$F per minute, where H represents the temperature of the coffee and t is time in minutes. If the coffee was at 120°F initially, what will its temperature be, to the nearest degree, 10 minutes later?

    (A) 73°F
    (B) 95°F
    (C) 100°F
    (D) 105°F

33. An investment of $4000 grows at the rate of $320e^{0.08t}$ dollars per year after t years. Its value after 10 years is approximately

    (A) $12902
    (B) $8902
    (C) $7122
    (D) $4902

34. The density of the population of a city is 10,000 people per square mile at the beginning of a certain year. We can model the time, in years (t), it will take until the population reaches a certain density, in people per square mile (D), by using the function $t = g(D)$. What are the units of $g'(D)$?

    (A) years
    (B) years per people per square mile
    (C) people per square mile per year
    (D) people per square mile

Questions 35 and 36. The graph below shows the velocity of an object during the interval $0 \le t \le 9$.

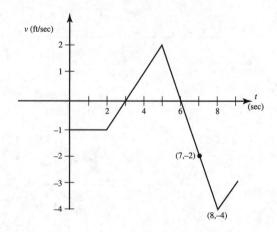

35. The object attains its greatest speed at $t =$

    (A) 3
    (B) 5
    (C) 6
    (D) 8

36. The object was at the origin at $t = 3$. It returned to the origin

    (A) during $6 < t < 7$
    (B) at $t = 7$
    (C) during $7 < t < 8$
    (D) at $t = 8$

37. When the region bounded by the y-axis, $y = e^x$, and $y = 2$ is rotated around the y-axis, it forms a solid with volume

    (A) 0.188
    (B) 0.386
    (C) 0.592
    (D) 1.214

38. If $\sqrt{x-2}$ is replaced by u, then $\int_3^6 \frac{\sqrt{x-2}}{x}\, dx$ is equivalent to

    (A) $\int_1^2 \frac{u\, du}{u^2 + 2}$

    (B) $\int_1^2 \frac{2u^2}{u^2 + 2}\, du$

    (C) $\int_3^6 \frac{2u^2}{u^2 + 2}\, du$

    (D) $\int_3^6 \frac{u\, du}{u^2 + 2}\, du$

39. The line tangent to the graph of function f at the point $(8,1)$ intersects the y-axis at $y = 3$. Find $f'(8)$.

    (A) $-\frac{1}{4}$
    (B) 1
    (C) $\frac{1}{4}$
    (D) -4

40. How many points of inflection does the function f have on the interval $0 \le x \le 6$ if $f''(x) = 2 - 3\sqrt{x}\,\cos^3 x$?

    (A) 1
    (B) 2
    (C) 3
    (D) 4

41. The graph above shows the rate at which tickets were sold at a movie theater during the last hour before showtime. Using the right-rectangle method, estimate the size of the audience.

    (A) 270
    (B) 300
    (C) 330
    (D) 360

42. Find the x-coordinate where $f(x) = 4^{\sin x}$ and $g(x) = \ln(x^2)$ intersect and their derivatives have the same sign.

    (A) -5.240
    (B) -3.961
    (C) -1.151
    (D) 2.642

43. Which statement is true?

    (A) If $f'(c) = 0$, then f has a local maximum or minimum at $(c, f(c))$.
    (B) If $f''(c) = 0$, then the graph of f has an inflection point at $(c, f(c))$.
    (C) If f is differentiable at $x = c$, then f is continuous at $x = c$.
    (D) If f is continuous on (a,b), then f attains a maximum value on (a,b).

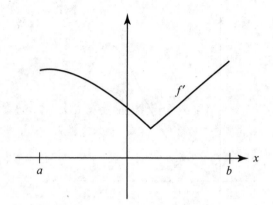

44. The graph of f' is shown above. Which statement(s) about f must be true for $a < x < b$?

   I. f is increasing.
  II. f is continuous.
  III. f is differentiable.

(A) I only
(B) II only
(C) I and II only
(D) I, II, and III

45. Given $\displaystyle\int_{-1}^{4} g(x)dx = 0$ and $g'(x) < 0$ for all x, select the table that could represent g on the interval $[-1,4]$.

(A)

x	y
-1	3
1	3
4	3

(B)

x	y
-1	7
1	-3
4	-1

(C)

x	y
-1	7
1	1
4	-3

(D)

x	y
-1	7
1	5
4	1

STOP

End of Section I

Section II

Part A

TIME: 30 MINUTES
2 PROBLEMS

A graphing calculator is required for some of these problems. See instructions on page 8.

1. A curve is defined by $x^2y - 3y^2 = 48$.

   (a) Verify that $\dfrac{dy}{dx} = \dfrac{2xy}{6y - x^2}$.

   (b) Write an equation of the line tangent to this curve at (5,3).

   (c) Using your equation from part (b), estimate the y-coordinate of the point on the curve where $x = 4.93$.

   (d) Show that this curve has no horizontal tangent lines.

2. The table shows the depth of water, W in a river, as measured at 4-hour intervals during a day-long flood. Assume that W is a differentiable function of time t.

t (hours)	0	4	8	12	16	20	24
$W(t)$ (feet)	32	36	38	37	35	33	32

   (a) Find the approximate value of $W'(16)$. Indicate units of measure.

   (b) Estimate the average depth of the water, in feet, over the time interval $0 \le t \le 24$ hours by using a trapezoidal approximation with subintervals of length $\Delta t = 4$ hours.

   (c) Scientists studying the flooding believe they can model the depth of the water with the function $F(t) = 35 - 3 \cos\left(\dfrac{t+3}{4}\right)$, where $F(t)$ represents the depth of the water, in feet, after t hours. Find $F'(16)$ and explain the meaning of your answer, with appropriate units, in terms of the river depth.

   (d) Use the function F to find the average depth of the water, in feet, over the time interval $0 \le t \le 24$ hours.

STOP   End of Part A, Section II

Part B

TIME: 60 MINUTES
4 PROBLEMS

> *No calculator is allowed for any of these problems.*
>
> *If you finish Part B before time has expired, you may return to work on Part A, but you may not use a calculator.*

3. The region **R** is bounded by the curves $f(x) = \cos(\pi x) - 1$ and $g(x) = x(2 - x)$, as shown in the figure.

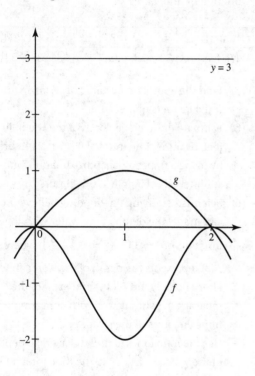

(a) Find the area of **R**.

(b) A solid has base **R**, and each cross section perpendicular to the *x*-axis is an isosceles right triangle whose hypotenuse lies in **R**. Set up, but do not evaluate, an integral for the volume of this solid.

(c) Set up, but do not evaluate, an integral for the volume of the solid formed when **R** is rotated around the line $y = 3$.

4. Two autos, *P* and *Q*, start from the same point and race along a straight road for 10 seconds. The velocity of *P* is given by $v_p(t) = 6(\sqrt{1 + 8t} - 1)$ feet per second. The velocity of *Q* is shown in the graph.

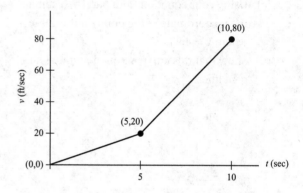

(a) At what time is *P*'s actual acceleration (in ft/sec²) equal to its average acceleration for the entire race?

(b) What is *Q*'s acceleration (in ft/sec²) then?

(c) At the end of the race, which auto was ahead? Explain.

GO ON TO THE NEXT PAGE

5. Given the differential equation $\dfrac{dy}{dx} = 2x(y^2 + 1)$

   (a) Sketch the slope field for this differential equation at the points shown in the figure.

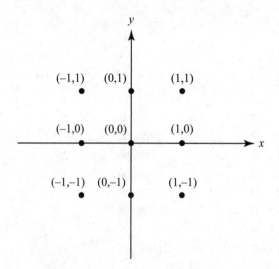

   (b) Let f be the particular solution to the differential equation whose graph passes through $(0,1)$. Express f as a function of x, and state its domain.

6. The graph shown is for $F(x) = \displaystyle\int_0^x f(t)\,dt$.

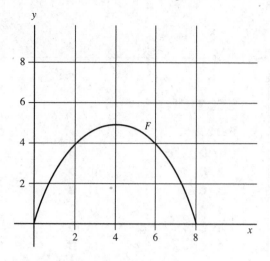

   (a) What is $\displaystyle\int_0^2 f(t)\,dt$?

   (b) What is $\displaystyle\int_2^6 f(t)\,dt$?

   (c) At what value of x does $f(x) = 0$?

   (d) Over what interval is $f'(x)$ negative?

   (e) Let $G(x) = \displaystyle\int_2^x f(t)\,dt$. Sketch the graph of G on the same axes.

STOP   End of Test

ANSWER KEY
AB Practice Test 3

SECTION I

Part A

1.	**D**	9.	**B**	17.	**D**	25.	**C**
2.	**C**	10.	**D**	18.	**B**	26.	**C**
3.	**C**	11.	**B**	19.	**D**	27.	**A**
4.	**A**	12.	**C**	20.	**D**	28.	**A**
5.	**C**	13.	**D**	21.	**D**	29.	**A**
6.	**C**	14.	**B**	22.	**D**	30.	**A**
7.	**A**	15.	**B**	23.	**B**		
8.	**D**	16.	**D**	24.	**C**		

Part B

31.	**B**	35.	**D**	39.	**A**	43.	**C**
32.	**C**	36.	**C**	40.	**A**	44.	**D**
33.	**B**	37.	**C**	41.	**C**	45.	**C**
34.	**B**	38.	**B**	42.	**B**		

Answer Explanations

Section I Multiple-Choice

Part A

1. **(D)** $\dfrac{dh}{dt}$ will increase above the half-full level (that is, the height of the water will rise more rapidly) as the area of the surface of the water diminishes.

2. **(C)** The given limit equals $f'\left(\dfrac{\pi}{2}\right)$, where $f(x) = \sin x$.

3. **(C)** Since $f(x) = x \ln x$,

$$f'(x) = 1 + \ln x, \qquad f''(x) = \frac{1}{x}, \qquad \text{and} \qquad f'''(x) = -\frac{1}{x^2}$$

4. **(A)** Differentiate implicitly to get $4x - 4y^3 \dfrac{dy}{dx} = 0$. Substitute $(-1,1)$ to find $\dfrac{dy}{dx} = -1$, the slope at this point, and write an equation of the tangent: $y - 1 = -1(x + 1)$.

5. **(C)** $f'(x) = 4x^3 - 12x^2 + 8x = 4x(x - 1)(x - 2)$. To determine the signs of $f'(x)$, inspect the sign at any point in each of the intervals $x < 0$, $0 < x < 1$, $1 < x < 2$, and $x > 2$. The function increases whenever $f'(x) > 0$.

6. **(C)** The integral is equivalent to $\dfrac{1}{2} \displaystyle\int \dfrac{2 \cos x \, dx}{4 + 2 \sin x} = \dfrac{1}{2} \displaystyle\int \dfrac{du}{u}$, where $u = 4 + 2 \sin x$.

7. **(A)** $\displaystyle\lim_{x \to 2^-} f(x) = \lim_{x \to 2^-} \sqrt{4x^2 + 9} = \sqrt{16 + 9} = 5$, $\displaystyle\lim_{x \to 2^+} f(x) = \lim_{x \to 2^+} \left(-\frac{1}{5}(4x^2 - 24x + 7)\right) = -\dfrac{1}{5}(16 - 48 + 7) = 5$.

Therefore, $\displaystyle\lim_{x \to 2} f(x) = 5 = f(2)$, so f is continuous at $x = 2$.

$$f'(x) = \begin{cases} \dfrac{4x}{\sqrt{4x^2 + 9}} & \text{if } x < 2 \\[2mm] -\dfrac{1}{5}(8x - 24) & \text{if } x > 2 \end{cases}$$

Notice that we removed the equal sign from the top piece—this is proper when taking derivatives of piecewise functions. The function f is continuous at $x = 2$, and the two derivatives are equal at $x = 2$.

$$\left.\frac{4x}{\sqrt{4x^2 + 9}}\right|_{x=2} = \frac{8}{5}; \qquad \left.-\frac{1}{5}(8x - 24)\right|_{x=2} = -\frac{1}{5}(-8) = \frac{8}{5}$$

Therefore, f is also differentiable at $x = 2$.

8. **(D)** Since $v = \dfrac{ds}{dt} = 5t^4 + 6t^2 = t^2(5t^2 + 6)$ is always positive, there are no reversals in motion along the line.

9. **(B)**

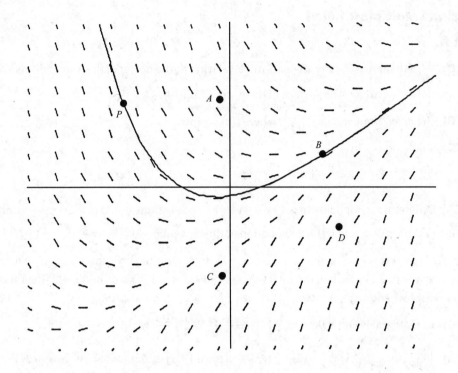

The slope field suggests the curve shown above as a particular solution.

10. **(D)** Since $f(x) = F'(x) = \dfrac{1}{1-x^2}$, f is discontinuous at $x=1$; the domain of F is therefore

$x > 1$. On [2,5] $f(x) < 0$, so $\displaystyle\int_5^2 f > 0$. $F''(x) = f'(x) = \dfrac{2x}{(1-x^2)^2}$, which is positive for $x > 1$.

11. **(B)**

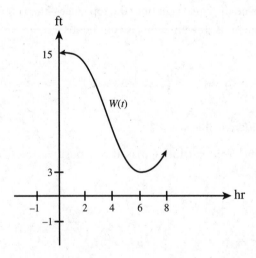

In the graph above, $W(t)$, the water level at time t, is a cosine function with amplitude 6 feet and period 12 hours:

$$W(t) = 6\cos\!\left(\tfrac{\pi}{6}t\right) + 9 \text{ ft}$$

$$W'(t) = -\pi \sin\!\left(\tfrac{\pi}{6}t\right) \text{ ft/hr}$$

Hence, $W'(4) = -\pi\sin\!\left(\dfrac{2\pi}{3}\right) = -\dfrac{\pi\sqrt{3}}{2}$ ft/hr.

12. **(C)** Solve the differential equation $\dfrac{dy}{dx} = x^2$, getting $y = \dfrac{x^3}{3} + C$. Use $x = -1$, $y = 2$ to determine $C = \dfrac{7}{3}$.

13. **(D)** $\displaystyle\int \dfrac{e^u\,du}{1 + (e^u)^2} = \tan^{-1}(e^u) + C$

14. **(B)** Since the denominator is a quadratic, there would have to be a linear function in the numerator for u-substitution to be a method we could try. Also, the denominator would have to factor for partial fraction decomposition (a BC topic) to be a method we could try. Here, we have to change the form of the quadratic by completing the square: $x^2 + 6x + 13 \Rightarrow (x^2 + 6x + 9) + 13 - 9 \Rightarrow (x+3)^2 + 4$. Then we rewrite the integrand and notice that the integrand is in the form of the derivative for $y = \tan^{-1}(u)$. Recall that $\dfrac{d}{dx}(\tan^{-1} u) = \dfrac{du}{1 + u^2}$.

$$\int_{-3}^{-1} \frac{1}{x^2 + 6x + 13}\,dx = \int_{-3}^{-1} \frac{1}{4 + (x+3)^2}\,dx = \int_{-3}^{-1} \frac{1}{4\left(1 + \left(\frac{x+3}{4}\right)^2\right)}\,dx$$

$$= \int_{-3}^{-1} \frac{1}{4\left(1 + \left(\frac{x+3}{2}\right)^2\right)}\,dx$$

$$= \frac{1}{4}\int_{-3}^{-1} \frac{1}{1 + \left(\frac{x+3}{2}\right)^2}\,dx = \frac{1}{2}\tan^{-1}\left(\frac{x+3}{2}\right)\Big|_{-3}^{-2}$$

$$= \frac{1}{2}\tan^{-1}(1) - \frac{1}{2}\tan^{-1}(0) = \frac{\pi}{8}$$

NOTE: In the integral we used $u = \dfrac{x+3}{2}$, $du = \dfrac{1}{2}\,dx$.

15. **(B)** $G'(2) = 4$, so $G(x) \approx 4(x - 2) + 5$.

16. **(D)** $A = 2\displaystyle\int_1^9 x\,dy = 2\int_1^9 \sqrt{y}\,dy = \dfrac{104}{3}$. See the figure below.

17. **(D)** Note that $\lim_{x \to 0} f(x) = f(0) = 1$.

18. **(B)** Note that $(0,0)$ is on the graph, as are $(1,2)$ and $(-1,-2)$. So only (B) and (D) are possible. Since $\lim_{x \to \infty} y = \lim_{x \to \infty} y = 0$, only (B) is correct.

19. **(D)** See the figure below.

20. **(D)** $\dfrac{x+3}{x^2-9} = \dfrac{x+3}{(x+3)(x-3)} = \dfrac{1}{x-3}$; $\lim_{x \to 3^+} \dfrac{1}{x-3} = +\infty$; $\lim_{x \to 3^-} \dfrac{1}{x-3} = -\infty$.

21. **(D)** In (D), $f(x)$ is not defined at $x = 0$. Verify that each of the other functions satisfies both conditions of the Mean Value Theorem.

22. **(D)** The signs within the intervals bounded by the critical points are given below.

Since f changes from increasing to decreasing at $x = -3$, f has a local maximum at -3. Also, f has a local minimum at $x = 0$, because it is decreasing to the left of zero and increasing to the right.

23. **(B)** Since $\ln \dfrac{x}{\sqrt{x^2+1}} = \ln x - \dfrac{1}{2}\ln(x^2+1)$, then

$$\frac{dy}{dx} = \frac{1}{x} - \frac{1}{2} \cdot \frac{2x}{x^2+1} = \frac{1}{x(x^2+1)}$$

24. **(C)** $\displaystyle\int_{-2}^{0} f = \int_{0}^{2} f = -\int_{0}^{4} f$, but $\displaystyle\int_{0}^{4} f = 0$.

25. **(C)** As seen from the figure, $\Delta V = \frac{1}{2}\pi r^2 \Delta x$, where $y = 2r$,

$$V = \frac{\pi}{2}\int_0^2 \left(\frac{y}{2}\right)^2 dx = \frac{\pi}{8}\int_0^2 \sqrt{4-2x}\ dx$$

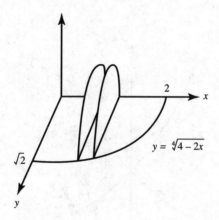

$y = \sqrt[4]{4-2x}$

26. **(C)** From the figure below, $\dfrac{\int_{-2}^{4} f}{6} = \dfrac{\dfrac{5+3}{2}\cdot 2 + \dfrac{3+7}{2}\cdot 4}{6} = \dfrac{28}{6}.$

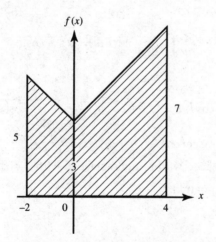

27. **(A)** Since the degrees of the numerator and the denominator are the same, the limit as $x \to \infty$ is the ratio of the coefficients of the terms of highest degree: $\dfrac{-2}{4}$.

28. **(A)** We see from the figure that $\Delta A = (y_2 - y_1)\Delta x$;

$$A = \int_0^1 (e^x - e^{-x})\ dx$$

$$= (e^x + e^{-x})\Big|_0^1 = e + \frac{1}{e} - 2$$

29. **(A)** Let $u = x^2 + 2$. Then

$$\frac{d}{du}\int_0^u \sqrt{1 + \cos t}\ dt = \sqrt{1 + \cos u}$$

and

$$\frac{d}{dx}\int_0^u \sqrt{1 + \cos t}\ dt = \sqrt{1 + \cos u}\ \frac{du}{dx} = \sqrt{1 + \cos(x^2 + 2)} \cdot (2x)$$

30. **(A)** From the integral we get $a = -1$, $b = 3$, so $\Delta x = \frac{3 - (-1)}{n} = \frac{4}{n}$ and $x_k = a + k \cdot \Delta x = -1 + \frac{4k}{n} = \frac{4k}{n} - 1$. Replace x with x_k and replace dx with Δx in the integrand to get the general term in the summation.

Part B

31. **(B)** Since $v = ks = \dfrac{ds}{dt}$, then $a = \dfrac{d^2 s}{dt^2} = k\dfrac{d^2 s}{dt^2} = k\dfrac{ds}{dt} = kv = k^2 s$.

32. **(C)** $\dfrac{dH}{H - 70} = -0.05\ dt.$ $\ln|H - 70| = -0.05t + C$

$$H - 70 = ce^{-0.05t}$$

$$H(x) = 70 + ce^{-0.05t}$$

The initial condition $H(0) = 120$ shows $c = 50$. Evaluate $H(10)$.

33. **(B)** Let P be the amount after t years. It is given that $P'(t) = 320\ e^{0.08t}$, so $P(10) = P(0)\int_0^{10} P'(t)\ dt = 8902.16$.

34. **(B)** Another way to write the derivative is $g'(D) = \dfrac{dt}{dD}$ since we are taking the derivative of the function t with respect to D. This suggests that the units of the derivative are the units of the numerator, which are in years, over the units of the denominator, which are in people per square mile. Thus, the units of the derivative are in years per people per square mile.

35. **(D)** Speed is the magnitude of velocity: $|v(8)| = 4$.

36. **(C)** For $3 < t < 6$ the object travels to the right $\dfrac{1}{2}(3)(2) = 3$ units. At $t = 7$ it has returned 1 unit to the left, and by $t = 8$, 4 units to the left.

37. **(C)**

Use disks: $\Delta V = \pi r^2 \Delta y = \pi x^2 \Delta y$, where $x = \ln y$. Use your calculator to evaluate

$$V = \pi \int_1^2 (\ln y)^2 \, dy.$$

38. **(B)** If $u = \sqrt{x-2}$, then $u^2 = x - 2$, $x = u^2 + 2$, $dx = 2u\, du$. When $x = 3$, $u = 1$; when $x = 6$, $u = 2$.

39. **(A)** The tangent line passes through points $(8,1)$ and $(0,3)$. Its slope, $\dfrac{1-3}{8-0}$, is $f'(8)$.

40. **(A)** Graph f'' in $[0,6] \times [-5,10]$. The sign of f'' changes only at $x = a$, as seen in the figure.

41. **(C)** In the graph below, the first rectangle shows 2 tickets sold per minute for 15 minutes, or 30 tickets. Similarly, the total is 2(15) + 6(15) + 10(15) + 4(15).

42. **(B)** Graph both functions in [−8,8] × [−5,5]. The curves intersect at *P*, *Q*, *R*, and *S* (at *T* the curves are close but do not intersect). At point of intersection *Q*, both are decreasing. Use a graphing calculator to find that the intersection point at *Q* reveals an *x*-coordinate of −3.961.

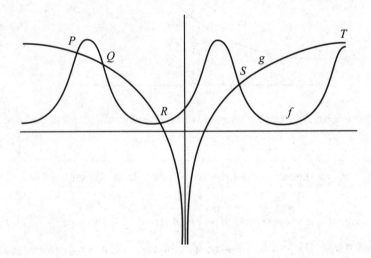

43. **(C)** Counterexamples are, respectively, for (A), $f(x) = x^3$, $c = 0$; for (B), $f(x) = x^4$, $c = 0$; for (D), $f(x) = x^2$ on (−1,1).

44. **(D)** $f'(x) > 0$; the curve shows that f' is defined for all $a < x < b$, so f is differentiable and therefore continuous.

45. **(C)** Since $g'(x) < 0$ for all x, the table must have decreasing values as x increases, and $\int_{-1}^{4} g(x)\,dx = 0$, so there must be at least one value above the *x*-axis and one value below the *x*-axis.

Section II Free-Response

Part A

AB 1.   (a) Since $x^2y - 3y^2 = 48$,

$$x^2\frac{dy}{dx} + 2xy - 6y\frac{dy}{dx} = 0$$

$$(x^2 - 6y)\frac{dy}{dx} = -2xy$$

$$\frac{dy}{dx} = \frac{2xy}{6y - x^2}$$

(b) At $(5,3)$, $\dfrac{dy}{dx} = \dfrac{2(5)(3)}{6(3) - 5^2}$, $= -\dfrac{30}{7}$ so an equation of the tangent line is

$$y - 3 = -\frac{30}{7}(x - 5)$$

(c) $y - 3 = -\dfrac{30}{7}(4.93 - 5) = 0.3$, so $y = 3.3$.

(d) Horizontal tangent lines have $\dfrac{dy}{dx} = \dfrac{2xy}{6y - x^2} = 0$. This could happen only if
$2xy = 0$, which means that $x = 0$ or $y = 0$.

If $x = 0$, $0y - 3y^2 = 48$, which has no real solutions.

If $y = 0$, $x^2 \cdot 0 - 3 \cdot 0^2 = 48$, which is impossible. Therefore, there are no horizontal tangents.

AB/BC 2.   (a) $W'(16) \approx \dfrac{W(20) - W(16)}{20 - 16} = \left(\dfrac{33 - 35}{4}\right) = -\dfrac{1}{2}$ ft/hr

$\left(\text{OR } \dfrac{W(16) - W(12)}{16 - 12} = \dfrac{35 - 37}{4} \text{ OR } \dfrac{W(20) - W(12)}{20 - 12} = \dfrac{33 - 37}{8}\right)$

(b) The average value of a function is the integral across the given interval divided by
the interval width. Here Avg $(W) = \dfrac{\displaystyle\int_0^{24} W(t)dt}{24} - 0$. Estimate the value of the integral using trapezoidal sum T with values from the table:

$$T = \left(\frac{32 + 36}{2}\right) \cdot 4 + \left(\frac{36 + 38}{2}\right) \cdot 4 + \left(\frac{38 + 37}{2}\right) \cdot 4 + \left(\frac{37 + 35}{2}\right) \cdot 4$$

$$+ \left(\frac{35 + 33}{2}\right) \cdot 4 + \left(\frac{33 + 32}{2}\right) \cdot 4 = 844$$

Hence

$$\text{Avg }(W) \approx \frac{844}{24} = 35.167 \text{ feet}$$

(c) For $F(t) = 35 - 3\cos\left(\dfrac{t + 3}{4}\right)$, use your calculator to evaluate $F'(16) \approx -0.749$.

After 16 hours, the river depth is dropping at the rate of 0.749 ft/hr.

(d) $\text{Avg }(F) = \dfrac{\displaystyle\int_0^{24} F(t)dt}{24 - 0} \approx 35.116 \text{ feet}$

Part B

AB/BC 3. (a) $\Delta A = (g(x) - f(x))\,\Delta x$, so:

$$A = \int_0^2 (g(x) - f(x))dx = \int_0^2 (x(2-x) - (\cos \pi x - 1))\,dx$$

$$= \int_0^2 (2x - x^2 - \cos \pi x + 1)dx = \left(x^2 - \frac{x^3}{3} - \frac{1}{\pi}\sin \pi x + x\right)\Big|_0^2$$

$$= \left(2^2 - \frac{2^3}{3} - \frac{1}{\pi}\sin 2\pi + 2\right) - \left(0^2 - \frac{0^3}{3} - \frac{1}{\pi}\sin 0 + 0\right)$$

$$= \frac{10}{3}$$

(b) Let h = the hypotenuse of an isosceles right triangle, as shown in the figure. Then each leg of the triangle is $\frac{h}{\sqrt{2}}$ and its area is $\frac{1}{2} \cdot \frac{h}{\sqrt{2}} \cdot \frac{h}{\sqrt{2}} = \frac{h^2}{4}$.

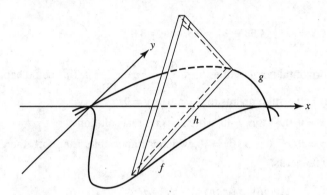

An element of volume is $\Delta V = \frac{h^2}{4}\,\Delta x = \frac{(g(x) - f(x))^2}{4}\,\Delta x$, and thus

$$V = \frac{1}{4}\int_0^2 (x(2-x) - (\cos \pi x - 1))^2\,dx.$$

(c) Washers; $\Delta V = \pi\left(r_1^2 - r_2^2\right)\Delta x$ where:

$$r_1 = 3 - f(x) = 3 - (\cos \pi x - 1)$$

$$r_2 = 3 - g(x) = 3 - x(2 - x)$$

So $V = \pi \displaystyle\int_0^{\pi 2} \left[(3 - (\cos \pi x - 1))^2 - (3 - x(2 - x))^2\right]dx.$

AB/BC 4. (a) $v_p(t) = 6(\sqrt{1 + 8t} - 1)$, so $v(0) = 0$ and $v(10) = 48$.

The average acceleration is $\dfrac{\Delta v}{\Delta t} = \dfrac{48 - 0}{10 - 0} = \dfrac{24}{5}$ ft/sec^2.

Acceleration $a(t) = v'(t) = 6 \cdot \frac{1}{2}(1 + 8t)^{-\frac{1}{2}}(8)$ ft/sec^2

$$\frac{24}{\sqrt{1 + 8t}} = \frac{24}{5} \text{ when } t = 3 \text{ seconds}$$

(b) Since Q's acceleration, for all t in $0 \le t \le 5$, is the slope of its velocity graph,

$$a = \frac{20 - 0}{5 - 0} = 4 \text{ ft/sec}^2.$$

(c) Find the distance each auto has traveled. For *P*, the distance is

$$\int_0^{10} 6(\sqrt{1+8t} - 1) =$$

$$6\left[\frac{1}{8} \int_0^{10} \sqrt{1+8t} \cdot 8dt - \int_0^{10} dt \right] =$$

$$6\left(\frac{1}{8} \cdot \frac{2}{3}(1+8t)^{\frac{3}{2}} - t \right)\Big|_0^{10} =$$

$$6\left(\frac{1}{12}\left(81^{\frac{3}{2}} - 1^{\frac{3}{2}}\right) - 10 \right) = 304 \text{ feet}$$

For auto *Q*, the distance is the total area of the triangle and trapezoid under the velocity graph shown below, namely,

$$\frac{1}{2}(5)(20) + \frac{1}{2}(20+80)(5) = 300 \text{ feet}$$

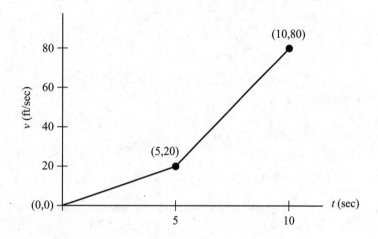

Auto *P* won the race.

AB 5. (a) Using the differential equation, evaluate the derivative at each point, then sketch a short segment having that slope. For example, at $(-1,-1)$, $\dfrac{dy}{dx} = 2(-1)((-1)^2 + 1) - 4$; draw a steeply decreasing segment at $(-1,-1)$.

Repeat this process at each of the other points. The result follows.

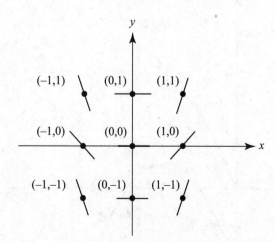

(b) The differential equation $\dfrac{dy}{dx} = 2x(y^2 + 1)$ is separable.

$$\int \frac{dy}{y^2 + 1} = \int 2x\, dx$$

$$\arctan(y) = x^2 + c$$

$$y = \tan(x^2 + c)$$

It is given that f passes through $(0,1)$, so $1 = \tan(0^2 + c)$ and $c = \dfrac{\pi}{4}$.

The solution is $f(x) = \tan\left(x^2 + \dfrac{\pi}{4}\right)$.

The particular solution must be differentiable on an interval containing the initial point $(0,1)$. The tangent function has vertical asymptotes at $x = \pm\dfrac{\pi}{2}$, hence:

$$-\frac{\pi}{2} < x^2 + \frac{\pi}{4} < \frac{\pi}{2} \ \left(\text{Since } x^2 \geq 0, \text{we ignore the left inequality.}\right)$$

$$x^2 < \frac{\pi}{4}$$

$$|x| < \frac{\sqrt{\pi}}{2}$$

AB 6. (a) $\displaystyle\int_0^2 f(t)\, dt = F(2) = 4$

(b) $\displaystyle\int_2^6 f(t)\,dt = F(6) - F(2) = 4 - 4 = 0.$

(c) $f(x) = F'(x); \ F'(x) = 0$ at $x = 4$

(d) $f'(x) = F''(x)$. F'' is negative when F is concave downward, which is true for the entire interval $0 < x < 8$.

(e) $G(x) = \displaystyle\int_2^x f(t)dt$

$\qquad = \displaystyle\int_0^x f(t)dt - \int_0^2 f(t)dt$

$\qquad = F(x) - 4$

Then the graph of G is the graph of F translated downward 4 units.

BC Practice Tests

ANSWER SHEET
BC Practice Test 1

SECTION I

Part A

1. Ⓐ Ⓑ Ⓒ Ⓓ
2. Ⓐ Ⓑ Ⓒ Ⓓ
3. Ⓐ Ⓑ Ⓒ Ⓓ
4. Ⓐ Ⓑ Ⓒ Ⓓ
5. Ⓐ Ⓑ Ⓒ Ⓓ
6. Ⓐ Ⓑ Ⓒ Ⓓ
7. Ⓐ Ⓑ Ⓒ Ⓓ
8. Ⓐ Ⓑ Ⓒ Ⓓ

9. Ⓐ Ⓑ Ⓒ Ⓓ
10. Ⓐ Ⓑ Ⓒ Ⓓ
11. Ⓐ Ⓑ Ⓒ Ⓓ
12. Ⓐ Ⓑ Ⓒ Ⓓ
13. Ⓐ Ⓑ Ⓒ Ⓓ
14. Ⓐ Ⓑ Ⓒ Ⓓ
15. Ⓐ Ⓑ Ⓒ Ⓓ
16. Ⓐ Ⓑ Ⓒ Ⓓ

17. Ⓐ Ⓑ Ⓒ Ⓓ
18. Ⓐ Ⓑ Ⓒ Ⓓ
19. Ⓐ Ⓑ Ⓒ Ⓓ
20. Ⓐ Ⓑ Ⓒ Ⓓ
21. Ⓐ Ⓑ Ⓒ Ⓓ
22. Ⓐ Ⓑ Ⓒ Ⓓ
23. Ⓐ Ⓑ Ⓒ Ⓓ
24. Ⓐ Ⓑ Ⓒ Ⓓ

25. Ⓐ Ⓑ Ⓒ Ⓓ
26. Ⓐ Ⓑ Ⓒ Ⓓ
27. Ⓐ Ⓑ Ⓒ Ⓓ
28. Ⓐ Ⓑ Ⓒ Ⓓ
29. Ⓐ Ⓑ Ⓒ Ⓓ
30. Ⓐ Ⓑ Ⓒ Ⓓ

Part B

31. Ⓐ Ⓑ Ⓒ Ⓓ
32. Ⓐ Ⓑ Ⓒ Ⓓ
33. Ⓐ Ⓑ Ⓒ Ⓓ
34. Ⓐ Ⓑ Ⓒ Ⓓ

35. Ⓐ Ⓑ Ⓒ Ⓓ
36. Ⓐ Ⓑ Ⓒ Ⓓ
37. Ⓐ Ⓑ Ⓒ Ⓓ
38. Ⓐ Ⓑ Ⓒ Ⓓ

39. Ⓐ Ⓑ Ⓒ Ⓓ
40. Ⓐ Ⓑ Ⓒ Ⓓ
41. Ⓐ Ⓑ Ⓒ Ⓓ
42. Ⓐ Ⓑ Ⓒ Ⓓ

43. Ⓐ Ⓑ Ⓒ Ⓓ
44. Ⓐ Ⓑ Ⓒ Ⓓ
45. Ⓐ Ⓑ Ⓒ Ⓓ

BC Practice Test 1

Section I

Part A

TIME: 60 MINUTES

The use of calculators is not permitted for this part of the examination.
There are 30 questions in Part A, for which 60 minutes are allowed. Because there is no deduction for wrong answers, you should answer every question, even if you need to guess.

DIRECTIONS: Choose the best answer for each question.

$$f(x) = \begin{cases} 3e^{x-1} - 2 & \text{for } x \leq 1 \\ 4x^2 + bx - 5 & \text{for } x > 1 \end{cases}$$

1. Consider the function $f(x)$, defined above, where b is a constant. What is the value of b for which the function f is continuous at $x = 1$?

   (A) -9
   (B) -4
   (C) 1
   (D) 2

x	$h(x)$
-4	-36
-3	-8
0	4
2	24

2. Function h is twice differentiable. The table above gives selected values of h. Which of the following must be true?

   (A) h has no critical points in the interval $-4 < x < 2$.
   (B) $h'(x) = 10$ for some value of x in the interval $-4 < x < 2$.
   (C) The graph of h has no points of inflection in the interval $-4 < x < 2$.
   (D) $h'(x) > 0$ for all values of x in the interval $-4 < x < 2$.

3. If $x = \sqrt{1 - t^2}$ and $y = \sin^{-1}t$, then $\dfrac{dy}{dx}$ equals

   (A) $-\dfrac{\sqrt{1 - t^2}}{t}$
   (B) $-t$
   (C) 2
   (D) $-\dfrac{1}{t}$

4. If $\displaystyle\sum_{n=1}^{\infty} b_n$ is a geometric series of all-positive terms with $b_1 = 90$ and $b_3 = 10$, then $\displaystyle\sum_{n=1}^{\infty} b_n$

   (A) diverges
   (B) $= 105$
   (C) $= 135$
   (D) converges to a sum that cannot be determined

GO ON TO THE NEXT PAGE

x	f	f'	g	g'
1	2	$\frac{1}{2}$	-3	5
2	3	1	0	4
3	4	2	2	3
4	6	4	3	$\frac{1}{2}$

5. The table above shows values of differentiable functions f and g. If $h(x) = g\,(f(x))$ then $h'(3) =$

(A) $\frac{1}{2}$

(B) 1

(C) 4

(D) 6

6. $\displaystyle\int_1^2 (3x - 2)^3 \, dx$ is equal to

(A) $\frac{64}{3}$

(B) $\frac{63}{4}$

(C) $\frac{85}{4}$

(D) $\frac{255}{4}$

7. Given the parametric equations $x(t) = 2t^2 - 1$ and $y(t) = 3 + 2t^{3/2}$, which expression gives the length of the curve from $t = 1$ to $t = 3$?

(A) $\displaystyle\int_1^3 \sqrt{1 + 9t} \, dt$

(B) $\displaystyle\int_1^3 \sqrt{1 + \frac{9}{16t}} \, dt$

(C) $\displaystyle\int_1^3 \sqrt{16\,t^2 + 9t} \, dt$

(D) $\displaystyle\int_1^3 \sqrt{(2\,t^2 - 1)^2 + (3 + 2\,t^{3/2})^2} \, dt$

8. What is the radius of convergence for the series $\displaystyle\sum_{n=0}^{\infty} \frac{(x - 7)^n}{3^n (n + 2)^4}$?

(A) $\frac{1}{3}$

(B) 1

(C) 3

(D) 7

9. Which equation has the slope field shown below?

(A) $\dfrac{dy}{dx} = \dfrac{5}{y}$

(B) $\dfrac{dy}{dx} = \dfrac{5}{x}$

(C) $\dfrac{dy}{dx} = \dfrac{x}{y}$

(D) $\dfrac{dy}{dx} = 5y$

10. Consider the function h that is continuous on the closed interval $[2,6]$ such that $h(4) = 5$ and $h(6) = 13$. Which of the following statements must be true?

(A) h is increasing on the interval $[4,6]$.

(B) $h'(x) = 4$ has at least one solution in the interval $(4,6)$.

(C) $5 \leq h(5) \leq 13$.

(D) $h(x) = 12$ has at least one solution in the interval $[4,6]$.

11. Which of the following series converges conditionally?

I. $\displaystyle\sum_{n=0}^{\infty} \frac{(-1)^n 3^n}{2^n}$

II. $\displaystyle\sum_{n=1}^{\infty} \frac{(-1)^n}{\sqrt[3]{n}}$

III. $\displaystyle\sum_{n=1}^{\infty} \frac{(-1)^n}{n^5}$

(A) I only

(B) II only

(C) I and II only

(D) II and III only

12. If $x = 2 \sin \theta$, $0 \le \theta \le \frac{\pi}{2}$, then $\int_0^2 \frac{x^2 \, dx}{\sqrt{4 - x^2}}$ is equivalent to:

    (A) $4 \int_0^2 \sin^2 \theta \, d\theta$

    (B) $\int_0^{\pi/2} 4 \sin^2 \theta \, d\theta$

    (C) $\int_0^{\pi/2} 2 \sin \theta \tan \theta \, d\theta$

    (D) $\int_0^2 \frac{2 \sin^2 \theta}{\cos \theta} \, d\theta$

13. $4 \int_{-1}^1 (1 - |x|) dx$

    (A) $= 0$

    (B) $= \frac{1}{2}$

    (C) $= 1$

    (D) does not exist

x	$f(x)$	$f'(x)$	$f''(x)$	$g(x)$	$g'(x)$	$g''(x)$
1	1	0	-7	1/3	-2	7

14. You are given two twice-differentiable functions, $f(x)$ and $g(x)$. The table above gives values for $f(x)$ and $g(x)$ and their first and second derivatives at $x = 1$. Find $\lim_{x \to 1} \frac{2f(x) - 6g(x)}{4x^2 - 4e^{3(x-1)}}$.

    (A) -3

    (B) 1

    (C) 2

    (D) nonexistent

15. What is the volume of the solid generated by rotating the region enclosed by the curve $y = \frac{3}{x}$ and the x-axis on the interval $[6, \infty)$ about the x-axis?

    (A) $\frac{\pi}{6}$

    (B) $\frac{3\pi}{6}$

    (C) 3π

    (D) divergent

16. If $A = \int_0^1 e^{-x} dx$ is approximated using various sums with the same number of subdivisions, and if L, R, and T denote, respectively, left Riemann Sum, right Riemann Sum, and trapezoidal sum, then it follows that

    (A) $R \le A \le T \le L$

    (B) $R \le T \le A \le L$

    (C) $L \le T \le A \le R$

    (D) $L \le A \le T \le R$

17. If $\frac{dy}{dx} = y \tan x$ and $y = 3$ when $x = 0$, then, when $x = \frac{\pi}{3}$, $y =$

    (A) $2\sqrt{3}$

    (B) $\frac{3}{2}$

    (C) $\frac{3\sqrt{3}}{2}$

    (D) 6

18. The parametric equations $x(t) = \sin(t^2 + 3)$ and $y(t) = 2\sqrt{t}$ give the position of a particle moving in the plane for $t \ge 0$. What is the slope of the tangent line to the path of the particle when $t = 1$?

    (A) 1

    (B) $\frac{1}{2 \cos 4}$

    (C) $\frac{2}{\sin 4}$

    (D) $\frac{1}{\cos 4}$

19. In which of the following series can the convergence or divergence be determined by using the Limit Comparison Test with $\sum_{n=1}^{\infty} \frac{1}{n^2}$?

    (A) $\sum_{n=1}^{\infty} \frac{5n}{2n + 4}$

    (B) $\sum_{n=1}^{\infty} \frac{5n}{2n^2 + 4}$

    (C) $\sum_{n=1}^{\infty} \frac{5n^2}{2n^3 + 4n}$

    (D) $\sum_{n=1}^{\infty} \frac{5n^2}{2n^4 + 4n}$

20. Find the slope of the curve $r = \cos 2\theta$ at $\theta = \frac{\pi}{6}$.

    (A) $\frac{\sqrt{3}}{7}$

    (B) $\frac{\sqrt{3}}{4}$

    (C) $-\frac{\sqrt{3}}{4}$

    (D) $-\sqrt{3}$

21. A particle moves along a line with velocity, in feet per second, $v = t^2 - t$. The total distance, in feet, traveled from $t = 0$ to $t = 2$ equals

    (A) $\frac{1}{3}$

    (B) $\frac{2}{3}$

    (C) 2

    (D) 1

Graph of f'

22. The graph of f' is shown in the figure above. Of the following statements, which one is true about f at $x = 1$?

    (A) f is not differentiable at $x = 1$.
    (B) f is not continuous at $x = 1$.
    (C) f attains an absolute maximum at $x = 1$.
    (D) There is an inflection point on the graph of f at $x = 1$.

23. The Maclaurin series for a function h is given by $\sum\limits_{n=1}^{\infty} \frac{(-1)^{n-1} x^{2n}}{3^{2n} \cdot n}$ and converges to h on the interval $-3 \le x \le 3$. If $h\left(\frac{1}{3}\right)$ is approximated with the sixth-degree Maclaurin polynomial, what is the Alternating Series Error Bound for this approximation?

    (A) $\frac{1}{3^{20} \cdot 5}$

    (B) $\frac{1}{3^{16} \cdot 4}$

    (C) $\frac{1}{3^{12} \cdot 3}$

    (D) $\frac{1}{3^8 \cdot 4}$

24. $\int x \cos x \, dx =$

    (A) $x \sin x + \cos x + C$
    (B) $x \sin x - \cos x + C$
    (C) $\frac{x^2}{2} \sin x + C$
    (D) $\frac{1}{2} \sin x^2 + C$

25. Which one of the following series converges?

    (A) $\sum\limits_{n=1}^{\infty} \frac{1}{\sqrt{n}}$

    (B) $\sum\limits_{n=1}^{\infty} \frac{1}{2n+1}$

    (C) $\sum\limits_{n=1}^{\infty} \frac{n}{n^2+1}$

    (D) $\sum\limits_{n=1}^{\infty} \frac{1}{n^2+1}$

26. The coefficient of the $(x-8)^2$ term in the Taylor polynomial for $y = x^{2/3}$ centered at $x = 8$ is

    (A) $-\frac{1}{144}$

    (B) $-\frac{1}{72}$

    (C) $-\frac{1}{9}$

    (D) $\frac{1}{144}$

27. If $f'(x) = h(x)$ and $g(x) = x^3$, then $\frac{d}{dx} f(g(x)) =$

    (A) $h(x^3)$
    (B) $3x^2 h(x)$
    (C) $3x^2 h(x^3)$
    (D) $h(3x^2)$

28. $\displaystyle\int_0^\infty e^{-x/2}\,dx =$

(A) -2

(B) 1

(C) 2

(D) ∞

30. Choose the integral that is the limit of the Riemann Sum: $\displaystyle\lim_{n\to\infty}\sum_{k=1}^{n}\left(\left(\frac{3k}{n}+2\right)^2\cdot\left(\frac{3}{n}\right)\right)$.

(A) $\displaystyle\int_2^5 (x+2)^2\,dx$

(B) $\displaystyle\int_0^3 (3x+2)^2\,dx$

(C) $\displaystyle\int_2^5 x^2\,dx$

(D) $\displaystyle\int_0^3 x^2\,dx$

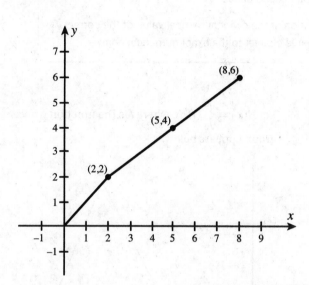

29. The graph of $f(x)$ consists of two line segments as shown above. If $g(x) = f^{-1}(x)$, the inverse function of $f(x)$, find $g'(4)$.

(A) $\frac{1}{5}$

(B) $\frac{2}{3}$

(C) $\frac{3}{2}$

(D) 5

STOP   End of Part A, Section I

Part B

TIME: 45 MINUTES

> *Some questions in this part of the examination require the use of a graphing calculator. There are 15 questions in Part B, for which 45 minutes are allowed. Because there is no deduction for wrong answers, you should answer every question, even if you need to guess.*
>
> **DIRECTIONS:** Choose the best answer for each question. If the exact numerical value of the correct answer is not listed as a choice, select the choice that is closest to the exact numerical answer.

31. Consider $h'(x) = 3e^{x-1} - 5\sin(x^3)$, the first derivative of the function h. The graph of h has a local maximum at which of the following values of x?

    (A) 0.318
    (B) 0.801
    (C) 1.116
    (D) 1.300

32. The graph of f', which consists of a quarter-circle and two line segments, is shown above. At $x = 2$, which of the following statements is true?

    (A) f is not continuous.
    (B) f is continuous but not differentiable.
    (C) f has a local maximum.
    (D) The graph of f has a point of inflection.

33. Let $H(x) = \displaystyle\int_0^x f(t)\,dt$, where f is the function whose graph appears below.

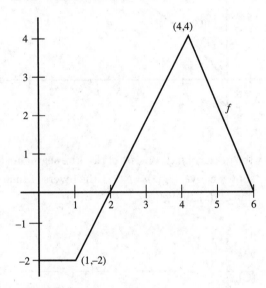

    The local linearization of $H(x)$ near $x = 3$ is $H(x) \approx$

    (A) $-2x + 8$
    (B) $2x - 4$
    (C) $-2x + 4$
    (D) $2x - 8$

34. Consider the velocity vector $\langle \ln(t^4 + 1), \cos(t^3) \rangle$ for a particle moving in the plane. The position vector of the particle at $t = 2$ is $\langle 4, -2 \rangle$. What is the x-coordinate of the position vector of the particle at $t = 0.5$?

    (A) 0.061
    (B) 1.941
    (C) 2.059
    (D) 5.941

GO ON TO THE NEXT PAGE

35. The function g is defined on the closed interval $[0,4]$ such that $g(0) = g(2) = g(4)$. The function g is continuous and strictly decreasing on the open interval $(0,4)$. Which of the following statements is true?

   (A) g has neither an absolute minimum nor an absolute maximum on $[0,4]$.

   (B) g has an absolute minimum but not an absolute maximum on $[0,4]$.

   (C) g has an absolute maximum but not an absolute minimum on $[0,4]$.

   (D) g has both an absolute minimum and an absolute maximum on $[0,4]$.

36. Find the volume of the solid generated when the region bounded by the y-axis, $y = e^x$, and $y = 2$ is rotated around the y-axis.

   (A) 0.592

   (B) 1.214

   (C) 2.427

   (D) 3.998

37. The path of a satellite is given by the parametric equations

$$x = 4 \cos t + \cos 12t$$
$$y = 4 \sin t + \sin 12t$$

   The upward velocity at $t = 1$ equals

   (A) 3.073

   (B) 3.999

   (C) 12.287

   (D) 12.666

38. If g is a differentiable function with $g(1) = 4$ and $g'(1) = 3$, which of the following statements could be false?

   (A) $\lim\limits_{x \to 1^+} g(x) = \lim\limits_{x \to 1^-} g(x)$

   (B) $\lim\limits_{x \to 1} g'(x) = 3$

   (C) $\lim\limits_{x \to 1} g(x) = 4$

   (D) $\lim\limits_{h \to 0} \dfrac{g(1+h) - g(1)}{h} = 3$

39. A particle is moving in the xy-plane. The position of the particle is given by $x(t) = t^2 + \ln(t^2 + 1)$ and $y(t) = 2t + 5\cos(t^2)$. What is the speed of the particle when $t = 3.1$?

   (A) 3.641

   (B) 3.807

   (C) 10.269

   (D) 12.041

40. Which definite integral represents the length of the first-quadrant arc of the curve defined by $x(t) = e^t$, $y(t) = 1 - t^2$?

   (A) $\displaystyle\int_{-1}^{1} \sqrt{1 + \dfrac{4t^2}{e^{2t}}}\, dt$

   (B) $\displaystyle\int_{1/e}^{e} \sqrt{1 + \dfrac{4t^2}{e^{2t}}}\, dt$

   (C) $\displaystyle\int_{-1}^{1} \sqrt{e^{2t} + 4t^2}\, dt$

   (D) $\displaystyle\int_{1/e}^{e} \sqrt{e^{2t} + 4t^2}\, dt$

41. For which function is $\displaystyle\sum_{n=0}^{\infty} \dfrac{(-1)^n x^{2n}}{(2n)!}$ the Taylor series about 0?

   (A) e^{-x}

   (B) $\sin x$

   (C) $\cos x$

   (D) $\ln(1 + x)$

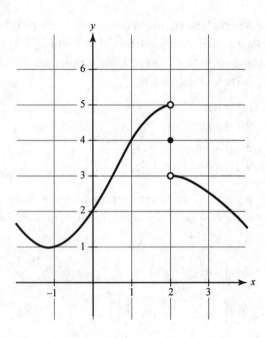

42. The graph of the function f is shown in the figure above. Find the value of $\lim\limits_{x \to 0} f(2x^2 + 2)$.

    (A) 3
    (B) 4
    (C) 5
    (D) nonexistent

43. At how many points on the interval $[0, \pi]$ does $f(x) = 2 \sin x + \sin 4x$ satisfy the Mean Value Theorem?

    (A) 1
    (B) 2
    (C) 3
    (D) 4

44. As a cup of hot chocolate cools, its temperature after t minutes is given by $H(t) = 70 + ke^{-0.4t}$. If its initial temperature was 120°F, what was its average temperature (in °F) during the first 10 minutes?

    (A) 79.1
    (B) 82.3
    (C) 95.5
    (D) 99.5

45. The rate at which a purification process can remove contaminants from a tank of water is proportional to the amount of contaminant remaining. If 20% of the contaminant can be removed during the first minute of the process and 98% must be removed to make the water safe, approximately how long will the decontamination process take?

    (A) 2 minutes
    (B) 5 minutes
    (C) 7 minutes
    (D) 18 minutes

STOP   End of Section I

SECTION II

PART A

TIME: 30 MINUTES

2 PROBLEMS

> *A graphing calculator is required for some of these problems. See instructions on page 8.*

1. The velocity vector of an object in motion in the plane for $0 \leq t \leq 6$ is given by its components $\frac{dx}{dt} = 3t \cdot \sin(t^2)$ and $\frac{dy}{dt} = 2\,e^{\cos(t)}$. At time $t = 5$, the position of the object is $(3, -2)$.

   (a) Find the speed of the object at time $t = 4$, and find the acceleration vector of the object at time $t = 4$.

   (b) Find the position of the object at time $t = 4$.

   (c) Find the total distance traveled by the object from $t = 0$ to $t = 4$.

2. The rate of sales of a new software product is given by the differentiable and increasing function $S(t)$, where S is measured in units per month and t is measured in months from the initial release date. The software company recorded these sales data:

t (months)	1	3	6	8	9
$S(t)$ (units/mo)	154	232	490	763	954

   (a) Use the data in the table to estimate $S'(7)$. Show the work that leads to your answer. Indicate units of measure.

   (b) Use a left Riemann Sum with the four subintervals indicated in the table to estimate the total number of units of software sold during the first 9 months of sales. Is this an underestimate or an overestimate of the total number of software units sold? Give a reason for your answer.

   (c) For $1 \leq t \leq 9$, must there be a time t when the rate of sales is increasing at $100\frac{\text{units}}{\text{month}}$ per month? Justify your answer.

   (d) For $9 \leq t \leq 12$, the rate of sales is modeled by the function $R(t) = 120 \cdot (2)^{\frac{x}{3}}$. Given that the actual sales in the first 9 months is 3636 software units, use this model to find the number of units of the software sold during the first 12 months of sales. Round your answer to the nearest whole unit.

STOP   End of Part A, Section II

Part B

TIME: 60 MINUTES

4 PROBLEMS

> No calculator is allowed for any of these problems.
>
> If you finish Part B before time has expired, you may return to work on Part A, but you may not use a calculator.

3. The graph of function $y = f(x)$ passes through the point (1,1) and satisfies the differential equation $\dfrac{dy}{dx} = \dfrac{6x^2 - 4}{y}$.

   (a) Sketch the slope field for the differential equation at the 12 indicated points on the axes provided.

   (b) Find an equation of the line tangent to $f(x)$ at the point (1,1) and use the linear equation to estimate $f(1.2)$.

   (c) Solve the differential equation, and find the particular solution for $y = f(x)$ that passes through the point (1,1).

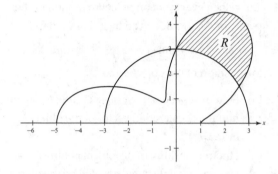

4. Shown above are the graphs of the polar curves $r = 3$ and $r = 3 - 2\cos(3\theta)$ for $0 \le \theta \le \pi$. Region R is the shaded region inside the graph of $r = 3 - 2\cos(3\theta)$ and outside the graph of $r = 3$. The two curves intersect at $\theta = \dfrac{\pi}{6}$, $\theta = \dfrac{\pi}{2}$, and $\theta = \dfrac{5\pi}{6}$.

   (a) Write, but do not evaluate, an integral expression for the area of region R.

   (b) For the graph of $r = 3 - 2\cos(3\theta)$, write expressions for $\dfrac{dx}{d\theta}$ and $\dfrac{dy}{d\theta}$ in terms of θ.

   (c) Write an equation, in terms of x and y, for the tangent line to the curve $r = 3 - 2\cos(3\theta)$ when $\theta = \dfrac{\pi}{2}$. Show the computations that lead to your answer.

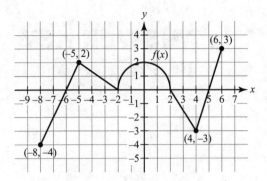

5. The graph of the function f is given above. f is twice differentiable and is defined on the interval $-8 \leq x \leq 6$. The function g is also twice differentiable and is defined as
$$g(x) = \int_{-2}^{x} f(q)dq.$$

(a) Write an equation for the line tangent to the graph of $g(x)$ at $x = -8$. Show the work that leads to your answer.

(b) Using your tangent line from part (a), approximate $g(-7)$. Is your approximation greater than or less than $g(-7)$? Give a reason for your answer.

(c) Evaluate: $\int_{-5}^{4} (2f'(x) - 3)dx$. Show the work that leads to your answer.

(d) Find the absolute minimum value of g on the interval $-8 \leq x \leq 6$. Justify your answer.

6. (a) Write the first four nonzero terms and the general term of the Maclaurin series for $f(x) = \ln(e + x)$.

(b) What is the radius of convergence?

(c) Use the first three terms of that series to write an expression that estimates the value of
$$\int_{0}^{1} \ln(e + x^2)dx.$$

STOP   **End of Test**

ANSWER KEY
BC Practice Test 1

SECTION I

Part A

1. **D**	9. **A**	17. **D**	25. **D**
2. **B**	10. **D**	18. **B**	26. **A**
3. **D**	11. **B**	19. **D**	27. **C**
4. **C**	12. **B**	20. **A**	28. **C**
5. **B**	13. **C**	21. **D**	29. **C**
6. **C**	14. **A**	22. **D**	30. **C**
7. **C**	15. **B**	23. **B**	
8. **C**	16. **A**	24. **A**	

Part B

31. **B**	35. **A**	39. **C**	43. **D**
32. **D**	36. **A**	40. **C**	44. **B**
33. **D**	37. **C**	41. **C**	45. **D**
34. **C**	38. **B**	42. **A**	

Answer Explanations

The explanations for questions not given below will be found in the answer explanation section for AB Practice Test 1 on pages 517–526. Identical questions in Section I of Practice Tests AB 1 and BC 1 have the same number. For example, an explanation of the answer for Question 5, not given below, will be found in Section I of AB Practice Test 1, Answer 5, page 517.

Section I Multiple-Choice

Part A

1. **(D)** Since we want the function continuous at $x = 1$, then $\lim_{x \to 1^-} f(x) = \lim_{x \to 1^+} f(x) = f(1)$. First find the two one-sided limits, and then solve for b.

    Left limit: $\lim_{x \to 1^-} f(x) = 3e^{1-1} - 2 = 1$

    Right limit: $\lim_{x \to 1^+} f(x) = 4(1)^3 + b \cdot 1 - 5 = b - 1$

    Set the one-sided limits equal and solve for b: $b - 1 = 1 \Rightarrow b = 2$. Therefore, since $f(1) = 1$, when $b = 2$, both one-sided limits are equal and are equal to $f(1)$.

2. **(B)** Since the function h is twice differentiable, it is continuous on the interval $[-4,2]$ so the Mean Value Theorem applies. Therefore, there must be a value of x in the interval $(-4,2)$ such that $h'(x) = \dfrac{h(2) - h(-4)}{2 - (-4)} = \dfrac{24 - (-36)}{6} = \dfrac{60}{6} = 10$.

    NOTE: h must have a critical point since the Intermediate Value Theorem applies on $h'(x)$. There is no way to determine if there is an inflection point since we don't know the behavior of h between the selected x-values in the table. Although the values of h that we see on the interval $[-4,2]$ are increasing, we don't know the function values between these selected values of x. So h could be decreasing on some interval.

3. **(D)** Here,
$$\frac{dy}{dx} = \frac{\dfrac{dy}{dt}}{\dfrac{dx}{dt}} = \frac{\dfrac{1}{\sqrt{1-t^2}}}{\dfrac{1}{2}\dfrac{(-2t)}{\sqrt{1-t^2}}} = -\frac{1}{t}$$

4. **(C)** If $b_1 = 90$ and $b_3 = 10$, this geometric series has a common ratio of $\frac{1}{3}$. Thus the sum of the series is $\dfrac{90}{1 - \frac{1}{3}} = 90 \cdot \dfrac{3}{2} = 135$. Note that the sum cannot be 105 because the sum of the first and third terms is already 100 and the second term must be greater than 10.

6. **(C)** $\dfrac{1}{3}\displaystyle\int_1^2 (3x-2)^3 (3dx) = \dfrac{1}{12}(3x-2)^4 \Big|_1^2$

7. **(C)** The length of the curve is $\displaystyle\int_a^b \sqrt{(x'(t))^2 + (y'(t))^2}\, dt$. Since $x'(t) = 4t$ and $y'(t) = 3\sqrt{t}$, the length is $\displaystyle\int_1^3 \sqrt{(4t)^2 + (3\sqrt{t})^2}\, dt = \int_1^3 \sqrt{16t^2 + 9t}\, dt$.

8. **(C)** Use the Ratio Test to find the radius of convergence.

    The ratio is $\left| \dfrac{(x-7)^{n+1}}{3^{n+1}(n+3)^4} \cdot \dfrac{3^n(n+2)^4}{(x-7)^n} \right| = \left| \left(\dfrac{n+2}{n+3}\right)^4 \cdot \dfrac{x-7}{3} \right|$. Find the limit:

    $\lim_{x \to \infty} \left| \left(\dfrac{n+2}{n+3}\right)^4 \cdot \dfrac{x-7}{3} \right| = \left| \dfrac{x-7}{3} \right|$. So for convergence, $\left| \dfrac{x-7}{3} \right| < 1 \Rightarrow |x-7| < 3$. Therefore, the radius of convergence is 3.

10. **(D)** Since the function h is continuous on the interval [4,6], $h(4) = 5$, and $h(6) = 13$, the Intermediate Value Theorem tells us that all function values between 5 and 13 must exist on that interval. Therefore, there must be an x-value on that interval where $h(x) = 12$. *NOTE:* Choice (A) doesn't have to be true—the function could be both increasing and decreasing on some subintervals of [4,6]. Choice (B) may not be true if h is not differentiable on (4,6). Choice (C) may not be true—the function may dip below 5 and then rise back up to 13 on the interval [4,6].

11. **(B)** Series I is a divergent geometric series with a common ratio of $-\frac{3}{2}$. Series II passes the Alternating Series Test and thus converges. However, the absolute value of series II is a p-series with $p = \frac{1}{3}$. Since $p < 1$, that p-series diverges. Thus, series II is conditionally convergent. Series III passes the Alternating Series Test and thus converges, and the absolute value of series III is a p-series with $p = 5$. Since $p > 1$, that p-series converges. Thus, series III is absolutely convergent.

12. **(B)** Note that, when $x = 2 \sin \theta$, $x^2 = 4 \sin^2\theta$, $dx = 2 \cos \theta \, d\theta$, and $\sqrt{4 - x^2} = 2 \cos\theta$. Also, when $x = 0$, $\theta = 0$, and when $x = 2$, $\theta = \frac{\pi}{2}$.

13. **(C)** The given integral is equivalent to $\displaystyle\int_{-1}^{0} (1 + x)\, dx + \int_{0}^{1} (1 - x)\, dx$.

    The figure shows the graph of $f(x) = 1 - |x|$ on $[-1,1]$.

    The area of triangle PQR is equal to $\displaystyle\int_{-1}^{1} (1 - |x|)\, dx$.

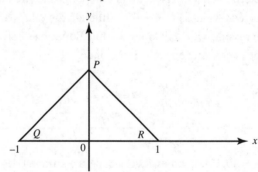

15. **(B)** This integral is an improper integral, so rewrite it as a limit and evaluate.

$$\int_{6}^{\infty} \pi \left(\frac{3}{x}\right)^2 dx = \lim_{b\to\infty}\int_{6}^{b} \pi \left(\frac{3}{x}\right)^2 dx = 9\pi \cdot \lim_{b\to\infty}\int_{6}^{b} \left(\frac{1}{x}\right)^2 dx = 9\pi \cdot \lim_{b\to\infty}\left(-\frac{1}{x}\right)\Big|_{6}^{b}$$

$$= 9\pi \cdot \lim_{b\to\infty}\left(-\frac{1}{b} + \frac{1}{6}\right) = 9\pi \cdot \frac{1}{6} = \frac{3\pi}{2}$$

17. **(D)** Separating variables yields $\dfrac{dy}{y} = \tan x \, dx$, so $\ln y = -\ln \cos x + C$. With $y = 3$ when $x = 0$, $C = \ln 3$.

    The general solution is therefore $y = \dfrac{3}{\cos x}$. When $x = \dfrac{\pi}{3}$, $y = \dfrac{3}{1/2} = 6$.

18. **(B)** The slope of the tangent line is $\dfrac{dy}{dx}$. Use the Chain Rule to find the slope with

$$\frac{dy}{dx} = \frac{dy/dt}{dx/dt} \cdot \frac{dy}{dt} = \frac{d}{dt}(2t^{1/2}) = t^{-1/2} = \frac{1}{\sqrt{t}} \text{ and } \frac{dx}{dt} = \cos(t^2 + 3) \cdot 2t.$$

The slope at $t = 1$ is $\dfrac{dy}{dx}\Big|_{t=1} = \dfrac{1/\sqrt{t}}{2t\cos(t^2 + 3)}\Big|_{t=1} = \dfrac{1}{2\cos 4}$.

19. **(D)** If we look at the end behavior of series (D), the numerator behaves like n^2 and the denominator behaves like n^4 as n gets larger. So the nth term of the series behaves like $\frac{n^2}{n^4} = \frac{1}{n^2}$. Thus, we can try and compare to $\sum_{n=1}^{\infty} \frac{1}{n^2}$, a known convergent series.

Using the Limit Comparison Test: $\lim_{n\to\infty} \dfrac{\dfrac{5n^2}{2n^4 + 4n}}{\dfrac{1}{n^2}} = \lim_{n\to\infty} \dfrac{5n^4}{2n^4 + 4n} = \dfrac{5}{2} > 0$. So both series

converge. If you use the series $\sum_{n=1}^{\infty} \frac{1}{n^2}$ and the Limit Comparison Test for series (A), (B), or (C), the limit is infinite. So the Limit Comparison Test is inconclusive since the comparison is with a known convergent series.

20. **(A)** Represent the coordinates parametrically as $(r \cos\theta,\ r \sin\theta)$. Then

$$\frac{dy}{dx} = \frac{\dfrac{dy}{d\theta}}{\dfrac{dx}{d\theta}} = \frac{r\cos\theta + \dfrac{dr}{d\theta}\cdot\sin\theta}{-r\sin\theta + \dfrac{dr}{d\theta}\cdot\cos\theta}$$

Note that $\frac{dr}{d\theta} = -2\sin 2\theta$, and evaluate $\frac{dy}{dx}$ at $\theta = \frac{\pi}{6}$. $\Big($Alternatively, write $x = \cos 2\theta \cos\theta$ and $y = \cos 2\theta \sin\theta$ to find $\frac{dy}{dx}$ from $\frac{dy/d\theta}{dx/d\theta}.\Big)$

21. **(D)** Note that v is negative from $t = 0$ to $t = 1$ but positive from $t = 1$ to $t = 2$. Thus the distance traveled is given by

$$-\int_0^1 (t^2 - t)\,dt + \int_1^2 (t^2 - t)\,dt$$

22. **(D)** The graph of f has an inflection point when the graph of f' goes from increasing to decreasing as it does at $x = 1$ on the given figure. The other statements are all false. Choice (A) is false because f is differentiable at $x = 1$; notice that $f'(1) = 3$. The error in choosing this statement could be thinking you are looking at the graph of f instead of f'. Choice (B) is false because f is differentiable at $x = 1$. Therefore, f is continuous at $x = 1$. Choice (C) is false because f' appears to attain an absolute maximum at $x = 1$. Again, the error in choosing this statement could be thinking you are looking at the graph of f instead of f'.

23. **(B)** $h(x) = \dfrac{x^2}{3^2 \cdot 1} - \dfrac{x^4}{3^4 \cdot 2} + \dfrac{x^6}{3^6 \cdot 3} - \dfrac{x^8}{3^8 \cdot 4} + \ldots$ and the sixth-degree Maclaurin polynomial is $M(x) = \dfrac{x^2}{3^2 \cdot 1} - \dfrac{x^4}{3^4 \cdot 2} + \dfrac{x^6}{3^6 \cdot 3}$.

$h\!\left(\frac{1}{3}\right) \approx M\!\left(\frac{1}{3}\right) = \dfrac{1}{3^2}\left(\frac{1}{3}\right)^2 - \dfrac{1}{3^4 \cdot 2}\left(\frac{1}{3}\right)^4 + \dfrac{1}{3^6 \cdot 3}\left(\frac{1}{6}\right)^6$. Since the Maclaurin series for $h(x)$ is an alternating series that converges by the Alternating Series Test at $x = \frac{1}{3}$, the Alternating Series Error Bound is the absolute value of the first omitted term from the Maclaurin series.

So $\left|h\!\left(\frac{1}{3}\right) - M\!\left(\frac{1}{3}\right)\right| \le \dfrac{1}{3^8 \cdot 4}\left(\frac{1}{3}\right)^8 = \dfrac{1}{3^{16} \cdot 4}$.

24. **(A)** Use parts; then $u = x$, $dv = \cos x\,dx$; $du = dx$, $v = \sin x$. Thus,

$$\int x \cos x\,dx = x \sin x - \int \sin x\,dx.$$

25. **(D)** (A), a p-series with $p = 1/2$, diverges. We would like to compare (B) to $\displaystyle\sum_{n=1}^{\infty} \frac{1}{n}$, but $\dfrac{1}{2n+1} < \dfrac{1}{n}$, so we use the Limit Comparison Test $\displaystyle\lim_{n\to\infty} \frac{\frac{1}{2n+1}}{1/n} = \lim_{n\to\infty} \frac{n}{2n+1} = \frac{1}{2}$; since $\displaystyle\sum_{n=1}^{\infty} \frac{1}{n}$ diverges, (B) diverges. For (C), we need to use the LCT for the same reason as (B), $\displaystyle\lim_{n\to\infty} \frac{\frac{n}{n^2+1}}{1/n} = \lim_{n\to\infty} \frac{n^2}{n^2+1} = 1$; again since $\displaystyle\sum_{n=1}^{\infty} \frac{1}{n}$ diverges, (C) diverges. For (D), we would like to compare to $\displaystyle\sum_{n=1}^{\infty} \frac{1}{n^2}$ and $\dfrac{1}{n^2+1} < \dfrac{1}{n^2}$, so by the Comparison Test (D) converges.

26. **(A)** By Taylor's Theorem, the coefficient is $\dfrac{f''(8)}{2!}$. For $f(x) = x^{\frac{2}{3}}, f'(x) = \frac{2}{3}x^{-\frac{1}{3}}$ and $f''(x) = -\frac{2}{9}x^{-\frac{4}{3}}$; hence $f''(8) = -\frac{2}{9}(8)^{-\frac{4}{3}} = -\frac{2}{9} \cdot \frac{1}{16} = -\frac{1}{72}$, making the coefficient $-\dfrac{1}{72} \cdot \dfrac{1}{2!}$.

27. **(C)** $\dfrac{d}{dx}f(g(x)) = f'(g(x))g'(x) = h(g(x))g'(x) = h(x^3) \cdot 3x^2$

28. **(C)** Evaluate:
$$\lim_{b\to\infty} \int_0^b e^{-x/2}\,dx = -\lim_{b\to\infty} 2e^{-x/2}\Big|_0^b = -2(0-1)$$

Part B

31. **(B)** The graph of $h'(x)$ on the interval $-1 \leq x \leq 3$ is shown below.

A local maximum occurs when the graph of the first derivative changes sign from positive to negative. This happens at the point (0.801,0) as shown in the figure.

34. **(C)** Use the Fundamental Theorem of Calculus to find the new position:
$$x(0.5) = x(2) + \int_2^{0.5} x'(t)\,dt = 2.059, \text{ where } x'(t) \text{ is the } x\text{-component of the velocity vector.}$$

35. **(A)** An example of such a function is shown below.

36. **(A)**

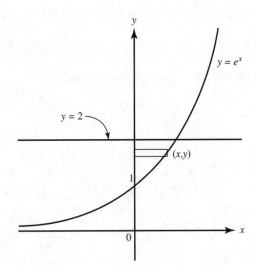

Using disks, $\Delta V = \pi R^2 \Delta y = \pi (\ln y)^2 \Delta y$. Evaluate:

$$V = \int_1^2 \pi (\ln y)^2 \, dy$$

The required volume is 0.592.

37. **(C)** The vertical component of velocity is

$$\frac{dy}{dt} = 4 \cos t + 12 \cos 12t$$

Evaluate at $t = 1$.

38. **(B)** Although most, if not all, of the functions studied in AP Calculus have continuous derivatives, there is no guarantee that the derivative of a function is continuous. *NOTE:* Choices (A) and (C) must be true since g is continuous because it is differentiable. Choice (D) is the definition of the derivative at $x = 3$ and thus must be true.

39. **(C)** The speed of the particle is the magnitude of the velocity vector. At $t = 3.1$, the velocity vector is $\langle 6.784, 7.709 \rangle$ and the magnitude is $\sqrt{6.784^2 + 7.709^2} = 10.269$.

40. **(C)** In the first quadrant, both x and y must be positive; $x(t) = e^t$ is positive for all t, but $y(t) = 1 - t^2$ is positive only for $-1 < t < 1$. The arc length is

$$\int_{t_1}^{t_2} \sqrt{\left(\frac{dx}{dt}\right)^2 + \left(\frac{dy}{dt}\right)^2} \, dt = \int_{-1}^{1} \sqrt{(e^t)^2 + (-2t)^2 \, dt}$$

41. **(C)** See series (2) on page 418.

42. **(A)** The condition for using the Composition Rule for limits is not met. In this problem, $f(x)$ is not continuous at $x = \lim_{x \to 0}(2x^2 + 2) = 2$. Therefore, we look at the one-sided limits.

Left limit: $\lim_{x \to 0^-} f(2x^2 + 2) = 3$. Note that as $x \to 0^-$, $(2x^2 + 2) \to 2$ from values above 2 (because $x^2 > 0$).

So to find $\lim_{x \to 0^-} f(2x^2 + 2)$, we must look at $\lim_{x \to 2^+} f(x)$.

Right limit: $\lim_{x \to 0^+} f(2x^2 + 2) = 3$. Note that as $x \to 0^+$, $(2x^2 + 2) \to 2$ again from values above 2.

So to find $\lim_{x \to 0^+} f(2x^2 + 2)$, we must look at $\lim_{x \to 2^+} f(x)$.

Since both one-sided limits of $f(2x^2 + 2)$ equal 3, then $\lim_{x \to 0} f(2x^2 + 2) = 3$.

44. **(B)** At $t = 0$, we know $H = 120$, so $120 = 70 + ke^{-0.4(0)}$, and thus $k = 50$. The average temperature for the first 10 minutes is $\dfrac{1}{10 - 0} \displaystyle\int_0^{10} (70 + 50e^{-0.4t}) \, dt$.

Section II Free-Response

Part A

1. (a) Note: $\frac{dx}{dt}$ and $x'(t)$ are equivalent notations for the x-component for velocity.

    Likewise, $\frac{dy}{dt}$ and $y'(t)$ are equivalent notations for the y-component for velocity.

    Speed $= \sqrt{(x'(4))^2 + (y'(4))^2} = 3.608$

    Acceleration $= \langle x''(4), y''(4) \rangle = \langle -92.799, 0.787 \rangle$

    (b) $x(4) = x(5) + \int_5^4 x'(t)dt = 3 + \int_5^4 x'(t)dt = 5.923$

    $y(4) = y(5) + \int_5^4 y'(t)dt = -2 + \int_5^4 y'(t)dt = -3.697$

    At time $t = 4$, the object is at position $\langle 5.923, -3.697 \rangle$

    (c) Total Distance $= \int_0^4 \sqrt{(x'(t))^2 + (y'(t))^2}\ dt = 20.261$

2. See the solution for AB/BC 2, page 523.

Part B

3. See the solution for AB/BC 3, page 523.

4. (a) $\frac{1}{2}\int_{\pi/6}^{\pi/2} ((3 - 2\cos(3\theta))^2 - 3^2)d\theta$

    (b) $\frac{dr}{d\theta} = 6\sin(3\theta)$, $x = r\cos\theta$, and $y = r\sin\theta$

    $\frac{dx}{d\theta} = \frac{dr}{d\theta}\cos\theta - r\sin\theta \Rightarrow \frac{dx}{d\theta} = 6\sin(3\theta)\cos\theta - (3 - 2\cos(3\theta))\sin\theta$

    $\frac{dy}{d\theta} = \frac{dr}{d\theta}\sin\theta + r\cos\theta \Rightarrow \frac{dy}{d\theta} = 6\sin(3\theta)\sin\theta + (3 - 2\cos(3\theta))\cos\theta$

    (c) The slope of the tangent line is $\frac{dy}{dx} = \frac{dy/d\theta}{dx/d\theta}$.

    When $\theta = \frac{\pi}{2}$, then $r\left(\frac{\pi}{2}\right) = 3 - 2\cos\left(\frac{3\pi}{2}\right) = 3 \Rightarrow x = 3\cos\left(\frac{\pi}{2}\right) = 0 \Rightarrow y = 3\sin\left(\frac{\pi}{2}\right) = 3$.

    $\left.\frac{dy}{dx}\right|_{\theta=\pi/2} = \left.\frac{dy/d\theta}{dx/d\theta}\right|_{\theta=\pi/2} = \frac{6\sin\left(\frac{3\pi}{2}\right)\sin\left(\frac{\pi}{2}\right) + \left(3 - 2\cos\left(\frac{3\pi}{2}\right)\right)\cos\left(\frac{\pi}{2}\right)}{6\sin\left(\frac{3\pi}{2}\right)\cos\left(\frac{\pi}{2}\right) - \left(3 - 2\cos\left(\frac{3\pi}{2}\right)\right)\sin\left(\frac{\pi}{2}\right)} = \frac{-6}{-3} = 2$

    The tangent line is $y = 3 + 2(x - 0) = 3 + 2x$.

5. See the solution for AB/BC 5, page 525.

6. (a) To write the Maclaurin series for $f(x) = \ln(e + x)$, use Taylor's Theorem at $x = 0$.

n	$f^{(n)}(x)$	$f^{(n)}(0)$	$a_n = \dfrac{f^{(n)}(0)}{n!}$
0	$\ln(e + x)$	1	1
1	$\dfrac{1}{e + x}$	$\dfrac{1}{e}$	$\dfrac{1}{e}$
2	$-(e + x)^{-2}$	$-\dfrac{1}{e^2}$	$-\dfrac{1}{2e^2}$
3	$2(e + x)^{-3}$	$\dfrac{2}{e^3}$	$\dfrac{1}{3e^3}$

$$f(x) = 1 + \frac{x}{e} - \frac{x^2}{2e^2} + \frac{x^3}{3e^3} \cdots + \frac{(-1)^{n+1}x^n}{ne^n} + \cdots \text{(for } n \geq 1)$$

(b) By the Ratio Test, the series converges when

$$\lim_{n \to \infty} \left| \frac{x^{n+1}}{(n+1)e^{n+1}} \cdot \frac{ne^n}{x^n} \right| < 1$$

$$|x| \lim_{n \to \infty} \frac{1}{e} < 1$$

$$|x| < e$$

Thus, the radius of convergence is e.

(c) $$\int_0^1 \ln(e + x^2)\, dx \approx \int_0^1 1 + \frac{x^2}{e} - \frac{(x^2)^2}{2e^2}\, dx$$

$$\int_0^1 \ln(e + x^2)\, dx \approx \left(x + \frac{x^3}{3e} - \frac{x}{5 \cdot 2e^2} \right) \Bigg|_0^1 = 1 + \frac{1}{3e} - \frac{1}{10e^2}$$

ANSWER SHEET
BC Practice Test 2

SECTION I

Part A

1. Ⓐ Ⓑ Ⓒ Ⓓ
2. Ⓐ Ⓑ Ⓒ Ⓓ
3. Ⓐ Ⓑ Ⓒ Ⓓ
4. Ⓐ Ⓑ Ⓒ Ⓓ
5. Ⓐ Ⓑ Ⓒ Ⓓ
6. Ⓐ Ⓑ Ⓒ Ⓓ
7. Ⓐ Ⓑ Ⓒ Ⓓ
8. Ⓐ Ⓑ Ⓒ Ⓓ

9. Ⓐ Ⓑ Ⓒ Ⓓ
10. Ⓐ Ⓑ Ⓒ Ⓓ
11. Ⓐ Ⓑ Ⓒ Ⓓ
12. Ⓐ Ⓑ Ⓒ Ⓓ
13. Ⓐ Ⓑ Ⓒ Ⓓ
14. Ⓐ Ⓑ Ⓒ Ⓓ
15. Ⓐ Ⓑ Ⓒ Ⓓ
16. Ⓐ Ⓑ Ⓒ Ⓓ

17. Ⓐ Ⓑ Ⓒ Ⓓ
18. Ⓐ Ⓑ Ⓒ Ⓓ
19. Ⓐ Ⓑ Ⓒ Ⓓ
20. Ⓐ Ⓑ Ⓒ Ⓓ
21. Ⓐ Ⓑ Ⓒ Ⓓ
22. Ⓐ Ⓑ Ⓒ Ⓓ
23. Ⓐ Ⓑ Ⓒ Ⓓ
24. Ⓐ Ⓑ Ⓒ Ⓓ

25. Ⓐ Ⓑ Ⓒ Ⓓ
26. Ⓐ Ⓑ Ⓒ Ⓓ
27. Ⓐ Ⓑ Ⓒ Ⓓ
28. Ⓐ Ⓑ Ⓒ Ⓓ
29. Ⓐ Ⓑ Ⓒ Ⓓ
30. Ⓐ Ⓑ Ⓒ Ⓓ

Part B

31. Ⓐ Ⓑ Ⓒ Ⓓ
32. Ⓐ Ⓑ Ⓒ Ⓓ
33. Ⓐ Ⓑ Ⓒ Ⓓ
34. Ⓐ Ⓑ Ⓒ Ⓓ

35. Ⓐ Ⓑ Ⓒ Ⓓ
36. Ⓐ Ⓑ Ⓒ Ⓓ
37. Ⓐ Ⓑ Ⓒ Ⓓ
38. Ⓐ Ⓑ Ⓒ Ⓓ

39. Ⓐ Ⓑ Ⓒ Ⓓ
40. Ⓐ Ⓑ Ⓒ Ⓓ
41. Ⓐ Ⓑ Ⓒ Ⓓ
42. Ⓐ Ⓑ Ⓒ Ⓓ

43. Ⓐ Ⓑ Ⓒ Ⓓ
44. Ⓐ Ⓑ Ⓒ Ⓓ
45. Ⓐ Ⓑ Ⓒ Ⓓ

BC PRACTICE TEST 2

BC Practice Test 2

Section I

Part A

TIME: 60 MINUTES

1. A function $f(x)$ equals $\dfrac{x^2 - x}{x - 1}$ for all x except $x = 1$. If $f(1) = k$, for what value of k would the function be continuous at $x = 1$?

    (A) 0
    (B) 1
    (C) 2
    (D) No such k exists.

2. $\displaystyle\lim_{x \to 0} \dfrac{\sin \frac{x}{2}}{x}$ is

    (A) 2
    (B) 0
    (C) $\dfrac{1}{2}$
    (D) nonexistent

3. The first four terms of the Taylor series about $x = 0$ of $\sqrt{1 + x}$ are

    (A) $1 - \dfrac{x}{2} + \dfrac{x^2}{4 \cdot 2} - \dfrac{3x^3}{8 \cdot 6}$

    (B) $x + \dfrac{x^2}{2} + \dfrac{x^3}{8} + \dfrac{x^4}{48}$

    (C) $1 + \dfrac{x}{2} - \dfrac{x^2}{8} + \dfrac{x^3}{16}$

    (D) $-1 + \dfrac{x}{2} - \dfrac{x^2}{8} + \dfrac{x^3}{16}$

4. Using the line tangent to $f(x) = \sqrt{9 + \sin(2x)}$ at $x = 0$, an estimate of $f(0.06)$ is

    (A) 0.02
    (B) 2.98
    (C) 3.01
    (D) 3.02

5. What is the radius of convergence for the Maclaurin series for $\dfrac{5x^2}{1 + 8x^3}$?

    (A) $\dfrac{1}{2}$
    (B) 1
    (C) 5
    (D) 8

6. The motion of a particle in a plane is given by the pair of equations $x = \cos 2t$, $y = \sin 2t$. The magnitude of its acceleration at any time t equals

    (A) 1
    (B) 2
    (C) 4
    (D) 16

7. Let $f(x) = (x-1) + \dfrac{(x-1)^2}{4} + \dfrac{(x-1)^3}{9} + \dfrac{(x-1)^4}{16} + \cdots$

The interval of convergence of $f'(x)$ is

(A) $0 \le x \le 2$

(B) $0 \le x < 2$

(C) $0 < x \le 2$

(D) $0 < x < 2$

8. A point moves along the curve $y = x^2 + 1$ so that the x-coordinate is increasing at the constant rate of $\frac{3}{2}$ units per second. The rate, in units per second, at which the distance from the origin is changing when the point has coordinates $(1,2)$ is equal to

(A) $\dfrac{7\sqrt{5}}{10}$

(B) $\dfrac{3\sqrt{5}}{2}$

(C) $\dfrac{3\sqrt{5}}{5}$

(D) $\dfrac{5}{2}$

9. $\displaystyle\lim_{h \to 0} \dfrac{\sqrt{25+h} - 5}{h}$

(A) $= 0$

(B) $= \dfrac{1}{10}$

(C) $= 1$

(D) does not exist

10. $\displaystyle\int_{-3}^{1} \dfrac{x^2 + 4x + 5}{x - 5} \, dx =$

(A) $-\dfrac{5}{9}$

(B) $\dfrac{98}{3} \ln 2$

(C) -8

(D) $32 - 50 \ln 2$

11. $\displaystyle\int_{1}^{e} \ln x \, dx$ equals

(A) $e - 1$

(B) $e + 1$

(C) 1

(D) -1

12. Consider the series $\displaystyle\sum_{n=0}^{\infty} \dfrac{(-1)^n x^{2n}}{(2n)!}$. To what value does the series converge when $x = \frac{\pi}{3}$?

(A) $\dfrac{1}{2}$

(B) $-\dfrac{\sqrt{3}}{3}$

(C) $\dfrac{\sqrt{3}}{2}$

(D) does not converge

13. $\displaystyle\int \dfrac{x - 6}{x^2 - 3x} \, dx =$

(A) $\ln |x^2(x-3)| + C$

(B) $-\ln |x^2(x-3)| + C$

(C) $\ln \left| \dfrac{x^2}{(x-3)} \right| + C$

(D) $\ln \left| \dfrac{x-3}{x^2} \right| + C$

GO ON TO THE NEXT PAGE

14. Given f' as graphed, which could be a graph of f?

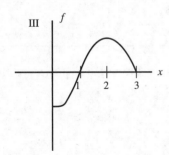

(A) I only
(B) II only
(C) III only
(D) I and III only

15. The first woman officially timed in a marathon was Violet Piercy of Great Britain in 1926. Her record of 3:40:22 stood until 1963, mostly because of a lack of women competitors. Soon after, times began dropping rapidly, but lately they have been declining at a much slower rate. Let $M(t)$ be the curve that best represents winning marathon times in year t. Which of the following is (are) positive for $t > 1963$?

I. $M(t)$

II. $M'(t)$

III. $M''(t)$

(A) I only
(B) I and II only
(C) I and III only
(D) I, II, and III

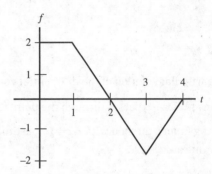

16. The graph of f is shown above. Let
$$G(x) = \int_0^x f(t)\,dt \text{ and } H(x) = \int_2^x f(t)\,dt.$$ Which of the following is true?

(A) $G(x) = H(x)$
(B) $G'(x) = H'(x+2)$
(C) $G(x) = H(x+2)$
(D) $G(x) = H(x) + 3$

17. Consider the series $\displaystyle\sum_{n=1}^{\infty} \frac{n^2}{3n^3 - n}$. Which of the following is true?

(A) The series converges by the nth Term Test.
(B) The series diverges by the nth Term Test.
(C) The series converges by the Limit Comparison Test with $\displaystyle\sum_{n=1}^{\infty} \frac{1}{n^3}$.
(D) The series diverges by the Limit Comparison Test with $\displaystyle\sum_{n=1}^{\infty} \frac{1}{n}$.

GO ON TO THE NEXT PAGE

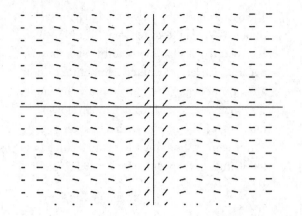

18. Which function could be a particular solution of the differential equation whose slope field is shown above?

(A) $y = \dfrac{2x}{x^2 + 1}$

(B) $y = \dfrac{x^2}{x^2 + 1}$

(C) $y = \sin x$

(D) $y = e^{-x^2}$

19. A particular solution of the differential equation $\dfrac{dy}{dx} = x + y$ passes through the point $(2,1)$. Using Euler's method with $\Delta x = 0.1$, estimate its y-value at $x = 2.2$.

(A) 1.30

(B) 1.34

(C) 1.60

(D) 1.64

Questions 20 and 21. Use the graph below, consisting of two line segments and a quarter-circle. The graph shows the velocity of an object during a 6-second interval.

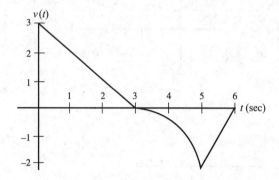

20. For how many values of t in the interval $0 < t < 6$ is the acceleration undefined?

(A) none

(B) one

(C) two

(D) three

21. During what time interval (in seconds) is the speed increasing?

(A) $0 < t < 3$

(B) $3 < t < 5$

(C) $5 < t < 6$

(D) never

22. If $\dfrac{dy}{dx} = \dfrac{y}{x}$ $(x > 0, y > 0)$ and $y = 3$ when $x = 1$, then

(A) $y = \sqrt{10 - x^2}$

(B) $y = x + 2$

(C) $y = \sqrt{8 + x^2}$

(D) $y = 3x$

23. A solid is cut out of a sphere of radius 2 by two parallel planes each 1 unit from the center. The volume of this solid is

(A) $\dfrac{10\pi}{3}$

(B) $\dfrac{32\pi}{3}$

(C) $\dfrac{25\pi}{3}$

(D) $\dfrac{22\pi}{3}$

GO ON TO THE NEXT PAGE

24. Which one of the following improper integrals converges?

    (A) $\int_{-1}^{1} \dfrac{dx}{(x+1)^2}$

    (B) $\int_{1}^{\infty} \dfrac{dx}{\sqrt{x}}$

    (C) $\int_{0}^{\infty} \dfrac{dx}{(x^2+1)}$

    (D) $\int_{1}^{3} \dfrac{dx}{(2-x)^3}$

25. The function $f(x) = x^5 + 3x - 2$ passes through the point (1,2). Let f^{-1} denote the inverse of f. Then $(f^{-1})'(2)$ equals

    (A) $\dfrac{1}{83}$

    (B) $\dfrac{1}{8}$

    (C) 8

    (D) 83

26. Find the domain of the particular solution of $\dfrac{dy}{dx} = 1 + y^2$ that passes through the origin.

    (A) all x

    (B) $x \geq 0$

    (C) $|x| < \dfrac{\pi}{2}$

    (D) $0 \leq x < \dfrac{\pi}{2}$

27. Which of the following statements is (are) true about the graph of $y = \ln(4 + x^2)$?

    I. It is symmetric to the y-axis.

    II. It has a local minimum at $x = 0$.

    III. It has inflection points at $x = \pm 2$.

    (A) I only

    (B) I and II only

    (C) II and III only

    (D) I, II, and III

28. $\int_{1}^{2} \dfrac{dx}{\sqrt{4 - x^2}}$ is

    (A) $-\dfrac{\pi}{3}$

    (B) $\dfrac{\pi}{6}$

    (C) $\dfrac{\pi}{3}$

    (D) nonexistent

29. Choose the integral that is the limit of the Riemann Sum: $\displaystyle\lim_{n\to\infty} \sum_{k=1}^{n} \left(\left(\sqrt{\dfrac{2k}{n}} + 3 \right) \cdot \left(\dfrac{1}{n} \right) \right)$.

    (A) $\int_{3}^{4} \sqrt{2x}\ dx$

    (B) $\int_{0}^{1} \sqrt{2x + 3}\ dx$

    (C) $\int_{0}^{1} \sqrt{2x}\ dx$

    (D) $\int_{3}^{4} \sqrt{2x + 3}\ dx$

30. Which infinite series converge(s)?

    I. $\displaystyle\sum_{n=1}^{\infty} \dfrac{3^n}{n!}$

    II. $\displaystyle\sum_{n=1}^{\infty} \dfrac{3^n}{n^3}$

    III. $\displaystyle\sum_{n=1}^{\infty} \dfrac{n^2}{n^3 + 1}$

    (A) I only

    (B) II only

    (C) III only

    (D) I and III only

STOP End of Part A, Section I

Part B

TIME: 45 MINUTES

> *Some questions in this part of the examination require the use of a graphing calculator. There are 15 questions in Part B, for which 45 minutes are allowed. Because there is no deduction for wrong answers, you should answer every question, even if you need to guess.*
>
> **DIRECTIONS:** Choose the best answer for each question. If the exact numerical value of the correct answer is not listed as a choice, select the choice that is closest to the exact numerical answer.

31. Find the area bounded by the spiral $r = \ln \theta$ on the interval $\pi \le \theta \le 2\pi$.

    (A) 2.405
    (B) 3.743
    (C) 4.810
    (D) 7.487

32. Write an equation for the line tangent to the curve defined by $\mathbf{F}(t) = \langle t^2 + 1, 2^t \rangle$ at the point where $y = 4$.

    (A) $y - 4 = (\ln 2)(x - 2)$
    (B) $y - 4 = (4 \ln 2)(x - 2)$
    (C) $y - 4 = (\ln 2)(x - 5)$
    (D) $y - 4 = (4 \ln 2)(x - 5)$

33. Bacteria in a culture increase at a rate proportional to the number present. An initial population of 200 triples in 10 hours. If this pattern of increase continues unabated, then the approximate number of bacteria after 1 full day is

    (A) 1056
    (B) 1440
    (C) 2793
    (D) 3240

34. When the substitution $x = 2t - 1$ is used, the definite integral $\int_3^5 t\sqrt{2t - 1}\ dt$ may be expressed in the form $k \int_a^b (x + 1)\sqrt{x}\ dx$, where $\{k, a, b\} =$

    (A) $\left\{ \frac{1}{4}, 2, 3 \right\}$
    (B) $\left\{ \frac{1}{4}, 5, 9 \right\}$
    (C) $\left\{ \frac{1}{2}, 2, 3 \right\}$
    (D) $\left\{ \frac{1}{2}, 5, 9 \right\}$

35. The curve defined by $x^3 + xy - y^2 = 10$ has a vertical tangent line when $x =$

    (A) 1.037
    (B) 1.087
    (C) 2.074
    (D) 2.096

GO ON TO THE NEXT PAGE

Questions 36 and 37. Use the graph of f shown on [0,7]. Let $G(x) = \int_2^{3x-1} f(t)\,dt.$

36. $G'(1)$ is

   (A) 1
   (B) 2
   (C) 3
   (D) 6

37. G has a local maximum at $x =$

   (A) 1
   (B) $\frac{4}{3}$
   (C) 2
   (D) 8

x	$f(x)$	$f'(x)$	$f''(x)$	$g(x)$	$g'(x)$	$g''(x)$
2	6	−1	−2	−2	1/3	−4/3

38. You are given two thrice-differentiable functions, $f(x)$ and $g(x)$. The table above gives values for $f(x)$ and $g(x)$ and their first and second derivatives at $x = 2$. Find $\lim_{x \to 2} \dfrac{f(x) + 3g(x)}{\frac{1}{2}x^2 - 2e^{x-2}}$.

   (A) 0
   (B) 1
   (C) 6
   (D) nonexistent

39. Using the left rectangular method and four subintervals of equal width, estimate $\int_0^8 |f(t)|\,dt$, where f is the function graphed below.

   (A) 4
   (B) 8
   (C) 12
   (D) 16

40. Given the function $f(x) = 2\cos(3x - 1) - 0.5e^{0.5x}$, find all values of x that satisfy the result of the Mean Value Theorem for the function f on the interval $[-3, -1]$. *NOTE*: The derivative of f is $f'(x) = -6\sin(3x - 1) - 0.25e^{0.5x}$.

   (A) −2.800 and −1.772
   (B) −2.242 and −1.296
   (C) −2.812 and −1.760
   (D) −2.843 and −1.729

41. The base of a solid is the region bounded by $x^2 = 4y$ and the line $y = 2$, and each plane section perpendicular to the y-axis is a square. The volume of the solid is

   (A) 8
   (B) 16
   (C) 32
   (D) 64

42. An object initially at rest at $(3,3)$ moves with acceleration $a(t) = \langle 2, e^{-t} \rangle$. Where is the object at $t = 2$?

    (A) $(4, -0.865)$
    (B) $(4, 1.135)$
    (C) $(7, 2.135)$
    (D) $(7, 4.135)$

43. Find the length of the curve $y = \ln x$ between the points where $y = \frac{1}{2}$ and $y = 1$.

    (A) 0.531
    (B) 0.858
    (C) 1.182
    (D) 1.356

44. Using the first two terms in the Maclaurin series for $y = \cos x$ yields accuracy to within 0.001 over the interval $|x| < k$ when $k =$

    (A) 0.394
    (B) 0.707
    (C) 0.786
    (D) 0.788

45. Consider the power series $\sum\limits_{n=0}^{\infty} b_n (x - 2)^n$. It is known that at $x = 4$, the series converges conditionally. Of the following, which is true about the convergence of the power series at $x = 1$?

    (A) There is not enough information.
    (B) At $x = 1$, the series diverges.
    (C) At $x = 1$, the series converges conditionally.
    (D) At $x = 1$, the series converges absolutely.

STOP   End of Section I

Section II

Part A

TIME: 30 MINUTES

2 PROBLEMS

> *A graphing calculator is required for some of these problems. See instructions on page 8.*

1. Let function f be continuous and decreasing, with values as shown in the table:

x	2.5	3.2	3.5	4.0	4.6	5.0
$f(x)$	7.6	5.7	4.2	3.8	2.2	1.6

   (a) Use a trapezoidal sum to estimate the area between f and the x-axis on the interval $2.5 \leq x \leq 5.0$.

   (b) Find the average rate of change of f on the interval $2.5 \leq x \leq 5.0$.

   (c) Estimate the instantaneous rate of change of f at $x = 2.5$.

   (d) If $g(x) = f^{-1}(x)$, estimate the slope of g at $x = 4$.

2. An object starts at point $(1,3)$ and moves along the parabola $y = x^2 + 2$ for $0 \leq t \leq 2$, with the horizontal component of its velocity given by $\dfrac{dx}{dt} = \dfrac{4}{t^2 + 4}$.

   (a) Find the object's position at $t = 2$.

   (b) Find the object's speed at $t = 2$.

   (c) Find the distance the object traveled during this interval.

STOP _____ End of Part A, Section II

Part B

TIME: 60 MINUTES

4 PROBLEMS

3. Given a function f such that $f(3) = 1$ and
$$f^{(n)}(3) = \frac{(-1)^n n!}{(2n+1)2^n}.$$

(a) Write the first four nonzero terms and the general term of the Taylor series for f around $x = 3$.

(b) Find the radius of convergence of the Taylor series.

(c) Show that the third-degree Taylor polynomial approximates $f(4)$ to within 0.01.

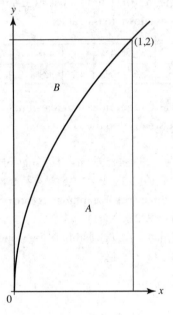

4. The curve $y = \sqrt{8 \; \sin\left(\frac{\pi x}{6}\right)}$ divides a first-quadrant rectangle into regions A and B, as shown in the figure above.

(a) Region A is the base of a solid. Cross sections of this solid perpendicular to the x-axis are rectangles. The height of each rectangle is 5 times the length of its base in region A. Find the volume of this solid.

(b) The other region, B, is rotated around the y-axis to form a different solid. Set up, but do not evaluate, an integral for the volume of this solid.

BC PRACTICE TEST 2

5. A bungee jumper has reached a point in her exciting plunge where the taut cord is 100 feet long with a 1/2-inch radius and is stretching. She is still 80 feet above the ground and is now falling at 40 feet per second. You are observing her jump from a spot on the ground 60 feet from the potential point of impact, as shown in the diagram below.

   (a) Assuming the cord to be a cylinder with volume remaining constant as the cord stretches, at what rate is its radius changing when the radius is 1/2 inch?

   (b) From your observation point, at what rate is the angle of elevation to the jumper changing when the radius is 1/2 inch?

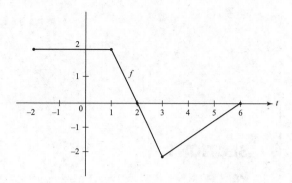

6. The figure above shows the graph of *f*, whose domain is the closed interval [−2,6]. Let

$$F(x) = \int_1^x f(t)\, dt.$$

   (a) Find $F(-2)$ and $F(6)$.

   (b) For what value(s) of *x* does $F(x) = 0$?

   (c) On what interval(s) is *F* increasing?

   (d) Find the maximum value and the minimum value of *F*.

   (e) At what value(s) of *x* does the graph of *F* have points of inflection? Justify your answer.

100'

80'

60'

θ

STOP ———————— **End of Test** ————————

ANSWER KEY
BC Practice Test 2

SECTION I

Part A

1. **B**	9. **B**	17. **D**	25. **B**
2. **C**	10. **D**	18. **A**	26. **C**
3. **C**	11. **C**	19. **D**	27. **D**
4. **D**	12. **A**	20. **C**	28. **C**
5. **A**	13. **C**	21. **B**	29. **B**
6. **C**	14. **D**	22. **D**	30. **A**
7. **B**	15. **C**	23. **D**	
8. **B**	16. **D**	24. **C**	

Part B

31. **B**	35. **C**	39. **D**	43. **C**
32. **C**	36. **D**	40. **A**	44. **A**
33. **C**	37. **B**	41. **C**	45. **D**
34. **B**	38. **C**	42. **D**	

Answer Explanations

The explanations for questions not given below will be found in the answer explanation section for AB Practice Test 2 on pages 541–552. Identical questions in Section I of Practice Tests AB 2 and BC 2 have the same number. For example, an explanation of the answer for Question 4, not given below, will be found in Section I of AB Practice Test 2, Answer 4, page 541.

Section I Multiple-Choice

Part A

1. **(B)** Since $\lim\limits_{x \to 1} f(x) = 1$, to render $f(x)$ continuous at $x = 1$, define $f(1)$ to be 1.

2. **(C)** The limit $\lim\limits_{x \to 0} \dfrac{\sin\left(\frac{x}{2}\right)}{x}$ is of the form $\dfrac{0}{0}$ so after applying L'Hospital's Rule, we

   get $\lim\limits_{x \to 0} \dfrac{\frac{1}{2}\cos\left(\frac{x}{2}\right)}{1} = \dfrac{1}{2}$. Or, using the substitution $\theta = \dfrac{x}{2}$, we can rewrite the limit

   as $\lim\limits_{x \to 0} \dfrac{\sin\left(\frac{x}{2}\right)}{x} = \lim\limits_{x \to 0} \dfrac{\sin \theta}{2\theta} = \dfrac{1}{2}\lim\limits_{x \to 0} \dfrac{\sin \theta}{\theta} = \dfrac{1}{2} \cdot 1 = \dfrac{1}{2}$.

3. **(C)** Obtain the first few terms of the Maclaurin series generated by $f(x) = \sqrt{1+x}$:

   $$f(x) = \sqrt{1+x} \qquad\qquad f(0) = 1$$

   $$f'(x) = \tfrac{1}{2}(1+x)^{-1/2} \qquad\qquad f'(0) = \tfrac{1}{2}$$

   $$f''(x) = -\tfrac{1}{4}(1+x)^{-3/2} \qquad\qquad f''(0) = -\tfrac{1}{4}$$

   $$f'''(x) = \tfrac{3}{8}(1+x)^{-5/2} \qquad\qquad f'''(0) = \tfrac{3}{8}$$

   So $\sqrt{1+x} = 1 + \dfrac{x}{2} - \dfrac{1}{4} \cdot \dfrac{x^2}{2} + \dfrac{3}{8} \cdot \dfrac{x^3}{6} - \cdots$.

5. **(A)** The function is in the form of the sum of a convergent geometric series; the series converges when the absolute value of the common ratio is less than 1. Here, the common ratio is $-8x^3$. So the series converges when $\left|-8x^3\right| < 1 \Rightarrow \left|x^3\right| < \dfrac{1}{8} \Rightarrow |x| < \dfrac{1}{2}$. Thus, the radius of convergence is $\dfrac{1}{2}$.

6. **(C)** Here,

   $$\frac{dx}{dt} = -2\,\sin 2t, \frac{dy}{dt} = 2\,\cos 2t$$

   and

   $$\frac{d^2x}{dt^2} = -4\,\cos 2t, \frac{d^2y}{dt^2} = -4\,\sin 2t$$

   and the magnitude of the acceleration, $|\mathbf{a}|$, is given by

   $$|\mathbf{a}| = \sqrt{\left(\frac{d^2x}{dt^2}\right)^2 + \left(\frac{d^2y}{dt^2}\right)^2}$$

   $$= \sqrt{(-4\,\cos 2t)^2 + (-4\,\sin 2t)^2}$$

   $$= \sqrt{16\left(\cos^2 2t + \sin^2 2t\right)} = 4$$

7. **(B)** $f'(x) = 1 + \dfrac{(x-1)}{2} + \dfrac{(x-1)^2}{3} + \dfrac{(x-1)^3}{4} + \cdots$

By the Ratio Test (page 407) the series converges when

$$\lim_{n\to\infty} \left| \frac{(x-1)^{n+1}}{n+2} \cdot \frac{n+1}{(x-1)^n} \right| < 1$$

$$|x-1| \lim_{n\to\infty} \frac{n+1}{n+2} < 1$$

$$|x-1| < 1$$

Checking the endpoints, we find:

$f'(0) = 1 - \dfrac{1}{2} + \dfrac{1}{3} - \dfrac{1}{4} + \cdots$ is the alternating harmonic series, which converges.

$f'(2) = 1 + \dfrac{1}{2} + \dfrac{1}{3} + \dfrac{1}{4} + \cdots$ is the harmonic series, which diverges.

Hence the interval of convergence is $0 \le x < 2$.

10. **(D)** When we have a rational function in the integrand and the degree of the numerator is greater than the degree of the denominator, we need to rewrite the integrand using long polynomial division.

$$
\begin{array}{r}
x + 9 \\
x - 5 \overline{)\, x^2 + 4x + 5} \\
\underline{-(x^2 - 5x)} \\
9x + 5 \\
\underline{-(9x - 45)} \\
50
\end{array}
$$

$$\int_{-3}^{1} \left(x + 9 + \frac{50}{x-5} \right) dx = \left(\frac{x^2}{2} + 9x + 50 \ln|x-5| \right) \Big|_{-3}^{1}$$

$$= \left(\frac{1}{2} + 9 + 50 \ln 4 \right) - \left(\frac{9}{2} - 27 + 50 \ln 8 \right) = 32 - 50 \ln 2$$

11. **(C)** We integrate by parts using $u = \ln x$, $dv = dx$; then $du = \frac{1}{x} dx$, $v = x$, and

$$\int_1^e \ln x \, dx = \left(x \ln x - \int x \frac{1}{x} dx \right) \Big|_1^e = (x \ln x - x) \Big|_0^2$$

$$= e \ln e - e - (1 \ln 1 - 1) = e - e - (0 - 1) = 1$$

12. **(A)** The given series is the Maclaurin series for $f(x) = \cos(x)$, which converges for all x. Thus, the series converges to $\cos\left(\frac{\pi}{3}\right) = \frac{1}{2}$.

13. **(C)** Use the method of partial fractions, letting

$$\frac{x-6}{x(x-3)} = \frac{A}{x} + \frac{B}{x-3}$$

$$x - 6 = A(x-3) + Bx$$

Letting $x = 0$, we find $A = 2$, and letting $x = 3$ yields $B = -1$.

Now $\int \left(\frac{2}{x} - \frac{1}{x-3} \right) dx = 2 \ln|x| - \ln|x-3| + C$.

17. **(D)** Check the nth Term Test: $\lim_{n\to\infty} \dfrac{n^2}{3n^3 - n} = 0$. Since the limit is zero, we must try another test to determine the convergence or divergence of the series. (*NOTE:* The nth Term Test cannot conclude convergence.) To determine which p-series to use, we can look at the end behavior (the largest-degree term) of the two polynomials. The end behavior of the numerator is n^2, and the end behavior of the denominator is $3n^3$. So the end behavior of

the general term is $\frac{n^2}{3n^3} = \frac{1}{3n}$. This series is behaving like the harmonic series (p-series with $p = 1$). If we perform the Limit Comparison Test using the harmonic series, we

get $\lim\limits_{n\to\infty} \dfrac{\frac{n^2}{3n^3-n}}{\frac{1}{n}} = \lim\limits_{n\to\infty} \dfrac{n^3}{3n^3-n} = \frac{1}{3}$. Since the limit is finite and the series $\sum\limits_{n=1}^{\infty} \frac{1}{n}$ diverges,

so does the series $\sum\limits_{n=1}^{\infty} \dfrac{n^2}{3n^3 - n}$.

19. **(D)** At $(2,1)$, $\dfrac{dy}{dx} = 3$. Use $\Delta x = 0.1$; then Euler's method moves to $(2.1, 1 + 3(0.1))$.

At $(2.1, 1.3)$, $\dfrac{dy}{dx} = 3.4$, so the next point is $(2.2, 1.3 + 3.4(0.1))$.

22. **(D)** Separate variables to get $\dfrac{dy}{y} = \dfrac{dx}{x}$, and integrate to get $\ln y = \ln x + C$.

Since $y = 3$ when $x = 1$, $C = \ln 3$. Then $y = e^{(\ln x + \ln 3)} = e^{\ln x} \cdot e^{\ln 3} = 3x$.

23. **(D)** The generating circle has the equation $x^2 + y^2 = 4$. Using disks, the volume, V, is given by

$$V = \pi \int_{-1}^{1} x^2\, dy = 2\pi \int_0^1 (4 - y^2) dy = 2\pi \left(4y - \frac{y^3}{3}\right)\Big|_0^1$$

24. **(C)** $\int_0^\infty \dfrac{dx}{x^2 + 1} = \lim\limits_{b\to\infty} \tan^{-1} x \Big|_0^b = \dfrac{\pi}{2}$. The integrals in (A), (B), and (D) all diverge to infinity.

26. **(C)** Using separation of variables:

$$\frac{dy}{dx} = 1 + y^2$$
$$\int \frac{dy}{1 + y^2} = \int dx$$
$$\arctan y = x + C$$
$$y = \tan(x + C)$$

Given initial point $(0,0)$, we have $0 = \tan(0 + C)$; hence $C = 0$ and the particular solution is $y = \tan(x)$. Because this function has vertical asymptotes at $x = \pm\frac{\pi}{2}$ and the particular solution must be differentiable in an interval containing the initial point $x = 0$, the domain is $|x| < \frac{\pi}{2}$.

28. **(C)** $\int_1^2 \dfrac{1}{\sqrt{4-x^2}}dx = \lim\limits_{h\to2^-}\int_1^h \dfrac{1}{\sqrt{4-x^2}}dx = \lim\limits_{h\to2^-}\int_1^h \dfrac{\left(\frac{1}{2}dx\right)}{\sqrt{1-\left(\frac{x}{2}\right)^2}} = \lim\limits_{h\to2^-} \sin^{-1}\frac{x}{2}\Big|_1^h$

$\qquad = \lim\limits_{h\to2^-}\left(\sin^{-1}\frac{h}{2} - \sin^{-1}\frac{1}{2}\right) = \frac{\pi}{2} - \frac{\pi}{6} = \frac{\pi}{3}$

29. **(B)** From the Riemann Sum, we see $\Delta x = \frac{1}{n}$, then $k \cdot \Delta x = \frac{k}{n}$. Notice that the term involving k in the Riemann Sum is not equal to $\frac{k}{n}$ but $2\left(\frac{k}{n}\right)$. Thus, the only choice for x_k is $x_k = \frac{k}{n}$, so $a = 0$ and $\Delta x = \frac{b-0}{n} = \frac{1}{n}$, so $b = 1$. Since x_k replaces x, $f(x) = \sqrt{2x+3}$ giving the integral $\int_0^1 \sqrt{2x+3}\ dx$.

30. **(A)** I. $\displaystyle\sum_{n=1}^{\infty} \frac{3^n}{n!}$ converges by the Ratio Test: $\displaystyle\lim_{n\to\infty}\left|\frac{3^{n+1}}{(n+1)!}\cdot\frac{n!}{3^n}\right| = \lim_{n\to\infty}\left|\frac{3}{n+1}\right| = 0 < 1$.

II. $\displaystyle\sum_{n=1}^{\infty} \frac{3^n}{n^3}$ diverges by the nth Term Test: $\displaystyle\lim_{n\to\infty}\frac{3^n}{n^3} = \infty$.

III. $\displaystyle\sum_{n=1}^{\infty} \frac{n^2}{n^3+1}$ diverges by the Limit Comparison Test: $\displaystyle\lim_{n\to\infty}\frac{\frac{n^2}{n^3+1}}{1/n} = \lim_{n\to\infty}\frac{n^3}{n^3+1} = 1$;

therefore both series diverge since we know that the harmonic series, $\displaystyle\sum_{n=1}^{\infty} \frac{1}{n}$, diverges.

Note that the Comparison Test is not appropriate for this series because $\dfrac{n^2}{n^3+1} < \dfrac{1}{n}$.

Part B

31. **(B)** Since the equation of the spiral is $r = \ln\theta$, use the polar mode. The formula for area in polar coordinates is

$$\frac{1}{2}\int_{\theta_1}^{\theta_2} r^2\, d\theta$$

Therefore, calculate

$$0.5\int_{\pi}^{2\pi} \ln^2\theta\, d\theta$$

The result is 3.743.

32. **(C)** When $y = 2^t = 4$, we have $t = 2$, so the line passes through point $F(2) = (5,4)$.

Also $\dfrac{dy}{dx} = \dfrac{\frac{dy}{dt}}{\frac{dx}{dt}} = \dfrac{2^t\ln 2}{2t}$, so at $t = 2$ the slope of the tangent line is $\dfrac{dy}{dx} = \ln 2$.

An equation for the tangent line is $y - 4 = \ln 2(x - 5)$.

40. **(A)** $f(x)$ is differentiable on $[-3,-1]$ so MVT guarantees that there exists at least one x-value such that $f'(x) = \dfrac{f(-1) - f(-3)}{-1 - (-3)} = 0.08957778$; solving this equation on your graphing calculator yields $x = -2.800, -1.772$ on $[-3,-1]$.

41. **(C)** See the figure below.

$$\Delta V = (2x)^2\,\Delta y = 4x^2\,\Delta y$$

$$= 16y\,\Delta y$$

$$V = \int_0^2 16y\, dy$$

42. **(D)** At $t = 0$, $\mathbf{R}(0) = \langle 3,3 \rangle$ and $\mathbf{v}(0) = \langle 0,0 \rangle$. Given $a(t) = \langle 2, e^{-t} \rangle$, using the FTC

$$\mathbf{v}(t) = \left\langle 0 + \int_0^t 2\,du, 0 + \int_0^t e^{-u}\,du \right\rangle = \left\langle 2u \Big|_0^t, -e^{-u}\Big|_0^t \right\rangle = \left\langle 2t, -e^{-t} + 1 \right\rangle \text{ and}$$

$$\mathbf{R}(2) = \left\langle 3 + \int_0^2 2t\,dt, 3 + \int_0^2 (-e^{-t} + 1)\,dt \right\rangle = \left\langle 7, 4.135 \right\rangle.$$

43. **(C)** The endpoints of the curve are $\left(\sqrt{e}, \frac{1}{2} \right)$ and $(e, 1)$. The length of the curve is given by

$$\int_{x_1}^{x_2} \sqrt{1 + \left(\frac{dy}{dx} \right)^2}\,dx = \int_{\sqrt{e}}^{e} \sqrt{1 + \left(\frac{1}{x} \right)^2}\,dx \text{ or } \int_{y_1}^{y_2} \sqrt{1 + \left(\frac{dx}{dy} \right)^2}\,dy = \int_{1/2}^{2} \sqrt{1 + (e^y)^2}\,dy.$$

44. **(A)** Find k such that $\cos x$ will differ from $\left(1 - \frac{x^2}{2} \right)$ by less than 0.001 at $x = k$.

Solve

$$0 = \cos x - \left(1 - \frac{x^2}{2} \right) - 0.001$$

which yields x or $k = 0.394$.

45. **(D)** The Ratio Test is used to find the radius of convergence for a power series. All points inside the resulting open interval must be absolutely convergent because the Ratio Test requires positive terms. For this power series, we are told that it is conditionally convergent at $x = 4$. This means that $x = 4$ is an endpoint of the interval of convergence because the endpoints are the only places in the interval of convergence where the series could be conditionally convergent. Thus, the radius of convergence is 2 and the open interval of convergence is $0 < x < 4$ because the series is centered at $x = 2$. Since $x = 1$ is inside the open interval of convergence, the power series is absolutely convergent at $x = 1$.

Section II Free-Response

Part A

1. See the solution for AB/BC 1, page 548.

2. (a) Position is the antiderivative of: $\dfrac{dx}{dt} = \dfrac{4}{t^2 + 4}$:

$$x = \int \frac{4}{t^2 + 4}\,dt = \frac{4}{4} \int \frac{1}{\left(\frac{t}{2} \right)^2 + 1}\,dt = 2 \int \frac{\frac{1}{2}\,dt}{\left(\frac{t}{2} \right)^2 + 1}$$

$$= 2 \arctan \frac{t}{2} + c$$

To find c, substitute the initial condition that $x = 1$ when $t = 0$:

$$1 = 2 \arctan \frac{0}{2} + c \text{ shows } c = 1, \text{ and } x = 2 \arctan \frac{t}{2} + 1$$

At $t = 2$, $x = 2 \arctan \dfrac{2}{2} + 1 = 2 \cdot \dfrac{\pi}{4} + 1 = \dfrac{\pi}{2} + 1$, and the position of the object

is $\left(\dfrac{\pi}{2} + 1, \left(\dfrac{\pi}{2} + 1 \right)^2 + 2 \right)$.

(b) At $t = 2$, $\dfrac{dx}{dt} = \dfrac{4}{2^2 + 4} = \dfrac{1}{2}$. Since $y = x^2 + 2$, $\dfrac{dy}{dt} = 2x \dfrac{dx}{dt}$, and at

$t = 2$, $\dfrac{dy}{dt} = 2 \left(\dfrac{\pi}{2} + 1 \right) \cdot \dfrac{1}{2} = \dfrac{\pi}{2} + 1$, speed $= \sqrt{ \left(\dfrac{dx}{dt} \right)^2 + \left(\dfrac{dy}{dt} \right)^2 } = \sqrt{ \left(\dfrac{1}{2} \right)^2 + \left(\dfrac{\pi}{2} + 1 \right)^2 }.$

(c) The distance traveled is the length of the curve of $y = x^2 + 2$ in the interval $1 < x < \frac{\pi}{2} + 1$:

$$L = \int \sqrt{1 + \left(\frac{dy}{dx}\right)^2}\, dx = \int_1^{\pi/2 + 1} \sqrt{1 + (2x)^2}\, dx = 5.839$$

Part B

3. (a)

n	$f^{(n)}(3) = \dfrac{(-1)^n n!}{(2n+1)2^n}$	$a_n = \dfrac{f^{(n)}(3)}{n!}$
0	1	1
1	$\dfrac{-1}{3 \cdot 2}$	$\dfrac{-1}{3 \cdot 2}$
2	$\dfrac{2!}{5 \cdot 2^2}$	$\dfrac{1}{5 \cdot 2^2}$
3	$\dfrac{-3!}{7 \cdot 2^3}$	$\dfrac{-1}{7 \cdot 2^3}$

$$f(x) = 1 - \frac{1}{6}(x-3) + \frac{1}{20}(x-3)^2 - \frac{1}{56}(x-3)^3 + \cdots + \frac{(-1)^n (x-3)^n}{(2n+1) \cdot 2^n} + \cdots$$

(b) By the Ratio Test, the series converges when

$$\lim_{x \to \infty} \left| \frac{(x-3)^{n+1}}{(2n+3) \cdot 2^{n+1}} \cdot \frac{(2n+1) \cdot 2^n}{(x-3)^n} \right| < 1$$

$$|x - 3| \lim_{x \to \infty} \left| \frac{(2n+1)}{(2n+3) \cdot 2} \right| < 1$$

$$|x - 3| < 2$$

Thus, the radius of convergence is 2.

(c) $f(4) = \displaystyle\sum_{n=0}^{\infty} \frac{(-1)^n}{(2n+1)2^n}$ is an alternating series. Since $\dfrac{1}{(2n+3)2^{n+1}} < \dfrac{1}{(2n+1)2^n}$

and $\displaystyle\lim_{n \to \infty} \frac{1}{(2n+1)2^n} = 0$ it converges by the Alternating Series Test. Therefore the error is

less than the magnitude of the first omitted term:

$$\text{error} < \frac{(4-3)^4}{(2 \cdot 4 + 1) \cdot 2^4} = \frac{1}{144} < 0.01$$

4. See the solution for AB 2/BC 4, page 548.

5. See the solution for AB/BC 5, page 550.

6. See the solution for AB/BC 6, page 551.

ANSWER SHEET
BC Practice Test 3

SECTION I

Part A

1. Ⓐ Ⓑ Ⓒ Ⓓ   9. Ⓐ Ⓑ Ⓒ Ⓓ   17. Ⓐ Ⓑ Ⓒ Ⓓ   25. Ⓐ Ⓑ Ⓒ Ⓓ
2. Ⓐ Ⓑ Ⓒ Ⓓ   10. Ⓐ Ⓑ Ⓒ Ⓓ   18. Ⓐ Ⓑ Ⓒ Ⓓ   26. Ⓐ Ⓑ Ⓒ Ⓓ
3. Ⓐ Ⓑ Ⓒ Ⓓ   11. Ⓐ Ⓑ Ⓒ Ⓓ   19. Ⓐ Ⓑ Ⓒ Ⓓ   27. Ⓐ Ⓑ Ⓒ Ⓓ
4. Ⓐ Ⓑ Ⓒ Ⓓ   12. Ⓐ Ⓑ Ⓒ Ⓓ   20. Ⓐ Ⓑ Ⓒ Ⓓ   28. Ⓐ Ⓑ Ⓒ Ⓓ
5. Ⓐ Ⓑ Ⓒ Ⓓ   13. Ⓐ Ⓑ Ⓒ Ⓓ   21. Ⓐ Ⓑ Ⓒ Ⓓ   29. Ⓐ Ⓑ Ⓒ Ⓓ
6. Ⓐ Ⓑ Ⓒ Ⓓ   14. Ⓐ Ⓑ Ⓒ Ⓓ   22. Ⓐ Ⓑ Ⓒ Ⓓ   30. Ⓐ Ⓑ Ⓒ Ⓓ
7. Ⓐ Ⓑ Ⓒ Ⓓ   15. Ⓐ Ⓑ Ⓒ Ⓓ   23. Ⓐ Ⓑ Ⓒ Ⓓ
8. Ⓐ Ⓑ Ⓒ Ⓓ   16. Ⓐ Ⓑ Ⓒ Ⓓ   24. Ⓐ Ⓑ Ⓒ Ⓓ

Part B

31. Ⓐ Ⓑ Ⓒ Ⓓ   35. Ⓐ Ⓑ Ⓒ Ⓓ   39. Ⓐ Ⓑ Ⓒ Ⓓ   43. Ⓐ Ⓑ Ⓒ Ⓓ
32. Ⓐ Ⓑ Ⓒ Ⓓ   36. Ⓐ Ⓑ Ⓒ Ⓓ   40. Ⓐ Ⓑ Ⓒ Ⓓ   44. Ⓐ Ⓑ Ⓒ Ⓓ
33. Ⓐ Ⓑ Ⓒ Ⓓ   37. Ⓐ Ⓑ Ⓒ Ⓓ   41. Ⓐ Ⓑ Ⓒ Ⓓ   45. Ⓐ Ⓑ Ⓒ Ⓓ
34. Ⓐ Ⓑ Ⓒ Ⓓ   38. Ⓐ Ⓑ Ⓒ Ⓓ   42. Ⓐ Ⓑ Ⓒ Ⓓ

BC Practice Test 3

Section I

Part A

TIME: 60 MINUTES

The use of calculators is not *permitted for this part of the examination.*
There are 30 questions in Part A, for which 60 minutes are allowed. Because there is no deduction for wrong answers, you should answer every question, even if you need to guess.

DIRECTIONS: Choose the best answer for each question.

1. A cylindrical tank, shown in the figure above, is partially full of water at time $t = 0$, when more water begins flowing in at a constant rate. The tank becomes half full when $t = 4$ and is completely full when $t = 12$. Let h represent the height of the water at time t. During which interval is $\dfrac{dh}{dt}$ increasing?

   (A) $0 < t < 4$
   (B) $0 < t < 8$
   (C) $0 < t < 12$
   (D) $4 < t < 12$

2. $\displaystyle\lim_{h \to 0} \dfrac{\sin\left(\frac{\pi}{2} + h\right) - 1}{h}$ is

   (A) 1
   (B) -1
   (C) 0
   (D) nonexistent

3. $\displaystyle\sum_{n=0}^{\infty} \dfrac{3^n}{4^{n+1}}$

   (A) $= 1$
   (B) $= 3$
   (C) $= 4$
   (D) diverges

4. An equation of the tangent to the curve $2x^2 - y^4 = 1$ at the point $(-1,1)$ is

   (A) $x + y = 0$
   (B) $x + y = 2$
   (C) $x - y = 0$
   (D) $x - y = -2$

5. The nth term of the Taylor series expansion about $x = 0$ of the function $f(x) = \dfrac{1}{1 + 2x}$ is

(A) $(2x)^n$

(B) $2x^n$

(C) $(-1)^{n-1}(2x)^{n-1}$

(D) $(-1)^n (2x)^{n-1}$

6. $\displaystyle\int \dfrac{2x + 3}{x^2 - 3x + 2}\, dx =$

(A) $\dfrac{1}{7}\ln|x - 1| - \dfrac{1}{5}\ln|x - 2| + C$

(B) $-5\ln|x - 1| + 7\ln|x - 2| + C$

(C) $-\dfrac{1}{5}\ln|x - 1| + \dfrac{1}{7}\ln|x - 2| + C$

(D) $7\ln|x - 1| - 5\ln|x - 2| + C$

7. Suppose the function f is defined as

$$f(x) = \begin{cases} \sqrt{4x^2 + 9} & \text{if } x \le 2 \\ -\dfrac{1}{5}(4x^2 - 24x + 7) & \text{if } x > 2 \end{cases}$$

Which of the following is true?

(A) The function f is both continuous and differentiable at $x = 2$.

(B) The function f is continuous but not differentiable at $x = 2$.

(C) The function f is differentiable but not continuous at $x = 2$.

(D) The function f is neither continuous nor differentiable at $x = 2$.

8. Which of the following series converge?

I. $\displaystyle\sum_{n=1}^{\infty} n^2 e^{-n}$

II. $\displaystyle\sum_{n=1}^{\infty} (-1)^n\, 3^{1/n}$

III. $\displaystyle\sum_{n=1}^{\infty} \dfrac{n^2 - 1}{n^3 + 1}$

(A) I only

(B) I and II only

(C) I and III only

(D) II and III only

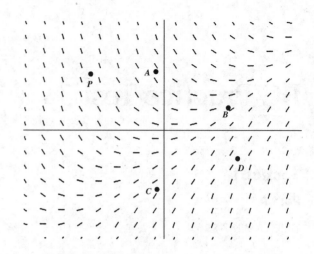

9. A particular solution of the differential equation whose slope field is shown above contains point P. This solution may also contain which other point?

(A) A

(B) B

(C) C

(D) D

10. $\displaystyle\int_0^x e^{t^3}\, dt =$

(A) $x^3 + \dfrac{x^6}{2!} + \dfrac{x^9}{3!} + \dfrac{x^{12}}{4!} + \cdots + \dfrac{x^{3(n+1)}}{(n+1)!} + \cdots$

(B) $x + \dfrac{x^4}{4} + \dfrac{x^7}{7 \cdot 2!} + \dfrac{x^{10}}{10 \cdot 3!} + \cdots$
$+ \dfrac{x^{3n+1}}{(3n+1) \cdot n!} + \cdots$

(C) $x + \dfrac{x^4}{4} + \dfrac{x^7}{14} + \dfrac{x^{10}}{30} + \cdots + \dfrac{x^{3n+1}}{n \cdot (3n+1)} + \cdots$

(D) $x + \dfrac{x^2}{2} + \dfrac{x^3}{6} + \dfrac{x^4}{12} + \cdots + \dfrac{x^{n+1}}{n \cdot (n+1)} + \cdots$

11. As the tides change, the water level in a bay varies sinusoidally. At high tide today at 8 A.M., the water level was 15 feet; at low tide, 6 hours later at 2 P.M., it was 3 feet. How fast, in feet per hour, was the water level dropping at noon today?

(A) $\dfrac{\pi}{2}$

(B) $\dfrac{\pi\sqrt{3}}{2}$

(C) $3\sqrt{3}$

(D) $6\sqrt{3}$

12. Let $\int_0^x f(t)dt = x \sin \pi x$. Then $f(3) =$

   (A) -3π

   (B) -1

   (C) 1

   (D) 3π

13. $\int \dfrac{e^u}{1+e^{2u}}\,du$ is equal to

   (A) $\ln(1+e^{2u})+C$

   (B) $\frac{1}{2}\ln|1+e^{2u}|+C$

   (C) $\frac{1}{2}\tan^{-1}e^{2u}+C$

   (D) $\tan^{-1}e^u+C$

14. $\int_{-3}^{-1} \dfrac{1}{x^2+6x+13}\,dx =$

   (A) $\dfrac{\pi}{16}$

   (B) $\dfrac{\pi}{8}$

   (C) $\dfrac{\pi}{4}$

   (D) $\ln 2$

15. Consider the polar curve $r = 2\cos(3\theta)$. What is the slope of the line tangent to the curve when $\theta = \dfrac{\pi}{4}$?

   (A) $-3\sqrt{2}$

   (B) $-\sqrt{2}$

   (C) -4

   (D) 2

16. The area bounded by the parabola $y = x^2$ and the lines $y = 1$ and $y = 9$ equals

   (A) $\dfrac{20}{3}$

   (B) $\dfrac{52}{3}$

   (C) 36

   (D) $\dfrac{104}{3}$

17. The first-quadrant region bounded by $y = \dfrac{1}{\sqrt{x}}$, $y = 0$, $x = q$, where $0 < q < 1$, and $x = 1$ is rotated about the x-axis. The volume obtained as $q \to 0^+$

   (A) $= \dfrac{2\pi}{3}$

   (B) $= \dfrac{4\pi}{3}$

   (C) $= 2\pi$

   (D) diverges

18. A curve is given parametrically by the equations $x = 3 - 2\sin t$ and $y = 2\cos t - 1$. The length of the curve from $t = 0$ to $t = \pi$ is

   (A) $\pi\sqrt{2}$

   (B) π

   (C) 2π

   (D) 4π

19. Suppose the graph of f is both increasing and concave up on $a \le x \le b$. If $\int_a^b f(x)dx$ is approximated using various sums with the same number of subintervals, and if L, R, M, and T denote, respectively, left Riemann Sum, right Riemann Sum, midpoint Riemann Sum, and trapezoidal sum, then it follows that

   (A) $R \le T \le M \le L$

   (B) $L \le T \le M \le R$

   (C) $R \le M \le T \le L$

   (D) $L \le M \le T \le R$

20. Which of the following statements about the graph of $y = \dfrac{x^2}{x-2}$ is (are) true?

   I. The graph has no horizontal asymptote.

   II. The line $x = 2$ is a vertical asymptote.

   III. The line $y = x + 2$ is an oblique asymptote.

   (A) I only

   (B) II only

   (C) I and III only

   (D) I, II, and III

21. The only function that does not satisfy the Mean Value Theorem on the interval specified is

   (A) $f(x) = x^2 - 2x$ on $[-3,1]$

   (B) $f(x) = \frac{1}{x}$ on $[1,3]$

   (C) $f(x) = \dfrac{x^3}{3} - \dfrac{x^2}{2} + x$ on $[-1,2]$

   (D) $f(x) = x + \frac{1}{x}$ on $[-1,1]$

22. $\int x^2 e^x \, dx =$

    (A) $\frac{1}{3}x^3 e^x + C$

    (B) $x^2 e^x - 2\int xe^x \, dx$

    (C) $2xe^x - \int x^2 e^x \, dx$

    (D) $\frac{1}{3}x^3 e^x - x^2 e^x + C$

23. The velocity vector of a particle moving in the plane is given by $\langle e^{4t+2}, \sin(9t+1) \rangle$ for $t \geq 0$. What is the acceleration vector of the particle?

    (A) $\langle e^{4t+2}, 9\cos(9t+1) \rangle$

    (B) $\langle \frac{1}{4}e^{4t+2}, -\frac{1}{9}\cos(9t+1) \rangle$

    (C) $\langle 4e^{4t+2}, 9\cos(9t+1) \rangle$

    (D) $\langle 4e^{4t+2}, -9\cos(9t+1) \rangle$

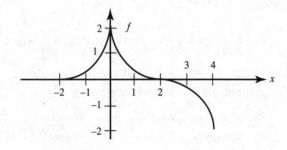

24. The graph of function f shown above consists of three quarter-circles. Which of the following is (are) equivalent to $\int_0^2 f(x) \, dx$?

    I. $\frac{1}{2}\int_{-2}^2 f(x)dx$

    II. $\int_4^2 f(x)dx$

    III. $\frac{1}{2}\int_0^4 f(x)dx$

    (A) I only

    (B) II only

    (C) I and II only

    (D) I, II, and III

25. The base of a solid is the first-quadrant region bounded by $y = \sqrt[4]{4-2x}$, and each cross section perpendicular to the x-axis is a semicircle with a diameter in the xy-plane. The volume of the solid is

    (A) $\frac{\pi}{2}\int_0^2 \sqrt{4-2x} \, dx$

    (B) $\frac{\pi}{4}\int_0^2 \sqrt{4-2x} \, dx$

    (C) $\frac{\pi}{8}\int_0^2 \sqrt{4-2x} \, dx$

    (D) $\frac{\pi}{8}\int_{-2}^2 \sqrt{4-2x} \, dx$

26. To determine whether the series $\sum_{n=1}^{\infty} \dfrac{n^3}{5n^4 - 2n^2 + 1}$ converges or diverges, we will use the Limit Comparison Test. Which of the following series should we use?

    (A) $\sum_{n=1}^{\infty} \dfrac{1}{n}$

    (B) $\sum_{n=1}^{\infty} \dfrac{1}{n^4}$

    (C) $\sum_{n=1}^{\infty} \dfrac{1}{n^3 + 1}$

    (D) $\sum_{n=1}^{\infty} \dfrac{n^2}{n^2 + 1}$

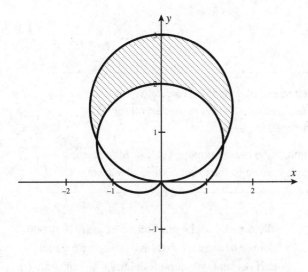

27. The area inside the circle $r = 3 \sin \theta$ and outside the cardioid $r = 1 + \sin \theta$, shaded in the figure above, is given by

(A) $\displaystyle\int_{\pi/6}^{\pi/2} [9 \sin^2 \theta - (1 + \sin \theta)^2] d\theta$

(B) $\displaystyle\int_{\pi/6}^{\pi/2} (2 \sin \theta - 1)^2 d\theta$

(C) $\displaystyle\frac{1}{2} \int_{\pi/6}^{5\pi/6} (8 \sin^2 \theta - 1) d\theta$

(D) $\displaystyle\frac{9\pi}{4} - \frac{1}{2} \int_{\pi/6}^{5\pi/6} (1 + \sin \theta)^2 d\theta$

28. Let $f(x) = \begin{cases} \dfrac{x^2 - 36}{x - 6} & \text{if } x \neq 6 \\ 12 & \text{if } x = 6 \end{cases}$

Which of the following statements is (are) true?

   I. f is defined at $x = 6$.
   II. $\displaystyle\lim_{x \to 6} f(x)$ exists.
   III. f is continuous at $x = 6$.

(A) I only

(B) II only

(C) I and II only

(D) I, II, and III

29. The table below shows values of $f''(x)$ for various values of x:

x	-1	0	1	2	3
$f''(x)$	-4	-1	2	5	8

The function f could be

(A) a linear function

(B) a quadratic function

(C) a cubic function

(D) an exponential function

30. Choose the Riemann Sum whose limit is the integral $\displaystyle\int_{-1}^{3} \ln (x^2 + 2) dx$.

(A) $\displaystyle\lim_{n \to \infty} \sum_{k=1}^{n} \left(\ln\left(\left(\frac{4k}{n} - 1 \right)^2 + 2 \right) \cdot \left(\frac{4}{n} \right) \right)$

(B) $\displaystyle\lim_{n \to \infty} \sum_{k=1}^{n} \left(\ln\left(\left(\frac{4k}{n} - 1 \right)^2 + 2 \right) \cdot \left(\frac{1}{n} \right) \right)$

(C) $\displaystyle\lim_{n \to \infty} \sum_{k=1}^{n} \left(\ln\left(\left(\frac{k}{n} - 1 \right)^2 + 2 \right) \cdot \left(\frac{1}{n} \right) \right)$

(D) $\displaystyle\lim_{n \to \infty} \sum_{k=1}^{n} \left(\ln\left(\left(\frac{k}{n} - 1 \right)^2 + 2 \right) \cdot \left(\frac{4}{n} \right) \right)$

STOP   End of Part A, Section I

Part B

TIME: 45 MINUTES

> *Some questions in this part of the examination require the use of a graphing calculator. There are 15 questions in Part B, for which 45 minutes are allowed. Because there is no deduction for wrong answers, you should answer every question, even if you need to guess.*
>
> **DIRECTIONS:** Choose the best answer for each question. If the exact numerical value of the correct answer is not listed as a choice, select the choice that is closest to the exact numerical answer.

31. Where, in the first quadrant, does the rose $r = \sin 3\theta$ have a vertical tangent?

    (A) $\theta = 0.393$
    (B) $\theta = 0.468$
    (C) $\theta = 0.598$
    (D) $\theta = 0.659$

32. A cup of coffee placed on a table cools at a rate of $\frac{dH}{dt} = -0.05(H - 70)°$F per minute, where H represents the temperature of the coffee and t is time in minutes. If the coffee was at 120°F initially, what will its temperature be, to the nearest degree, 10 minutes later?

    (A) 73°F
    (B) 95°F
    (C) 100°F
    (D) 105°F

33. Consider the power series $\sum_{n=0}^{\infty} b_n(x - 4)^n$. It is known that at $x = -1$, the series converges conditionally. Of the following, which is true about the convergence of the power series at $x = 10$?

    (A) There is not enough information.
    (B) At $x = 10$, the series diverges.
    (C) At $x = 10$, the series converges conditionally.
    (D) At $x = 10$, the series converges absolutely.

34. The density of the population of a city is 10,000 people per square mile at the beginning of a certain year. We can model the time, in years (t), it will take until the population reaches a certain density, in people per square mile (D), by using the function $t = g(D)$. What are the units of $g'(D)$?

    (A) years
    (B) years per people per square mile
    (C) people per square mile per year
    (D) people per square mile

Questions 35 and 36. The graph below shows the velocity of an object during the interval $0 \le t \le 9$.

35. The object attains its greatest speed at $t =$

    (A) 3
    (B) 5
    (C) 6
    (D) 8

GO ON TO THE NEXT PAGE

36. The object was at the origin at $t = 3$. It returned to the origin

    (A) during $6 < t < 7$
    (B) at $t = 7$
    (C) during $7 < t < 8$
    (D) at $t = 8$

---

37. An object in motion in the plane has acceleration vector $\mathbf{a}(t) = \langle \sin t, e^{-t} \rangle$ for $0 \leq t \leq 5$. It is at rest when $t = 0$. What is the maximum speed it attains?

    (A) 1.022
    (B) 1.414
    (C) 2.217
    (D) 3.162

38. If $\sqrt{x-2}$ is replaced by u, then $\displaystyle\int_3^6 \frac{\sqrt{x-2}}{x} \, dx$ is equivalent to

    (A) $\displaystyle\int_1^2 \frac{u \, du}{u^2 + 2}$

    (B) $\displaystyle\int_1^2 \frac{2u^2}{u^2 + 2} \, du$

    (C) $\displaystyle\int_3^6 \frac{2u^2}{u^2 + 2} \, du$

    (D) $\displaystyle\int_3^6 \frac{u \, du}{u^2 + 2}$

39. The set of all x for which the power series $\displaystyle\sum_{n=0}^{\infty} \frac{x^n}{(n+1) \cdot 3^n}$ converges is

    (A) $\{-3,3\}$
    (B) $|x| < 3$
    (C) $-3 \leq x < 3$
    (D) $-3 < x \leq 3$

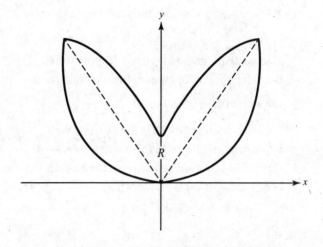

40. The graph above shows region R, which is bounded above by the polar graph of $r = 3 + 2\sin(3\theta)$ and below by the polar graph of $r = 5\sin\theta$. The intersection of the graphs occur at $\theta = 0.893$ and $\theta = 2.249$ as indicated by the dashed line segments. What is the area of region R?

    (A) 3.301
    (B) 4.155
    (C) 6.141
    (D) 8.667

41. The definite integral $\displaystyle\int_1^{10} \sqrt{1 + \frac{9}{x^2}} \, dx$ represents the length of a curve of an increasing function $f(x)$. If one end of the curve is at the point $(1,2)$, then an equation describing the curve is

    (A) $y = 3\ln x + 2$
    (B) $y = 3\ln x$
    (C) $y = 11 - \dfrac{9}{x}$
    (D) $y = 9 - \dfrac{9}{x}$

42. The Maclaurin series for the function $f(x)$ is given by $f(x) = \sum_{n=2}^{\infty} \frac{(-1)^n x^n}{2^n \cdot (3n-5)}$. If the kth-degree Maclaurin polynomial is used to approximate the values of the function in the interval of convergence, then $f(x) \approx P_k(x) = \sum_{n=2}^{k} \frac{(-1)^n x^n}{2^n \cdot (3n-5)}$. If we desire an error of less than 0.01 when approximating $f(2)$ with $P_k(2)$, what is the least degree, k, we would need so that the Alternating Series Error Bound guarantees $|f(2) - P_k(2)| < 0.01$?

(A) 31

(B) 33

(C) 35

(D) 37

43. Which statement is true?

(A) If $f'(c) = 0$, then f has a local maximum or minimum at $(c, f(c))$.

(B) If $f''(c) = 0$, then the graph of f has an inflection point at $(c, f(c))$.

(C) If f is differentiable at $x = c$, then f is continuous at $x = c$.

(D) If f is continuous on (a,b), then f maintains a maximum value on (a,b).

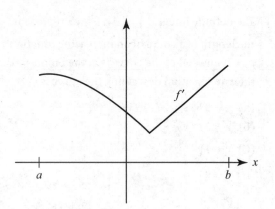

44. The graph of f' is shown above. Which statement(s) about f must be true for $a < x < b$?

I. f is increasing.

II. f is continuous.

III. f is differentiable.

(A) I only

(B) II only

(C) I and II only

(D) I, II, and III

45. Given $\int_{-1}^{4} g(x)\,dx = 0$ and $g'(x) < 0$ for all x, select the table that could represent g on the interval $[-1,4]$.

(A)

x	y
-1	3
1	3
4	3

(B)

x	y
-1	7
1	-3
4	-1

(C)

x	y
-1	7
1	1
4	-3

(D)

x	y
-1	7
1	5
4	1

STOP   End of Section I

Section II

Part A

TIME: 30 MINUTES

2 PROBLEMS

A graphing calculator is required for some of these problems. See instructions on page 8.

---

1. The Boston Red Sox play in Fenway Park, notorious for its Green Monster, a wall 37 feet tall and 315 feet from home plate at the left-field foul line. Suppose a batter hits a ball 2 feet above home plate, driving the ball down the left-field line at an initial angle of 30° above the horizontal, with initial velocity of 120 feet per second. (Since Fenway is near sea level, assume that the acceleration due to gravity is -32.172 ft/sec^2.)

   (a) Write the parametric equations for the location of the ball t seconds after it has been hit.

   (b) At what elevation does the ball hit the wall?

   (c) How fast is the ball traveling when it hits the wall?

2. The table shows the depth of water, W, in a river, as measured at 4-hour intervals during a day-long flood. Assume that W is a differentiable function of time t.

t (hours)	0	4	8	12	16	20	24
$W(t)$ (feet)	32	36	38	37	35	33	32

   (a) Find the approximate value of $W'(16)$. Indicate units of measure.

   (b) Estimate the average depth of the water, in feet, over the time interval $0 \le t \le 24$ hours by using a trapezoidal approximation with subintervals of length $\Delta t = 4$ hours.

   (c) Scientists studying the flooding believe they can model the depth of the water with the function $F(t) = 35 - 3 \cos\left(\dfrac{t+3}{4}\right)$, where $F(t)$ represents the depth of the water, in feet, after t hours. Find $F'(16)$ and explain the meaning of your answer, with appropriate units, in terms of the river depth.

   (d) Use the function F to find the average depth of the water, in feet, over the time interval $0 \le t \le 24$ hours.

STOP   **End of Part A, Section II**

Part B

TIME: 60 MINUTES

4 PROBLEMS

No calculator is allowed for any of these problems.

If you finish Part B before time has expired, you may return to work on Part A, but you may not use a calculator.

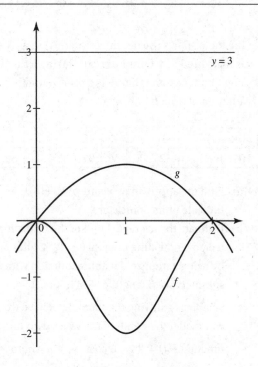

3. The region **R** is bounded by the curves $f(x) = \cos(\pi x) - 1$ and $g(x) = x(2 - x)$, as shown in the figure.

(a) Find the area of **R**.

(b) A solid has base **R**, and each cross section perpendicular to the x-axis is an isosceles right triangle whose hypotenuse lies in **R**. Set up, but do not evaluate, an integral for the volume of this solid.

(c) Set up, but do not evaluate, an integral for the volume of the solid formed when **R** is rotated around the line $y = 3$.

4. Two autos, P and Q, start from the same point and race along a straight road for 10 seconds. The velocity of P is given by $v_p(t) = 6(\sqrt{1 + 8t} - 1)$ feet per second. The velocity of Q is shown in the graph.

(a) At what time is P's actual acceleration (in ft/sec^2) equal to its average acceleration for the entire race?

(b) What is Q's acceleration (in ft/sec^2) then?

(c) At the end of the race, which auto was ahead? Explain.

5. Given that a function f is continuous and differentiable throughout its domain and that $f(5) = 2, f'(5) = -2, f''(5) = -1$, and $f'''(5) = 6$.

   (a) Write a Taylor polynomial of degree 3 that approximates f around $x = 5$.

   (b) Use your answer to estimate $f(5.1)$.

   (c) Let $g(x) = f(2x + 5)$. Write a cubic Maclaurin polynomial approximation for g.

6. Let f be the function that contains the point $(-1, 8)$ and satisfies the differential equation $\dfrac{dy}{dx} = \dfrac{10}{x^2 + 1}$.

   (a) Write an equation of the line tangent to f at $x = -1$.

   (b) Using your answer to part (a), estimate $f(0)$.

   (c) Using Euler's method with a step size of 0.5, estimate $f(0)$.

   (d) Estimate $f(0)$ using an integral.

STOP   **End of Test**

ANSWER KEY
BC Practice Test 3

SECTION I

Part A

1. **D**	9. **B**	17. **D**	25. **C**
2. **C**	10. **B**	18. **C**	26. **A**
3. **A**	11. **B**	19. **D**	27. **A**
4. **A**	12. **A**	20. **D**	28. **D**
5. **C**	13. **D**	21. **D**	29. **C**
6. **B**	14. **B**	22. **B**	30. **A**
7. **A**	15. **D**	23. **C**	
8. **A**	16. **D**	24. **C**	

Part B

31. **B**	35. **D**	39. **C**	43. **C**
32. **C**	36. **C**	40. **D**	44. **D**
33. **B**	37. **C**	41. **A**	45. **C**
34. **B**	38. **B**	42. **C**	

Answer Explanations

The explanations for questions not given below will be found in the answer explanation section for AB Practice Test 3 on pages 567–578. Identical questions in Section I of Practice Tests AB 3 and BC 3 have the same number. For example, explanations of the answers for Questions 1 and 2, not given below, will be found in Section I of AB Practice Test 3, Answers 1 and 2, page 567.

Section I Multiple-Choice

Part A

3. **(A)** The series $\frac{1}{4} + \frac{3}{16} + \frac{9}{64} + \cdots$ is geometric with $a = \frac{1}{4}$ and $r = \frac{3}{4}$; it converges to $\dfrac{\frac{1}{4}}{1 - \frac{3}{4}}$.

5. **(C)** $\dfrac{1}{1 + 2x}$ is the sum of an infinite geometric series with first term 1 and common ratio $-2x$.
   The series is $1 - 2x + 4x^2 - 8x^3 + 16x^4 - \cdots$.

6. **(B)** Assume that $\dfrac{2x + 3}{(x - 1)(x - 2)} = \dfrac{A}{x - 1} + \dfrac{B}{x - 2}$. Then $2x + 3 = A(x - 2) + B(x - 1)$.
   If we let $x = 1$, then we get $2(1) + 3 = A(1 - 2)$ and $A = -5$. If we let $x = 2$, then we get
   $2(2) + 3 = B(2 - 1)$ and $B = 7$. So $\displaystyle\int \dfrac{2x + 3}{x^2 - 3x + 2}\,dx = \int \left(\dfrac{-5}{x - 1} + \dfrac{7}{x - 2}\right) dx = $
   $-5 \ln |x - 1| + 7 \ln |x - 2| + C$.

8. **(A)** I. $\displaystyle\sum_{n=1}^{\infty} n^2 e^{-n} = \sum_{n=1}^{\infty} \dfrac{n^2}{e^n}$. Using the Ratio Test, we get

   $$\lim_{n \to \infty} \left| \dfrac{a_{k+1}}{a_k} \right| = \lim_{n \to \infty} \left| \dfrac{(n+1)^2}{e^{n+1}} \cdot \dfrac{e^n}{n^2} \right| = \lim_{n \to \infty} \left| \left(\dfrac{n+1}{n}\right)^2 \cdot \dfrac{1}{e} \right| = 1^2 \cdot \dfrac{1}{e} < 1, \text{ convergent.}$$

   II. $\displaystyle\lim_{n \to \infty} 3^{1/n} = 3^0 = 1$, so $\displaystyle\lim_{n \to \infty} (-1)^n 3^{1/n}$ does not exist and thus the series

   $\displaystyle\sum_{n=1}^{\infty} (-1)^n 3^{1/n}$ diverges because it doesn't pass the nth Term Test.

   III. We would like to compare this series with $\displaystyle\sum_{n=1}^{\infty} \dfrac{1}{n}$; however, we cannot use the
   Comparison Test because $\dfrac{n^2 - 1}{n^3 + 1} < \dfrac{1}{n}$ (see note below). So we must use the Limit

   Comparison Test: $\displaystyle\lim_{n \to \infty} \dfrac{\frac{n^2 - 1}{n^3 + 1}}{1/n} = \lim_{n \to \infty} \dfrac{n^3 - n}{n^3 + 1} = 1$.

   Since $\displaystyle\sum_{n=1}^{\infty} \dfrac{1}{n}$ diverges, $\displaystyle\sum_{n=1}^{\infty} \dfrac{n^2 - 1}{n^3 + 1}$ also diverges by the Limit Comparison Test.

   NOTE: We can rewrite $\dfrac{n^2 - 1}{n^3 + 1} = \dfrac{1 - \frac{1}{n^2}}{n + \frac{1}{n^2}}$; comparing to $\dfrac{1}{n}$ we see that the numerator

   $1 - \dfrac{1}{n^2} < 1$ and the denominator $n + \dfrac{1}{n^2} > n$; therefore $\dfrac{n^2 - 1}{n^3 + 1} < \dfrac{1}{n}$.

10. **(B)** The Maclaurin series for $e^u = 1 + u + \dfrac{u^2}{2!} + \dfrac{u^3}{3!} + \cdots + \dfrac{u^n}{n!} + \cdots$. The Maclaurin series for
    e^{t^3} is found by substitution with $u = t^3$:

    $$e^{t^3} = 1 + t^3 + \dfrac{t^6}{2!} + \dfrac{t^9}{3!} + \cdots + \dfrac{t^{3n}}{n!} + \cdots$$

    So, $\displaystyle\int_0^x e^{t^3}\,dt = \int_0^x \left(1 + t^3 + \dfrac{t^6}{2!} + \dfrac{t^9}{3!} + \cdots + \dfrac{t^{3n}}{n!} + \cdots\right) dt = x + \dfrac{x^4}{4} + \dfrac{x^7}{7 \cdot 2!} + \dfrac{x^{10}}{10 \cdot 3!} + \cdots$

12. **(A)** $f(x) = \dfrac{d}{dx}(x \sin \pi x) = \pi x \cos \pi x + \sin \pi x$

15. **(D)** Like working with parametric equations, to find the slope of the tangent line to a polar curve, we must use the Chain Rule with $\dfrac{dy}{dx} = \dfrac{dy/d\theta}{dx/d\theta}$. First we will need $\dfrac{dr}{d\theta} = -6\sin(3\theta)$ to compute $\dfrac{dy}{d\theta}$ and $\dfrac{dx}{d\theta}$. Next convert the polar curve to parametric equations with the identities $x = r\cos\theta$ and $y = r\sin\theta$. Then find the derivatives with respect to θ:

$\dfrac{dx}{d\theta} = r \cdot (-\sin\theta) + \dfrac{dr}{d\theta} \cdot \cos\theta = -2\sin\theta\cos(3\theta) - 6\cos\theta\sin(3\theta)$ and

$\dfrac{dy}{d\theta} = r \cdot (\cos\theta) + \dfrac{dr}{d\theta} \cdot \sin\theta = 2\cos\theta\cos(3\theta) - 6\sin\theta\sin(3\theta)$

At $\theta = \dfrac{\pi}{4}$,

$\dfrac{dy}{dx}\Big|_{\theta=\pi/4} = \dfrac{dy/d\theta}{dx/d\theta}\Big|_{\theta=\pi/4} = \dfrac{2\left(\frac{\sqrt{2}}{2}\right)\left(-\frac{\sqrt{2}}{2}\right) - 6\left(\frac{\sqrt{2}}{2}\right)\left(\frac{\sqrt{2}}{2}\right)}{-2\left(\frac{\sqrt{2}}{2}\right)\left(-\frac{\sqrt{2}}{2}\right) - 6\left(\frac{\sqrt{2}}{2}\right)\left(\frac{\sqrt{2}}{2}\right)} = \dfrac{-1-3}{1-3} = 2$

17. **(D)**

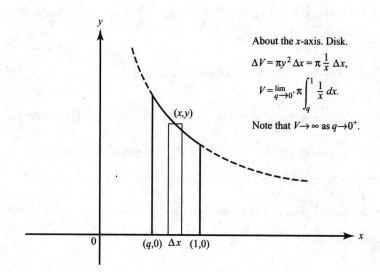

About the x-axis. Disk.

$\Delta V = \pi y^2 \Delta x = \pi \frac{1}{x} \Delta x$,

$V = \lim_{q\to 0^+} \pi \int_q^1 \frac{1}{x}\, dx$.

Note that $V \to \infty$ as $q \to 0^+$.

18. **(C)** See the figure below, which shows that the length of a semicircle of radius 2 is needed here. The answer can, of course, be found by using the formula for length of curve:

$$s = \int_0^\pi \sqrt{\left(\frac{dx}{dt}\right)^2 + \left(\frac{dy}{dt}\right)^2}\, dt = \int_0^\pi \sqrt{(-2\,\cos t)^2 + (-2\,\sin t)^2}\, dt = 2\int_0^\pi dt$$

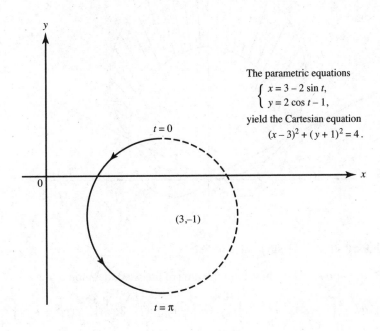

The parametric equations
$$\begin{cases} x = 3 - 2\sin t, \\ y = 2\cos t - 1, \end{cases}$$
yield the Cartesian equation
$$(x-3)^2 + (y+1)^2 = 4.$$

20. **(D)** $\lim\limits_{x\to\infty} f(x) = \infty$; $\lim\limits_{x\to 2} f(x) = -\infty$; using long division, $y = x + 2 + \dfrac{4}{x-2}$.

22. **(B)** Using parts we let $u = x^2$, $dv = e^x dx$; then $du = 2x\, dx$, $v = e^x$, and

$$\int x^2 e^x\, dx = x^2 e^x - 2\int x e^x\, dx$$

23. **(C)** The components of the acceleration vector are the derivatives of the components of the velocity vector.

26. **(A)** When we have a series that has a general term in the form of a ratio of two polynomials, we can use the Limit Comparison Test with a p-series. To determine the p-series to use, we can look at the end behavior (the largest-degree term) of the two polynomials. The end behavior of the numerator is n^3, and the end behavior of the denominator is $5n^4$. So the end behavior of the general term is $\dfrac{n^3}{5n^4} = \dfrac{1}{5n}$. This series is behaving like the harmonic series (p-series with $p = 1$). If we perform the Limit Comparison Test using the harmonic series, we get:

$$\lim_{n\to\infty} \frac{\dfrac{n^3}{5n^4 - 2n^2 + 1}}{\dfrac{1}{n}} = \lim_{n\to\infty} \frac{n^4}{5n^4 - 2n^2 + 1} = \frac{1}{5}$$

Since the limit is finite and the series $\sum\limits_{n=1}^{\infty} \dfrac{1}{n}$ diverges, so does the series $\sum\limits_{n=1}^{\infty} \dfrac{n^3}{5n^4 - 2n^2 + 1}$.

27. **(A)** The required area is lined in the figure below.

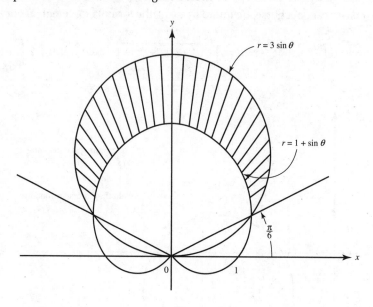

Intersections: $3 \sin \theta = 1 + \sin \theta$; $\sin \theta = \frac{1}{2}$; $\theta = \frac{\pi}{6}, \frac{5\pi}{6}$

Let $r_1 = 3 \sin \theta$ and $r_2 = 1 + \sin \theta$, then the area of the shaded region is

$$\int_{\pi/6}^{5\pi/6} \frac{1}{2} r_1{}^2 \, d\theta - \int_{\pi/6}^{5\pi/6} \frac{1}{2} r_2{}^2 \, d\theta = \frac{1}{2} \int_{\pi/6}^{5\pi/6} \left(r_1{}^2 - r_2{}^2\right) \, d\theta = 2 \cdot \frac{1}{2} \int_{\pi/6}^{\pi/2} \left(r_1{}^2 - r_2{}^2\right) \int_{\pi/6}^{\pi/6} d\theta$$

28. **(D)** Note that $f(x) = x + 6$ if $x \neq 6$, that $f(6) = 12$, and that $\lim\limits_{x \to 6} f(x) = 12$. So f is defined and continuous at $x = 6$.

29. **(C)** The table indicates both positive and negative values for $f''(x)$. $f''(x)$ for a linear function would be zero for all x. Both quadratic and exponential functions have no change in concavity for all x. A cubic function would include a change in concavity, and $f''(x)$ would be linear. Both of these properties are indicated in the table with the given points. Therefore, the function could be a cubic function.

30. **(A)** From the integral we get $a = -1$, $b = 3$, so $\Delta x = \dfrac{3 - (-1)}{n} = \dfrac{4}{n}$ and $x_k = a + k \cdot \Delta x = -1 + \dfrac{4k}{n} = \dfrac{4k}{n} - 1$. Replace x with x_k and replace dx with Δx in the integrand to get the general term in the summation.

Part B

31. **(B)** Expressed parametrically, $x = \sin 3\theta \cos \theta$, $y = \sin 3\theta \sin \theta$. $\dfrac{dy}{dx}$ is undefined where $\dfrac{dx}{d\theta} = -\sin 3\theta \sin \theta + 3 \cos 3\theta \cos \theta = 0$. Use your calculator to solve for θ.

CALCULATOR TIP: Graph in function mode: $y = -\sin(3x) \cdot \sin(x) + 3\cos(3x) \cdot \cos(x)$, and find the roots (zeros) of the function on the interval $\left[0, \frac{\pi}{2}\right]$.

33. **(B)** The Ratio Test is used to find the radius of convergence for a power series. All points inside the resulting open interval must be absolutely convergent because the Ratio Test requires positive terms. For this power series, we are told that it is conditionally convergent at $x = -1$. This means that $x = -1$ is an endpoint of the interval of convergence because the endpoints are the only places in the interval of convergence where the series could be conditionally convergent. Thus the radius of convergence is 5. The open interval of convergence is $-1 < x < 9$ because the series is centered at $x = 4$. Since $x = 10$ is outside the open interval of convergence, the power series is divergent at $x = 10$.

37. **(C)** It is given that $\mathbf{a}(t) = \langle \sin t, e^{-t} \rangle$.

$$v(t) = \left\langle 0 + \int_0^t \sin u \, du, 0 + \int_0^t e^{-u} \, du \right\rangle = \left\langle -\cos u \Big|_0^t, -e^{-u} \Big|_0^t \right\rangle = \left\langle -\cos t + 1, -e^{-t} + 1 \right\rangle$$

The object's speed is $|v(t)| = \sqrt{(-\cos t + 1)^2 + (-e^{-t} + 1)^2}$.

Use a calculator to find that the object's maximum speed is 2.217.

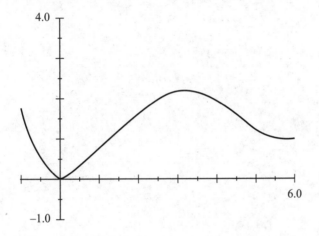

39. **(C)** Use the Ratio Test:

$$\lim_{n \to \infty} \left| \frac{x^{n+1}}{(n+2) \cdot 3^{n+1}} \cdot \frac{(n+1) \cdot 3^n}{x^n} \right| = \lim_{n \to \infty} \frac{n+1}{n+2} \cdot \frac{1}{3}|x| = \frac{|x|}{3}$$

which is less than 1 if $-3 < x < 3$. When $x = -3$, the convergent alternating harmonic series is obtained. When $x = 3$, the divergent harmonic series is obtained.

40. **(D)** Let the top curve be $r_1 = 3 + 2\sin(3\theta)$ and the bottom curve be $r_2 = 5\sin\theta$. The total area consists of three pieces: the area inside the r_2 curve from $\theta = 0$ to $\theta = 0.893$, the area inside the r_1 curve from $\theta = 0.893$ to $\theta = 2.249$, and the area inside the r_2 curve from $\theta = 2.249$ to $\theta = \pi$. So the total area is

$$\int_1^{0.893} \frac{1}{2}(5\sin\theta)^2 \, d\theta + \int_{0.893}^{2.249} \frac{1}{2}(3 + 2\sin(3\theta))^2 \, d\theta + \int_{2.249}^{\pi} \frac{1}{2}(5\sin\theta)^2 \, d\theta = 8.667.$$

NOTE: Because of symmetry, the area can also be found by doubling the area from

$$\theta = 0 \text{ to } \theta = \frac{\pi}{2} : 2 \cdot \left(\int_0^{0.893} \frac{1}{2}(5\sin\theta)^2 \, d\theta + \int_{0.893}^{\pi/2} \frac{1}{2}(3 + 2\sin(3\theta))^2 \, d\theta \right) = 8.667.$$

41. **(A)** Curve length is given by $\int \sqrt{1 + \left(\frac{dy}{dx}\right)^2} \, dx$. Here the integrand $\sqrt{1 + \frac{9}{x^2}}$ implies that $\frac{dy}{dx} = \frac{3}{x}$, hence, $y = 3\ln x + C$. Since the curve contains $(1,2)$, $2 = 3\ln 1 + C$, which yields $C = 2$.

42. **(C)** The approximation is $f(2) \approx P_k(2) = \displaystyle\sum_{n=2}^{k} \frac{(-1)^n}{3n-5}$. The Alternating Series Error Bound is the first omitted term of the alternating series. So the error bound is the absolute value of the term in the series with $n = k + 1$. If $k = 31$, the error term is the term with $n = 32$, $\left| \dfrac{(-1)^{32}}{3 \cdot 32 - 5} \right| = \dfrac{1}{91} = 0.01099 > 0.01$. For $k = 33$, the error term is the term with $n = 34$, $\left| \dfrac{(-1)^{34}}{3 \cdot 34 - 5} \right| = \dfrac{1}{97} = 0.01031 > 0.01$. For $k = 35$, the error term is the term with $n = 36$, $\left| \dfrac{(-1)^{36}}{3 \cdot 36 - 5} \right| = \dfrac{1}{103} = 0.00971 < 0.01$. *NOTE:* For $k = 37$, the error will also be less than 0.01. However, $k = 35$ is the least value of k for which the error is less than 0.01.

Section II Free-Response

Part A

1.

Green
Monster

120 ft/sec

30γ

37'

2'

315'

(a) The following table shows x- and y-components of acceleration, velocity, and position:

	Horizontal	Vertical
acceleration components:	$a_x = 0$	$a_y = -32.172$
velocity $= \int a\, dt$:	$v_x = c_1$	$v_y = -32.172t + c_2$
initial velocities:	$v_x(0) = 60\sqrt{3} = c_1$	$v_y(0) = 60 = c_2$
velocity components:	$v_x(t) = 60\sqrt{3}$	$v_y(t) = -32.172t + 60$
position $= \int v\, dt$:	$x(t) = 60\sqrt{3}t + c_3$	$y(t) = -16.086t^2 + 60t + c_4$
initial position (0,2):	$x(0) = 0 = c_3$	$y(0) = 2 = c_4$
position components:	$x(t) = 60\sqrt{3}t$	$y(t) = -16.086t^2 + 60t + 2$

The last line in the table is the answer to part (a).

(b) To determine how far above the ground the ball is when it hits the wall, find out when $x = 315$, and evaluate y at that time.

$$60\sqrt{3}t = 315 \text{ yields } t = \frac{315}{60\sqrt{3}} \approx 3.03109 \text{ seconds}$$

$$y\left(\frac{315}{60\sqrt{3}}\right) \approx 36.075 \text{ feet}$$

(c) The ball's speed at the moment of impact in part (b) is $|v(t)|$ evaluated at $t = \frac{315}{60\sqrt{3}}$.

$$|v(t)| = \sqrt{(v_x(t))^2 + (v_y(t))^2}$$

$$= \sqrt{(60\sqrt{3})^2 + (-32.172t + 60)^2} \text{ when } x = 315$$

$$\left|v\left(\frac{315}{60\sqrt{3}}\right)\right| \approx 110.487 \text{ ft/sec}$$

2. See the solution for AB/BC 2, page 575.

Part B

3. See the solution for AB/BC 3, page 576.

4. See the solution for AB/BC 4, pages 576–577.

5. (a) The table below is constructed from the information given.

n	$f^{(n)}(5)$	$a_n = \dfrac{f^{(n)}(5)}{n!}$
0	2	2
1	-2	-2
2	-1	$-\dfrac{1}{2}$
3	6	1

$$f(x) \approx 2 - 2(x-5) - \frac{1}{2}(x-5)^2 + (x-5)^3$$

(b) $f(5.1) \approx 2 - 2(5.1-5) - \dfrac{1}{2}(5.1-5)^2 + (5.1-5)^3$

$\approx 2 - 0.2 - 0.005 + 0.001 = 1.796$

(c) Use Taylor's Theorem around $x = 0$.

n	$g^{(n)}(x)$	$g^{(n)}(0)$	$a_n = \dfrac{g^{(n)}(0)}{n!}$
0	$f(2x+5)$	$f(5) = 2$	2
1	$2f'(2x+5)$	$2f'(5) = 2(-2) = -4$	-4
2	$4f''(2x+5)$	$4f''(5) = 4(-1) = -4$	-2
3	$8f'''(2x+5)$	$8f'''(5) = 8(26) = 48$	8

$$g(x) \approx 2 - 4x - 2x^2 + 8x^3$$

6. (a) At $(-1,8)$, $\dfrac{dy}{dx} = \dfrac{10}{x^2+1} = \dfrac{10}{2} = 5$, so an equation of the tangent line is

$y - 8 = 5(x - (-1))$. Therefore $f(x) \approx 8 + 5(x+1)$.

(b) $f(0) \approx 8 + 5(0+1) = 13$

(c) At $(-1,8)$, $\dfrac{dy}{dx} = 5 \approx \dfrac{\Delta y}{\Delta x}$. For $\Delta x = 0.5$, $\Delta y = 0.5(5) = 2.5$, so move to

$(-1 + 0.5, 8 + 2.5) = (-0.5, 10.5)$.

At $(-0.5, 10.5)$, $\dfrac{dy}{dx} = \dfrac{10}{\left(-\dfrac{1}{2}\right)^2 + 1} = \dfrac{10}{\dfrac{5}{4}} = 8 \approx \dfrac{\Delta y}{\Delta x}$. For $\Delta x = 0.5$, $\Delta y = 0.5(8) = 4$,

so move to $(-0.5 + 0.5, 10.5 + 4)$.

Thus $f(0) \approx 14.5$.

(d) $\displaystyle\int_{-1}^{0} \frac{10}{x^2+1}\,dx = f(0) - f(-1)$, so $f(0) = 8 + \displaystyle\int_{-1}^{0} \frac{10}{x^2+1}\,dx$

$= 8 + \arctan(x)\Big|_{-1}^{0}$

$= 8 + 10(\arctan(0) - \arctan(-1))$

$= 8 + \dfrac{5\pi}{2}$

Appendix: Formulas and Theorems for Reference

Algebra

1. QUADRATIC FORMULA.   The roots of the quadratic equation

$$ax^2 + bx + c = 0 \ (a \neq 0)$$

   are given by

$$x = \frac{-b \pm \sqrt{b^2 - 4ac}}{2a}$$

2. BINOMIAL THEOREM.   If n is a positive integer, then

$$\left(a + b\right)^n = a^n + na^{n-1}b + \frac{n(n-1)}{1 \cdot 2}a^{n-2}b^2 + \frac{n(n-1)(n-2)}{1 \cdot 2 \cdot 3}a^{n-3}b^3$$
$$+ \cdots + nab^{n-1} + b^n$$

3. REMAINDER THEOREM.   If the polynomial $Q(x)$ is divided by $(x - a)$ until a constant remainder R is obtained, then $R = Q(a)$. In particular, if a is a root of $Q(x) = 0$, then $Q(a) = 0$.

Geometry

The sum of the angles of a triangle is equal to a straight angle (180°).

PYTHAGOREAN THEOREM

In a right triangle,
$c^2 = a^2 + b^2$

In the following formulas,

A	is	area	B	is	area of base
S		surface area	r		radius
V		volume	C		circumference
b		base	l		arc length
h		height or altitude	θ		central angle (in radians)
s		slant height			

4. Triangle: $A = \dfrac{1}{2}bh$

5. Trapezoid: $A = \left(\dfrac{b_1 + b_2}{2}\right)h$

6. Parallelogram: $A = bh$

7. Circle: $C = 2\pi r;\ A = \pi r^2$

8. Circular sector: $A = \dfrac{1}{2}r^2\theta$

9. Circular arc: $l = r\theta$

10. Cylinder:

$$V = \pi r^2 h = Bh$$
$$S\,(\text{lateral}) = 2\pi rh$$
$$\text{Total surface area} = 2\pi r^2 + 2\pi rh$$

11. Cone:

$$V = \tfrac{1}{3}\,\pi r^2 h = \tfrac{1}{3}\,Bh$$
$$S\,(\text{lateral}) = \pi r\sqrt{r^2 + h^2}$$
$$\text{Total surface area} = \pi r^2 + \pi r\sqrt{r^2 + h^2}$$

12. Sphere: $V = \dfrac{4}{3}\pi r^3$

$$S = 4\pi r^2$$

Trigonometry

BASIC IDENTITIES

13. $\sin^2\theta + \cos^2\theta = 1$

14. $1 + \tan^2\theta = \sec^2\theta$

15. $1 + \cot^2\theta = \csc^2\theta$

SUM AND DIFFERENCE FORMULAS

16. $\sin(\alpha \pm \beta) = \sin\alpha\cos\beta \pm \cos\alpha\sin\beta$

17. $\cos(\alpha \pm \beta) = \cos\alpha\cos\beta \mp \sin\alpha\sin\beta$

18. $\tan(\alpha \pm \beta) = \dfrac{\tan\alpha \pm \tan\beta}{1 \mp \tan\alpha\tan\beta}$

DOUBLE-ANGLE FORMULAS

19. $\sin 2\alpha = 2\sin\alpha\cos\alpha$

20. $\cos 2\alpha = \cos^2\alpha - \sin^2\alpha = 2\cos^2\alpha - 1 = 1 - 2\sin^2\alpha$

21. $\tan 2\alpha = \dfrac{2\tan\alpha}{1 - \tan^2\alpha}$

HALF-ANGLE FORMULAS

22. $\sin\dfrac{\alpha}{2} = \pm\sqrt{\dfrac{1 - \cos\alpha}{2}};\ \sin^2\alpha = \dfrac{1}{2} - \dfrac{1}{2}\cos 2\alpha$

23. $\cos\dfrac{\alpha}{2} = \pm\sqrt{\dfrac{1 + \cos\alpha}{2}};\ \cos^2\alpha = \dfrac{1}{2} + \dfrac{1}{2}\cos 2\alpha$

REDUCTION FORMULAS

24. $\sin(-\alpha) = -\sin\alpha$ $\qquad\qquad$ $\cos(-\alpha) = \cos\alpha$

25. $\sin\left(\dfrac{\pi}{2} - \alpha\right) = \cos\alpha$ $\qquad$ $\cos\left(\dfrac{\pi}{2} - \alpha\right) = \sin\alpha$

26. $\sin\left(\dfrac{\pi}{2} + \alpha\right) = \cos\alpha$ $\qquad$ $\cos\left(\dfrac{\pi}{2} + \alpha\right) = -\sin\alpha$

27. $\sin(\pi - \alpha) = \sin\alpha$ $\qquad\quad$ $\cos(\pi - \alpha) = -\cos\alpha$

28. $\sin(\pi + \alpha) = -\sin\alpha$ $\qquad$ $\cos(\pi + \alpha) = -\cos\alpha$

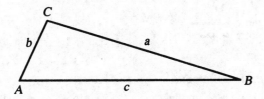

If a, b, c are the sides of triangle ABC, and A, B, C are, respectively, the opposite interior angles, then:

29. Law of Cosines: $c^2 = a^2 + b^2 - 2ab\cos C$

30. Law of Sines: $\dfrac{a}{\sin A} = \dfrac{b}{\sin B} = \dfrac{c}{\sin C}$

31. The area $A = \dfrac{1}{2}ab\sin C$

GRAPHS OF TRIGONOMETRIC FUNCTIONS

The four functions sketched above, sin, cos, csc, and sec, all have period 2π.

The two functions tan and cot have period π.

$y = \tan x$ ———
$y = \cot x$ - - - -

INVERSE TRIGONOMETRIC FUNCTIONS

$y = \sin^{-1} x = \arcsin x$ implies $x = \sin y,$ where $-\dfrac{\pi}{2} \leq y \leq \dfrac{\pi}{2}.$

$y = \cos^{-1} x = \arccos x$ implies $x = \cos y,$ where $0 \leq y \leq \pi.$

$y = \tan^{-1} x = \arctan x$ implies $x = \tan y,$ where $-\dfrac{\pi}{2} < y < \dfrac{\pi}{2}.$

Analytic Geometry

Rectangular Coordinates

DISTANCE

32. The distance d between two points, $P_1(x_1, y_1)$ and $P_2(x_2, y_2)$, is given by

$$d = \sqrt{\left(x_2 - x_1\right)^2 + \left(y_2 - y_1\right)^2}$$

EQUATIONS OF THE STRAIGHT LINE

33. POINT-SLOPE FORM. Through $P_1(x_1, y_1)$ and with slope m:

$$y - y_1 = m\left(x - x_1\right)$$

34. SLOPE-INTERCEPT FORM. With slope m and y-intercept b:

$$y = mx + b$$

35. TWO-POINT FORM. Through $P_1\left(x_1, y_1\right)$ and $P_2\left(x_2, y_2\right)$:

$$y - y_1 = \frac{y_2 - y_1}{x_2 - x_1}\left(x - x_1\right)$$

36. INTERCEPT FORM. With x- and y-intercepts of a and b, respectively:

$$\frac{x}{a} + \frac{y}{b} = 1$$

37. GENERAL FORM. $Ax + By + C = 0$, where A and B are not both zero. If $B \neq 0$, the slope is $-\dfrac{A}{B}$; the y-intercept, $-\dfrac{C}{B}$; the x-intercept, $-\dfrac{C}{A}$.

DISTANCE FROM POINT TO LINE

38. Distance d between a point $P\left(x_1, y_1\right)$ and the line $Ax + By + C = 0$ is

$$d = \left|\frac{Ax_1 + By_1 + C}{\sqrt{A^2 + B^2}}\right|$$

EQUATIONS OF THE CONICS

CIRCLE

39. With center at $(0,0)$ and radius r: $x^2 + y^2 = r^2$
40. With center at (h,k) and radius r: $(x - h)^2 + (y - k)^2 = r^2$

PARABOLA

41. With vertex at $(0,0)$ and focus at $(p,0)$: $y^2 = 4px$
42. With vertex at $(0,0)$ and focus at $(0,p)$: $x^2 = 4py$

With vertex at (h,k) and axis

43. parallel to x-axis, focus at $(h + p,k)$: $(y - k)^2 = 4p(x - h)$
44. parallel to y-axis, focus at $(h,k + p)$: $(x - h)^2 = 4p(y - k)$

ELLIPSE

With major axis of length $2a$, minor axis of length $2b$, and distance between foci of $2c$:

45. Center at $(0,0)$, foci at $(\pm c,0)$, and vertices at $(\pm a,0)$:

$$\frac{x^2}{a^2} + \frac{y^2}{b^2} = 1$$

46. Center at $(0,0)$, foci at $(0,\pm c)$, and vertices at $(0,\pm a)$:

$$\frac{y^2}{a^2} + \frac{x^2}{b^2} = 1$$

47. Center at (h,k), major axis horizontal, and vertices at $(h \pm a,k)$:

$$\frac{(x - h)^2}{a^2} + \frac{(y - k)^2}{b^2} = 1$$

48. Center at (h,k), major axis vertical, and vertices at $(h,k \pm a)$:

$$\frac{(y - k)^2}{a^2} + \frac{(x - h)^2}{b^2} = 1$$

For the ellipse, $a^2 = b^2 + c^2$, and the eccentricity $e = \frac{c}{a}$, which is *less* than 1.

HYPERBOLA

With real (transverse) axis of length $2a$, imaginary (conjugate) axis of length $2b$, and distance between foci of $2c$:

49. Center at $(0,0)$, foci at $(\pm c,0)$, and vertices at $(\pm a,0)$:

$$\frac{x^2}{a^2} - \frac{y^2}{b^2} = 1$$

50. Center at $(0,0)$, foci at $(0,\pm c)$, and vertices at $(0,\pm a)$:

$$\frac{y^2}{a^2} - \frac{x^2}{b^2} = 1$$

51. Center at (h,k), real axis horizontal, vertices at $(h \pm a,k)$:

$$\frac{(x - h)^2}{a^2} - \frac{(y - k)^2}{b^2} = 1$$

52. Center at (h,k), real axis vertical, vertices at $(h,k \pm a)$:

$$\frac{(y - k)^2}{a^2} - \frac{(x - h)^2}{b^2} = 1$$

For the hyperbola, $c^2 = a^2 + b^2$, and the eccentricity $e = \frac{c}{a}$, which is *greater* than 1.

Polar Coordinates

RELATIONS WITH RECTANGULAR COORDINATES

53. $x = r \cos \theta$
     $y = r \sin \theta$
     $r^2 = x^2 + y^2$
     $\tan \theta = \dfrac{y}{x}$

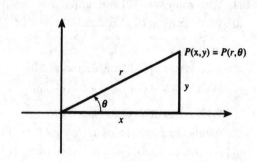

SOME POLAR EQUATIONS

54. $r = a$ circle, center at pole, radius a
55. $r = 2a \cos \theta$ circle, center at $(a,0)$, radius a
56. $r = 2a \sin \theta$ circle, center at $(0,a)$, radius a
57. $\left.\begin{array}{l} r = a \sec \theta \\ \text{or } r \cos \theta = a \end{array}\right\}$ line, $x = a$
58. $\left.\begin{array}{l} r = b \csc \theta \\ \text{or } r \sin \theta = b \end{array}\right\}$ line, $y = b$

roses (four leaves)

59. $r = \cos 2\theta$ 60. $r = \sin 2\theta$

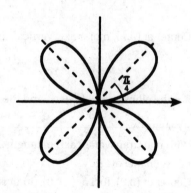

cardioids (specific examples below)

61. $r = a(1 \pm \cos \theta)$ 62. $r = a(1 \pm \sin \theta)$

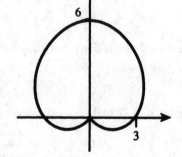

$r = 4(1 - \cos \theta)$ $r = 3(1 + \sin \theta)$

63. $r^2 = \cos 2\theta$, lemniscate, symmetric to the x-axis

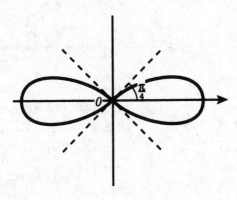

($r^2 = \sin 2\theta$ is a lemniscate symmetric to $y = x$.)

64. $r = \theta$, (double) spiral of Archimedes

For $\theta > 0$, the curve consists only of the solid spiral.

65. $r\theta = a$ $(\theta > 0)$, hyperbolic (or reciprocal) spiral

This curve is for $r\theta = 2$.
Note that $y = 2$ is an asymptote.

Exponential and Logarithmic Functions

PROPERTIES

e^x	$\ln x \,(x > 0)$
$e = 1$	$\ln 1 = 0$
$e = e$	$\ln e = 1$
$e^{x_1} \cdot e^{x_2} = e^{x_1 + x_2}$	$\ln(x_1 \cdot x_2) = \ln x_1 + \ln x_2$
$\dfrac{e^{x_1}}{e^{x_2}} = e^{x_1 - x_2}$	$\ln \dfrac{x_1}{x_2} = \ln x_1 - \ln x_2$
$e^{-x} = \dfrac{1}{e^x}$	$\ln x^r = r \ln x \,(r \text{ real})$

INVERSE PROPERTIES

$f(x) = e^x$ and $f^{-1}(x) = \ln x$ are inverses of each other:

$$f^{-1}\big(f(x)\big) = f\big(f^{-1}(x)\big) = x$$

$$\ln e^x = e^{\ln x} = x \,(x > 0)$$

GRAPHS

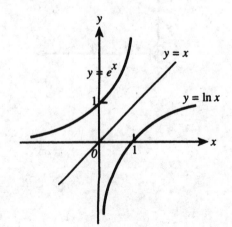

Index

A

Absolute convergence, 408–409
Absolute maximum/minimum, 164
Absolute-value function, 66, 100
Acceleration, 180–183
Acceleration vector, 183
Alternating harmonic series, 408
Alternating series, 408–410
Alternating Series Error Bound, 411
Alternating Series Test, 408–411
Amplitude, 70
Antiderivatives, 213–214
Antidifferentiation, 213–214
Approximating sums, 252–253
Approximation(s)
    of definite integral, 248–249
    a limit of an alternating series, 410–411
    local linear, 186
    tangent-line, 186–187
Arc length, 297–299
Area
    bounded by polar curve, 288–290
    calculating, 283–285
    between curves, 285
    Fundamental Theorem of Calculus (FTC), 247–248
    using symmetry, 286–287
    of volume, 290–292
    in x interpreted as, 260–261
Asymptotes, 94–95
Average rate of change, 113, 160, 262
Average value, 261–268
Axis of revolution, 292

B

Best linear approximation, 186, 419

C

Calculator use, 8–12
Calculus AB Exam, 1–8
Calculus BC Exam, 1–8
Candidates Test, 171
Carrying capacity, 376
Chain Rule, 116, 245
Closed Interval Test, 171
Common Maclaurin series, 417–418
Comparison test, 305–308, 405, 408–409
Complex numbers, 429
Components, 182
Composition (composite) function, 64, 116
Composition of Continuous Functions Theorem, 104
Composition Rule, 96

Concavity, 164
Conditional convergence, 409
Constant Multiple Rule, 96
Constant of integration, 213
Constant Rule, 96
Continuity, 99–104
Continuous Functions Theorem, 104
Converge, 300
Converge conditionally, 409
Convergence
    of infinite series, 403–404
    of nonnegative series, 408–410
    in power series, 411
    theorems, 402–403
    topic outline for, 7
Convergent, 401
Converges, 399
Critical points, in slope, 159–160
Cross section, area of a, 290–292
Cubic functions, 69
Curve(s)
    area between, 289
    fields, 358–362
    logistic growth, 377
    motion along, 183–184
    parametrically defined, 73–75, 159–163
    points in slope of a, 159–160
    sketching, 165–168
    slope of, 159–161
    slope of polar, 190–192
    tangents to, 161
Cylindrical shell, 295–297

D

Decay, 369
Definite Integral(s)
    approximations of, 248–253
    definition, 241–242
    definition as sum, 246–247
    evaluation of, 241
    properties of, 241–242
    of a rate is net change, 344–345
Degree and order, 419
Derivatives
    of a composite function, 116
    continuity of, 121–125
    definition, 113
    differentiability, 121–125
    formulas for, 115
    graphically estimating, 125–126
    graphing a function from its, 253–260
    of implicit defined functions, 128–129, 362–363

    of inverse function, 130–131
    and L'Hospital's Rule, 133
    numerically estimating, 122–123
    parametric equations, 126–127
    Product Rule, 115
    Quotient Rule, 115
    recognizing limits as, 136
    symmetric difference quotient, 124
    topic outline for, 2–4
    See also Differential equations
Difference quotient, 113
Difference Rule, 96
Differential equations
    definition, 357
    Euler's Method, 363–367
    exponential growth, 369–373
    logistic growth, 376–380
    motion problems, 224–225
    restricted growth, 373–376
    slope fields, 358–362
    solving first-order, 367–369
Differentiation. *see* Derivatives
Discontinuities, 100–101
Disks, 292–293
Displacement, 337
Distance, 337
    See also Motion
Diverge, 300
Divergence, theorems, 402–403
Divergent, 399
Domains, 63, 357

E

Equations
    parametric, 73–74
    of the tangent, 161
    See also Differential equations
Euler's method, 363–367
Exercises
    antiderivatives, 227–235
    applications of integration, 346–350
    definite integrals, 270–276
    derivatives
    differential calculus applications, 193–203
    differential equations, 381–389
    functions, 80–82
    integration to geometry, 309–316
    limits and continuity, 105–109
    sequences and series, 431–436
Exponential decay, 372
Exponential functions, 72
Exponential growth, 369–373
Extreme Value Theorem, 104, 171

F

First-order differential equations, 367–369
Formulas
  for antiderivatives, 213–214
  derivatives, 115
  integration by parts, 221
Function(s)
  absolute-value, 66
  average value of a, 261–262
  composite, 64, 116
  continuity of, 2
  with continuous derivatives, 163
  cubic, 69
  defined by power series, 413–415
  definition, 63
  derivative discontinuity, 163–164
  derivative of implicitly defined, 362–363
  derivatives graphically, 177–179
  differentiable, 165–168, 171
  domain of, 63
  even, 64
  exponential, 72
  graphing from its derivative, 253–260
  greatest-integer, 66
  increasing/decreasing, 163
  inverse, 65
  inverse trigonometric, 71
  linear, 69
  logarithmic, 72–73
  Maclaurin series, 415–417
  not everywhere differentiable, 169, 171
  odd, 64
  one-to-one, 65
  parametrically defined, 73–76
  polar, 76–78
  polynomial, 69
  quadratic, 69
  range of, 63
  rational, 69
  special, 66
  Taylor series, 415–417
  that generate no series, 418
  trigonometric, 69–73
Fundamental Theorem of Calculus (FTC), 4–6, 241–242, 247–248, 337

G

General solution, 357
General term, 400
Geometric series, 401, 403–404
Global maximum/minimum, 164, 171
Greatest-integer function, 66
Growth
  exponential growth, 369–372
  logistic growth, 376–380
  restricted growth, 373–376
  unrestricted vs. restricted, 376

H

Half-life, 369
Harmonic series, 401, 402
Horizontal asymptote, 94
Horizontal components, 182

I

Implicit differentiation, 128–129
Improper integrals
  classes of, 299–305
  comparison tests, 305–308
Indefinite integral, 213
Indeterminate forms, limits, 133
infinite discontinuity, 101
Infinite sequence, 399
Infinite series
  alternating, 408–409
  convergence tests, 403–404
  definitions, 400–401
  geometric series test, 403–404
  nth term test, 403
  tests for convergence of nonnegative series, 404
  theorems about convergence/divergence, 402–403
Inflection points, 165
Inhibited model, 376
Initial-value problem, 224
Inputs, 63
Instantaneous rate of change, 113, 160
Integrable, 241
Integral test, 404
Integrals
  improper, 299–308
  Mean Value Theorem, 242
  topic outline for, 4–6
Integrands, 213
Integration
  differential equations, 224
  by partial fractions, 220–221
  by parts, 221–222
  tic-tac-toe method, 223
Intermediate Value Theorem, 104
Interval of convergence, 412
Inverse function, 65, 130–131
Inverse trigonometric function, 71
Inverses, 71

J

Jump discontinuity, 101

L

Lagrange error bound, 423
Lagrange remainder, 423
Law of exponential growth, 369
Left sum, 248
Length of a curve, 297–299
L'Hospital's Rule, 133–136, 404
Limit Comparison Test, 406–408
Limit definition of e, 99
Limit(s)
  Composition Rule, 96
  Constant Multiple Rule, 96
  definition, 89–93
  Difference Rule, 96
  one-sided, 89
  other basic limits, 99
  Product Rule, 96
  quotient of polynomials, 98
  Quotient Rule, 96
  Squeeze (Sandwich) Theorem, 96
  Sum Rule, 96

  theorems on, 96–97
  topic outline for, 1–2
  trigonometric, 99
Linear functions, 69
Lines, motion and, 180–182, 337–339
Local extreme value, 164
Local extremum, 164
Local linear approximation, 186
Local maximum/minimum, 164
Logarithmic functions, 72–73
Logarithmic response, 358
Logistic curve, 376
Logistic differential equation, 376
Logistic growth, 376–380

M

Maclaurin polynomial, 418–419
Maclaurin series, 415–417
Magnitude, 182
Marginal cost of production, 344
Maxima, 173–177
Maximum sustainable population, 376
Mean Value Theorem, 131, 242
Midpoint sum, 248
Minima, 173–177
Motion
  along a curve, 182–185, 183–185
  along a line, 180–182
  along a plane curve, 339–342
  along a straight line, 337–339
  differential equations and, 224–225

N

Natural logarithm, 260
Net change, 337
Nonnegative series
  Comparison Test, 405
  convergence tests of, 404
  integral tests, 404
  Limit Comparison Test, 406–408
  p-Series test, 405
  Ratio Test, 408–409
  Root Test, 408
  tests for, 404–408
nth Term Test, 403

O

One-sided limit, 89
One-to-one functions, 65
Optimization, 173–177
Order and degree, 419
Oscillation, 302, 400
Outputs, 63

P

Parametric differentiation, 126–127
Parametric equations, 73–74
Parametrically defined curve, tangents to, 161–163, 339–342
Parametrically defined functions, 73–76
Partial fractions, integration by, 220–221
Particular solution, 357
Parts, integration by, 221–222

Parts Formula, 221–222
Period, 70
Periodicity, 70
Plane curves, 292–293
Point of inflection, 165
Polar curve, slope of, 190–191
Polar functions, 76–78
Polynomial(s)
  end behavior, 93–94
  function, 69
  Maclaurin series, 418
  Taylor, 419–420
Position vector, 182
Power rule, 115
Power series
  Common Maclaurin series,
    417–418
  complex numbers, 429–430
  computations with, 425–429
  convergence, 411
  functions defined by, 413–415
  Maclaurin, 415–417
  Taylor, 415–417
Product Rule, 96, 115
p-Series Test, 401, 405, 408

Q
Quadratic functions, 69
Quotient Rule, 96, 115

R
Radium of convergence, 411
Rate of change
  average, 113, 160
  differential equations for, 358
  of exponential growth and decay,
    369–372
  instantaneous, 113, 180
  of logistic growth, 376–378
  of restricted growth, 373–375
Ratio Test, 407–408, 412
Rational functions, 69
Real numbers, sequence of, 399–400
Related rates, 188–190

Removable discontinuity, 101
Restricted growth, 373–376
Restricted model, 376
Riemann Sums, 4–6, 246, 248–249,
    342–344
Right sum, 248
Rolle's Theorem, 132–133
Root Test, 408

S
Second Derivative Test, 165
Separable, 367
Sequence of real numbers, 399–400
Series
  infinite, 400–402
  power, 411–416
  topic outline for, 6–7
Shells, 295–297
Sketching, 165, 171
Slope
  of a curve, 113, 159–160
  fields, 358–362
  graphically, 177
Solids
  area of a cross section of a,
    290–292
  disks, 292–293
  of revolution, 292–294
  shells, 295–297
  washers, 294–295
Solution, 357
Solution curve, 358
Speed, 180, 339
Squeeze (Sandwich) Theorem, 96
Straight line, motion along, 180–182
Sum Rule, 96
Sums, 248–253
Symmetric Difference Quotient, 124
Symmetry
  for area, 286–287
  of functions, 64–65

T
Tangent-line approximation, 186–188

Taylor polynomial, 418–419
Taylor series, 415–417
Taylor's Theorem, 423–425
Terms, 400
Theorem(s)
  Composition of Continuous
    Functions, 104
  Continuous Functions, 104
  Extreme Value, 104
  Fundamental Theorem of
    Calculus, 241
  Intermediate Value, 104
  Mean Value, 131
  Rational Function, 98
  Rolle's, 132–133
  Squeeze (Sandwich), 96
  Taylor's, 423
Tic-tac-toe method, 223
Trapezoidal sum, 250
Trapezoids, 252–253
Trigonometric functions, 69–73
Trigonometric limits, 99

U
Unchecked growth, 376
Uninhibited growth, 376
Unrestricted growth, versus restricted
    growth, 376

V
Variables, separating, 367–369
Vector, 182
Velocity, 180
Velocity vector, 182
Vertical asymptote, 95
Vertical components, 182
Volume, area of a cross section,
    290–292

W
Washers, 294–295

Z
Zeros, of functions, 66